AF323126

WATER QUALITY

Edited by **Hlanganani Tutu**
and **Bronwyn Patricia Grover**

Water Quality

http://dx.doi.org/10.5772/62562
Edited by Hlanganani Tutu and Bronwyn Patricia Grover

Contributors

Salmiati Salmiati, Mohd Razman Salim, Nor Zaiha Arman, Matjaž Glavan, Marina Pintar, Andrej Jamšek, Cezara Voica, Roxana Ionete, Andreea Maria Iordache, Ioan Ştefanescu, Zati Sharip, Saim Suratman, Benias Nyamunda, Mahmut Çetin, Hyriye Ibrikci, Ebru Karnez, Hande Sağir, Muhammet Said Gölpinar, Mehmet Ali Akgul, Bloodless Dzwairo, Munyaradzi Mujuru, Oghenekaro Nelson Odume, Pablo Fierro, Claudio Valdovinos, Luis Vargas-Chacoff, Carlos Bertran, Ivan Arismendi, Jacinto Elías Sedeño-Díaz, Ricardo A Ruiz-Picos, Eugenia López López, Lydia Bondareva, Valerii Rakitskii, Ivan Tananaev, Stefania Gheorghe, Catalina Stoica, Maurício Gustavo Coelho Emerenciano, Marcel Martínez-Porchas, Anselmo Miranda-Baeza, Luis Rafael Martínez-Córdoba, Marek Ruman, Bogdan Zygmunt, Żaneta Polkowska, Meilin Wu, Isaac Mwangi, Joshua Edokpayi, John Odiyo, Olatunde Durowoju, Gulnihal Ozbay, Chunlei Fan, Zhiming Yang

CBS, Edition **2018**

Published by InTech

Janeza Trdine 9, 51000 Rijeka, Croatia

Notice

Statements and opinions expressed in the chapters are these of the individual contributors and not necessarily those of the editors or publisher. No responsibility is accepted for the accuracy of information contained in the published chapters. The publisher assumes no responsibility for any damage or injury to persons or property arising out of the use of any materials, instructions, methods or ideas contained in the book.

Publishing Process Manager Dajana Pemac

Technical Editor SPi Global

Cover Designer InTech Design Team

Additional hard copies can be obtained from orders@intechopen.com

Water Quality, Edited by Hlanganani Tutu and Bronwyn Patricia Grover
p. cm.
Print ISBN 978-953-51-2881-6
Online ISBN 978-953-51-2882-3

Contents

Preface

Water is a very valuable resource, supporting a wide variety of life. There is an increasing demand on this resource as populations around the world continue to increase. Industrial development has continued to increase, further exerting increased demand for this resource.

Surface and groundwater are the main sources of fresh water for drinking, household purposes, irrigation, manufacturing and other uses. As such, these have to be conserved and protected as much as possible. Increased population growth and industrial activity have led to increased anthropogenic contamination of water resources through sewage effluents, agricultural runoff, wastewater discharges and industrial and mining discharge. This has resulted in elevated levels of eutrophic nutrients, pathogenic microorganisms, toxic elements, pesticides, health care products, persistent organic pollutants and other emerging pollutants.

Most legislative bodies and frameworks, from the United States Environmental Protection Agency (USEPA) to the European Union, across the world seek to protect water resources from overexploitation and contamination and to promote sustainable water use.

This book entitled *Water Quality* covers issues related to water chemistry and quality, biomonitoring, modelling and some approaches to water treatment. It is divided into sections addressing those particular issues, viz:

Biomonitoring

This section consists of five chapters that cover biofloc technology, aquatic ecosystems, biomonitoring and the use of bioindicators.

Modelling

This section consists of four chapters that cover chemometrics, modelling nutrient leaching and remote sensing.

Water treatment

This section consists of two chapters that cover cyanide removal and the use of algal biomass for removal of elements.

Water quality

This section consists of seven chapters that straggle trace elements, radionuclides, wastewater and water quality monitoring.

We would like to convey our gratitude to the authors who contributed to this book, to the reviewers for their valuable input as well as to the organisers and the staff at InTech Open-Access Publisher, especially Dajana Pemac, Ana Simčić and Ivona Lovrić, for their help and guidance in putting the book together. Each book chapter was reviewed by three reviewers who gave suggestions for improvement, extension or reduction of aspects.

Prof. Hlanganani Tutu
Molecular Sciences Institute, School of Chemistry
University of the Witwatersrand, Johannesburg
South Africa

Dr. Bronwyn Patricia Grover (Associated Editor)
Geostratum Groundwater and Geochemistry Consulting
Johannesburg, South Africa

Reviewers

Prof. Elvis Fosso-Kankeu, School of Chemical Engineering, North West University, South Africa

Prof. Luke Chimuka, School of Chemistry, University of the Witwatersrand, South Africa

Dr. Bronwyn P. Grover, Geostratum Groundwater and Geochemistry Consulting, South Africa

Prof. Hlanganani Tutu, School of Chemistry, University of the Witwatersrand, South Africa

Biomonitoring

Chapter 1

Ecosystem Approach to Managing Water Quality

Oghenekaro Nelson Odume

Additional information is available at the end of the chapter

http://dx.doi.org/10.5772/106111

Abstract

This chapter argues for the ecosystem approach to managing water quality, which advocates the management of water, land and the associated living resources at the catchment scale as complex social-ecological systems and proactively defend and protect the ecological health of the ecosystem for the continuing supply of ecosystem services for the benefit of society. It argues for a shift from the engineering-driven command and control approach to water resource management. Environmental water quality (EWQ) is discussed as a holistic and integrated tripod ecosystem approach to managing water quality. Water physico-chemistry, biomonitoring and aquatic ecotoxicology are discussed as and their application and limitation with respect to water quality management, particularly in South Africa, is critically evaluated. The chapter concludes with a case study illustrating the application of biomonitoring for the assessment of ecosystem health in the Swartkops River, Eastern Cape, South Africa. The macroinvertebrates-based South African Scoring System version 5 was applied at three impacted sites and one control site. Two of the three impacted sites downstream of an effluent discharge point had very poor health conditions. The urgent need for ecological restoration was recommended.

Keywords: aquatic ecosystems, biomonitoring, ecotoxicology, macroinvertebrates, pollution, water chemistry

1. Introduction

The sustainability of freshwater ecosystems is being threatened globally [1]. A growing human population, coupled with changing demography, increasing socio-economic development as well as urbanisation and industrialisation of freshwater ecosystems catchments are the major drivers of change, resulting in deteriorating freshwater quality and depleting quantity. Climate change and other human-induced influences will, in the foreseeable future, exacerbate the conditions of the already stressed freshwater ecosystems [2]. Globally, there is a growing recognition that the typical hard-engineering informed 'command-and-control' approach to

managing freshwater ecosystems, particularly water quality, is no longer sustainable [3, 4]. The hard engineering command and control approach (CCP) arises out of the insatiable quest for humans to tame, control and command everything in the environment, including nature [4]. Its primacy is the development of water resources for the socio-economic benefits of human with little or no attention to the ecosystems that provide the resource base. It is, however, becoming increasingly clear that an alternative approach that takes account of both ecosystem sustainability and socio-economic development is needed for managing water resources, including water quality.

The ecosystem approach is a holistic and integrated management strategy with an appreciation of the ecosystem as the source of water as well as a water user with specific requirements in terms of water quality, quantity, in-stream ecological and riparian conditions as well as the overall health and functionality of the ecosystem [5]. It advocates the management of water, land and the associated living resources at the catchment scale as complex social-ecological systems [6]. It proactively defends and protects the ecological health of the ecosystem. It is becoming the preferred approach for managing water quality, for example, in Europe [7], Australia [8] and South Africa [9]. For example, the European Union Water Framework Directive (WFD, 2000/60/EC) explicitly recognises and consciously advocates the ecosystem approach to managing the surface water quality of water bodies within the EU member states. It mandates all EU members to maintain surface water quality in 'good status' and to restore degraded systems to 'good conditions'.

2. The ecosystem approach and water quality in South Africa

South Africa's ground-breaking water law provides for an ecosystem approach to managing water resources (National Water Act No. 36 of 1998). The strategies for achieving the ecosystem-oriented objectives of the Act are designed in the National Water Resource Strategy 2 (NWRS2) [5]. The NWRS2 provides for two complementary approaches, the Resource Directed Measures (RDM) and the Source Directed Controls (SDC).

The RDM are directed at protecting and using the water resources sustainably, in terms of water quality, ecological and riparian habitat conditions [5]. The RDM are composed of the national water resource-classification system, the ecological reserve, and the Resource Quality Objectives (RQOs). In South Africa, water resources are classified into three management classes: Class I (a resource with no noticeable or with minimal human impacts); Class II (a resource slightly or moderately impacted by human activities with little deviation from natural conditions); and Class III (a resource with significant impacts resulting in serious deviation from natural conditions) [5, 10]. Water resources in Classes I and II are given high management priority to keep them in good condition; while depending on the scenarios, efforts are made to restore the conditions of those in management Class III. The ecological reserve provides the legal basis for assessing and protecting the quality, quantity and reliability of water needed for the functioning and maintaining the aquatic ecosystem [9]. The RQO provides measurable quantitative and qualitative descriptions/objectives for the physical, biological and chemical attributes that should be protected. The RQOs thus capture the management class and the ecological requirements, giving directions on how a water resource should be managed to protect key ecosystem attributes and functionalities [11]. The determination of the ecological reserve involves derivation of the present ecological state (PES) of the water resource. The PES is determined by integrating biological, physical and chemical information, including fish, macroinvertebrates, geomorphology, vegetation, riparian condition as well as hydrological and physico-chemical variables.

The NWRS2 also provides for measures to control the use of water resources to protect the water quality and ecological conditions needed to ensure the functionality of the aquatic ecosystems. Human activities impacting water quality in terms of abstraction and discharges are regulated through the SDC, which are used in combination with the RDM. The SDC define, and then impose limits, and restrict the use of water resources to achieve the desired levels of protection. Licensing, registration, authorisation and special permit are the tools used to achieve the control of water use impact on water quality. Guidelines and limits, discharges of effluent as well as water abstractions are used to impose limit on water use activities. The combined process of the RDM and the SDC involves integrating biophysical information from multiple components of the ecosystems, and in terms of water quality, environmental water quality (EWQ) provides a sound ecosystem-based methodology for managing aquatic ecosystems in South Africa.

3. Environmental water quality (EWQ)

Environmental water quality is an integrated approach that links the chemical, physical and radiological characteristics of a water resource to the responses of the in-stream assemblage structure, function and processes [12, 13]. The EWQ combines water physico-chemistry, biomonitoring and ecotoxicology. The conventional approach to managing water quality is physico-chemistry, which involves measuring and analysing physical and chemical variables to indicate water quality without taking into account their effects on biological organisms. Biomonitoring is the systematic deployment of resident biota to provide information on aquatic ecosystem health with limited capacity for a cause-effect relationship, while ecotoxicology is the experimental evaluation of the effects of specific toxicants on aquatic biota, adding the potential for causal linkages.

3.1. Water physico-chemistry

Human activities such as agriculture, domestic and industrial wastewater discharges, environmental engineering, and natural factors including geology and soils, hydrology, seasonal patterns, geomorphology, climate and weather, influence the physico-chemical conditions of the aquatic ecosystems. The physico-chemical variable analysis is the traditional approach to controlling pollution and managing water quality. It helps water-resource managers to measure and analyse the concentrations of pollutants, determine their fate and transport, as well as their persistence in the aquatic environment. In South Africa, for example, the National Physico-Chemical Monitoring Programme (NCPM) uses analyses of physico-chemical variables to provide the water quality status of rivers and streams [14].

The physico-chemical approach forms an important component of the EWQ in terms of managing water quality. However, its drawbacks include (i) high analytical costs of monitoring physico-chemical variables, (ii) inexhaustible numbers of both dissolved and suspended chemicals and pollutants, making the choice of variables for analysis difficult and also making it impossible to measure all variables, (iii) lack of spatial and temporal representativeness of water quality conditions, as results are only reflective of the time and place of sampling and (iv) provision of very little or no insights into ecological response of aquatic biota and overall biophysical health of the system. Given that conserving biodiversity and protecting the ecosystem health are critical objectives of the ecosystem approach, the physico-chemical analysis alone is inadequate. The second pillar of the EWQ, biological monitoring also known as biomonitoring, provides the opportunity for detecting ecological impairments and measuring both taxonomic and functional diversity, which are important components of the aquatic ecosystem.

3.2. Biomonitoring

Biomonitoring integrates multiple effects of stressors including chemical (e.g. salinisation), physical (e.g. sedimentation) and biological (e.g. parasitism) to evaluate aquatic ecosystem health [15]. It relies on the sound ecological understanding that in-stream biota, for example, plants, algae, animals and microorganisms integrate the conditions of their environment and are therefore able to provide an indication of the health of the ecosystem in which they live [16]. Biomonitoring can be applied at multiple biological organisations including sub-organismal (e.g. gene mutation and cell alteration), individual species composition, population, community and ecosystem levels. In South Africa, for example, the science of biomonitoring is well developed compared to the rest of sub-Saharan African countries. The design of the National Aquatic Ecosystem Health Monitoring Programme (NAEHMP) is met to generate information needed regarding the ecological conditions of aquatic ecosystems in South Africa [17]. The NAEHMP utilises the responses of in-stream biota and system drivers to characterise the impacts of disturbances in aquatic ecosystems and to determine present ecological states of the systems. The NAEHMP uses fish, macroinvertebrate and riparian vegetation as its primary biological indicators, while abiotic indicators such as habitat, geomorphology, hydrology and water chemistry form the framework for the interpretation of the biotic results. In terms of the NAEHMP, assessment models such as the fish response assessment index (FRAI), vegetation response assessment index (VEGRAI) and macroinvertebrate response assessment index (MIRAI) have been developed for assessing the ecological states of riverine ecosystems [18–20].

At the core of biomonitoring is the search for and identification of suitable biological indicators (i.e. bioindicators), whose presence or absence, abundance and diversity, and behaviour reflect environmental conditions. Over the years, many studies have used bioindicators such as fish, diatoms, algae and macroinvertebrates to assess ecological water quality [21]. However, among the bioindicators, macroinvertebrates are arguably the most widely used groups [22]. Their wide application in biomonitoring can be attributed to their ubiquitous occurrence, abundance and diversity in the aquatic ecosystems. In addition, they can be easily collected and identified to the family level, though species-level identification requires more time and for some taxa may not be possible especially in the Afrotropical region. They have a huge species richness that offers a wide spectrum of environmental responses and they are relatively sedentary, representing local conditions. They provide an indication of environmental conditions over varying times and are differentially sensitive to a variety of pollutants and, consequently, capable of a graded response to stress. They also serve as a critical pathway for transporting and utilising energy and matter in the aquatic ecosystem.

Freshwater macroinvertebrates spend at least part of their lifecycles in the aquatic environment and are large enough to be seen unaided [23]. Depending on the goal of the biomonitoring, they can be monitored for changes in population, community, growth rate and cohorts. They can also be monitored for bioaccumulation of pollutants, as well as for morphological and biochemical changes in cells, tissues, organs and systems. Macroinvertebrates-based biomonitoring approaches include single biotic indices such as the Biological Monitoring Working Party (BMWP) and the South African Scoring System version 5 (SASS5) [24, 25]; multimetric indices, for example, the Index of Biotic Integrity 12 (IBI 12) and the Serra dos Órgãos Multimetric Index (SOMI) [26]; multivariate predictive techniques, for example, the Australian River Assessment System (AUSRIVAS) and the United Kingdom's River Invertebrate Prediction and Classification System (RIVPACS, UK) [27] and finally the traits-based techniques.

A multivariate predictive technique evaluates aquatic ecosystem condition by comparing biota at a site to those expected to occur in the absence of human disturbances [16]. A predictive model is constructed using reference sites' biotic communities and correlating the community to natural environmental variables using multivariate statistics to predict expected communities at the impacted sites. A multimetric approach on the other hand combines metrics representing several aspects of macroinvertebrate attributes (e.g. structure, function and processes) to indicate river health. Bonada et al. [16] assessed the utilities, strengths and weaknesses of both approaches using a set of 12 criteria in 3 categories: rationale, implementation and performance. Out of the 12 criteria evaluated, the multivariate approach satisfies 9, while the multimetric fulfils 10.

3.3. Aquatic ecotoxicology

Protecting water resources requires a thorough understanding of the mechanisms by which pollutant(s) or toxicant(s) influence the aquatic ecosystems. This often involves experimental manipulation to establish an evidence-based cause-effect relationship between the toxicant and the observed effects on the organism. Aquatic ecotoxicology is the third pillar of the EWQ, and it provides data needed to explore a cause-effect relationship between stressors and biota [28]. The traditional approach to aquatic ecotoxicology is the single-species tests in the laboratory. Depending on the duration of the exposure and the endpoints measured, these tests are termed acute or chronic. Acute toxicity tests are short term, usually lasting between 48 and 96 h, measuring mortality as an endpoint [29]. Chronic toxicity tests last longer, and in addition to long-term mortality, sub-lethal effects on organismal attributes such as growth, reproduction, behaviour, enzymatic activities and histology are also measured. Many of these single-species acute and chronic toxicity tests have been standardised and are widely use in the ecosystem-based approach to managing water quality [30]. The strengths of the laboratory single-species tests include (i) precision: they are conducted in a highly regulated environment, where external influences are isolated, so that there is a high level of precision with regard to the toxicant effects on the organism; (ii) repeatability: single-species, laboratory-based experiments are easily reproducible and repeatable, provided that sets, guidelines and protocols are followed; (iii) high level of acceptance in the regulatory circle: these tests still form the cornerstone of risk assessments of harmful chemicals in the environment; (iv) simplicity: these test are usually very simple to undertake, hence their appeal in regulatory circles.

Although the single-species laboratory-based tests are widely used in managing aquatic ecosystems, they are unable to provide direct community or ecosystem-level effects. They rely heavily on laboratory to field extrapolations by applying safety assessment factors or the species sensitivity distribution (SSD) approach [31]. Reducing uncertainties requires using more ecologically relevant and realistic assessments that employ multi-species in experimental settings that are closer to the natural field conditions. The multi-species model-stream ecosystem approach occupies an intermediate space between field biomonitoring studies and the traditional single-species laboratory-based approach. If reasonably controlled, manipulated and replicated, they can simulate community and even ecosystem effects [32]. While the single-species approach offers high degree of precision, repeatability and simplicity, model-stream ecosystems represent a compromise between these factors, and their high environmental realisms [32].

Model-stream ecosystems are termed mesocosms or microcosms depending on their sizes and locations [33–37]. For example, Odum [33] defined mesocosms as outdoor experimental streams bounded and partially closed, which closely simulate the natural conditions. Buikema and Voshell [34] use volume as a factor for differentiating between microcosms and mesocosms, referring to microcosms as experimental streams (usually indoor) with a volume equal or less than 10 m^3 and mesocosms as those (usually outdoor) having a volume greater than 10 m^3. Hill et al. [35] defined mesocosms as experimental streams that are more than 15 m long and microcosms as those that are shorter. However, Belanger [36] review revealed that increased physical sizes of experimental streams did not correspond to increased biological complexity. Since the goal of a multi-species model-stream ecosystem is to achieve an adequate ecological realism irrespective of size, the terms 'microcosm' and 'mesocosm' are actually inappropriate. Instead, the appropriate terminologies should be a 'model-stream ecosystem', 'experimental streams or artificial streams'.

Model-stream ecosystems have some advantages over conventional single-species toxicity tests. They enable the simulation of natural conditions, offering a high degree of environmental realism and enabling complex biophysical interactions. They enable the researcher to evaluate direct effects of pollutants at higher biological organisation such as population, community and even ecosystem levels [37]. Moreover, they enable the study of biotic interaction and community dynamics and measurement of indirect ecosystem effects. Their shortcomings are that they are not easily reproducible, have low precision and are not simple to undertake.

4. Case study of the application of biomonitoring for the assessment of the ecosystem health in the Swartkops River

This case study illustrates the application of biomonitoring in the Swartkops River using the South African Scoring System version 5.

4.1. The South African Scoring System version 5

The South African Scoring System version 5 (SASS5) is a rapid bioassessment index based on the presence or absence of selected families of aquatic macroinvertebrates and their perceived sensitivity or tolerance to deteriorating water quality [24]. In SASS5, macroinvertebrate families are awarded scores in the range of 1–15 in increasing order of sensitivity to deteriorating water quality. Families considered sensitive are awarded high scores and those considered tolerant low scores. The results are expressed both as an index score, that is, SASS5 score, and as an average score per recorded taxon (ASPT) value. The SASS5 score is calculated by summing the scores of all recorded families, while the ASPT value is obtained by dividing the total SASS5 score by the number of families recorded. In addition to being a useful water quality assessment index, SASS5 is used to assess emerging water quality problems, development impacts, ecological state and spatio-temporal trends of biological assemblages.

4.2. The study area

The Swartkops River originates in the foothills of the Groot Winterhoek Mountains and then meanders through the towns of Uitenhage, Despatch and Perseverance before discharging into the Indian Ocean at Algoa Bay, near the city of Port Elizabeth (**Figure 1**). Climate in the catchment is warm and temperate, and rainfalls vary between the upper and lower regions. The upper region usually receives higher rainfall than the lower region. The catchment geology is mainly of marine, estuarine and fluvial origin. Soils in the upper catchment are not deep and are unsuitable for agriculture. Those in the low-lying floodplain region are deep and well suited for agriculture. The dominant vegetation in the catchment is bushveld and succulent thicket.

Although the river is an important ecological and socio-economic asset, serving as a home to important bird and fish species, and providing water for small-scale irrigation, the health and functionality of the entire system are being threatened by deteriorating water quality. Several sources of pollution including raw sewage run-off from informal settlements, treated waste-water effluent discharges from municipal treatment works, agricultural farmlands, surrounding road and rail networks, and industrial sites were all influencing the water quality of the river and hence the need to assess its health using the SASS5.

4.3. Sampling sites and macroinvertebrates sampling

Four sites within the same ecoregion were selected for the study. Site 1 (33°45′08.4″ S, 25°20′ 32.6″ E), situated in the upper reaches of the river was the least impacted and thus was chosen as the control site. It has a well diverse range of macroinvertebrate sampling habitats. Site 2

(33°47′29.0″ S, 25°24′26.4″ E) was in the industrial town of Uitenhage, where surrounding impacts include run-off from roads and informal settlements, free-ranging livestock and other agricultural practices. All macroinvertebrate sampling biotopes were adequately represented at the site. Site 2 is situated upstream of the discharge point of the Kelvin Jones wastewater treatment work (WWTW) in the town of Uitenhage. Site 3 (33°47′11.8″ S, 25°25′53.97″ E) is further downstream, but also within the industrial town of Uitenhage, where surrounding impacts include industrial and wastewater effluent discharges, run-off from road and rail networks, and agricultural activities. The Kelvin Jones WWTW is the main pollution source at Site 3. Macroinvertebrate sampling biotopes at Site 3 were also adequate. Site 4 (33°47′34.0″ S, 25°27′58.7″ E) further downstream of Site 3 was situated in the residential town of Despatch. Municipal run-off, sand and gravel mining on the riparian zone were the main impacts at Site 4. Although Site 4 was not as polluted as Site 3, it would have been good to select another site further downstream to monitor for potential system recovery. However, the tidal limit at Perseverance between the estuary and the freshwater section is only a short distance downstream of Site 4. Consequently, it was not possible to select a fifth site further downstream because of likely estuarine effects.

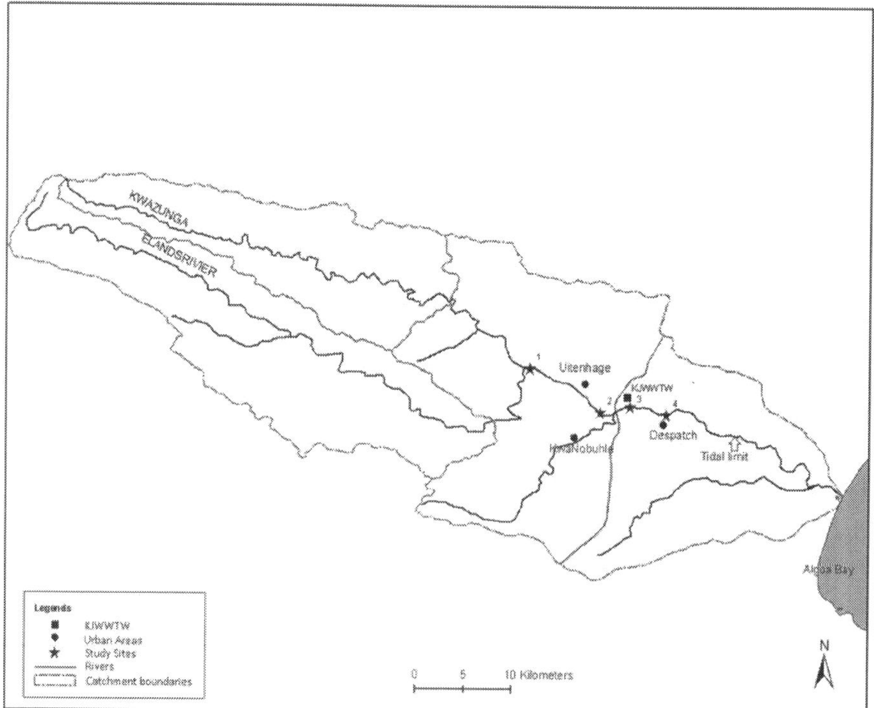

Figure 1. Map of the Swartkops River showing the sampling sites and the relative position of the Kelvin Jones Wastewater Treatment Works.

Macroinvertebrates were sampled using the SASS5 protocol. At each site, over a period of three years, between late August 2009 and September 2012, samples were collected seasonally. A total of eight sampling events were conducted over the sampling period. Macroinvertebrates were collected using a kick net (300 × 300 mm frame, 1000 µm mesh) from three distinct biotope groups: stones (stones-in-and-out-of-current), vegetation (marginal and aquatic vegetation) and sediment (gravel, sand and mud, GSM) as prescribed in the SASS5 protocol.

Sampled macroinvertebrates were tipped into a white rectangular tray, half-filled with river water, and macroinvertebrate families identified on site using identification keys by Gerber and Gabriel [38]. The identified families were recorded on a SASS5 sheet together with their abundance estimates. SASS5 scores, number of taxa and ASPT values were calculated and then interpreted as described in the following section. Time spent on field identification adhered strictly to recommendation in the SASS5 protocol.

4.4. Interpretation of macroinvertebrate data based on the SASS5 protocol for river health

Guidelines developed by Dallas [39] were used for the interpretation of the macroinvertebrate data. The guidelines stipulate range of SASS5 scores and ASPT values indicative of different ecological categories reflective of water quality/river health conditions for the upper and lower areas of each geo-morphological zone in South Africa. The Swartkops River is within the southern eastern coastal belt (lower zone) and the ranges of SASS5 scores and ASPT values for this zone were applied in this study to interpret the SASS5 data in order to determine the Swartkops River health condition (**Table 1**).

4.5. Water sampling and physico-chemical analyses

Basic water physico-chemical analyses were undertaken at each site at the same time when macroinvertebrates were sampled. Dissolved oxygen (DO), electrical conductivity (EC), turbidity, temperature and pH were measured using CyberScan DO 300, CyberScan Con 300, Orbeco-Hellige 966, mercury-in-glass thermometer and CyberScan pH 300 m, respectively. Five-day biochemical oxygen demand (BOD_5) was analysed according to APHA [40].

Ecological category	Water quality category name	Description	Range of SASS5 scores	Range of ASPT values
E/F	Very poor	Seriously/critically modified	<62.9	<5
D	Poor	Largely modified	63–81.9	5.1–5.3
C	Fair	Moderately modified	82–99.9	5.4–5.9
B	Good	Largely natural with few modifications	100–148.9	6.0–7.0
A	Natural	Unmodified	149–180	7.1–8

Table 1. Range of SASS5 scores and ASPT values indicative of the different ecological categories and water quality for the southern eastern coastal belt lower zone ecoregion [39].

4.6. Statistical analysis

One-way analysis of variance (ANOVA) was used to test for differences ($p < 0.05$) in the means of the analysed physico-chemical variables between the four sampling sites. When ANOVA indicated significant differences, a post hoc test, the Tukey's Honestly Significant Different (HSD) test was computed to indicate sites that differed. The basic assumptions of normality and homogeneity of variance were investigated using the Shapiro-Wilk test and the Levene's test, respectively. The nonparametric Kruskal-Wallis multiple comparison test was used to evaluate whether SASS5 scores, ASPT values and the number of taxa differed significantly between the biotope groups. ANOVA and Kruskal-Wallis multiple comparison tests were undertaken using the Statistica software package version 9.

4.7. Results

4.7.1. Water physico-chemical variables

Table 2 shows the mean, standard deviation and range of physico-chemical variables measured during the study period. With the exception of pH and temperature, the measured variables were statistically significantly different between the sampling sites ($p < 0.05$). The lowest value of DO and highest turbidity and BOD_5 values were recorded at Site 3. The Tukey's HSD post hoc test revealed that the mean DO concentration was significantly lower at Site 3 than at Sites 1 and 2. Although pH and temperature were not statistically significantly different between the sampling sites, the highest mean pH and temperature values were at Sites 2 and 3, respectively, and the lowest at Sites 1 and 2, respectively. The Tukey's HSD post hoc test showed that the mean EC concentration was significantly lower at Site 1 than at the rest of the sampling sites and turbidity significantly higher at Site 3. The mean BOD_5 concentrations were significantly higher at Sites 3 and 4 than at Site 1 (**Table 2**).

Variable	Site 1	Site 2	Site 3	Site 4	p value	F value
Dissolved oxygen (mg/l)	6.99 ± 1.15^a (4.73–9.5)	7.4 ± 1.52^a (5.53–9.48)	3.19 ± 1.47^b (1.81–6.36)	4.81 ± 3.01^{ab} (0.9–8.31)	0.001	7.18
pH	6.53 ± 1.11 (4.69–7.75)	7.37 ± 1.11 (5.69–8.99)	7.29 ± 0.42 (6.56–7.9)	7.27 ± 0.56 (6.31–8.01)	0.201	1.65
Temperature (°C)	17.48 ± 5.46 (7.31–24.0)	17.27 ± 7.17 (6.11–27.3)	20.88 ± 3.29 (14.3–25.2)	18.9 ± 4.14 (12.2–24.0)	0.415	0.98
Electrical conductivity (mS/m)	32.45 ± 17.74^a (8.23–62.0)	160.75 ± 146^b (30–460)	262.51 ± 76.14^b (154.8–333)	259.63 ± 56.28^b (171–354)	0.000	22.57
Turbidity (NTU)	5.3 ± 2.22^a (3.0–10.1)	6.33 ± 2.44^a (3.0–11.2)	72.7 ± 102.36^b (10.5–320)	7.08 ± 8.06^a (2.2–26)	0.000	15.67
BOD_5 (mg/l)	4.62 ± 1.45^a (2.16–6.86)	8.25 ± 4.33^{ab} (4.58–16.68)	14.54 ± 3.57^c (8.32–20.62)	11.77 ± 5.28^{bc} (2.24–22.94)	0.002	13.50

Table 2. Mean ± standard deviation and range (in parenthesis) of the physico-chemical variables ($n = 8$) in the Swartkops River during the study period (August 2009–September 2012). p and F values are indicated by ANOVA. Different superscript letters per variable across sites indicate significant differences ($p < 0.05$) revealed by Tukey's HSD post hoc test. The same superscript letter between sites per variable indicates no significant differences ($p > 0.05$).

4.7.2. Assessing the Swartkops River health using the South African Scoring System version 5 (SASS5)

The interpretation of the SASS5 results were based on the range of SASS5 scores and ASPT values reflecting ecological categories A, B, C, D and E/F indicative of natural, good, fair, poor and very poor water quality conditions, respectively (**Table 1**). The SASS5 scores and ASPT values revealed that the Swartkops river health conditions differed between the sampling sites. Seasonally, with the exception of the autumn and spring (2012) collections, SASS5 scores at Site 1 indicated the B ecological category indicative of good water quality condition (**Figure 2**). The ASPT values on the other hand, in all the sampling seasons, indicated the C ecological category for Site 1, suggesting that the water quality at Site 1 was fair (**Figure 3**). The numbers of taxa vary slightly between the sampling seasons at Site 1 with more taxa occurring in spring (2012) (**Figure 4**). Overall, the SASS5 score showed good water quality (ecological category B) for Site 1, but the ASPT value indicated that the water quality condition at the site was fair (ecological category C) (**Figure 5**).

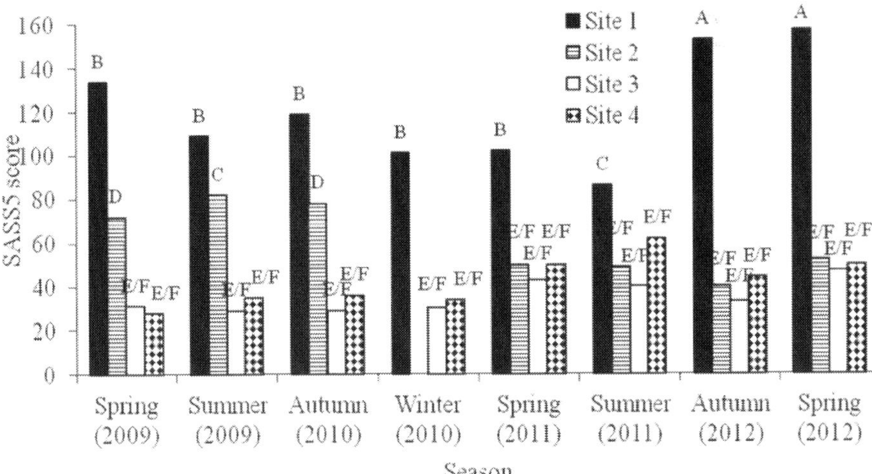

Figure 2. Seasonal variations for the South African Scoring System version 5 (SASS5) score at the four sampling sites in the Swartkops River during the study period (August 2009–September 2012). The ecological categories: A (natural water quality), B (good water quality), C (fair water quality), D (poor water quality) and E/F (very poor water quality) are indicated on the bars.

At Site 2, SASS5 scores indicated the D ecological category, that is, poor water quality in spring (2009) and in autumn (2010), while in summer (2009), it revealed the C category indicative of fair water quality (**Figure 2**). During the rest of the sampling events, SASS5 scores revealed the E/F ecological category indicating very poor water quality. Although the SASS5 scores reflected other ecological categories in addition to the E/F for Site 2, the ASPT values consistently showed that Site 2 was in the E/F ecological category (**Figure 3**). Although the number of taxa did not vary significantly between the sampling seasons at Site 2, the highest number of taxa (20) was recorded during autumn (2010). At Sites 3 and 4, SASS5 scores and ASPT values revealed the E/F ecological category (very poor water quality) throughout the sampling seasons. The overall lowest number of taxa (8) in the river was recorded at Site 3 in winter 2010 (**Figure 4**).

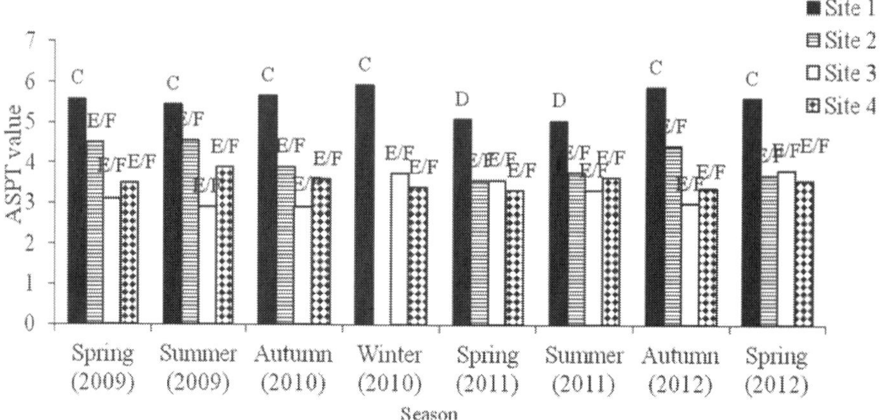

Figure 3. Seasonal variations for the average score per recorded taxon (ASPT) at the four sampling sites in the Swartkops River during the study period (August 2009–September 2012). The ecological categories: C (fair water quality), D (poor water quality) and E/F (very poor water quality) are indicated on the bars.

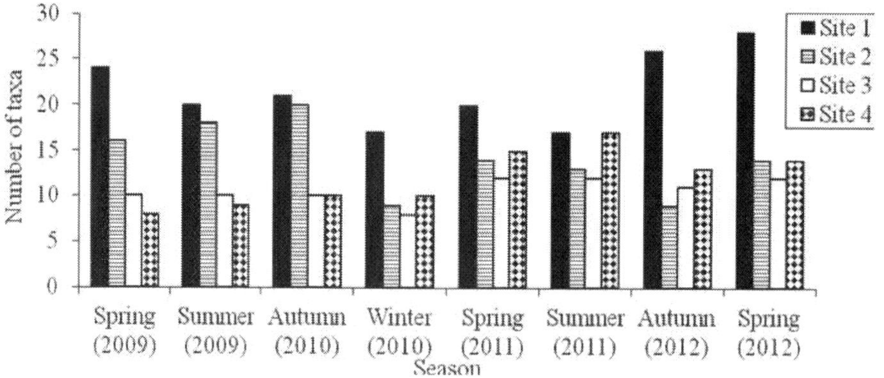

Figure 4. Seasonal variations for the number of taxa at the four sampling sites in the Swartkops River during the study period (August 2009–September 2012).

4.7.3. Comparing SASS5 scores, ASPT values and the numbers of taxa between the sampling biotopes (stone, vegetation and GSM)

The vegetation and stone biotope had higher SASS5 scores, ASPT values and numbers of taxa than the GSM biotope at Site 1 (**Figure 6**). The Kruskal-Wallis multiple comparison test revealed that SASS5 scores were significantly higher for the vegetation than for the GSM biotope at Site 1 ($p < 0.05$; KW-H = 7.21). Similarly, at Site 2, SASS5 scores were significantly higher for the vegetation than for the GSM biotope ($p < 0.05$; KW-H = 10.13), and though the stone had higher SASS5 scores, they were not statistically higher than the scores recorded for the GSM biotope. The pattern described for Site 2 was similar to those observed for Sites 3 and 4 where the SASS5 scores were significantly higher for the vegetation biotope than the stone and GSM biotopes (Site 3: $p < 0.05$; KW-H = 40.44), (Site 4: $p < 0.05$; KW-H = 18.14).

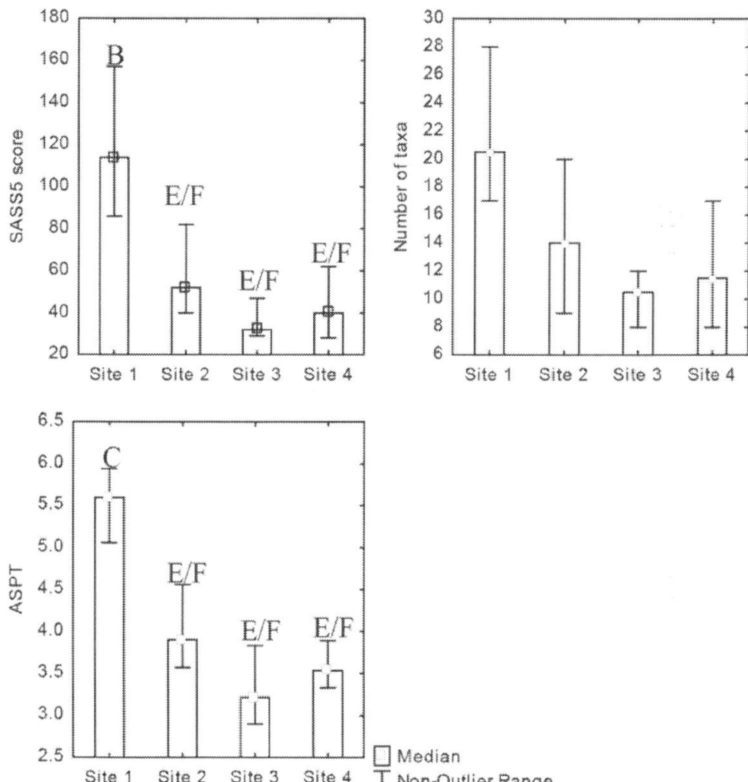

Figure 5. Summary of the SASS5 scores, number of taxa and ASPT values at the four sampling sites in the Swartkops River during the study period (August 2009–September 2012). The overall ecological categories: B (good water quality), C (fair water quality), D (poor water quality) and E/F (very poor water quality) are indicated on the bars.

The average score per recorded taxon (ASPT) values were similar between the three biotopes at Site 1, but at Site 2, the ASPT values were significantly higher for the vegetation than the GSM biotope ($p < 0.05$; KW-H = 9.45). The vegetation had significantly higher ASPT values than the stone and GSM biotopes at Sites 3 ($p < 0.05$; KW-H = 26.9) and 4 ($p < 0.05$; KW-H = 14.25).

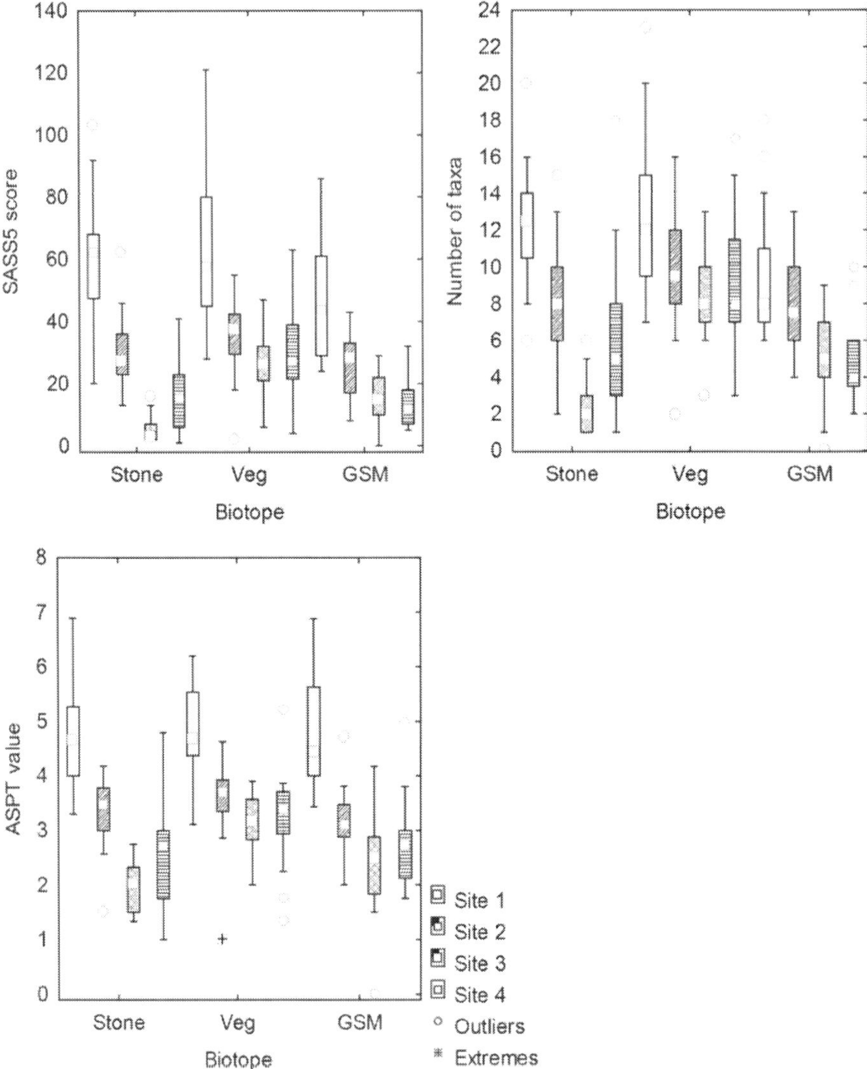

Figure 6. Median (small square), inter-quartile ranges (box), non-outlier ranges (bars) for SASS5 scores, numbers of taxa and ASPT values recorded per biotope at the four sampling sites in the Swartkops River during the study period (August 2009–September 2012).

Stone and vegetation biotopes supported significantly higher numbers of taxa than the GSM biotope at Site 1 ($p < 0.05$; KW-H = 11.89), but at Site 2, only the vegetation supported significantly higher numbers of taxa than the GSM ($p < 0.05$; KW-H = 7.23). More taxa were recorded on the vegetation and GSM biotopes than on the stone biotopes at Site 3. The Kruskal-Wallis multiple comparison test indicated that the numbers of taxa for the stone biotope were significantly lower than the taxa recorded for the vegetation and GSM ($p < 0.05$; KW-H = 40.44) at Site 3. At Site 4, the stone and vegetation supported more taxa, but only the numbers of taxa supported by the vegetation biotope were significantly higher than the values recorded for the GSM ($p < 0.05$; KW-H = 16.27).

4.8. Discussion

The ecosystem approach takes into account biodiversity conservation and therefore prioritises the protection of biodiversity as well as the sustainable use of water resources and the associated ecosystems. In the case study provided, the South African Scoring System version 5 (SASS5) was used to evaluate the health of the Swartkops River. In South Africa, SASS5 is one of the tools that contribute ecological information for the determining the ecological reserve and setting Resource Quality Objectives (RQOs). The SASS5 results indicated that water quality in the Swartkops River was critically modified at Sites 3 and 4 throughout the sampling period and the numbers of taxa occurring at these sites were significantly reduced compared to those occurring at Sites 1 and 2. Sites 3 and 4 were situated downstream of a WWTWs, which influenced the health of the river. The values of the measured physico-chemical variables at these sites, that is, Sites 3 and 4, provided evidence for negative impact arising from the discharges of wastewater effluents. For example, at Sites 3 and 4, higher values of turbidity and EC concentrations and lower DO concentrations were recorded. Since highly sensitive taxa have higher scores in the SASS5 sheet, oxygen depletion could easily affect the occurrence and distribution of these taxa. Therefore, it was expected that sites with low concentration of DO would experience the disappearance of sensitive taxa and the dominance of tolerant taxa, and hence, the critically modified health conditions recorded at Sites 3 and 4. In addition to lower DO concentrations at the downstream sites, the elevated turbidity level recorded at Site 3 could be detrimental to oxygen-sensitive biota as decomposition of solids with high organic content could lead to oxygen depletion, as was evident at Sites 3 and 4. The majority of the highly sensitive taxa on the SASS5 sheet use external gills for respiration. Highly turbid water is likely to impact on the breathing apparatus of external gill-bearing organisms, which can then lead to clogging [41]. The river health condition at Site 2, which was upstream of the effluent discharge point, but still situated within the urban and industrial town of Uitenhage, was mostly in the range of fair and very poor conditions. Diffuse pollution sources on the river catchments were the main contributors to deteriorating river health recorded at this site. Site 1, which was used as the control site, had conditions mostly in the good and fair categories. The implication was that the control site had some sensitive taxa, which had disappeared from the impacted sites.

The number of taxa, SASS5 scores and ASPT values were highest mostly in the stone and vegetation biotopes and differed significantly between the three biotopes. These differences could be attributed to differences in hydraulic, substrate and thermal conditions between the three biotopes. The stone and vegetation biotopes are morphologically complex and more stable than the GSM biotope and are therefore more likely to support more food and space resources, and thus more macroinvertebrate families leading to increased SASS5 scores and ASPT values. These results are in agreement with those of Dallas [42] who reported that the stone and vegetation biotopes supported more macroinvertebrate families and higher SASS5 score and ASPT values than the GSM biotope. It is therefore important to sample all available biotopes to capture a wider range of biodiversity when undertaken aquatic biomonitoring.

In summary, the deteriorating environmental water quality in the Swartkops River has impacted on the macroinvertebrate assemblages particularly at the downstream sites. This was expected because of the ranges of impacts these sites receive which include industrial and sewage effluent discharges, run-off from informal settlement and agricultural activities such as livestock farming. Water quality at Site 1 which was used as the control site in this study was indicated as good and fair by the SASS5 score and ASPT value, respectively. This is a cause for concern as the results showed that macroinvertebrates at this site were experiencing noticeable impacts. Overall, both the physico-chemical variable analysis and the biotic index results revealed that the Swartkops River was deteriorating in quality as it flowed down-stream, indicating the need for an urgent management intervention.

5. Conclusion

In this chapter, the ecosystem-based approach to managing water quality was critically reviewed with a clear focus on environmental water quality (EWQ). The three pillars to EWQ were discussed and their contributions and limitation highlighted. Of particular interest is that, in this chapter, the relevance of the EWQ approach was discussed with respect to its application to water resources management in South Africa. It is argued that the EWQ is an integrative approach for sound and sustainable management of water quality. The biomonitoring case study illustrated the utility of one of the three pillars of the EWQ approach.

Acknowledgements

The South African National Research Foundation (NRF) is acknowledged for providing post-doctoral grant (Grant no.: 88517) to Dr. O.N. Odume for this research. The Carnegie Corporation of New York through the Regional Initiative in Science and Education (RISE) is also acknowledged for doctoral bursary for this work. Rhodes University Research Committee (RC) is acknowledged for a research grant.

Author details

Oghenekaro Nelson Odume

Address all correspondence to: nelskaro@yahoo.com

Unilever Centre for Environmental Water Quality, Institute for Water Research, Rhodes University, Grahamstown, South Africa

References

[1] Bunn, S.E. (2016) Grand challenge for the future of freshwater ecosystems. *Frontiers in Environmental Science* **4(21)**: 1–4.

[2] Woznicki, S.A., Nedadhashemi, A.P., Tang, Y., and Wang, L. (2016) Large-scale climate change vulnerability assessment of stream health. *Ecological Indicators* **69**: 578–598.

[3] Masese, F.O., Omokoto, J.O., and Nyakeya, K. (2013) Biomonitoring as a prerequisite for sustainable water resources: a review of current status, opportunities and challenges to scaling up in East Africa. *Ecohydrology and Hydrobiology* **13(3)**: 173–191.

[4] Brown, P.G., and Schmidt, J.J. (2010) An ethic of compassionate retreat. In: Brown, P.G. and Schmidt, J.J. editors *Water ethics — foundational readings for students and professionals*. Island Press, Washington. p. 265–286.

[5] Department of Water Affairs (2013) National water resource strategy. Second edition. Department of Water Affairs, Pretoria, South Africa.

[6] Pollard, S. and du Toit, D. (2008) Integrated water resource management in complex systems: how the catchment management strategies seek to achieve sustainability and equity in water resources in South Africa. *Water SA* **34(6)**: 671–679.

[7] Doulgeris, C., Georgiou, P., Papadimos, D., and Papamichail, D. (2012) Ecosystem approach to water resources management using the Mike 11 modeling system in the Strymonas River and Lake Kerkini. *Journal of Environmental Management* **94**: 132–143.

[8] Thoms, M.C., and Sheldon, F. (2002) An ecosystem approach for determining environmental water allocations in Australian dryland river systems: the roe of geomorphology. *Geomorphology* **47**: 153–168.

[9] King, J., and Pienaar, H. (2011) Sustainable use of South Africa's inland waters. WRC Report No. TT 491/11. Water Research Commission, Pretoria, South Africa.

[10] DWAF — (Department of Water Affairs and Forestry) (2008b) Draft regulations for the establishment of a water resource classification system. Government Gazette No. 31417, Pretoria, South Africa.

[11] Odume, O.N. (2014) Macroinvertebrates-based biomonitoring and ecotoxicological assessment of deteriorating environmental water quality in the Swartkops River, South Africa. PhD thesis, Rhodes University, Grahamstown, South Africa.

[12] Scherman, P.A., Muller, W.J., and Palmer, C.G. (2003) Links between ecotoxicology, biomonitoring and water chemistry in the integration of water quality into environmental flow assessments. *River Research and Applications* **19**: 1–11.

[13] Palmer, C.G., Berold R., and Muller W.J. (2004) Environmental water quality in water resource management. WRC Report No.TT 217/04, Water Research Commission, Pretoria, South Africa.

[14] Hohls, B.C., Silberbauer, M.J., Kühn, A.L., Kempster, P.L., and van Ginkel, C.E. (2002) National water resource quality status report: inorganic chemical water quality of surface water resources in SA—the big picture. Report No. N/0000/REQ0801. ISBN No. 0-621-32935-5. Institute for Water Quality Studies, Department of Water Affairs and Forestry, Pretoria, South Africa. http://www.dwaf.gov.za/iwqs/water_quality/NCMP/ReportNationalAssmt3c.pdf [Accessed 28 May, 2012]

[15] Extence, C.A., Chadd, R.P., England, J., Dunbar, M.J., Wood, P.J., and Taylor, E.D. (2013) The assessment of fine sediment accumulation in rivers using macroinvertebrate community response. *River Research and Applications* **29**: 17–55.

[16] Bonada, N., Prat, N., Resh, V.H., and Statzner, B. (2006) Development in aquatic insect biomonitoring: a comparative analysis of recent approaches. *Annual Review of Entomology* **51**: 495–523.

[17] Department of Water Affairs and Forestry (2003) National Aquatic ecosystem biomonitoring programme—compiling State of the rivers report and posters: a manual. NAEBP Report Series No. 17. DWAF, Pretoria, South Africa.

[18] Kleynhans, C.J., and Louw, M.D. (2008) River ecoclassification: manual for ecostatus determination (version 2)—Module A: EcoClassification and EcoStatus determination. Joint Water Research Commission and Department of Water Affairs and Forestry Report. WRC Report No TT 329/08. Water Research Commission, Pretoria, South Africa.

[19] Kleynhans, C.J. (2008) River ecoclassification: manual for ecostatus determination (version 2)—Module D volume 1: Fish Response Assessment Index. Joint Water Research Commission and Department of Water Affairs and Forestry Report. WRC Report No TT 330/08. Water Research Commission, Pretoria, South Africa.

[20] Thirion, C. (2008) River ecoclassification: manual for ecostatus determination (version 2) —Module E: macroinvertebrate response assessment index in. Joint Water Research Commission and Department of Water Affairs and Forestry Report. Water Research Commission, Pretoria, WRC Report No. TT 332/08.

[21] Pace, G., Bella, V.D., Barile, M., Andreani, P., Mancini, L., and Belfiore, C. (2012) A comparison of macroinvertebrate and diatom responses to anthropogenic stress in a small sized volcanic siliceous streams of central Italy (Mediterranean Ecoregion). *Ecological Indicators* **23**: 544–554.

[22] Murphy, J.F., Davy-Bowker, J., McFarland, B., and Ormerod, S.J. (2013) A diagnostic biotic index for assessing acidity in sensitive streams in Britain. *Ecological Indicators* **24**: 562–572.

[23] Rosenberg, D.M., and Resh, V.H. editors. (1993) *Freshwater biomonitoring and benthic macroinvertebrates*, p. 488. Chapman and Hall One Penn Plaza, New York, NY.

[24] Dickens, C.W.S., and Graham, P.M. (2002) The South African Scoring System (SASS) version 5 rapid bioassessment method for rivers. *African Journal of Aquatic Science* **27**: 1–10.

[25] Walley, W.J., and Hawkes, H.A. (1996) A computer-based reappraisal of the biological monitoring working party scores using data from the 1990 river quality survey of England and Wales. *Water Research* **30**: 2086–2094.

[26] Baptista, D.F., Bus, D.F., Egler, M., Giovanelli, A., Silveira, M.P., and Nessimian, J.L. (2007) A multimetric index based on benthic macroinvertebrates for evaluation of Atlantic Forest streams at Rio de Janeiro State, Brazil. *Hydrobiologia* **575**: 83–94.

[27] Jorgensen S.E., Costanza, R., and Xu, F.L. (2005) Handbook of ecological indicators for assessment of ecosystem health. CRC Press Boca Raton, USA.

[28] Rand, G.M., (Ed.) (1995) Fundamentals of aquatic toxicology—effects, environmental fates, and risk assessment (2nd edn), p. 1148. Taylor and Francis, Washington D.C., USA.

[29] Mensah, P.K., Muller, W.J., and Palmer, C.G. (2011) Acute toxicity of Roundup® herbicide to three life stages of the freshwater shrimp *Caridina nilotica* (Decapoda: Atyidae). *Physics and Chemistry of the Earth* **36**: 905–909.

[30] Moiseenko, T.I. (2008) Aquatic ecotoxicology: theoretical principles and practical application. *Water Resources* **35(5)**: 530–541.

[31] Schmitt-Jansen, M., Veit, U., Dudel, G., and Altenburger, R. (2008) An ecological perspective in aquatic ecotoxicology: approaches and challenges. *Basic and Applied Ecology* **9**: 337–345.

[32] Ledger, M.E., Harris, R.M.L., Armitage, P.D., and Milner, A.M. (2009) Realism of model ecosystems: an evaluation of physicochemistry and macroinvertebrate assemblages in artificial streams. *Hydrobiologia* **617**: 91–99.

[33] Odum, E.P. (1984) The mesocosm. *Bioscience* **34**: 558–562.

[34] Buikema, A.L., and Voshell, J.R. (1993) Toxicity studies using freshwater benthic macroinvertebrates. In: Rosenberg, D.M., and Resh, V.H. (Eds.), *Freshwater biomonitoring and benthic macroinvertebrates*. Chapman and Hall one penn plaza, New York, NY. pp. 344–398.

[35] Hill, I.R., Heimbach, F., Leeuwangh, P., and Matthiessen, P. (Eds.) (1994) Freshwater field tests for hazard assessment of chemicals, p. 561. CRC Press, Boca Raton, FL.

[36] Belanger, S.E. (1997) Literature review and analysis of biological complexity in model stream ecosystems: influence of size and experimental design. *Ecotoxicology and Environmental Safety* **36**: 1–16.

[37] Choung, C.B., Hyne, R.V., Stevens, M.M., and Grant C.H. (2013) The ecological effects of a herbicide-insecticide mixture on an experimental freshwater ecosystem. *Environmental Pollution* **172**: 264–274.

[38] Gerber, A., and Gabriel, M.J.M. (2002) Aquatic invertebrates of South African Rivers—field guide. Resource quality services, Department of Water Affairs, Pretoria.

[39] Dallas, H.F. (2007) River health programme: South African scoring system (SASS) data interpretation guidelines. Institute of Natural Resources and Department of Water Affairs and Forestry, Pretoria, South Africa. Available online: http://safrass.com/reports/SASS% 20Interpretation%20Guidelines.pdf Accessed: 18 October, 2010.

[40] APHA—(American Public Health Association) (1992) Standard methods for the examination of water and waste water (18th edn). APHA, Washington DC, USA.

[41] Bilotta, G.S., and Brazier, R.E. (2008) Understanding the influence of suspended solids on water quality and aquatic biota. *Water Research* **42**: 2849–2861.

[42] Dallas, H.F. (2007) The influence of biotope availability on macroinvertebrate assemblages in South African rivers: implications for aquatic bioassessment. *Freshwater Biology* **52**:370–380.

Chapter 2

Macroinvertebrates and Fishes as Bioindicators of Stream Water Pollution

Pablo Fierro, Claudio Valdovinos,

Luis Vargas-Chacoff, Carlos Bertrán and

Ivan Arismendi

Additional information is available at the end of the chapter

http://dx.doi.org/10.5772/65084

Abstract

Freshwater ecosystems worldwide have been progressively deteriorated during the past decades due to an increasing human pressure that has lead to a decrease in aquatic biodiversity. Among the human activities of high impact on freshwater ecosystems is the land-use change, principally from native forests to agriculture. To evaluate the impacts of human activities on water quality, a traditional approach has considered the use of single physical-chemical parameters. However, this approach may be insufficient to fully assess the impact of these human activities on freshwaters. Therefore, there is a need for alternative tools such as the indices of biotic integrity that may provide a complement to traditional approaches. In the literature, there are several examples of biotic indicators that have shown promising results in evaluating water quality including the use of macroinvertebrates and fish diets. Here, we provide a review of the indicators of biotic integrity that included fish assemblages as well as macroinvertebrates as bioindicators. We identify pros and cons of using aquatic communities as indicators of water quality. Finally, we develop a procedure that combines fish and macroinvertebrate assemblages as bioindicators and discuss their effectiveness using illustrative examples from streams under several agricultural uses in the Mediterranean region of Chile.

Keywords: biological monitoring, biotic index, macroinvertebrates, fishes, Mediterranean, Chile

1. Introduction

Of all the water on earth, freshwater accounts for just 0.01% and covers only 0.8% of the planet's surface [1]. Freshwaters are among the most threatened ecosystems of the world, and thus, understanding their health statuses is of special relevance. Indeed, the physical, chemical, and biological integrities of water are highly important for successfully implementing conservation and management strategies before ecosystem health or biotic integrity are affected [2–4]. This chapter provides a review of known biotic integrity indicators, including benthic macroinvertebrate and fish communities that have been proposed to serve as water quality indicators. In addition, the pros and cons of using aquatic communities as water quality indicators are discussed. Finally, we present a research case study in which benthic macroinvertebrate and fish communities are used as bioindicators, in addition to discussing the effectiveness of using illustrative examples for streams subject to several agriculture uses in a region of Chile dominated by agricultural activities.

Worldwide, a primary threat to freshwater ecosystems is the rapid changes occurring in land uses (**Figure 1**), a situation that has intensified over the past decade [5, 6]. Most recent land use conversion has been for crop production, which notably impacts proximal ecosystems due to changes over extensive crop areas [7]. In particular, the fertilizers and pesticides used in agriculture negatively affect freshwater ecosystems by draining into rivers, where eutrophication and other negative effects, such as high sediment deposits and postsedimentation, subsequently occur. Furthermore, the extensive land use of farming many times results in landscape deforestation, which often arrives to the riverbank itself. This deforestation can increase the temperature of and quantity of light in river water. When coupled with eutrophication, the trophic changes within the aquatic ecosystem can be disturbed, causing, for example, a decreased quantity of aquatic taxa as compared to rivers with fewer alterations [8, 9].

Figure 1. Examples of land use in the central-south of Chile. Left: Stream nearby corn crops, right: Stream borderer by native forest of the Maule Region watershed (photographs by P. Fierro).

2. Indicators of aquatic ecosystem health

The definition of a healthy ecosystem has been widely debated in the literature. Nevertheless, the definition proposed by Rapport is one of the most widely accepted [10]. This definition states that a healthy ecosystem is defined by the "absence of danger signals in the ecosystem, the ability of the ecosystem to quickly and completely recover (resilience), and/or the lack of risks or threats that push the ecosystem composition, structure, and/or function." The purpose of monitoring aquatic ecosystem health is to identify physicochemical and biological changes arising from anthropogenic impacts [11]. This information is crucial for managers and policy makers to make informed decisions towards improving the environment and, consequently, human health [12].

Traditional techniques for measuring water quality and to establish aquatic health assess a number of physical and chemical parameters of the water. However, these measurements do not accurately account for the real impacts that physicochemical activities have on freshwater ecosystems [13]. Indeed, these parameters interact and evidence accumulative effects over time, the impacts of which can finally affect aquatic biota [14]. Due to this, other measurements that consider non-natural disturbing effects on ecological integrity should be used to calculate the quality of aquatic resources [15]. Indices based on aquatic biota have been widely successful in determining the integrity of aquatic ecosystems [16].

The use of indices that evaluate water quality through biological parameters, such as freshwater ecosystem structure and performance, has considerably increased in recent years and has gained recognition as an important measure for calculating the global integrity of freshwater ecosystems [17–19]. Biological monitoring is advantageous in that it can integrate and reflect accumulative changes over time, which is in contrast to a number of other methods, such as flow regimen, energetic resources, and biotic interactions [20, 21]. Another benefit is that the high fauna diversity found in aquatic ecosystems, which include microorganisms, algae, periphyton, phytoplankton, zooplankton, macroinvertebrates, fish, and mammals, can be included in evaluations of river health [4].

Among fauna, fish and macroinvertebrate assemblages have been highlighted as good bioindicators for monitoring ecosystem degradation related to farming and forestry, as well as to urban and industrial effluents [9, 22]. Diverse proxies are used to measure ecosystem condition, such as species density and the presence/absence of several species in assemblage structures [23]. A notable advantage of using these aquatic biota is the relative simplicity of their capture and sampling [24, 25]. In particular, the sampling of fish assemblages can be performed through electrofishing, a highly common tool, whereas macroinvertebrate sampling is facilitated and simplified by Surber, D-frame dip, and kick nets (**Figure 2**).

Furthermore, recent studies report that the stomach contents of salmonids (i.e., *Oncorhynchus mykiss* and *Salmo trutta*) contain a diversity of invertebrate prey present in the benthos of nonintervened (hereafter termed "native") basins, thereby reflecting anthropogenic impacts to the basin [26]. Related to this, Fierro et al. [6] reported similarities in stomach contents and prey diversity of the benthos in river sections with land use different than in the basin.

Likewise, similarities have been found between rivers with more local perturbation, such as through the effects of dams [27, 28]. Therefore, the *O. mykiss* diet might represent an effective bioindicator for evaluating environmental disturbances within the entire basin [6].

Figure 2. Left: Fish communities sampled using electrofishing. Right: Aquatic macroinvertebrates sampled using a Surber Net (photograph by P. Fierro).

Among the ecological indices commonly used to evaluate river health, three primary groups exist – biotic indices, multivariate methods, and multimetric indices [15, 19]. Of these, multimetric indices are the most recommended since a large quantity of data can be considered and since these indices may also identify the cause(s) of degradation. This information can then be applied to obtain better understandings of ecosystem status [4]. In turn, biotic indices evaluate river health based only on organism tolerance to organic pollution. One of the most well-known biotic indices is the Hilsenhoff Biotic Index [29], which has been widely used and adapted around the world (e.g., [30–32]). Continuing, multivariate methods require the use of models that relate physicochemical properties of rivers with observed organisms, which are represented under reference (relatively pristine) conditions. These models then compare the observed organisms with those that were "expected." This comparative method can ultimately detect potentially degraded areas. The most widely used multivariate index is the River Invertebrate Prediction and Classification System [33], which was first implemented in the UK and then adapted to other countries, including Australia [34]. Finally, multimetric indices capture broad characteristic of community structure and function (metric), thus providing a broader understanding of the events occurring in the river [35]. Multimetric indices are powerful tools for establishing the consequences of human activities. These effects may include a high amount of specific and blurred disturbances (nonpoint pollutant discharge), which encompass impacts arising from agriculture, grazing, deforestation, physical alterations of river or bank habitats, damps, sewage discharges, urban areas, and mining [36, 37]. These indices can be applied in several animal assemblages, plant communities, and ecosystems, including terrestrial, marine, and freshwater environments [35]. Corresponding indices of integrity are frequently performed and applied in fish [38] and macroinvertebrates [39]. A summary that contrasts among the three types of indices is presented in **Table 1**.

	Biotic indices	Multivariate methods	Multimetric indices
Examples	Hilsenhoff Biotic Index. Fish Species Biotic Index	River Invertebrate Prediction and Classification System. Australian River Assessment Scheme	Index of Biotic Integrity. Benthic Index of Biotic Integrity
Advantages	Simple, measure only one disturbance (e.g., organic pollution tolerance)	Model created to predict the species and number of organisms that would be expected to appear in a stream system	Include diverse disturbances. Applicable in several animal/plant groups. Incorporates temporal and spatial scale attributes
Disadvantages	Organisms do not respond to only one disturbance; many more stressors affect distribution in the wild	Created models can be easily changed, making the results uncertain. These methods were developed to find patterns and not establish impact	Limited by sampling technique efficiencies. Seasonal migration of biota influences results. Easy confusions with natural perturbations

Table 1. Summary of the characteristics considered with stream health indices (adapted from [4]).

3. Assessing the ecological integrity of streams

Ecological integrity, which is also referred to as river health or ecological status, is a measure of the global condition of an aquatic ecosystem. This measurement integrates physical, chemical, and biological integrity elements [15, 17]. Importantly, biological integrity is defined as the ability of aquatic ecosystems to support and maintain a balanced and integrated community with adapted organisms and a composition, diversity, and functional organization comparable to natural habitats within the same region [40–42]. Therefore, a loss of integrity indicates any human-induced positive or negative divergence of the system from a natural, model condition [43].

The Index of Biotic Integrity (IBI), which was initially developed for western USA rivers by [28], is the most used index based on fish assemblages. Consequently, the IBI has been adapted for use to numerous rivers on all continents to evaluate stream health [4, 28]. Indeed, since the creation of the IBI, over 2374 researchers, as of 2014, have used, modified, or mentioned the importance of the IBI (Google Scholar). Furthermore, the number of citations for the IBI grew exponential until 2005, at which point citations "stagnated" near 140 studies per year (**Figure 3**).

Worth highlighting, of the studies presented in this review, the most important milestone occurred from 1986 to 1990. During this period, researchers first began adapting and making modifications to indices based on fish, in addition to these indices being applied in reports to the US government. Between 1991 and 1995, integrity indices were developed for several groups, including macroinvertebrates, birds, and zooplankton. Furthermore, this period was witness to index adaptations to marine and estuary environments. Even terrestrial environments were assessed by the IBI to measure the environmental quality of forests. Between 1996 and 2000, the IBI continued to expand to other groups and environments, such as periphyton communities, macrophytes, corals, and wetlands. Corresponding adaptations of the IBI to other continents, including Africa, Europe, and South America (Brazil), also occurred [44, 45]. Since 2001, this index is in use on almost all continents and has been adapted several times to different ecoregions within the same countries.

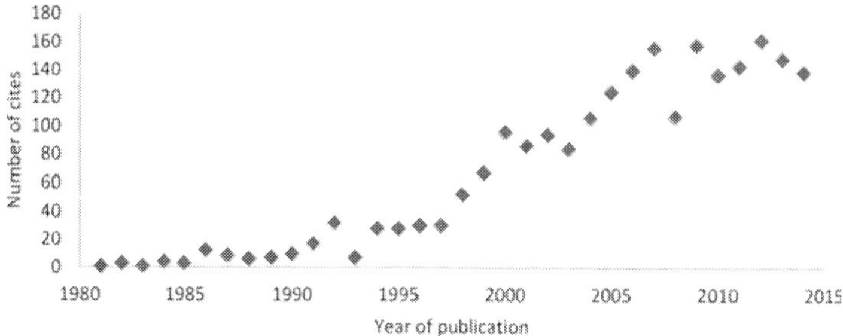

Figure 3. Accumulative number of worldwide publications on the index of biotic integrity around the world, starting with the first related publication by [28] (Source: own elaboration).

The advantage of establishing the biotic integrity of rivers based on fish arises as these organisms are present in all, or almost all, rivers, even those that are polluted. Additionally, extensive life history information is available for many species, and fish assemblages generally represent a variety of trophic levels. Indeed, fish are located within the top of the aquatic food chain and can thus help to provide an integrated view of basin environments. Other benefits of the IBI using fish are that fish populations are relatively stable in the summer, when most monitoring occurs; fish are easily identifiable; and the general public can relate to statements about the conditions of fish assemblages. On the other hand, a noted disadvantage of the IBI is that fish are highly mobile, making sampling difficult. Indeed, large groups of personnel, various tools, and an extended period in the field are needed to record daily and seasonal variations [31].

Although less used, the Benthic Index of Biotic Integrity (B-IBI) was developed by Kerans and Karr [46] for rivers of the Tennessee Valley (USA), using the IBI as an initial base [28]. The advantages of using macroinvertebrates as bioindicators are a great biodiversity and an extreme sensitivity and fast response of many taxa to pollution. This quick response is likely due to many macroinvertebrates being sessile and having aquatic life cycles, thus any alterations in environmental limits could lead to death [14]. One significant disadvantage of the B-IBI is that a taxonomic specialist is needed to identify the macroinvertebrate species, which takes a long time. To address this limitation, Rolls et al. [27] used higher levels of taxonomic identification (e.g., genus, family, or both) as a method for adequately describing taxa traits for B-IBI use. Through this technique, a greater cost-benefit might be obtained as less time will be required to taxonomically identify species. Indeed, in countries with few taxonomists and without access to species-level identification keys, application of the B-IBI is very important, as is the case in Chile. Other disadvantages include widespread ignorance about the life histories of many species. Furthermore, it is more difficult for the general public to feel connected to index results based on macroinvertebrates. Finally, [47] reported that B-IBI requires a large number of samples and multiple metrics to correctly establish the biological condition of a river.

4. Chile: a case study

Mediterranean-climate ecosystems are priority areas of conservation efforts; however, these ecosystems remain threatened globally due to environment degradation [48, 49]. Of the five regions worldwide that present this climate, Chile is the least studied in regards to aquatic ecology [50]. This is despite reporting high national endemism and being considered among the 34 biodiversity hotspots in the world [48, 51].

The Mediterranean-climate ecosystem basins of Chile are host to significant industrial activities. This constitutes an increasing problem for aquatic ecosystems due to severe site degradations. Of the various human activities that threaten this region, land use and land cover conversion are highly ranked [52]. Indeed, while many activities directly or indirectly influence aquatic ecosystems, land use is the principal determinant of water quality and of water quantity entering aquatic ecosystems [53]. Furthermore, land cover conversions for crop production or monoculture plantations directly affect freshwater fauna, decreasing, for example, aquatic insect densities and possibly inducing local extinction [9].

In Chile, the use of bioindicators to assess water quality is limited, with applications focused on benthic macroinvertebrate assemblages through a modified Hilsenhoff Biotic Index (e.g., [31, 32, 54]). Notably, these studies were conducted only as a part of basic scientific research as no regulations or laws in Chile stipulate the use of biological criteria for measuring water quality. In contrast, bioindicators are widely used in other countries for assessing and monitoring water quality, often times to meet governmental regulations. In the United States, for example, the Environmental Protection Agency established the "Use of Biological Assessments and Criteria in the Water Quality Program" [55], whereas the European Environment Agency has used biomarker-based monitoring in a number of countries (e.g., Austria 1968 and United Kingdom 1970 [56]).

5. Effects of agricultural land use on aquatic ecosystems

Agricultural land use can increase the delivery of several compounds, such as phosphorous and nitrogen, to fluvial ecosystems. In turn, this can produce eutrophication and, consequently, limit the presence of some macroinvertebrate and fish species. For example, when 22 streams were sampled across five Mediterranean-climate watersheds in the farming, central-south region of Chile, agricultural land use was found to be an important predictor of both macroinvertebrate and fish assemblages. Specifically, significant differences in the composition of macroinvertebrate (**Figure 4**; ANOSIM: r = 0.203, P = 0.01) and fish (**Figure 5**; ANOSIM: r = 0.563, P = 0.01) assemblages between land use types were found. In addition, taxonomic diversity of macroinvertebrates was higher in native streams than agricultural streams (average Shannon-Wiener index in native streams: 1.5, agricultural streams: 1.1).

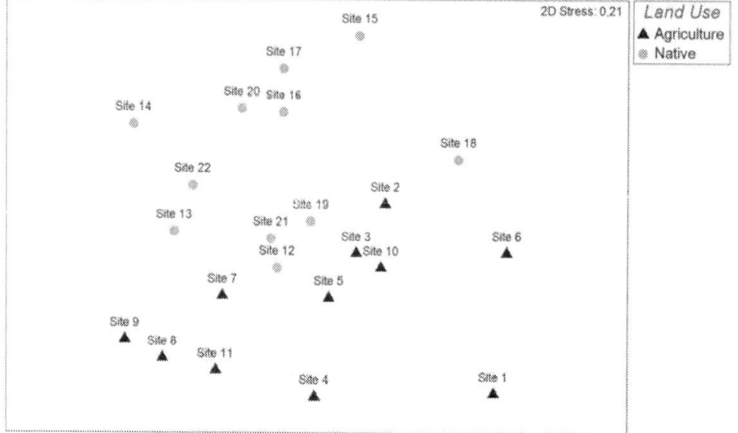

Figure 4. nMDS plot based on the composition of macroinvertebrates in 11 native streams and 11 agriculture streams in Mediterranean-climate ecosystems in the farming, central-south region of Chile. The data matrix was constructed using the Bray-Curtis Similarity Index with the square-root transformation of data (9999 restarts). Axes are relative scales and therefore appear without legends (personal data P. Fierro).

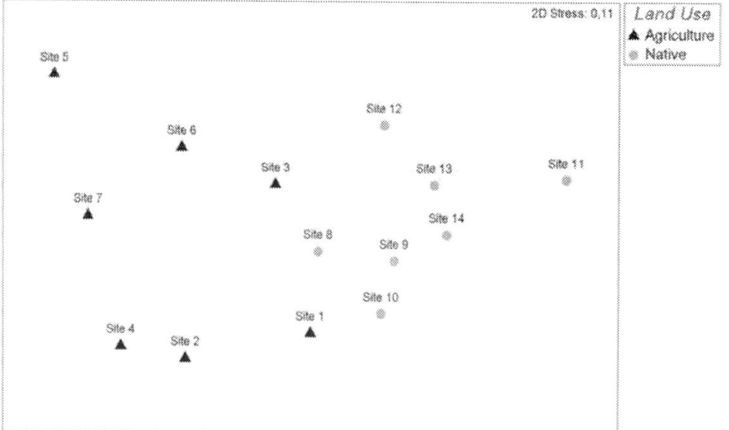

Figure 5. nMDS plot based on the composition of fish in seven native streams and seven agriculture streams in Mediterranean-climate ecosystems in the farming, central-south region of Chile. The data matrix was constructed using the Bray-Curtis Similarity Index with the square-root transformation of data (9999 restarts). Axes are relative scales and therefore appear without legends (personal data P. Fierro).

The principal difference in both assemblages was community heterogeneity, where native streams were constituted by greater abundances of Ephemeroptera larvae and presented Plecoptera larvae, while in agriculture streams, Diptera larvae and gastropods were more abundant (**Figure 6**). Regarding fish assemblages, a higher amount of taxa were recorded in native streams, and included exotic trout (e.g., *O. mykiss* and *S. trutta*; **Table 2**). These species are unique to environments with low temperatures and high oxygen content, indicators of good water quality. In contrast, the catfish *Trichomycterus areolatus* (**Figure 7**) was recorded at all native and agriculture sites, supporting the broad environmental tolerance of catfish species in general [57].

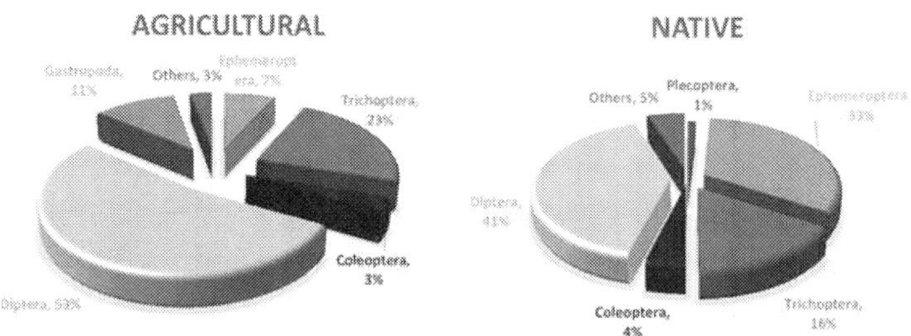

Figure 6. Macroinvertebrate classes found in agricultural dominated and reference streams (N = 22) (unpublished data P. Fierro).

	Agriculture	Native
Diplomystes nahuelbutaensis	0%	4.4%
Trichomycterus areolatus	20.9%	34.1%
Brachygalaxias bullocki	0.2%	0%
Cheirodon galusdae	3.5%	0.6%
Percilia gillisi	20.4%	28.7%
Basilicthys microlepidotus	0%	1.6%
Percichthys trucha	3.2%	0.8%
*Gambusia holbrooki**	50.3%	0%
*Cnesterodon decemmaculatus**	0.1%	0%
*Oncorhynchus mykiss**	1.8%	26.7%
*Salmo trutta**	0%	3.2%
*Cyprinus carpio**	0.5%	0%

*Exotic species (unpublished data P. Fierro).

Table 2. Species richness and relative abundances of fish species in agriculture and native streams in the farming, central-south region of Chile.

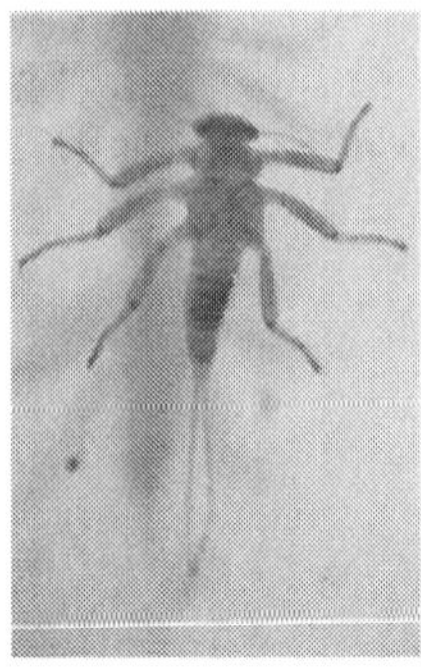

Figure 7. Left: Catfish, *Trichomycterus areolatus,* Siluriformes, 9 cm in total length. Center: *Andesiops torrens*, Ephemeroptera, 0.5 cm in total length. Right: *Antarctoperla michaelseni,* Plecoptera, 0.8 cm in total length. All individuals were collected from streams in the farming, central-south region of Chile (photographs by P. Fierro).

6. Conclusion

Macroinvertebrates and fish are used to evaluate the health of streams worldwide. The case results presented in this chapter evidence the importance of using one or more taxonomic groups in bioassessments, where both evaluated assemblages efficiently responded to pressures of human agricultural activities. These results suggest that macroinvertebrates and fish can be used as indicators of water pollution in monitoring programs. Using both assemblages as bioindicators presents several methodological advantages as compared to only assessing physicochemical parameters. These include low costs, easily identifiable fish, and, principally, the sensitivity of both assemblages to different stressors. For example, macroinvertebrates responded differently to substrate compositions than fish, which, in turn, responded to variables such as stream morphometry.

Rivers are increasingly affected by multiple physicochemical and biological stressors. Considering the ongoing rise in environmental management programs for aquatic communities, one related future goal is to develop appropriate indices, such as multimetric or biotic integrity indices, to differentiate between taxonomic groups, thereby facilitating assessments of stream health. However, the effectiveness of these indices will be highly dependent on applicability in different ecoregions.

Acknowledgements

This work was supported by Becas Doctorado Nacional CONICYT and funded in part by DAND Codelco-Andina, Fondecyt 1151375 and Fondap-Ideal 15150003 to L.V-C. We are grateful to Loretto Arriagada, Alfonso Jara, Gustavo Diaz, Jorge González, Cindy Cerna, and Aliro Manosalva for technical assistance in the field and laboratory.

Author details

Pablo Fierro[1,2*], Claudio Valdovinos[1], Luis Vargas-Chacoff[2,3], Carlos Bertrán[2] and
Ivan Arismendi[4]

*Address all correspondence to: pablofierro@udec.cl

1 Department of Aquatic Systems, Faculty of Environmental Sciences Center EULA,
University of Concepción, Concepción, Chile

2 Institute of Marine Science and Limnology, Austral University of Chile, Valdivia, Chile

3 Fondap High Latitudes Research Center (IDEAL), Austral University of Chile, Valdivia,
Chile

4 Department of Fisheries and Wildlife, Oregon State University, Corvallis, USA

References

[1] Dudgeon D, Arthington A, Gessner M, Kawabata Z, Knowler D, Lévêque C, Naiman
R, Prieur-Richard A, Soto D, Stiassny M, Sullivan A. Freshwater biodiversity: impor-
tance, threats, status and conservation challenges. Biological Reviews 2006; 81: 163–182.
DOI: 10.1017/S1464793105006950.

[2] Lyons J, Navarro-Pérez S, Cochran P, Santos E, Guzmán-Arroyo M. Index of biotic
integrity based on fish assemblages for the conservation of streams and rivers in West-
Central Mexico. Conservation Biology 1995; 9: 569–584. DOI: 10.1046/j.
1523-1739.1995.09030569.x.

[3] Butcher J, Stewart P, Simon T. A Benthic community index for streams in the Northern
Lakes and forests ecoregion. Ecological Indicators 2003; 3: 181–193. DOI: 10.1016/
S1470-160X(03)00042-6.

[4] Herman M, Nejadhashemi A. A review of macroinvertebrate- and fish-based stream
health índices. Ecohydrology & Hydrobiology 2015; 15: 53–67. DOI: 10.1016/j.ecohyd.
2015.04.001.

[5] Barletta M, Jaureguizar AJ, Baigun C, Fontoura NF, Agostinho AA, Almeida-Val VMF,
Val AL, Torres RA, Jimenes-Segura LF, Giarrizzo T, Fabré NN, Batista VS, Lasso C,
Taphorn DC, Costa MF, Chaves PT, Vieira JP, Correa MFM. Fish and aquatic habitat
conservation in South America: a continental overview with emphasis on neotropical
systems. Journal of Fish Biology 2010; 76: 2118–2176. DOI: 10.1111/j.
1095-8649.2010.02684.x.

[6] Fierro P, Quilodrán L, Bertrán C, Arismendi I, Tapia J, Peña-Cortés F, Hauenstein
E,Arriagada R, Fernández E, Vargas-Chacoff L. Rainbow Trout diets and macroinver-

tebrates assemblages responses from watersheds dominated by native and exotic plantations. Ecological Indicators 2016; 60: 655–667. DOI: 10.1016/j.ecolind.2015.08.018.

[7]   Allan JD. Landscapes and riverscapes: the influence of land use on stream ecosystems. Annual Review of Ecology, Evolution, and Systematics 2004; 35: 257–284. DOI: 10.1146/annurev.ecolsys.35.120202.110122.

[8]   Wang L, Robertson DM, Garrison PJ. Linkages between nutrients and assemblages of macroinvertebrates and fish in Wadeable streams: implication to Nutrient Criteria Development. Environmental Management 2007; 39: 194–212. DOI: 10.1007/s00267-006-0135-8.

[9]   Fierro P, Bertrán C, Mercado M, Peña-Cortés F, Tapia J, Hauenstein E, Caputo L, Vargas-Chacoff L. Landscape composition as a determinant of diversity and functional feeding groups of aquatic macroinvertebrates in southern rivers of the Araucanía, Chile. Latin American Journal of Aquatic Research 2015; 43: 186–200. DOI: 10.3856/vol43-issue1-fulltext-16.

[10]   Fu-Liu X, Shu T. On the study of ecosystem health: state of the art. Journal of Environmental Sciences 2000; 12: 33–38.

[11]   Hughes R, Whittier T, Thiele S, Pollard J, Peck D, Paulse S, Mcmullen D, Lazorchak J, Larsen D, Kinney W, Kaufmann P, Hedtke S, Dixit S, Collins G, Baker J. Lake and stream indicators for the United States Environmental Protection Agency's environmental monitoring and assessment program. In: McKenzie D, Hyatt D, McDonald V, editors. Ecological Indicator, I. Elsevier Applied Science, London, UK; 1992. p. 305–335. DOI: 10.1007/978-1-4615-4659-7_20.

[12]   Weigel B, Henne L, Martínez-Rivera L. Macroinvertebrate-based index of biotic integrity for protection of streams in west-central Mexico. Journal of the North American Benthological Society 2002; 21: 686–700. DOI: 10.2307/1468439.

[13]   Oberdoff T, Hughes RM. Modification of an index of biotic integrity based on fish assemblages to characterize rivers of the Seine Basin, France. Hydrobiologia 1992; 228: 117–130. DOI: 10.1007/BF00006200.

[14]   Roldan G. Macroinvertebrates and value as water quality indicators - Los Macroinvertebrados y su valor como indicadores de la calidad del agua. Revista Academica Colombiana de Ciencias 1999; 23: 375–387.

[15]   Oliveira SV, Cortes RMV. Environmental indicators of ecological integrity and their development for running waters in northern Portugal. Limnetica 2006; 25: 479–498.

[16]   Karr J. Biological monitoring and environmental assessment: a conceptual framework. Environmental Management 1987; 11: 249–256. DOI: 10.1007/BF01867203.

[17]   Barbour MT, Gerritsen J, Snyder BD, Stribling JB. Rapid bioassessment protocols for use in streams and Wadeable Rivers: periphyton, benthic macroinvertebrates and fish, second edition. EPA 841-B-99-002 U.S. Environmental Protection Agency; Office of Water; Washington, D.C.; 1999. 339 p.

[18] Karr J, Chu E. Sustaining living rivers. Hydrobiologia. 2000; 422/423: 1–14. DOI: 10.1023/A:1017097611303.

[19] Ollis D, Dallas H, Esler K, Boucher Ch. Bioassessment of the ecological integrity of river ecosystems using aquatic macroinvertebrates: an overview with a focus on South Africa. African Journal of Aquatic Science 2006; 31: 205–227. DOI: 10.2989/16085910609503892.

[20] Cairns J, Pratt JR. A history of biological monitoring using benthic macroinvertebrates. In: Rosenberg DM, Resh VH, editors. Freshwater Biomonitoring and Benthic Macroinvertebrates. Chapman & Hall, New York; 1993. p. 10–27. DOI: 10.1002/aqc.3270040110.

[21] Alba-Tercedor J. Aquatic Macroinvertebrates and water quality of streams - Macroinvertebrados acuáticos y calidad de las aguas de los ríos. IV Simposio del Agua en Andalucía, Almería 1996; 2: 203–213.

[22] Dos Santos DA, Molineri C, Reynaga MC, Basualdo C. Which index is the best to assess stream health? Ecological Indicators 2011; 11: 582–589. DOI: 10.1016/j.ecolind.2010.08.004.

[23] Hilty J, Merenlender A. Faunal indicator taxa selection for monitoring ecosystem health. Biological Conservation 2000; 92: 185 197. DOI: 10.1016/S0006-3207(99)00052-X.

[24] Li H, Li J. Role of fish assemblages in stream communities. In: Hauer FR, Lamberti GA, editors. Methods in Stream Ecology, second edition. Academic Press, San Diego, CA; 2007. p. 489–514.

[25] Merrit RW, Cummins KW. Trophic relationships of macroinvertebrates. In: Hauer FR, Lamberti GA, editors. Methods in Stream Ecology, second edition. Academic Press, San Diego, CA; 2007. p. 585–610.

[26] Vargas-Chacoff L, Quilodrán L, Bertrán C, Arismendi I, Fierro P, Tapia J, Peña-Cortés F, Hauenstein E. Food of rainbow trout and changes in soil use: the Chilean Example. In: Polakof S, Moon TW, editors. Trout: From Physiology to Conservation. Nova Science Publishers, New York; 2013. p. 65–80.

[27] Rolls R, Boulton A, Gowns I, Maxwell S, Ryder D, Westhorpe D. Effects of an experimental environmental flow release on the diet of fish in a regulated coastal Australian river. Hydrobiologia 2012; 686: 195–212. DOI: 10.1007/s10750-012-1012-5.

[28] Veloso C, Costa F, Santos P. Hydropeaking effects of on the diet of a Neotropical fish community. Neotropical Ichthyology. 2014; 12: 795–802. DOI: 10.1590/1982-0224-20130151.

[29] Hilsenhoff W. Rapid field assessment of organic pollution with a family-level biotic index. Journal of the North American Benthological Society. 1988; 71: 65–68. DOI: 10.2307/1467832.

[30] Lenat D. A biotic index for the southeastern United States: derivation and list of tolerance values, with criteria for assigning water-quality ratings. Journal of the North American Benthological Society. 1993; 12: 279–290. DOI: 10.2307/1467463.

[31] Figueroa R, Valdovinos C, Araya E, Parra O. Benthic Macroinvertebrates as indicators of water quality of southern Chile rivers - Macroinvertebrados bentónicos como indicadores de calidad de agua de ríos del sur de Chile. Revista Chilena de Historia Natural. 2003; 76: 275–285. DOI: 10.4067/S0716-078X2003000200012.

[32] Fierro P, Bertrán C, Mercado M, Peña-Cortés F, Tapia J, Hauenstein E, Vargas-Chacoff L. Benthic macroinvertebrate assemblages as indicators of water quality applying a modified biotic index in a spatio-seasonal context in a coastal basin of Southern Chile. Revista de Biología Marina y Oceanografía. 2012; 47: 23–33. DOI: 10.4067/S0718-19572012000100003.

[33] Wright JF, Furse MT, Moss D. River classification using invertebrates: RIVPACS applications. Aquatic Conservation: Marine and Freshwater Ecosystems. 1998; 8: 617–631. DOI: 10.1002/(SICI)1099-0755(199807/08)8:4<617::AID-AQC255>3.0.CO;2-#.

[34] Davies PE. Development of a National River Bioassessment System (AUSRIVAS) in Australia. In: Wright JF, Sutcliffe DW, Furse MT, editors. Assessing the Biological Quality of Fresh Waters: RIVPACS and Other Techniques. Freshwater Biological Association, Ambleside, UK, 2000. p. 113–124.

[35] Reynoldson TB, Norris RH, Resh VH, Day KE, Rosenberg DM. The reference condition: a comparison of multimetric and multivariate approaches to assess water-quality impairment using benthic macroinvertebrates. Journal of the North American Benthological Society, 1997; 16: 833–852. DOI: 10.2307/1468175.

[36] Barbour MT, Gerritsen J, Griffith GE, Frydenborg R, McCarron E, White JS, Bastian ML. A framework for biological criteria for Florida streams using benthic macroinvertebrates. Journal of the North American Benthological Society. 1996; 15: 185–211. DOI: 10.2307/1467948.

[37] Varandas S, Vitor R. Evaluating macroinvertebrate biological metrics for ecological assessment of streams in northern Portugal. Environmental Monitoring Assessment. 2010; 166: 201–221. DOI: 10.1007/s10661-009-0996-4.

[38] Karr J. Assessment of biotic integrity using fish communities. Fisheries. 1981; 6: 21–27. DOI: 10.1577/1548-8446(1981)006<0021:AOBIUF>2.0.CO;2.

[39] Griffith M, Hill B, McCormick F, Kaufmann P, Herlihy A, Selle A. Comparative application of indices of biotic integrity based on periphyton, macroinvertebrates, and fish to southern Rocky Mountain streams. Ecological Indicators. 2005; 5: 117–136. DOI: 10.1016/j.ecolind.2004.11.001.

[40] Karr J, Dudley D. Ecological perspectives in water quality goals. Environmental Management. 1981; 5: 55–68. DOI: 10.1007/BF01866609.

[41] Karr J. Biological integrity: a long neglected aspect of water resource management. Ecological Applications. 1991; 1: 66–84. DOI: 10.2307/1941848.

[42] Angermeier PL, Karr J. Applying an index of biotic integrity based on stream-fish communities: considerations in sampling and interpretation. North American Journal of Fisheries Management. 1986; 6: 418–429. DOI: 10.1577/1548-8659(1986)6<418:AAIO-BI>2.0.CO;2.

[43] Westra L, Miller P, Karr J, Rees W, Ulanowicz E. Ecological Integrity and the aims of the global integrity project. In: Pimentel D, Westra L, Noss RF, editors. Ecological Integrity: Integrating Environment, Conservation, and Health. Island Press, Washington, DC. p. 19–41. DOI: 10.1093/ije/31.3.704.

[44] De Freitas Terra B, Hughes R, Rocha Francelino M, Gerson Araújo F. Assessment of biotic condition of Atlantic Rain Forest streams: a fish-based multimetric approach. Ecological Indicators 2013; 34: 136–148. DOI: 10.1016/j.ecolind.2013.05.001.

[45] Tiku Mereta S, Boets P, De Meester L, Goethals P. Development of a multimetric index based on benthic macroinvertebrates for the assessment of natural wetlands in Southwest Ethiopia. Ecological Indicators 2013; 29: 510–521. DOI: 10.1016/j.ecolind. 2013.01.026.

[46] Kerans BL, Karr J. A benthic index of biotic integrity (B-IBI) for rivers of the Tennesse Valley. Ecological Applications 1994; 4: 768–785. DOI: 10.2307/1942007.

[47] Karr J, Chu E. Biological Monitoring and Assessment: Using Multimetric Indexes Effectively. EPA 235-R97-001. University of Washington, Seattle, WA; 1997. 149 p.

[48] Myers N, Mittermeier R, Mittermeier C, Da Fonseca G, Kent J. Biodiversity hotspots for conservation priorities. Nature 2000; 403: 853–858. DOI: 10.1038/35002501.

[49] García N, Cuttelod A. Biodiversity loss in the Mediterranean: causes and conservation proposals - Pérdida de biodiversidad en el Mediterráneo: causas y propuestas de conservación. Memorias Real Sociedad Española de Historia Natural 2013; 2: 41–54.

[50] Gasith A, Resh V. Streams in mediterranean climate regions: abiotic influences and biotic responses to predictable seasonal events. Annual Review of Ecology, Evolution, and Systematics 1999; 30: 51–81. DOI: 10.1146/annurev.ecolsys.30.1.51.

[51] Myers N. Biodiversity Hotspot Revisited. Bioscience 2003; 53: 916–917. DOI: 10.1641/0006-3568(2003)053[0916:BHR]2.0.CO;2.

[52] Aguayo M, Pauchard A, Azocar G, Parra O. Land use change in the south central Chile at the end of the 20th century. Understanding th spatio-temporal dynamics of the landscape - Cambio del uso del suelo en el centro sur de Chile a fines del siglo XX. Entendiendo la dinámica espacial y temporal del paisaje. Revista Chilena de Historia Natural. 2009; 82: 361–374. DOI: 10.4067/S0716-078X2009000300004.

[53] Cuevas J, Huertas J, Leiva C, Paulino L, Dörner J, Arumi J. Nutrient retention in a microcatchment with low levels of anthropogenic pollution. Bosque 2014; 35: 75–88. DOI: 10.4067/S0717-92002014000100008.

[54] Figueroa R, Ruiz VH, Encina-Montoya F, Palma A. Simplification in the use of macro-invertebrates in the evaluation of water quality in fluvial systems - Simplificación en el uso de macroinvertebrados en la evaluación de la calidad de las aguas en sistemas fluviales. Interciencia 2005; 30: 770–774.

[55] EPA. U.S. Environmental Protection Agency. Policy on the USE of Biological Assessments and Criteria in the water quality program. Office of Science and Technology, Washington, D.C.; 1991. 19 p.

[56] EEA Report. Surface water quality monitoring: 4.2 Biological assessments of river quality [Internet]. 2016. Available from: http://www.eea.europa.eu/publications/92-9167-001-4/page021.html [accessed: 2016-04-04].

[57] Habit E, Victoriano P, Campos H. Trophic ecology and reproductive aspects of *Trichomycterus areolatus* (Pisces, Trichomycteridae) in irrigation canal environments - Ecología trófica y aspectos reproductivos de *Trichomycterus areolatus* (Pisces, Trychomycteridae) en ambientes loticos artificiales. Revista Biología Tropical. 2005; 53: 195–210. DOI: 10.15517/rbt.v53i1-2.14416.

Chapter 3

Calibrating and Validating the Biomonitoring Working Party (BMWP) Index for the Bioassessment of Water Quality in Neotropical Streams

Ricardo Arturo Ruiz-Picos ,

Jacinto Elías Sedeño-Díaz and Eugenia López-López

Additional information is available at the end of the chapter

http://dx.doi.org/10.5772/66221

Abstract

The Biological Monitoring Working Party (BMWP) is among the most used bioassessment indices for aquatic ecosystems quality assessment, which assigns scores to each macroinvertebrate taxa according to their sensitivity to organic pollution. However, BMWP scores must be calibrated to each geographical and ecological conditions. In this study, we obtain statistically derived scores of sensitivity for macroinvertebrates taxa from Neotropical Mexican rivers, Apatlaco and Chalma-Tembembe rivers, Balsas Basin. We obtained water samples and aquatic macroinvertebrates in four sampling campaigns (dry and rainy seasons). Physicochemical parameters and the abundances of the aquatic macroinvertebrates were used for the BMWP index calibration, which was performed in steps obtaining: the physicochemical quality index (Pcq), incorporation of abundances classes of macroinvertebrates taxa in the corresponding Pcq interval and the determination of bioindication values for each macroinvertebrate family. The BMWP calibrated index was validated and tested for the geographical range extension. The BMWP scores for Chalma-Tembembe River (located in agricultural areas) showed bad polluted to regular and moderated polluted categories. The urban river zone of Apatlaco River showed: bad, very polluted to very bad categories. The BMWP calibrated is a suitable biomonitoring tool, allowing the detection of those zones that needs urgently a management and recovery plan.

Keywords: water quality assessment, biomonitoring tool, macroinvertebrate sensitivity, land use, bioindication values, water quality categories

INTECH

open science | open minds

1. Introduction

In the framework of the World Economic Forum, the "Water Crisis" is positioned as the highest concern global risk for the next 10 years [1]. In this sense, water quality and management of freshwater ecosystems are one of the main challenges worldwide [2]. However, these ecosystems face impacts and degradation that are result of human population increase and agricultural and industrial development [3]. Consequently, freshwater ecosystems and their biota are considered as the most endangered and threatened worldwide [4]. In developing countries, there is an extremely high population growth, increasing industrialization and urbanization processes, with severe and constant changes in land use, whereby the freshwater ecosystems are highly impaired [5]. Rivers crossing different land uses (urban, industrial and agricultural) are the most threatened by anthropogenic activities [5]. The threat to freshwater systems, in particular in developing countries, make evident an urgent need for developing tools for the assessment and classification of aquatic conditions in order to manage water resources and bring them a sustainable management.

Biomonitoring is considered as the most appropriate method for environmental studies and for the control of water quality, due to that living organisms are excellent biosensors of the physicochemical and biological characteristics of water [6]. The aquatic macroinvertebrates have been used as bioindicators because they have a wide range of habitats and sensitivity to environmental pollution and other types of stressors, including sediment [7]. Thus, the macroinvertebrate assemblages change in response to environmental disturbances in predictable ways including a strong reduction in species and abundance in impacted areas and more tolerant species predominate; whereas, sensitive species are only present in environments with the least impact or un-impaired conditions. Moreover, biomonitoring integrate information over longer periods of time and better represent the responses of aquatic habitats providing information concerning the present state and the past trends in environmental conditions [8].

The Biological Monitoring Working Party (BMWP) is among the most used bioassesment index in Europe, which was originally developed in the UK in 1976 [9] and it has been used by the regulatory authorities in the UK as the basis of their river invertebrate status classification system since 1980. This index assigns scores to each macroinvertebrate taxon according to their responses to oxygen deficits caused by organic pollution. The analysis of these pollution-induced responses allows the calculation of sensitivity values by the different groups of organisms. Because of its ease of use and low cost, the BMWP index has been used in many other countries in Africa, Asia, Oceania and Latin America [10]. Nevertheless, the BMWP scores for each taxon must be calibrated to each ecological region since the taxonomic composition, ecological, zoogeographic and anthropogenic conditions promote important geographical differences.

Additionally, the scale to ranking water conditions must be adapted for each particular condition. In Latin America, attempts have been made to develop regional indices [5]. In México, the water quality indicators used by the National Commission of Water are fecal coliforms, biochemical oxygen demand, chemical oxygen demand and total dissolved solids [11]; unfortunately, biomonitoring is not included in the current legislation, while information on bioindication is scarce [12]. However, main urban zones of Mexico exert a high rate of changes in land use and deforestation for agricultural, industrialization and urban expansion provoking serious damages in water bodies [13]. Consequently, there is a need for a tool that considers both, biotic and abiotic variables and their relationships to assess the river water quality.

The aim of this study is to obtain the statistically derived scores of sensitivity for aquatic macroinvertebrates taxon for Neotropical Mexican rivers (Apatlaco and Chalma-Tembembe rivers, Balsas Basin). This chapter presents the calibration of the BMWP index based on Riss et al. [14] with some modifications, using physical, chemical and biological data from Apatlaco and Chalma-Tembembe rivers. This index, besides being an easy-to-use tool, allows for the implementation of a permanent biomonitoring network.

2. Materials and methods

2.1. Study area

Apatlaco and Chalma-Tembembe rivers are located in the Balsas Basin (**Figure 1a**), one of the largest catchment areas in Mexico (area of 117,405 km^2) [15]. Both rivers are in the same zoogeographic region (Neotropical), which belongs to the Ecoregion Balsas Complex and belongs to the Biogeographic Province "Depresión del Balsas," particularly to the "Alto Balsas" [15]. The Chalma-Tembembe River is formed by the Chalma and Tembembe rivers; the first has a length of 70 km and the second of 50.72 km. The Chalma River joins the Tembembe River at its lower reaches. The Chalma-Tembembe subbasin has a mean annual rainfall around 600 mm. The area consists of tropical deciduous forest ($\approx$47% of landcover), with some areas of rain-dependent and irrigated agricultural use. The strong pressure from agricultural activities has favored land use changes and the loss of the original vegetation [15]. Six study sites were selected in Chalma-Tembembe: El Arco (I), La Loma (II), El Platanar (III), Casa de la Escuela (IV), Coatlán (V) and Hacienda de Cuautlita (VI) (**Figure 1a**). The Apatlaco River has a length of 63 km, with annual rainfall from 850 to 1500 mm. The natural vegetation has been highly fragmented and transformed, with only 27% of the original area of tropical deciduous, coniferous and oak forest remaining. Moreover, an important urban-industrial corridor (Cuernavaca corridor), runs alongside the river. The activities in the vicinity of the river include agriculture, lumber forestry, hunting and fishing [15]. Nine study sites along the main river channel (Apatlaco River) were selected: Las Truchas (VII), El Pollo (VIII), El Rayo (IX), El Encanto (X), Salida Panochera (XI), Xochitepec (XII), Alpuyeca (XIII), Xoxocotla (XIV) and Zacatepec (XV) and two more along the westerly tributary: Buenavista 1 (XVI) and Buenavista 2 (XVII), study sites located before and after the effluent of a wastewater treatment plant and three more along the easterly tributaries: El Texcal (XVIII), La Gachupina (XIX) and Las Juntas (XX), resulting in a total of 14 sampling sites (**Figure 1a**). In both rivers, four sampling campaigns were undertaken (the dry season in December 2012 and February-March 2013 and the rainy season in August-September 2012 and June 2013).

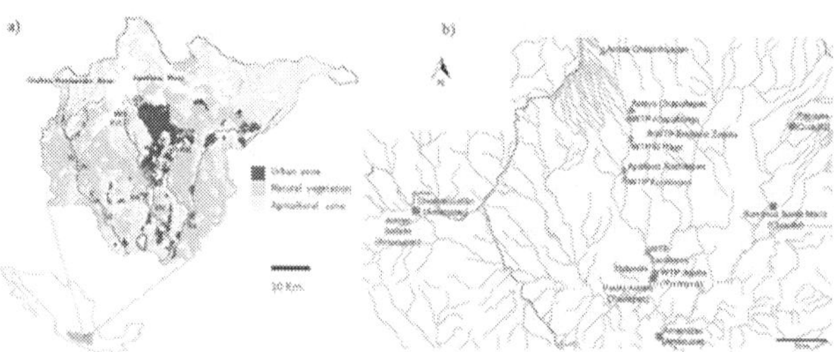

Figure 1. Study area location. (a) Study sites. (b) Validation and range extension study sites.

2.2. Water quality measurements and water quality index

For each site and sampling period, some variables were recorded: altitude (masl), water temperature (°C), conductivity (µS/cm), pH, salinity (PSU), dissolved oxygen (DO mg/L O_2) and turbidity (NTU), using a Quanta® multiparameter probe. Air temperature was recorded with a 45118 EXTECH anemometer. Water samples were transported to the laboratory in refrigerated and dark conditions. Biochemical oxygen demand (BOD_5 mg/L O_2), chlorides (mg/L Cl), alkalinity (mg/L $CaCO_3$) and total and fecal coliforms (MPN/100 mL) were determined according to [16]. Nitrite (mg/L NO_2), nitrate (mg/L NO_3), ammonia (mg/L NH_3), total nitrogen (mg/L TN), orthophosphate (mg/L PO_4), total phosphorus (mg/L TP), hardness (mg/L $CaCO_3$), sulfates (mg/L SO_4^{2-}) and color (U Pt-Co) were analyzed using a Hach DR 2500 spectrophotometer. Additionally, the DO saturation (%) was computed. The water quality index (WQI) proposed by Dinius [17] was calculated, which range from 0 to 100; 100 is excellent and 0 is strongly polluted.

2.3. Macroinvertebrate sampling

Aquatic macroinvertebrates were sampled at each sampling site and season with a multi-habitat monitoring system [18], along a section of 100 m length to incorporate local habitat variation (fast flowing riffles, pools, submerged vegetation and riparian vegetation). An area of 1 m² was sampled for each type of habitat. Triplicate samples for each type of habitat, with a 10-min collecting effort, were collected and pooled for analysis. The samples were taken using a kick net for fast-flowing riffles and pools, while type-D nets for submerged and riverine vegetation, all nets with a mesh size of 500 µm. Organisms collected were preserved with 70% alcohol. Taxonomic identification at the family level was conducted using stereomicroscopes (Nikon C-Leds) and with the use of keys [19, 20].

2.4. BMWP index calibration

2.4.1. Data processing

The BMWP index is calculated by adding up the individual tolerance scores of aquatic macroinvertebrates at family taxonomic level present at a sample site. We calibrate the tolerance scores of the aquatic macroinvertebrates in several steps [14]: (1) obtaining a physicochemical quality index (*Pcq*) for all the study sites; (2) assessing the bioindication values to each macroinvertebrate family according to the *Pcq* and their abundance class; and (3) with the scores calibrated for each family of macroinvertebrate from Apatlaco and Chalma Tembembe rivers, we assessed the BMWP. Additionally, we define the water quality categories following the procedure of Ref. [21].

2.4.2. Mathematical formulation

The physicochemical quality index (*Pcq*) is a value that describes water quality at each sample site on a scale from 0 to 10; 0 corresponds to a highly impacted site and 10 to a site with excellent water quality. Its calculation utilizes a data matrix with the mean values of each physicochemical parameters obtained from the four study periods for each sample site. Parameters recorded in situ, as well as quantified in the laboratory, were included. Values for each parameter were normalized: $Ci = $ Ln $(i +1)$ and for percentages $Ci = (2/\pi)^*$ arcsin $\sqrt{i}$. Where Ci is a physicochemical variable and i is the mean value of a physicochemical variable. The data matrix was subjected to a factor analysis with the software XLstat version 2013. Parameters that showed a significant correlation ($p < 0.05$), either positive or negative, in the first two factors were considered as qualifying variables or C_{i2}; these variables were arranged in a new data matrix of sampling sites *vs* the C_{i2}. The maximum (C_{i2max}) and minimum (C_{i2min}) values for each qualifying variable were determined and assigned taking into account the environmental legislation for the water management in México (Mexican Official Standards: [22, 23]), the USA [24], Canada [25], Central and South American countries [26, 27] and a worldwide level [28]. With C_{i2max} and C_{i2min} values for each qualifying variable the C_{i2} were standardized:

$$Ci_3 = \frac{Ci_2 - Ci_{2\min}}{Ci_{2\max} - Ci_{2\min}}. \tag{1}$$

Each C_{i3} value was adjusted to fit the BMWP scale that goes from 0 to 10, using the following formula: $C_{i4} = (1 - C_{i3})^*10$. In the case of the qualifying variables associated with good water quality (DO and DO% saturation), the inverse procedure was followed by applying the formula: $C_{i4} = C_{i3}{}^* 10$. For each study site, an average was assessed with the C_{i4} values for the selected variables to determine the physicochemical quality:

$$P_{cq} = \frac{\sum_{i=1}^{n} Ci_4}{ni}. \tag{2}$$

The *Pcq* values fluctuate from 0 to 10, for all 20 sampling sites and were incorporated into 10 categories of quality or *Pcq* intervals.

2.4.3. BMWP scores for macroinvertebrates families

To assign bioindication values to the different macroinvertebrates families, a data matrix of sampling sites *vs* abundance was constructed using the mean abundance values for each aquatic macroinvertebrate family from all sampling and study sites. The mean abundances were standardized to six abundance classes: class 0 (0 organisms), class 1 (1–3 organisms), class 2 (4–10 organisms), class 3 (11–33 organisms), class 4 (34–100 organisms) and class 5 (>100 organisms). This standardization was carried on following [14] to reduce the possible effect of the overvaluation by the local dominance of some groups due to the nonhomogeneous natural distribution of the macroinvertebrates.

For each family of macroinvertebrates, the abundance class data were pooled within each *Pcq* interval. In cases where a family appeared in more than one study site within the same *Pcq* interval, the abundance classes of such sites were averaged in order to obtain a single abundance value per family for each *Pcq* interval. The value of class abundance obtained by a *Pcq* interval indicates the number of times you have to replicate the value of the superior limit corresponding to the *Pcq* interval; for example, if a family obtained a value of 1 for its abundance class within the 0–1 *Pcq* interval, a 3 for the 3–4 *Pcq* interval, a 4 for the 5–6 *Pcq* interval and a 2 for the 6–7 *Pcq* interval, the data for calculating the fifth percentile would be 1, 4, 4, 4, 6, 6, 6, 6, 7 and 7. The bioindication values for each aquatic macroinvertebrate family were calculated by obtaining the fifth percentile of the abundance class distributions along the *Pcq* intervals where that family was present. This value represents the minimum tolerance value of this family in relation to organic pollution; in the case of this example, the fifth percentile would be 2, which is the family BMWP bioindication value.

2.5. Definition of BMWP water quality categories

The water quality category ranges for the BMWP values were assigned following [21]. The median value of the data set of the reference sites was calculated. Scores above this median value will correspond to the "Excellent" quality category, while values that fall between the median and the tenth percentile of that distribution are considered to be in the "Good, not sensible affected" quality category. Values below the tenth percentile were subdivided into four equal parts, which correspond to the categories "Regular,""Bad, polluted,""Bad, very polluted," and "Bad, extremely polluted." The names assigned to each of the water quality categories with some modifications were those proposed by Alba-Tercedor [29]. The selection of reference conditions included physical, chemical and biological criteria (WQI, *Pcq*, land use and macroinvertebrate community) following [30].

2.6. BMWP statistical validation

The validation process was performed using three approaches. First, a score prediction test, proposed by Armitage et al. [31], with the average per study site of BMWP$_{observed}$ *vs* BMWP$_{expected}$ values. The expected values were calculated with a multiple linear regression (best model procedure) with the qualifying variables for each study site. The observed and expected values were plotted and confidence intervals ($\alpha = 0.05$) were calculated using the XLSTAT software

2013. The goodness of fit of the model was evaluated using the coefficient of determination (R^2) and p values.

For the second validation approach, the degree of fit of the model for the BMWP$_{observed}$ *vs* BMWP$_{expected}$ was tested using two independent methods: the [32] efficiency model (NSE) (range $-\infty$ to 1.0), a widely used statistic for hydrological models

A third approach for the index validation and for the assessment of the geographical extension of the BMWP calibrated was performed with additional information that was obtained from the National Agency for Water in México (CONAGUA), data included aquatic macroinvertebrates collected in the county of Morelos. For the index validation, nine sites within the Apatlaco subbasin were considered: Arriba Chalchihuapan, Arroyo Chapultepec, water treatment plant (WTP) Acapatzingo, WTP Emiliano Zapata, WTP El Rayo, Apatlaco-Xochitepec, WTP Xochitepec, WTP Zacatepec and Tlaltenchi (**Figure 1b**). For the geographical extension, data from CONAGUA of seven sites in three subbasins were considered: Amacuzac subbasin, sites Chontalcoatlán, Amacuzac and Arroyo Salado; Cuautla subbasin, sites Barranca Santa María and Papayos; and Yautepec subbasin, sites Pedro Amaro and WTP Jojutla. These sites belong to the Balsas Basin and the last three subbasins are adjacent to the two rivers monitored in this study (**Figure 1b**). The monitoring team of CONAGUA used D-nets (mesh size of 500 µm), a multihabitat sampling and each habitat was sampled in 1 m², with three replicates. Based on the calibrated scores for aquatic macroinvertebrates, the BMWP scores were calculated for each site of the data set from CONAGUA and these scores were included in the previous model generated with *observed vs expected* data for the BMWP with confidence limits ($\alpha = 0.05$). A Pearson correlation analysis was also performed for the BMWP.

2.7. Statistical analysis

WQI and BMWP values are presented as the mean values of each study site for the four monitoring campaigns. Mean values were also calculated for each study season taking into account the values of all the studied sites. Significant differences between sites and seasons were detected with a bivariate analyses of variance (ANOVA), followed by Student-Neuman-Keuls multiple comparison tests (if the data were normally distributed as well as homoscedastic), or Kruskal-Wallis test for nonparametric data, both $p < 0.05$, using SigmaPlot version 11.0.

3. Results

3.1. Water quality index

For the Chalma-Tembembe River, mean values of WQI fluctuated from 52 to 74 (**Figure 2**), from slightly polluted to acceptable for human consumption; however, no significant differences between study sites ($p > 0.05$) were observed. In Apatlaco River, the study sites Las Truchas and El Texcal achieved excellence in water quality with mean values of 96 and 84, respectively (**Figure 2**) and were significantly different from other study sites ($p < 0.05$). The other study sites showed mean WQI values from 57 to 60 with no statistical differences ($p >$ 0.05) and water quality fluctuated from polluted to mildly polluted for human consumption. Additionally, the WQI values decreased during the dry season ($p < 0.05$), where August-September differed from February-March and June seasons; and December was statistically different from February-March and June seasons (**Figure 2**).

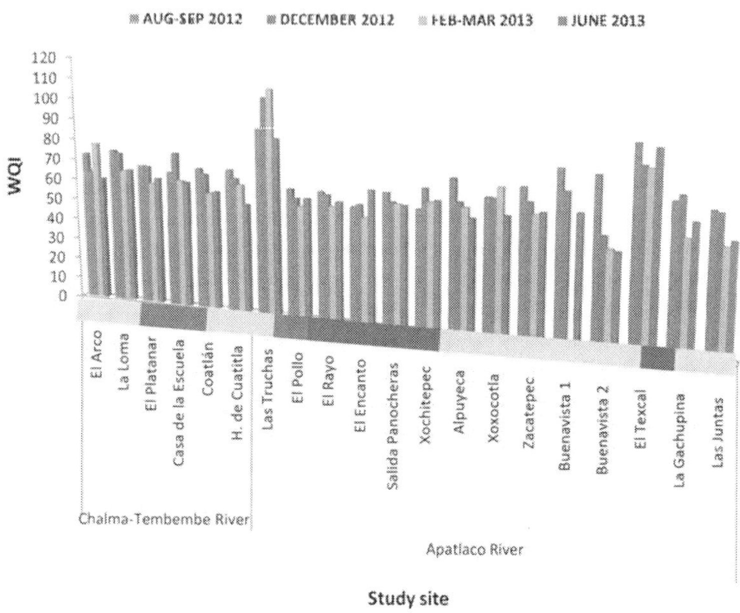

Figure 2. Water quality index.

3.2. Physicochemical quality and bioindication values

The factor analysis showed a total of 60.32% of explained variance for the first two axes. The parameters that showed a significant correlation, either positive or negative, in the first two axes of the factor analysis and which were considered as qualifying variables for this study, C_{i2}, were NO_3, NO_2, NO_3, TN, TP, BOD_5, DO, DO saturation (%), SO_4^{2-}, color, alkalinity, chlorides and conductivity (**Table 1**). Using these variables to assess the physicochemical quality, we found that five study sites of the Chalma-Tembembe River obtained a *Pcq*, corresponding to classes 6–7 and one site in classes 7–8. The Apatlaco River sites were distributed across a broad spectrum of physicochemical quality classes: one site in classes 0–1, another in classes 2–3, four sites in classes 3–4, six sites in classes 4–5, one in classes 6–7 and finally one in classes 9–10. A total of 66 taxa were taxonomically determined, distributed in 5 phyla, 7 classes, 21 orders and 63 families: Oligochaeta, Hirudinea and Turbellaria were identified only to class level. "Las Truchas" was the site that reached the maximum taxa (35).

The bioindication values for each aquatic macroinvertebrates family (obtaining the fifth percentile of the abundance class distributions along the *Pcq* intervals) are shown in **Table 2**.

Variable	F1 (37.24%)	F2 (23.08%)
Nitrates (mg/L)	**0.592**	0.207
Nitrites (mg/L)	**0.567**	0.195
Ammonium (mg/L)	0.297	**0.738**
Total N (mg/L)	**0.606**	0.600
Total P (mg/L)	**0.818**	0.392
Sulfates (mg/L)	**0.750**	−0.525
Color (Pt-Co)	**0.605**	0.472
Alkalinity (mg/L)	**0.756**	−0.203
Chlorides (mg/L)	**0.906**	0.166
BOD (mg/L)	0.516	**0.599**
DO (mg/L)	−0.152	**−0.777**
Conductivity (µS/cm)	**0.775**	−0.522
% DO	0.024	**−0.868**
% Explained variance	37.24	23.08
% Accumulated variance	37.24	60.32

Numbers in bold represent significant variables from the factor analyses.

Table 1. Variables used for the calculation of *Pcq* and their correlations with the first two factors F1 and F2 (explained variance).

3.3. BMWP index and BMWP water quality classes

The BMWP values assessed with the calibrated bioindication values fluctuated from 2 to 109, but showed no statistical differences ($p > 0.05$) between seasons. The study sites "Las Truchas" obtained the highest BMWP values ($p < 0.05$) during the four study periods. It was selected as the reference site and their scores were averaged. The BMWP scores above this median value will correspond to the "Excellent" quality category, from which range values for all water quality classes were assigned (**Figure 3a**). Seasonal fluctuations showed that this site fluctuate from "Regular" in August, "Good" in June to "Excellent" in December and February. The study site "El Encanto" had the lowest BMWP index value during its four study periods, from 2 to 6 points and thus qualifying as "Very bad, extremely polluted"; this was followed by La Gachupina, Buenavista 1, Buenavista 2, El Texcal and El Pollo, with values between 2 and 43 points and thus reaching the "Bad, polluted" and "Bad, very polluted" categories. The remaining sites qualified as "Regular, medium pollution" (**Figure 3a**). A comparison of BMWP and WQI scores for each study site is shown in **Figure 3b**.

Taxon		Bioindication value
Cordulegastridae	Lepidostomatidae	10
Heptageniidae	Perlidae	
Caenidae	Scathophagidae	8
Gyrinidae		
Aeshnidae	Philopotamidae	7
Blaberidae	Pseudothelphusidae	
Cambaridae	Ptilodactylidae	
Helicopsychidae	Saldidae	
Hydrobiosidae	Scirtidae	
Naucoridae		
Corbiculidae	Hydrobiidae	5
Dixidae	Hydroptilidae	
Dryopidae	Leptophlebiidae	
Ephydridae	Polycentropodidae	
Glossosomatidae	Pyralidae	
Gordiidae	Thiaridae	
Ancylidae	Leptohyphidae	4
Asellidae	Physidae	
Calopterygidae	Planorbidae	
Gomphidae	Psychodidae	
Hirudinea	Sphaeriidae	
Hyallelidae	Stratiomyidae	
Hydropsychidae	Turbellaria	
Corydalidae	Elmidae	3
Belostomatidae	Libellulidae	2
Coenagrionidae	Tabanidae	
Hebridae	Tipulidae	
Staphylinidae		
Baetidae	Muscidae	1
Chironomidae	Nepidae	
Corixidae	Notonectidae	
Culicidae	Oligochaeta	
Dytiscidae	Simuliidae	
Hydrophilidae	Syrphidae	
Lestidae		

Table 2. Calibrated bioindication values for the aquatic invertebrates of the Apatlaco and Chalma-Tembembe Rivers.

The multiple linear regression equation of the quality test for predicting the BMWP scores is presented below (with $r = 0.935$). The adjusted line of $BMWP_{observed}$ vs $BMWP_{expected}$ indices attained R^2 values of 0.874 (**Figure 4a**); furthermore, all points related to the study sites were distributed within the confidence interval ($\alpha = 0.05$). The Nash and Sutcliffe efficiency model was of 0.87, demonstrating a well-fitted model:

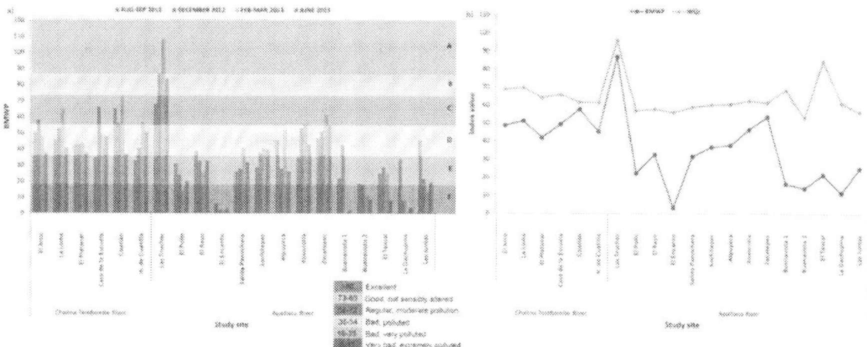

Figure 3. Scores of the BMWP. A: Excellent; B: Good, not sensible affected; C: Regular, moderate pollution; D: Bad, polluted; E: Bad, very polluted and F: Bad, extremely polluted. (a) Scores by period and study sites. (b) Scores of the BMWP and WQI.

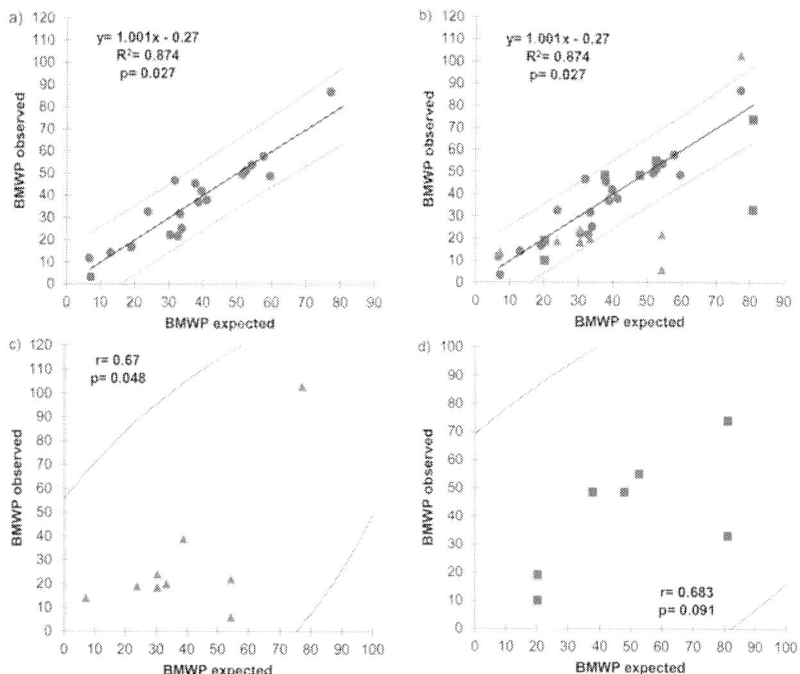

Figure 4. Validation model. (a) BMWP observed *vs* expected. (b) BMWP observed *vs* expected including validation data. (c) BMWP observed and expected values for the index validation. (d) BMWP observed and expected values for the regional extrapolation sites.

$$BMWP_{expected} = [138.854 - 7.69 \text{ Nitrates} + 123.51 \text{ Nitrites} - 0.18 \text{ Ammonia}$$
$$+2.17 \text{ Total N} - 33.11 \text{ Total P} + 0.80 \text{ Sulfates} - 1.18 \text{ Color} - 0.14 \text{ Alkalinity} \qquad (3)$$
$$+0.28 \text{ Chlorides} - 1.85 \text{ BOD}_5 - 24.96 \text{ DO} - 0.18 \text{ Conductivity} 2.19 \text{ \%DO}].$$

3.4. Index validation and regional extrapolation

The BMWP values for their validation and regional extrapolation (**Figure 5**) span the whole range of water quality classes: from "Very bad, extremely polluted" (Tlatenchi) to "Excellent" (Arriba Chalchihuapan).The observed and expected BMWP values for the nine index validation sites in the Apatlaco River and the seven regional in the neighboring river subbasins (Amacuzac, Cuautla and Yautepec) were calculated using the previously BMWP scores calibrated and the derived multiple linear regression model. Four sites of the BMWP values lay outside the confidence limits (α = 0.05) for the BMWP linear regression models (**Figure 4b**). Pearson's correlations were calculated (α = 0.05) for the BMWP observed and expected values for the index validation and regional extrapolation sites (**Figures 4c** and **d**). As in the previous analysis, acceptable (p = 0.048) values were calculated for the index validation, whereas the regional extrapolation showed weaker correlation (p = 0.091). Three out sites of the nine independent sites were outliers for the calculated multiple linear regressions BMWP model (**Figures 4c** and **d**).

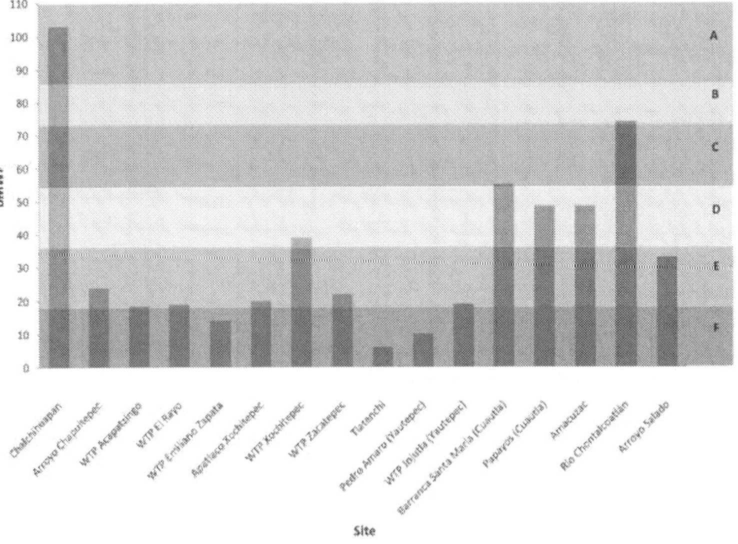

Figure 5. Scores of the BMWP for the validation and range extension study sites. A: Excellent; B: Good, not sensible affected; C: Regular, moderate pollution; D: Bad, polluted; E: Bad, very polluted and F: Bad, extremely polluted.

The index validation, adding nine independent sites, validated the regression model as a satisfactory indicator for river water quality in the Apatlaco River for the BMWP index ($r = 0.67$, $p = 0.048$). For the combined nine sites, there was a positive significant correlation for the BMWP index (**Figure 4a** and **b**).

4. Discussion

4.1. Water quality index

In Latin America, WQIs have been used to compare rivers in a country-wide dimension, the effect of a city discharge and also for a spatial and historical water quality assessment [13]. However, in Apatlaco and Chalma-Tembembe rivers, the WQI scores do not detect significant differences neither in the spatial nor in the temporal dimensions. In consequence, in this study, the WQI do not allow the detection of the most impaired portions of the rivers.

4.2. The BMWP index

Nowadays, the BMWP index is widely used in various countries of Europe (UK, Spain, Portugal, Turkey, Poland, among others) [8]. This index, also, has been used in some countries of Latin America [5, 14]. However, the procedure for calibrating the scores of the BMWP index has not been detailed. In the present study, we follow several steps for calibrating the BMWP values. The first step included the calculation of the minimum tolerance scores for each macroinvertebrate family from the study area. The *Pcq* index was used to obtain the bioindication values for each macroinvertebrate family. The factor analysis displayed the major explanatory variables, which are related with organic matter (N, P, BOD, DO, among others). These variables highlight the role of the different human activities that impinge on river water quality in the study area and can affect the abundance and distribution of aquatic macroinvertebrates families. Due to the particular conditions of each basin or biogeographical region, the relative importance of the physicochemical variables will vary, between study areas (basins), because of unique geological nature, land uses, as well as contamination and disturbance histories. Among changes in land use, those related to agricultural activities can affect several parameters of water quality (NO_3, NO_2, NH_3, TN, TP, among others), mainly because runoff contributes with the addition of agrochemicals into river, modifying water quality [33]. However, urban wastewater discharges can exert major impairments in the water quality due to their high load of organic matter that can deplete the DO and increase the DBO_5 [34]. The data set from the studied rivers included a wide spectrum of conditions: from those with low impact (clean) to those very affected (highly polluted), giving rise to a wide range of the *Pcq* scores and water quality, data very valuable for the calibration scores of the BMWP index.

4.3. The BMWP index and BMWP water quality classes

While the WQI scores showed small differences among study sites, the BMWP index showed a wider variation, making evident that the latter has a higher level of sensitivity, as it was able to register fine and important differences between study sites, which were not evidenced by the WQI (**Figure 5**). The sensitivity of the BMWP index is related with the procedure to assign the values for the water quality classes. For this step, the reference sites are indispensable. The BMWP index was able to detect the pristine condition of "Las Truchas," the clean river reference site, located inside an undisturbed oak forest, showing the highest richness score of this study site. Furthermore, this site showed the presence of the family Perlidae, a bioindicator of excellent water quality. Lakew and Moog [30] stated that a reference site must meet both abiotic and biotic requirements; the same authors consider a reference site as the least impaired site characterized by selected physical, chemical and biological characteristics. Las Truchas site reached the higher *Pcq* and WQI scores, the less impairment of adjacent land use and also meet the biological requirements of the higher richness, as well as the higher scores of the BMWP index. In consequence, the site "Las Truchas" meets the requirements of a reference site. The BMWP scores obtained in this study also showed a wide seasonal fluctuation, particularly for Las Truchas site. The BMWP fluctuation provided for Las Truchas showed scores from 69 to 110, these data conformed the set of values for obtaining the median value that correspond to the "Excellent" quality category, from which range values for all water quality classes were assigned. The method used for the allocation of scores to each water quality class was proposed by Pond and McMurray [21]. These procedure displayed a suitable assignation of the water quality categories and consequently reducing the probability of commit errors (Type I and II) [35] for the categorization of the study sites.

The bioindication values of the aquatic macroinvertebrate families do not always match completely from one country to another, which can be due to the variations in taxonomic tolerances of each basin and biogeographic region and to the method of assigning bioindication values, which in most cases is unknown, generating some uncertainty in the scores assigned to each family. However, there are families of aquatic macroinvertebrates characteristic of very healthy environments, as is the case of the Perlidae with a score of 10; or in extremely hostile conditions, the midges and lumbriculids with score values of 1 [14]. In the present study, the wide distribution of families in different intervals of *Pcq*, coupled with the abundance classes, allowed us to generate a bioindication value that is neither overestimated nor underestimated, as was evidenced by the degree of adjustment of the models described above.

Our results show that the BMWP index has a good discriminating capacity; nevertheless, doubtless, any index has to be adjusted, as demonstrated here, to the particular ecological conditions of each region in order to generate a powerful and representative biomonitoring tool.

4.4. Index validation and regional extrapolation

A third step in our procedure included the statistical index validation process, which produced multiple linear regression models for the BMWP index with good results in general. The obtained R^2 value in this study for the BMWP ($R^2 = 0.874$) indicates that the model included a great proportion of the variance.

For the index validation, we included a procedure with the addition of nine independent sites, validating the regression model as a satisfactory indicator for river water quality in the Apatlaco River for the BMWP index ($r = 0.67$, $p = 0.048$). For the combined nine sites, there was a positive significant correlation. Three out sites of the nine independent sites were outliers for the calculated multiple linear regressions (**Figure 6a** and **b**).

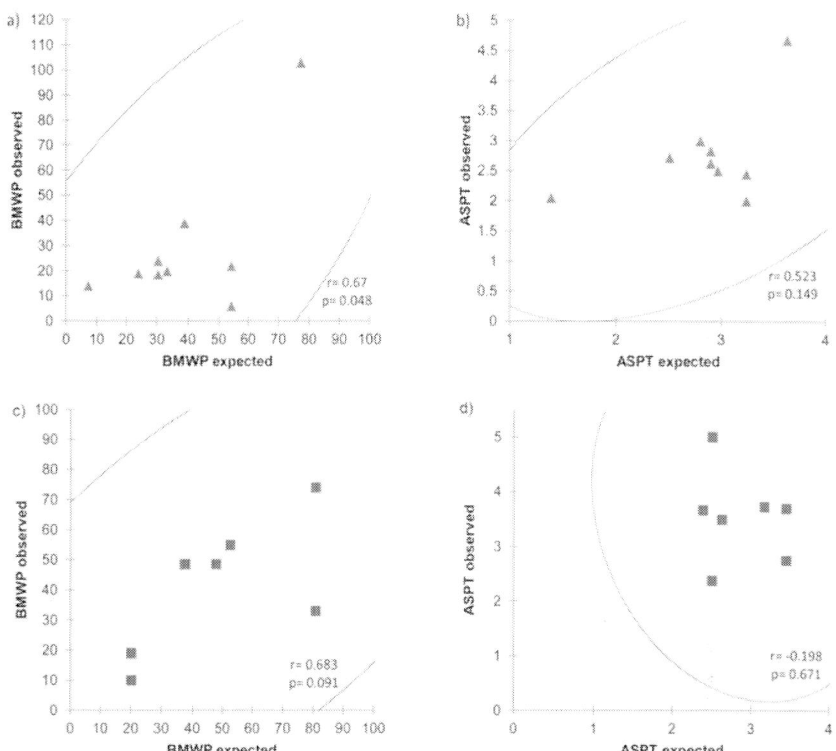

Figure 6. Validation and range extension data correlation. (a) Validation BMWP observed *vs* expected. (b) Validation ASP observed *vs* expected. (c) Extension BMWP observed *vs* expected. (d) Extension ASP observed *vs* expected.

The fourth step in our procedure included the range extrapolation analysis, where the multiple linear regression model was extended to study sites from Amacuzac, Cuautla and Yautepec subbasins for BMWP index, in this case we obtained lower correlation and p values ($r = 0.683$, $p = 0.091$). The scores accurately reflect the human impacts on the aquatic macroinvertebrates assemblage, with scores from "Bad, polluted" to "Very Bad, Extremely polluted." There were one outlier out of a total of seven sites for the BMWP, thus indicating that the BMWP regression model seemed to represent a more satisfactory geographical extension in this case (**Figure 6c** and **d**).

Therefore, the calibrated BMWP scores and the proposed water quality classes of this study can be used as a tool for the biomonitoring of water quality in the Apatlaco and Chalma-Tembembe rivers and even the subbasins Cuautla, Yautepec and Amacuzac. Furthermore, our ranges for water quality class showed also a good fit for the qualification of study sites and the spectrum of the land use conditions [36]. The studied rivers showed that a great portion of the rivers Apatlaco and Chalma-Tembembe (nine study sites), with agriculture as the main land use, is qualified as: bad polluted to regular, moderated polluted, while another great portion of the Apatlaco River mainly located in urban zones is qualified as bad, very polluted to very bad. The BMWP calibrated and the water quality assignations resulted to be suitable to assess water categories in the studied rivers. Our results make evident that Apatlaco River needs urgently a management and recovery plan.

5. Conclusions

The procedure followed in this study, included four steps, resulted to be efficient, reliable, repeatable and suitable for the development of a robust index to assess the water quality for rivers in the Neotropical region of México.

The tolerance values of the BMWP index developed in this study and their respective water quality classes can be applicable without modification to the adjacent river subbasins of the Apatlaco River, such as the subbasins Cuautla, Yautepec and Amacuzac.

The aquatic invertebrates and the BMWP index calibrated proved to be excellent indicators of water quality, being very sensitivity to differentiate the degree of pollution, different land uses and degrees of perturbation and thus, assign a water quality class that is also strongly related with the land use of their surrounding area.

These results make evident that the BMWP index calibrated is a suitable tool for the biomonitoring in the Neotropical region, where changes in land use have exerted strong impacts on aquatic resources and where the assessment of ecological conditions in freshwater ecosystems is urgently needed in a relatively simple, effective, reliable, fast and economical way. The procedure followed in this study to calibrate the BMWP is recommended and can be extended to other Neotropical rivers.

Acknowledgements

We thank to the join Fund of the National Science and Technology Council (CONACyT) and the Morelos Government, Project FOMIX CONACyT 173996, as well as the Research and Postgraduate Studies Secretariat (SIP Instituto Politécnico Nacional SIP20121087).

Author details

Ricardo Arturo Ruiz-Picos[1], Jacinto Elías Sedeño-Díaz[2] and Eugenia López-López[1*]

*Address all correspondence to: eulopez@ipn.mx

1 National Polytechnic Institute, National School of Biological Science, Laboratory of Aquatic Ecosystem Health Assessment, Laboratorio de Evaluación de la Salud de Ecosistemas Acuáticos, Departamento de Zoología, ENCB, Instituto Politécnico Nacional, Ciudad de México, México

2 Polytechnical Coordination for Sustainability, National Polytechnic Institute, Coordinación Politécnica para la Sustentabilidad, Instituto Politécnico Nacional, Ciudad de México, México

References

[1] World Economic Forum. World Economic Forum Annual Meeting 2016 [Internet]. 2016 [Updated: 2016]. Available from: https://www.weforum.org/events/world-economic-forum-annual-meeting-2016/ [Accessed: 2016-06-16].

[2] Balian EV, Segers H, Lévêque C, Martens K. The freshwater animal diversity assessment: an overview of the results. Hydrobiologia. 2008; 595: 627–637. DOI: 10.1007/s10750-007-9246-3.

[3] Ferreira WR, Paiva RT, Callisto M. Development of a benthic multimetric index for biomonitoring a Neotropical watershed. Brazilian Journal of Biology. 2011; 71(1): 15–25. DOI: 10.1590/S1519-69842011000100005.

[4] Dudgeon D, Arthington AH, Gessner MO, Kawabata ZI, Knowler D J, Lévêque C, Naiman RJ, Prieur-Richard AH, Soto D, Stiassny MLJ and Sullivan CA. Freshwater biodiversity: importance, threats, status and conservation challenges Biological Reviews. 2006; 81:163–182. DOI: 10.1017/S1464793105006950.

[5] Thorne RS, Williams WP. The response of benthic macroinvertebrates to pollution in developing countries: a multimetric system of bioassessment. Freshwater Biology. 1997; 37: 671–686. DOI: 10.1046/j.1365-2427.1997.00181.x.

[6] Kholodkevich SV, Ivanov AV, Kurakin AS, Kornienko EL, Fedotov VP. Real time biomonitoring of surface water toxicity level at water supply stations. Journal of Environmental Bioindicators. 2008; 3(1): 23–34. DOI: 10.1080/15555270701885747.

[7] Resh VH. Multinational freshwater biomonitoring programs in the developing world: lessons learned from South African and South East Asian river surveys. Environmental Management. 2007; 39(5): 737–748. DOI: 10.1007/s00267-006-0151-8.

[8] Oertel N, Salánki J. Biomonitoring and bioindicators in aquatic ecosystems. In: Ambasht RS, Ambasht NK (eds), Modern Trends in Applied Aquatic Ecology. Kluwer Academic/Plenum Publishers, New York, 2003; pp 219–246.

[9] Hawkes HA. Origin and development of the biological monitoring working party score system. Water Research 1997; 32: 964–968. DOI: 10.1016/S0043-1354(97)00275-3.

[10] Chang F, Lawrence JE, Ríos-Touma B, Resh, VH. Tolerance values of benthic macroin-vertebrates for stream biomonitoring: assessment of assumptions underlying scoring systems worldwide. Environmental Monitoring and Assessment. 2014; 186(4): 2135–2149. DOI: 10.1007/s10661-013-3523-6.

[11] Comisión Nacional del Agua. Indicadores de calidad del agua. 2014. National Commission of water. Indicators of water quality. Available at: http://www.cona-gua.gob.mx/Contenido.aspx?n1=3&n2=63&n3=98&n4=98 [Accessed: 2016-06-17].

[12] Weigel BM, Henne LJ, Martínez-Rivera LM. Macroinvertebrate-based index biotic integrity for protection of stream in west-central México. Journal of the North American Benthological Society. 2002; 21: 686–700. DOI: 10.2307/1468439.

[13] Sedeño-Díaz JE and López-López E. Water quality in the río Lerma, México: an overview of the last quarter of the twentieth century. Water Resources Management. 2007; 21: 1797–1812. DOI: 10.1007/s11269-006-9128-x.

[14] Riss W, Ospina R, Gutiérrez JD. Establecimiento de valores de bioindicación para macroinvertebrados acuáticos de la sabana de Bogotá. Construction of a biological indication system for aquatic macroinvertebrates of the Savanna of Bogotá. Caldasia. 2002; 24(1): 135–156.

[15] CONAGUA. Comisión Nacional del Agua. Atlas del agua en México. Secretaría de Medio Ambiente y Recursos Naturales. México, DF. 2012.

[16] APHA. Standard Methods for Examination of Water and Waste-water, 21st ed. APHA, Washington. 2005.

[17] Dinius SH. Design of an index of water quality. Water Resources Bulletin. 1987; 23: 833–843.

[18] Barbour MT, Gerritsen J, Snyder BD, Stribling JB. Rapid Bioassessment Protocols for Use in Streams and Wadeable Rivers: Periphyton, Benthic Macroin-vertebrates and Fish, 2nd ed. EPA 841-0B-99-002 U. S Environmental Protection Agency Office of Water, Washington, DC. 1999.

[19] Thorp JH and Covich AP. Ecology and Classification of North American Freshwater Invertebrates, 3rd ed. Academic Press, Elsevier, USA. 2001. 173 p.

[20] Merrit RW, Cummins KW, Berg MB. An Introduction to the Aquatic Insects of North America. Kendall/Hunt Publishing Company, USA. 2008. 1158 pp.

[21] Pond GJ, McMurray SE. A macroinvertebrate Bioassesment Index for Headwater Streams of the Eastern Coalfield Region, Kentucky. Kentucky Department for Environmental Protection, USA. 2002. 48 pp.

[22] Norma Oficial Mexicana NOM-127-SSA1-1994. Salud ambiental, agua para uso y consumo humano-límites permisibles de calidad y tratamientos a que debe someterse el agua para su potabilizacion. México. 1994. Mexican oficial normativity NOM-127-SSA1-1994. Environmental health, water for use and human consume, threshold and water treatment for the water potabilization, Mexico. 1994.

[23] Norma Oficial Mexicana NOM-001-SEMARNAT-1996. Establece los límites máximos permisibles de contaminantes en las descargas de aguas residuales en aguas y bienes nacionales. México. 1996.

[24] Canadian Drinking Water Guideliness. http://www.hc-sc.gc.ca/ewh-semt/watereau/ 2012. [Accessed: 2016-06-16].

[25] EPA. Environmental Protection Agency, www.epa.gov. 2003. [Accessed: 2014-04-18].

[26] Diario Oficial de Costa Rica, Reglamentan calidad de agua potable. Official newspaper of Costa Rica. Regulate water potable quality. http://www.tecdigital.itcr.ac.cr/file/ 5972860/Reglamento-de-Calidad-del-Agua-Potable-32327-S-La-Gaceta-84.pdf. 2005. [Accessed: 2016-06-16].

[27] Normas Técnicas de Calidad del Agua Potable. Technical standards for drinking water quality. DEC475/98. Decreto 475 de 1998. Diario Oficial de Colombia. Colombia. 1998.

[28] WHO – World Health Organization. Guidelines for Drinking-Water Quality, 3rd ed. WHO, Geneva. http://www.who.int/water_sanitation_health/dwq/ GDWQ2004web.pdf. 2004. [Accessed: 2016-06-16].

[29] Alba-Tercedor J. Macroinvertebrados acuáticos y calidad de las aguas de los ríos. Aquatic macroinvertebrates and water quality. IV Simposio del agua en Andalucía. 1996; 2: 203–213.

[30] Lakew A and Moog O. A multimetric index based on benthic macroinvertebrates for assessing the ecological status of streams and rivers in central and southeast highlands of Ethiopia. Hydrobiologia 2015; 751: 229–242.

[31] Armitage PD, Moss D, Wright JF and Furse MT. The performance of a new biological water quality score system based on macroinvertebrates over a wide range of unpolluted running-water sites. Water Research. 1983; 17: 333–347.

[32] Nash JE and Sutcliffe JV. River flow forecasting through conceptual models Part I- A discussion of principles. Journal of Hydrology. 1970; 10: 282–290.

[33] Ding J, Jiang Y, Fu L, Liu Q, Peng Q and Kang M. Impacts of land use on surface water quality in a subtropical river basin: acase study of the Dongjiang River Basin, Southeastern China. Water 2015; 7: 4427–4445 DOI:10.3390/w7084427.

[34] Daniel MHB, Montebelo AA, Bernardes MC, Ometto JPHB, De Camargo PB, Krusche AV, Ballester MV, Victoria RL and Martinelli LA. Effects of urban sewage on dissolved oxygen, dissolved inorganic and organic carbon and electrical conductivity of small streams along a gradient of urbanization in the Piracicaba river basin. Water, Air and Soil Pollution. 2002; 136: 189–206. DOI: 10.1023/A:1015287708170.

[35] Hawkins CP, Norris RH, Gerritsen J, Hughes RM, Jackson SK, Johnson RK and Stevenson RJ. Evaluation of the use of landscape classifications for the prediction of freshwater biota: synthesis and recommendations. Journal of the North American Benthological Society. 2000; 19: 541–556. DOI: 10.2307/1468113.

[36] Ruiz-Picos RA, Sedeño-Díaz JE and López-López E. Ensambles de macroinvertebrados acuáticos relacionados con diversos usos del suelo en los ríos Apatlaco y Chalma-Tembembe (cuenca del Río Balsas). Aquatic macroinvertebrate assemblages related with different land uses in the rivers Apatlaco and Chalma-Tembembe. In press. Hidrobiologica. 2016.

Chapter 4

Metals Toxic Effects in Aquatic Ecosystems: Modulators of Water Quality

Stefania Gheorghe, Catalina Stoica,

Gabriela Geanina Vasile, Mihai Nita-Lazar,

Elena Stanescu and Irina Eugenia Lucaciu

Additional information is available at the end of the chapter

http://dx.doi.org/10.5772/65744

Abstract

The topic of this work was based on the assessment of aquatic systems quality related to the persistent metal pollution. The use of aquatic organisms as bioindicators of metal pollution allowed the obtaining of valuable information about the acute and chronic toxicity on common Romanian aquatic species and the estimation of the environment quality. Laboratory toxicity results showed that Cd, As, Cu, Zn, Pb, Ni, Zr, and Ti have toxic to very toxic effects on *Cyprinus carpio*, and this observation could raise concerns because of its importance as a fishery resource. The benthic invertebrates' analysis showed that bioaccumulation level depends on species, type of metals, and sampling sites. The metal analysis from the shells of three mollusk species showed that the metals involved in the metabolic processes (Fe, Mn, Zn, Cu, and Mg) were more accumulated than the toxic ones (Pb, Cd). The bioaccumulation factors of metals in benthic invertebrates were subunitary, which indicated a slow bioaccumulation process in the studied aquatic ecosystems. The preliminary aquatic risk assessment of Ni, Cd, Cr, Cu, Pb, As, and Zn on *C. carpio* revealed insignificant to moderate risk considering the measured environmental concentrations, acute and long-term effects and environmental compartment.

Keywords: metals, fish, crustaceans, benthic invertebrates, toxicity, LC50, MATC, bioaccumulation, risk

1. Introduction

Metals are constantly released in aquatic systems from natural and anthropic sources such as industrial and domestic sewage discharges, mining, farming, electronic waste, anthropic accidents, navigation traffic as well as climate change events like floods (**Figure 1**) [1, 2]. Moreover, metals are easily dissolved in water and are subsequently absorbed by aquatic organisms such as fish and invertebrates inducing a wide range of biological effects, from being essential for living organisms to being lethal, respectively. In spite of the fact that some metals are essential at low concentrations for living organisms, such as (i) micronutrients (Cu, Zn, Fe, Mn, Co, Mo, Cr, and Se) and (ii) macronutrients (Ca, Mg, Na, P, and S); at higher concentrations, they could induce toxic effects disturbing organisms' growth, metabolism, or reproduction with consequences to the entire trophic chain, including on humans [3]. In addition, the non-essential metals such as Pb, Cd, Ni, As, and Hg enhance the overall toxic effect on organisms even at very low concentrations. High levels of metals in the environment could be a hazard for functions of natural ecosystems and human health, due to their toxic effects, long persistence, bioaccumulative proprieties, and biomagnification in the food chain [4, 5]. In this context, metal pollution is a global problem; therefore, the international regulations demanded for water quality compliance with the quality standards both in surface water or groundwater and in biota [6–9]. Currently, in accordance with the European Water Framework Directive (EU-WFD, 2000/60/EC), the ecological status of water bodies is assessed based on five biological indicators such as phytoplankton, macrophytes, phytobenthos, benthic invertebrates, and fish alongside with chemical and hydromorphological quality elements. Due to the fact that biota has the ability to accumulate various chemicals, it has been extensively used to measure the effects of metals on aquatic organisms as an essential indicator of water quality [10]. The mollusks [11–14] and fish [3, 15] are the most used organisms as bioindicators of metal pollution in water or sediment.

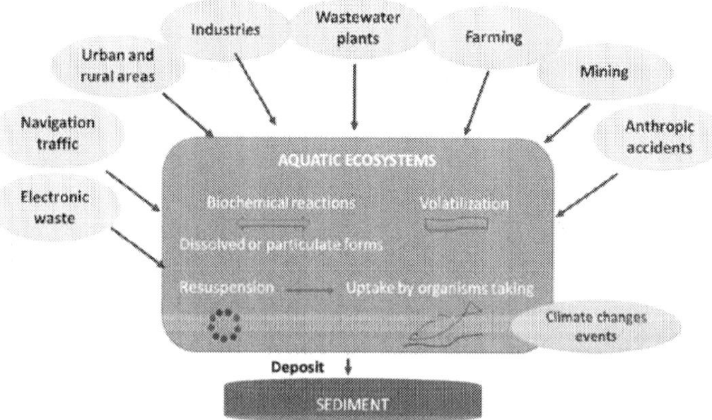

Figure 1. Sources of metal contamination affecting aquatic ecosystems.

The proposed topic of this chapter is based on the assessment of aquatic systems quality linked to persistent metal pollution. The chapter includes an extensive literature review concerning the impact of heavy metals on aquatic systems followed by an experimental part based on metal distribution and toxicity effect on the Romanian surface waters. Due to the European economic and strategic importance of Danube Delta, the final receptor of Danube's flow, the toxic effect of various metal concentrations (Ni, Zn, Cu, Cd, As, Cr, Pb, Co, Ti, Zr, Fe, Mn, etc.) was analyzed.

2. Metal ecotoxicology

2.1. Heavy metal bioavailability to aquatic organisms

Unlike organic chemicals, the majority of metals cannot be easily metabolized into less toxic compounds, a characteristic of them being the lack of biodegradability. Once introduced into the aquatic environment, metals are redistributed throughout the water column, accumulated in sediments or consumed by biota [16]. Due to desorption and remobilization processes of metals, the sediments constitute a long-term source of contamination to the food chain. Metal residues in contaminated habitats have the ability to bioaccumulate in aquatic ecosystems — aquatic flora and fauna [17], which, in turn, may enter into human food chain and result in health problems [18]. Metal accumulation in sediments occurs through processes of precipitation of certain compounds, binding fine solid particles, association with organic molecules, co-precipitation with Fe or Mn oxides or species bounded as carbonates — according to the physical and chemical conditions existing between the sediment and the associated water column [19, 20].

Metal bioavailability is defined as the fraction of the total concentration of metal which has the potential to accumulate in the body. The factors that control the bioavailability of metals (**Figure 2**) are the following: the organism biology (metals assimilation efficiency, feeding strategies, size or age, reproductive stage); metal geochemistry (distribution in water — sediment, suspended matters, and metal speciation) [21, 22]; physical and chemical factors (temperature, salinity, pH, ionic strength, concentration of dissolved organic carbon, total suspended solids) [23, 24].

Metal bioavailability controls their accumulation in aquatic organisms. The metals uptaken paths are through the permeable epidermis if metals are in dissolved forms or through the food ingestion if metals are in particulate forms. Metal speciation, the presence of organic or inorganic complexes, pH, temperature, salinity, and redox conditions [24] are the main factors that could modulate metal toxicity. The ingestion uptake depends on similar factors, plus the rate of feeding, intestinal transit time, and the digestion efficiency [25].

Many studies have shown that the free hydrated metallic ion is the most bioavailable form for Cu, Cd, Zn [26], and Pb [27], but some exceptions have been reported [28]. Thus, the importance of other chemical forms of dissolved metals and complexes formed with suitable organic ligands with low molecular weight should not be neglected. It has been found that the presence of organic binders increases the bioavailability of Cd in mussels and fish, by facilitating the diffusion of the hydrophobic compound in the lipid membrane. The organic compounds of metals could be more bioavailable than the ionic forms [29]. For instance, the organic mercurial compounds are lipid-soluble and penetrate quickly the lipid membranes, increasing the toxicity compared to mercuric chloride which is not lipid-soluble [30].

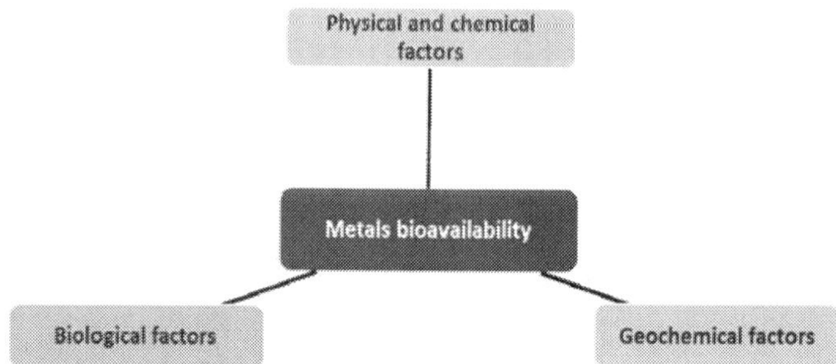

Figure 2. The main control factors that influence metal bioavailability.

The adsorption on suspended solids affects the total concentration of metals present in water. The association between solid particles and metals is also critical for the metal uptake into organisms through food ingestion [31]. The suspended solids accumulate the insoluble metal compounds, but under certain conditions, the metal reached the interstitial water being dissolved. Heavy metal concentrations from sediments or suspended solids are much higher than in water, so a small fraction of them could be an important source for bioaccumulation in planktonic and benthic organisms [32]. The dynamics of different forms of metals in the aquatic environment is not fully understood, so new studies are required to analyze the different accumulation/bioaccumulation pathways based on dissolved or suspended metal forms.

Other studies highlighted that bioavailability of metals in bivalve mollusks depends on sediment particle size due to their filter feeding behavior. If the particles were coated with bacterial extracellular polymers or fulvic acids, the Cd, Zn, and Ag bioavailability was significantly increased. Overall, the binding of metal decreased the bioavailability of metals from the sediment [28, 33].

2.2. Metal toxicity effects

After metal ingestion, this is specifically transported by lipoproteins into different body compartments (organs, blood, or other physiological structures) where they can be specifically oriented to different centers: (i) *action centers* where the toxic metal interacts with an endogen macromolecule (protein or ADN) or a certain cellular structure inducing toxic effects for all body; (ii) *metabolism centers* where the metal is processed by detoxified enzymes; (iii) *storage centers* where the metals are collected in a toxic inactive state; and (iv) *excretion centers* where the metals are disposed.

The heavy metal overload has inhibitory effects on the development of aquatic organisms (phytoplankton, zooplankton, and fish) [34, 35]. The metallic compounds could disturb the oxygen level and mollusks development, byssus formation, as well as reproductive processes. Several histological changes such as gill necrosis or fatty degeneration of the liver occur in the fish and crustaceans [36, 37]. Assessments at the cellular level enable to understand the action of toxic metals on the enzymatic metabolism and physiology of the aquatic organisms.

The lethal effects of metals in crustaceans were induced by the inhibition of enzymes involved in cellular respiration. The histological changes observed in fish and crustaceans after chronic exposure to metals are the result of antioxidant enzymes inhibition [38–41]. The effects on organisms' growth and development were triggered by the inhibition of enzymatic systems involved in protein synthesis and cell division. The metal type modulates the bioaccumulation level and enzymatic systems vulnerability generating a multitude of effects, toxic or not [42, 43].

In order to understand the interaction mechanism between the toxic metals and the aquatic organisms and how organisms answer to metal contamination, more information on bioavailability is needed [44].

At the present, many studies on the assessment of acute and chronic toxicity of metals mentioned the following parameters: survival, growth, development, reproduction, behavior, accumulation, effects on enzyme systems, etc. In **Table 1**, the values of acute (LC50) and chronic (MATC/NOEC/LOEC) toxic concentration for fish and planktonic crustaceans according to PAN Pesticide Database—Chemical Toxicity Studies on Aquatic Organisms [45] are exemplified. The studies highlighted that the toxic concentration intervals depend on the species, exposure time, age of specimens, type of toxicity test type, and laboratory conditions.

| Metal | Fish (*Cyprinus carpio*) | | Crustaceans (*Daphnia magna*) EC50 |
	LC50 (96 h)	MATC/NOEC/LOEC	(48 h)
Ni	1.3–10.4 mg/L	NOEC 50 µg/L LOEC 3.50; 13 µg/L; 1.97 mg/L	1 g/L
Zn	0.45–30 mg/L	NOEC 2.60 µg/L; 0.43; 4.20 mg/L	0.35–3.29 mg/L
Cd	2–240 µg/L 3–17.05 mg/L	NOEC 0.02–37 µg/L LOEC 6.7–440 µg/L	24–355.4 µg/L
As	0.49 mg/L (*Carassius sp.*) 0.9 mg/L (*Pimephales sp.*)	LOEC 25 µg/L	3.8 mg/L
Cr	14.3–93 mg/L	NOEC 0.19–17 µg/L LOEC 25 µg/L	22–160 µg/L
Pb	0.44–2 mg/L	NOEC 0. 07; 128 µg/L LOEC 0.03–128 µg/L	4.4; 5.7 mg/L (*Daphnia sp.*)
Sb	6.2–8.3 mg/L (*Cypridon sp.*)	NOEC 6.2 mg/L	>1 g/L
Mn	0.1–15.61 mg/L (*Oncorhynchus sp.*)	–	40 mg/L

LC (EC) 50—lethal concentrations for 50% of tested organisms after 96 or 48 h; MATC—maximum acceptable toxicant concentration in aquatic systems; NOEC—no observed effect concentration; LOEC—low observed effect concentration.

Table 1. Literature acute and chronic toxicity values [45–47].

2.3. Metal bioaccumulation and bioamplification

Monitoring the toxicity and accumulation of metals into the aquatic biota or sediment is mainly performed for assessing both the surface water quality and ensure food safety, respectively, as well as for the compliance with the directives. The toxicity and accumulation parameters are used for various environment monitoring programs such as wastewater discharges or various risk assessments of natural and anthropogenic events (floods or dredging activities) and also for identifying the source of metal contamination [48, 49].

According to the European document COM (2011)—876 final—2011/0429 (COD) (2012/C 229/22) amending the Directive 2000/60/EC and 2008/105/EC on priority substances in the field of water policy, new concentration limits of a number of harmful chemical compounds were allowed for the aquatic biota (fish, mollusks, or crustaceans). For instance, diphenyl brominated, fluoranthene, hexachlorobenzene, hexachlorobutadiene, benzene compounds, dicofol, perfluorooctane sulfonic acid and its derivatives, dioxins and dioxin-type compounds, cyclohexa-bromo-dodecane, heptachlor epoxide, and heptachlor have a concentration values in the range 6.7×10^{-3} to 167 mg/kg wet weight. The rest of the chemical compounds were not yet amended, which represents a considerable research opportunity to assess their chronic toxicity, bioaccumulation, and subsequently to set their maximum permissible concentration limits in aquatic organisms. Furthermore, the Directive 2008/105/EC of Environmental Quality Standards (EQS) entail values for various chemicals in biota.

More and more studies of various organic and inorganic chemical bioaccumulation/bioconcentration in freshwater organisms revealed induced harmful effects, especially of heavy metals (Hg, As, Cd, Zn, Fe, Pb, Fe, Mn, etc.) [10, 50, 51] and metal nanoparticles [52]. Bioaccumulation remains to be an ongoing highly debated subject. According to United States Geological Survey (USGS) Toxic Substances Hydrology Program, the bioaccumulation represents *"the biological sequestering of a substance at a higher concentration than that at which it occurs in the surrounding environment or medium."* Pollutants can be uptaken by organism directly from the environment or through ingestion of particles [53], and the accumulation occurs when an organism absorbs toxic chemical with a rate faster than the chemical is metabolized. On the contrary, the bioconcentration refers to the chemical uptake from the water only, which could be assessed in the laboratory conditions. The value of the concentration factor index gives information if there is a bioaccumulation (the concentration factor of <1) or a bioconcentration (concentration factor >1) [54]. The understanding of the bioaccumulation process is important because persistent pollutants (such as metals) could increase the toxic potential risk by bioaccumulation in the ecosystem, triggering a long-term effect on the ecosystem which cannot be assessed by laboratory toxicity tests [54]. It is considered that a high bioaccumulation potential does not necessarily imply a high potential for toxicity, and as a result, the toxic effects should be estimated separately. In addition, it was made a distinction between accumulation in a small concentrations range, which occurs due to physiological needs (e.g., Zn) and apparently uncontrolled accumulation (e.g., Cd) [55].

It was observed that the mollusks from the Black Sea have shown a great tendency to accumulate in high concentration Cd and Cu from sediments as well as Cd, Ni, and Cu from water. Data showed that the highest concentrations of heavy metals were found in the digestive tract of fish [56]. Also, the fish *Cyprinus carpio* could differentially bioaccumulate metals inform one organ to another: Zn > Cr > Pb > Cu in muscle; Pb > Cr > Zn > Cu in gills; Pb > Cr > Zn in liver [10]. Moreover, for the same species, it has been shown that gills and liver or kidney were accumulating the following metals: Pb > Cd > Cr > Ni and Pb > Cd > Ni > Cr. On the other hand, bioaccumulation of Pb and Cd was significant in all *C. carpio* tissues [57].

Metal transfer in the aquatic food chain is another interesting environmental topic for many reasons such as the accumulation of metals in aquatic organisms that could transfer up to humans, leading to a potential risk of public health through consumption of contaminated fish [58, 59]. It is known that aquatic organisms can be exposed to high or low concentrations of metals as a result of continuous or accidental release, causing long-term effects. The main uptake pathways of metals in aquatic organisms are direct through the food or sediment particles ingestion and water via epidermis and gills then they are transported inside the cells through biological membranes and ionic channels [60]. Bioconcentration and bioaccumulation of metals into the trophic chain occur if metals are excreted into the water or the contaminated organisms are food for some predator's organisms [61, 62].

The study named "*Ecotoxicology of heavy metals in the Danube meadow*" [55] revealed that the amplification of metal concentrations in the food chains of ecosystems depends on the type of metal and the food chain. The metal accumulation in plants depends on the species, metal type, and ecosystem, especially for species which predominantly take metals from soil/sediment. Benthic gastropods tend to differentially accumulate the metals. The populations which are using the seston as energy source concentrate in many cases metals from different sources: Bivalves shell accumulate Pb, the tissue—Mn, Zn, Cd, and sometimes the amphibians in young stages accumulate Cd, Mn, and Zn.

Metal concentrations such as Fe, Mn, Cu, Cr, and Pb were not amplified in the food chain (benthic fauna-fish-birds), but they were amplified for Zn and Cd. Concentrations of metals were greater at the end of the trophic chains, as follows: vegetation/detritus—terrestrial invertebrates phytophase/detritophage—terrestrial invertebrates' predators—amphibians (Cd, Cr, Pb, and Cu in case of detritus chain and Zn in case of vegetation). The fish always accumulate metals, with some exceptions in the case of Cd, for which the transfer coefficient indicates accumulation in muscle and liver. The transfer of metals from benthic invertebrates to omnivorous fish revealed concentration of Zn and Cu in the liver and Zn in muscle. The Mn, Cr, and Cd metals transfer from omnivorous fish (muscle) to predatory fish, more specifically in their muscles and liver. At the end, the birds that are using contaminated fish as food source will accumulate Fe, Mn, Zn, Cu, and Cd in muscle and all metals (except Cr) in the liver [55].

3. Experimental part

3.1. Occurrence of metals in surface water and sediments

Several monitoring studies performed by INCD ECOIND Bucharest researchers during 2003–2013 in the Danube Delta—Sfantu Gheorghe Branch (sampling points: Mahmudia,

Murighiol, and Uzlina) emphasized some heavy metal concentration patterns in the study area (**Table 2**). The metal concentrations in water were within the limits of Romanian legislations, for class I and class II quality (according to the EU-WFD and the requirements set by the Romanian Law 310/2004 which amends the Law 17/1996). Cu and Ni showed the highest total concentration (**Table 2**, marked lines) among the determined metals.

Metal	Mahmudia (2009–2013)				Murighiol (2003–2013)				Uzlina (2003–2013)			
	Min	Max	Average	SD	Min	Max	Average	SD	Min	Max	Average	SD
Ni	<1.0	24.0	**4.02**	5.79	<1.00	68.1	**12.4**	18.6	<1.00	10.3	**2.60**	2.85
Fe	<20	880	**300**	270	112	3400	**710**	750	80.0	1040	**350**	280
Mn	<2.0	30.0	**10.0**	10.0	3.00	290	**70.0**	80.0	5.00	50.0	**20.0**	10.0
Cd	0.40	0.40	**0.40**	0.00	<0.10	0.50	**0.36**	0.14	<0.10	0.50	**0.37**	0.13
Cr	<0.5	6.00	**2.09**	1.74	<0.50	21.0	**5.38**	6.00	<0.50	21.0	**3.52**	5.33
Cu	2.50	10.5	**5.74**	2.67	0.012	55.3	**12.9**	17.8	0.03	123	**14.5**	26.7
Pb	<2.0	3.20	**2.08**	0.31	<2.00	5.00	**2.15**	1.29	<2.00	5.00	**2.17**	1.27
As	<2.0	2.20	**2.01**	0.05	<2.00	3.90	**1.82**	0.88	<2.00	2.64	**1.73**	0.59
Hg	<0.1	0.24	**0.32**	0.06	<0.10	0.77	**0.15**	0.20	<0.10	0.14	**0.22**	0.10
Zn	<2.0	24.7	**10.1**	7.07	<2.00	56.0	**9.58**	11.4	<2.00	57.0	**8.15**	11.7
Co	<0.5	1.30	**0.66**	0.33	<0.50	5.00	**1.17**	1.65	<0.50	5.00	**1.18**	1.65

Min—the minimum detected concentration; Max—the maximum detected concentration; SD—standard deviation.

Table 2. Occurrence of metals in water of Danube Delta—Sf. Gheorghe (2003–2013) in µg/L [63, 64].

The studies revealed that metals Cu, Pb, Zn, Cr, Ni, Cd, Mn, and Fe were the most abundant in the sediments of the Danube Delta—Sf. Gheorghe Branch sampling sites. The concentrations of these metals ranged with the sampling location and seasonal or natural events, as follows: Cu 4.65–194 mg/kg d.m (dry matter), Pb 4.76–51.3 mg/kg d.m., Zn 17.7–218 mg/kg d.m., Cr 7.5–61.9 mg/kg d.m., Ni 10.8–111 mg/kg d.m., Cd <0.01–1.5 mg/kg d.m. (**Figure 3**). The average value in the period 2009–2013 for Mn was 614.03 mg/kg d.m. and for Fe, it was 20 987 mg/kg d.m. [64, 65].

Alongside metal concentration, several chemical (nutrients, oxygen and pH regime, pesticides, petroleum products, polychlorinated biphenyls) and biological (phytoplankton, zooplankton, and benthic macroinvertebrates) elements were investigated, showing that the organochlorine pesticides and petroleum products exceeded the maximum allowed limits [63, 66].

In addition, other studies performed along Romanian rivers showed that the mining activities had a great impact on sediment ecosystems due to metal pollution. For instance, a study performed during 2003 in Baia Mare (in North Vest of Romania) mining area after a pollution accident showed a high content of heavy metals in Somes River sediment (Cu 104–339 mg/kg, Pb 59–465 mg/kg, Zn 56–2060 mg/kg, Cd 0.05–14.14 mg/kg, CN 0.33–15.86 mg/kg). The detected concentration affected the aquatic ecosystem where the microalgae species disappeared and the number of fish species decreased dramatically compared to the period before the incident. Also many species of mollusks disappeared because their capacity to accumulate large amount of heavy metals was exceeded [67]. In addition, in Rosia Montana area (in the West part of Romania), significant water contamination with heavy metals occurred due to the mining acidic waters from area on two water courses: Rosia and Corna stream. The results showed exceedances of Cu, Cd, Fe, Ni, and Cr, in particular in the Rosia Montana water stream [68]. Along Jiu River (in south of Romania) sediments, heavy metal pollution in most sampling points was recorded according to the pollution load index (PLI) [69].

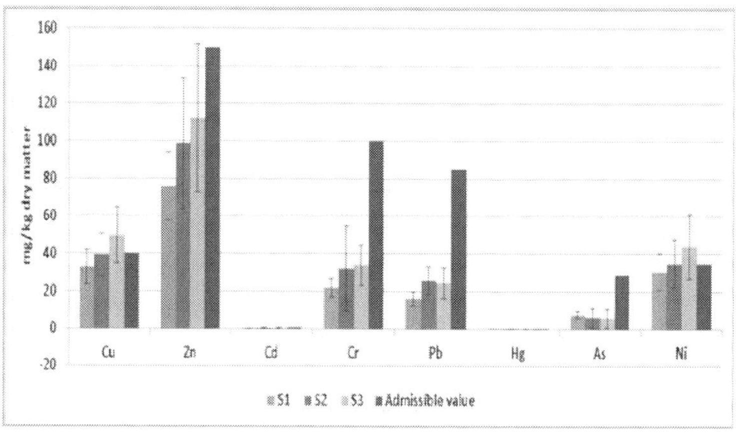

Figure 3. Occurrence of metals in sediments of Danube Delta—Sf. Gheorghe (2003–2013) (average values). S1—Mahmudia, S2—Murighiol, and S3—Uzlina [65].

In the above context, in the following sections will be presented some data concerning the metals effects on freshwater organisms (fish, planktonic crustacean, and mollusks), obtained through laboratory testing or by biological samples collected from contaminated fields.

3.2. Laboratory tests: acute and chronic effects of metals

3.2.1. Materials and methods

The assessment of metals acute and chronic effects was based on fish (*C. carpio*) and planktonic crustacean (*Daphnia magna*) laboratory data. The tested organisms were those recommended by the international ecotoxicology protocols (OECD or ISO), and they are frequently found in Romanian surface waters, easily to acclimatize in laboratory and sensitive to various contaminants. *C. carpio* are in particular the most affected organisms due to the fact they ingest both planktonic and benthic organisms, respectively, and thus, they especially accumulate the contamination from water and sediment. The tests were performed on the following metals: Ni, Zn, Cu, Cd, As, Cr, Pb, Sb, Mn, Ti, and Zr, which were usually detected in the aquatic systems.

3.2.1.1. Sample preparation

For stock solution preparation, a known quantity of metals test as $NiSO_4$, $ZnSO_4$, $CuSO_4$, $CdCl_2/CdSO_4$, As_2O_3, $K_2Cr_2O_7$, $Pb(NO_3)_2$, $SbCl_5$, $MnCl_2x4H_2O$, TiO_2, $ZrCl_4$ was dissolved into the specified volume of dilution water or growth medium. No added solvents have been used, and all substances have been tested under their maximum solubility. The solutions were stirred for 24 h, in the dark at 25°C. The testing solutions were prepared by mixing the appropriate volumes of stock solution with dilution water or growth medium in order to obtain the final concentrations used for testing. Finally, the pH values of tested solutions were situated between 6.5 and 8.5 units.

3.2.1.2. Fish toxicity test procedure

Using OECD methodologies for acute toxicity, the lethal concentrations for 50% of tested organisms were estimated. Metals' long-term toxicities on fish were conducted using an in-house methodology based on the changes in some physiological indicators such as growth rate, mortality, biomass, production, food use and biochemical indicators, hepatic enzyme activity, respectively. **Table 3** presents the technical parameters of fish toxicity tests.

3.2.1.3. Crustacean toxicity test procedure

The toxicity test determined the metal concentration that immobilizes or kills 50% (LC50) of *D. magna* crustacean, after chemical exposure at 20°C ± 2°C in the dark for 24 or 48 h. The test procedure was performed according to OECD 202 using the microbiotest Daphoxkit F Magna provided by MicroBioTests Inc., Belgium. Briefly, the test was performed in three replicates, in multiwall test plates (six rinsing wells and 24 wells for toxicant dilutions) using 20 organisms per each concentration (at least five different concentrations for each metal) and control (untreated standard freshwater). The mortality/immobility percentage of organisms was registered after 24 and 48 h.

3.2.1.4. Data processing and statistics

The acute effect concentration values in the fish and crustacean tests were calculated using probity analysis method, based on exponential regression relationship between cumulative percentages of mortality (expressed as probity units) for each exposure period against logarithmic concentrations of test substance. For each result, standard deviations were calculated.

3.2.2. Results and discussion

3.2.2.1. Fish toxicity

Acute toxicity tests provide a measure of toxicity for a target species under specific environmental situations and could suggest a rapid and severe effect of contaminants. Acute and chronic toxicity test mimicked the metals accidental release or long-term accumulation in sediment [67–70]. The carp fish LC50-96h values showed different responses in direct correlation with the metals type and concentration. The LC50-96h values were 0.16, 0.28, 0.31, and 0.40 mg/L for Cd, Ti, Zr, and As (**Figure 4**), 2.17, 12.2, 30.10, and 65.8 mg/L for Cu, Zn, Pb, and Ni (**Figure 5**), 120, and 758 mg/L for Cr and Sb, obtained from two replicates for each metal (**Table 4**).

Test conditions	OECD 203 (acute tests)	In-house procedure (chronic tests)
Holding of fish	Acclimatization of fish in laboratory tanks for 3 weeks	
Limit test	One concentration selected according to scientific literature	MATC estimated = LC50-96 h × 0.1
Definitive test		
Test concentrations in definitive test	Five concentrations in a geometric series	Two concentrations (under or over the estimated MATC)
Type of test	Static	Discontinuous (renewal solutions at 24 h)
Time of exposure	96 h	60 days
Fish species and characteristics	*Cyprinus carpio* (1 year) 10 exemplars/test solution, 5–7 cm, 10–15 g/exemplary	*Cyprinus carpio* (2 years) 20–30 exemplars/test solution, 12–14 cm, 25–30 g/exemplary
Fish source	Romanian specialized fish farm	
Testing vessels	10 L	100 L
Temperature, oxygen concentration, pH, light	18–25°C, ≥4 mgO$_2$/L, pH 6.5–8.5 (daily measuring), 12- to 16-h photoperiod daily. Mean of water total hardness 13 mg/L CaCO$_3$	
Feeding	Not food	2% from the surviving lot weight/day
Control test	All toxicity tests were carried out in the same time with a control test	
Replicates	Two replicates/test/metal	
Analytical control in test solutions	Inductively coupled plasma atomic emission spectrometry (ICP-OES)	
Toxicity criteria	Organisms mortalities and visible abnormalities (at 24, 48, 72, and 96 h)	Growth instant rate, mortality rate, biomass mean, production, used food rate, and biochemical indicators—hepatic enzymes activity—GOT and GPT
Results treatment	Probity analysis method based on the exponential regression model between the mortality (probity units) and the log of concentrations of the metal	Comparative analyses with the controls
End points	Lethal concentrations for 50% of tested fish after 96 h of exposure (LC50-96 h)	Maximum acceptable toxicant concentration in aquatic systems (MATC)

Table 3. Test conditions of acute and chronic toxicity tests.

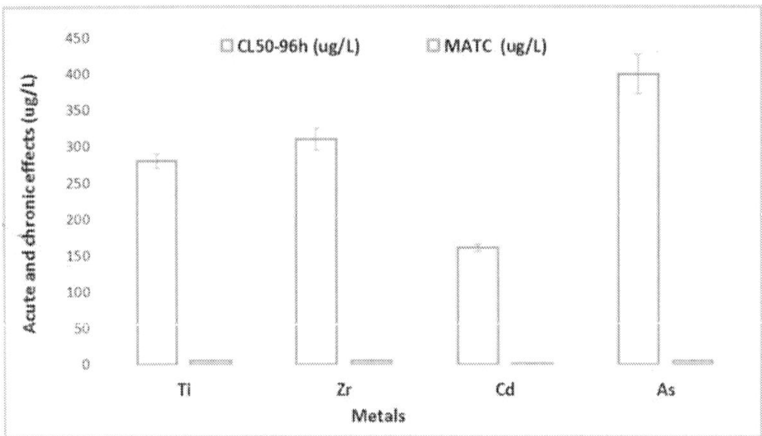

Figure 4. Acute and chronic toxicity of Ti, Zr, Cd, and As classified in very toxic class for *Cyprinus carpio*.

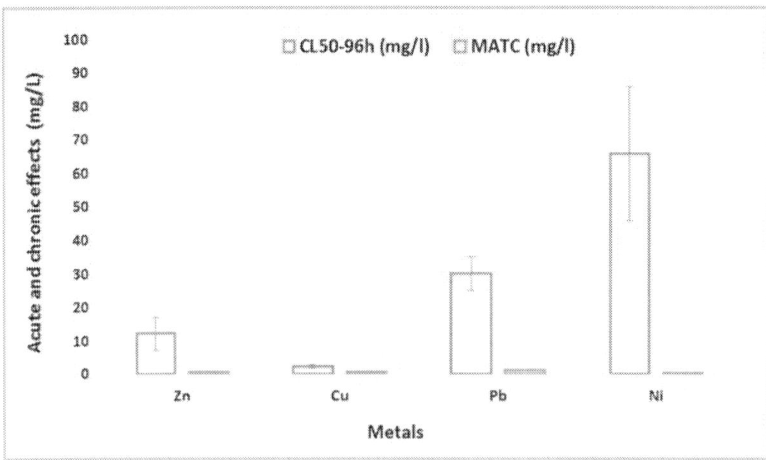

Figure 5. Acute and chronic toxicity of Zn, Cu, Pb, and Ni classified in toxic class for *Cyprinus carpio*.

According to Global Harmonization System for chemical classification and labeling, Cd, Ti, Zr, and As were the most toxic metals for fish. Cd, Ti, Zr, and As showed to be very toxic compared with the other analyzed metals. Research studies revealed similar acute toxicity intervals: 6.16–47.58 mg/L for Ni, 0.15–21.4 mg/L for Zn, 0.28–34.5 mg/L for Cu, 0.005–7.92 mg/L for Cd and 90 to >139 mg/L for Cr [71]. The maximum acceptable toxicant concentration (MATC) is a value calculated from chronic toxicity tests [72] in order to set water quality norms for aquatic life protection.

Metals	Cyprinus carpio		Daphnia magna	G.D. 351/2005[a] (μg/L)	Directive 105[b] (μg/L)	National plan[c] (μg/L)	Toxicity class[d]
	LC50-96h (mg/L)	MATC (mg/L)	LC50-48h (mg/L)				
Ti (TiO$_2$)	0.28 ± 0.01	0.005 ± 0.001	5.56 ± 0.8	–	–	–	Very toxic—fish
Zr (ZrCl$_4$)	0.31 ± 0.01	0.005 ± 0.002	91.20 ± 10	–	–	–	Very toxic—fish
Ni (NiSO$_4$)	65.8 ± 20.0	0.10 ± 0.02	–	20	20	4–34	Toxic—fish
Zn (ZnSO$_4$)	12.2 ± 5.0	0.60 ± 0.01	–	5	–	11.80–73	Toxic—fish
Cu (CuSO$_4$)	2.17 ± 0.50	0.05 ± 0.01	–	100	–	1.22–10	Toxic—fish
Cd (CdCl$_2$/ CdSO$_4$)	0.16 ± 0.001	0.001 ± 0.0005	0.14 ± 0.01	5	0.2	–	Very toxic—fish and daphnia
As (As$_2$O$_3$)	0.40 ± 0.02	0.005 ± 0.001	–	10	–	49	Very toxic—fish
Cr (K$_2$Cr$_2$O$_7$)	120 ± 22	1.00 ± 0.01	0.81 ± 0.02	50	–	8.8	Very toxic—daphnia
Pb (Pb(NO$_3$)$_2$)	30.1 ± 5.0	1.00 ± 0.02	–	10	7.2	–	Toxic—fish
Sb (SbCl$_5$)	758 ± 24	0.060 ± 0.001	148 ± 21	5	–	–	Non-toxic—fish and daphnia
Mn (MnCl$_2$×4H$_2$O)	>53 ± 8	–	–	–	–	–	Non-toxic—fish

[a] Governmental Decision no. 351/2005 concerning the hazard chemical discharge.
[b] Directive 2008/105/EC on environmental quality standards in the field of water policy.
[c] National Plan of River Basin Management (2016-2021)—Annex 6.1.3B.
[d] According to REACH 1907/2006; Regulation (EC) 1272/2008; Regulation (EU) 286/2011; Global Harmonization System for chemical classification and labeling (GHS) Revision 2011. The toxicity class was decided on the highest toxicity of target organisms.

Table 4. In-house toxicity data of metals for fish and crustacean in relation with the national and international norms for metals limits in surface water.

Experimental exposure of fish for 60 days to different concentrations of metals revealed different long-term effects. The final results showed no effects concentrations on target organisms, assessment of environmentally safe concentrations, respectively. The calculation of MATC values started by multiplication of the LC50-96h of each metal with an application factor of 0.1 (**Table 2**). The monitored physiological parameters from chronic test revealed that Cd is non-toxic at 0.001 mg/L, Ti, Zr, and As were safety to 0.005 mg/L, Cu at 0.05 mg/L, Sb at 0.06 mg/L, Ni at 0.10 mg/L, Zn at 0.60 mg/L, Cr and Pb at 1.00 mg/L, comparative with the controls (**Figures 4** and **5, Table 4**). Similar values for Cu (0.012 mg/L) and Zn (0.5 mg/L) were also obtained in other studies [73].

3.2.2.2. Crustaceans toxicity

Toxicity tests on *D. magna* crustaceans showed various toxicities of metals; the LC50-48h showed 0.14 mg/L for Cd, 0.81 mg/L for Cr, 5.56 mg/L for Ti, 91.2 mg/L for Zr, and 148 mg/L for Sb. Cd and Cr showed the highest toxicities and were classified in very toxic chemicals class for *Daphnia* sp. (**Figure 6**). Similar literature values were reported for Cr between 0.02 and 0.05 mg/L [71] and for Cd between 0.024 and 0.355 mg/L [45].

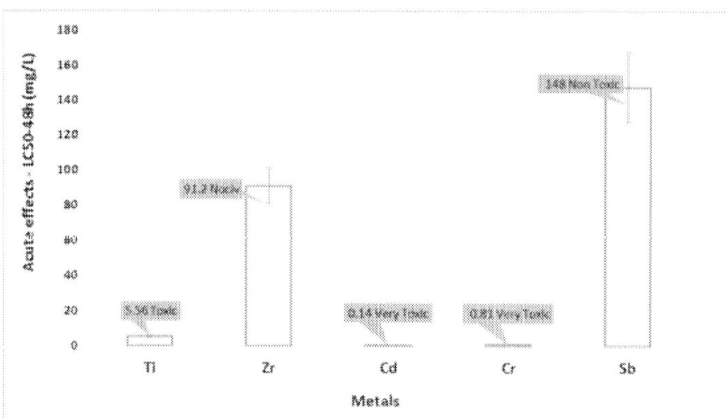

Figure 6. Acute toxicity of Ti, Zr, Cd, Cr, and Sb for *Daphnia magna*.

The surface water quality norms require specific limits only for few very toxic and toxic metals. For example, Ti, Zr, Cd, and Pb norms are not established by the National Plan of River Basin Management—Annex 6.1.3B, despite of their acute toxic effects at very low concentrations (**Table 4**). Also the Directive 2008/105/EC on environmental quality standards in the field of water policy sets limits only for Ni, Cd, and Pb. The present limits assure the protection of aquatic organisms, especially for fish and planktonic crustaceans.

3.3. Field test: bioaccumulation

In order to assess the impact of metals in the field, the following sections present some preliminary data concerning the metal bioaccumulation into benthic invertebrates (mollusks).

3.3.1. Materials and methods

3.3.1.1. Studied area characterization

The studied area was focused on a highly sinuous channel, located on the southeast area of the Danube Delta (Sf. Gheorghe Branch) receiving 22% of Danube's water flow. The Sf. Gheorghe Branch has a width varying between 150 and 550 m, and the water depth varies between 3 and 27 m. The sampling sites location was selected taking into consideration the changes in the Sf. Gheorghe Branch morphology as a result of the pressure from anthropic and environmental factors. Iron Gates I dam construction on Danube River led to a 10% decrease in the suspended sediment amount at Isaccea station. Moreover, the Iron Gates II dam building induced a 50% decrease in suspended sediment at Isaccea. These constructions alongside meander modification (during the years 1984–1988) have produced major changes in sediment distribution. The establishing of space location was performed using GPS type system map 60CSx—Garmin [74].

In addition, the anthropic activities undertaken to strength the banks against coastal erosion led to meanders cutoff, which in turn caused continuous biotope degradation. These changes negatively impacted the ecosystem functions by reducing the structure of the main and constant ecological communities, the benthic invertebrates. So that, to characterize metal bioaccumulation (in benthic invertebrates), two representative sampling sites were selected considering the pressure resulted from anthropic and environmental factors (Murighiol and Uzlina)—**Figure 7**. At temporal scale, this study was conducted during summer and autumn of 2013.

Figure 7. Location of sampling sites in Danube Delta (Sfantu Gheorghe Branch) (St 1—Murighiol; St 2—Uzlina).

3.3.1.2. Sampling collection

The sediment samples for both benthic invertebrates and metal analysis were collected in two replicates using a Van Veen grab, according to the following methodologies: EN ISO 5667-1:2008, ISO SR 5667-6:2009, SR ISO 5667-12:2001 and EN ISO 9391:2000. Surface sample unit was of 255 cm², and the sampling depth was of 10 cm. The analysis of benthic invertebrates was performed according to SR EN ISO 8689-1:2003. The species identification was performed using a Motic stereomicroscope. The results were calculated taking into consideration the wet biomass.

3.3.1.3. Sample preparation

The biota samples were dried at 40°C (24 h) and crushed then about three grams of biological sample were dissolved in aqua regia (a mixture of suprapure acids HCl 30 and 65% HNO_3 in the report 21–7 mL). The mixture was mineralized using a sand bath until complete dissolution. After cooling, the samples were filtered on paper filter (porosity <45 µm) in a 50-mL volumetric flask and filled with ultrapure water. The metal content in the samples was determined by inductively coupled plasma optical emission spectrometry. A calibration curve in the range of 0.1–0.5 mg/L (As, Se, Sb, Cd, Cr, Cu, Co, Fe, Mn, Mg, Ni, Pb, Zn) was performed using a

Certified Reference Material solution (100 mg/L Multi Element Standard Solution, Certipur, Merck). The quality control of the data was carried out according to Quality Control Standards 21A, 100 mg/L, produced by PerkinElmer. A reagent blank in order to estimate the metal contents from acids was prepared.

The mollusks (two bivalves' species: *Unio pictorum* and *Anodonta cygnea*) and one gastropod species (*Viviparus viviparus*) were selected in this study (**Figure 8**) as they prevail in the total biomass of benthic invertebrate community structure, and they are widely used as bioindicators for water quality. Their shells were subjected to metals detection, because they are formed throughout mollusks life and their chemical composition is an integral index to describe the composition of the aquatic environment over time [75]. Bioaccumulation factors of metals were calculated for each tested species.

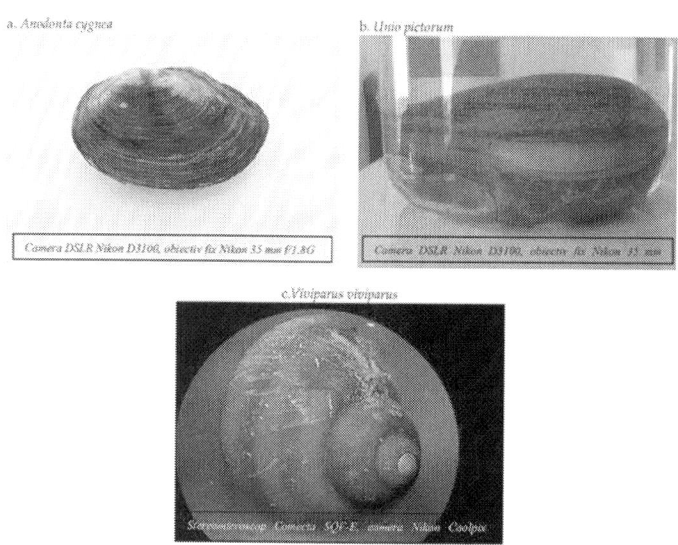

Figure 8. The analyzed benthic macroinvertebrates species.

3.3.2. Results and discussion

3.3.2.1. Metal accumulation in benthic organisms

This study included metal analysis results in the bivalves and gastropod shells from Murighiol and Uzlina sampling site. Other researchers [76–79] performed their studies as well using the same biological model, mollusk shells, for the metal accumulation analysis.

The mollusks have the largest representation and are the most valuable groups among the benthic invertebrates' communities due to the fact they are dominant in the total benthic community biomass and represent a basic food for the next trophic level (e.g., fish).

Two types of bivalve species identified at Uzlina and Murighiol were selected for metal analysis, respectively: *U. pictorum, A. cygnea*. Also from Gasteropoda, the species *V. viviparus* were selected (**Figure 8**). For each species, the dry and wet biomasses were determined (**Table 5**).

Sampling point/month	Species	Wet biomass (g)	Dry biomass (g)
Murighiol/July	*Unio pictorum*	25.38	24.49
Uzlina/July	*Viviparus viviparous*	10.96	0.88
	Unio pictorum	39.64	35.44
Uzlina/September	*Anodonta cygnea*	32.73	30.16
	Unio pictorum	19.68	18.58

Table 5. Dry and wet biomass values of the selected species.

The Biota Sediment Accumulation Factor (BSAFsed) was calculated using the equation: $BSAFsed = Cb/Csed$, where Cb is the metal concentration in biota/organism and $Csed$ is the metal concentration in the sediment sample [80].

At Murighiol sampling site, in the *U. pictorum*, shells (collected in July) were recorded the highest values for Cu, Ni, and Zn. Moreover, the Cu concentration in sediment was 47 mg/kg d.m. over the set limit. It was estimated that 4% of the Cu concentration, 2% of Zn, and 1% of Ni were found in *U. pictorum* shell species. The BSAFsed values were <0.05 (**Table 6**). Also, various metals were detected in the *U. pictorum* and *V. viviparous* shells from Uzlina, in July (**Table 7**).

Metal	Cb* *Unio pictorum*	Csed*	Csed*	BSAFsed 2009–2013**
As	<0.05	12.2	9.61	0.004
Cd	<0.01	–	0.50	–
Cu	1.97	47.0	35.1	0.04
Cr	0.12	27.6	31.6	0.004
Co	0.05	9.41	8.37	0.005
Fe	73.2	–	14,895	–
Mn	230	–	464	–
Ni	0.60	35.0	30.8	0.02
Pb	<0.05	25.6	22.3	0.002
Se	0.44	–	–	–
Sb	<0.05	–	–	–
Zn	1.17	91.7	88.5	0.01
Mg	62.5	–	–	–

* Average of metal concentrations (for two replicates) expressed in mg/kg d.m.
** Csed 2009–2013—average of the metal concentration detected in sediment from 2009 to 2013 [63, 65].

Table 6. Metal concentration (mg/kg d.m) in the shell of *Unio pictorum* at Murighiol in July 2013.

Metal	Cb* *Unio pictorum*	BSAFsed	Cb* *Viviparus viviparus*	BSAFsed	Csed*	Csed* 2009–2013**
As	<0.05	0.006	<0.05	0.006	7.75	9.30
Cd	<0.01	–	<0.01	–	–	0.51
Cu	2.61	0.05	2.60	0.05	54.7	47.0
Cr	<0.01	0.0003	0.42	0.01	29.6	29.2
Co	0.11	0.01	0.19	0.02	10.8	9.84
Fe	140	–	279	–	–	20987
Mn	58.7	–	30.0	–	–	614
Ni	0.34	0.0085	0.58	0.015	40.0	39.3
Pb	<0.05	0.002	0.15	0.006	26.7	21.3
Se	<0.09	–	<0.09	–	–	–
Sb	<0.05	–	<0.05	–	–	–
Zn	0.90	0.006	3.87	0.02	158	96.9
Mg	154	–	211	–	–	–

* Average of metal concentrations (for two replicates) expressed in mg/kg d.m.
** Csed 2009–2013 — average of the metal concentration detected in sediment from 2009 to 2013 [63, 65].

Table 7. Metal concentration (mg/kg d.m) in the shell of *Unio pictorum* and *Viviparus viviparus* at Uzlina in July 2013.

Concentrations of As, Cd, Cr, Fe, Pb, Se, Sb, Mg did not showed significant values in shells of analyzed benthic organisms. The metals Cu, Ni, and Zn were present in sediment over the set limits of national norms inducing their accumulation in shells. The highest values of Cu and Zn were both in *U. pictorum* and *V. viviparus* (**Table 7**). Similar concentrations of Zn, Cu, Pb, Cd, and Co in *V. viviparus* were found in the River Dnieper in the same gastropod shells [81].

Metal concentrations showed a lower magnitude in mollusk shells than in their bodies, and this result could be explained by the fact that metals were accumulated in shell only after they were absorbed by the body. The bioaccumulation selectivity of metals in gastropod shells follows the next order: Fe > Mn > Zn > Cu > Pb > Co > Cd. Thus, the quantitative distribution of metals in mollusk shells is considered by the level of biochemical involvement, metabolic processes, their toxicity degree as well as the bioavailability to aquatic organisms [81].

Some studies [82] revealed that Fe belongs to metals which play an important role in body metabolism and is not toxic. The Mn, Mg, Co, Cu, Zn, and Ni are involved in growth, development, and reproduction process, but in high concentrations can show toxic effects (see the above section "*Laboratory tests—acute and chronic effects*"). Pb and Cd are not involved in metabolic processes; thus, they are highly toxic at low concentrations and have a great storage capacity in the organisms at long-term exposure. The results on metal concentrations in *U. pictorum* and *A. cygnea* shells, metal detection in sediment samples (2013), average of metals detection in sediment in period of 2009–2013 at Uzlina and BSAFsed values are presented in

Table 8. The metals were determined in both bivalve species *U. pictorum* and *A. cygnea* shells, in September 2013 (**Table 8**).

| Metal | Cb* | BSAFsed | Cb* | BSAFsed | Csed* | Csed* |
	Unio pictorum		*Anodonta cygnea*			2009–2013**
As	<0.05	0.007		0.008	6.60	9.30
Cd	<0.01	–	<0.01	–	–	0.51
Cu	2.57	0.05	4.63	0.09	48.9	47.0
Cr	<0.01	0.0004	0.14	0.005	28.2	29.2
Co	0.13	0.01	<0.01	0.001	9.47	9.84
Fe	153	–	97.8	–	–	20,987
Mn	248	–	157	–	–	614
Ni	0.29	0.008	0.24	0.007	35.6	39.3
Pb	<0.05	0.002	<0.05	0.001	30.2	21.3
Se	<0.09	–	<0.09	–	–	–
Sb	<0.05	–	<0.05	–	–	–
Zn	0.59	0.006	1.27	0.01	91.2	96.9
Mg	42.8	–	79.6	–	–	–

* Average of metal concentrations (for two replicates) expressed in mg/kg d.m.
** Csed 2009–2013—average of the metal concentration detected in sediment from 2009 to 2013 [63, 65].

Table 8. Metal concentration (mg/kg d.m) in the shell of *Unio pictorum* and *Anodonta cygnea* at Uzlina in September 2013.

In this case, the highest metal concentration was recorded for Cu, Ni, and Zn. The Cu and Ni concentrations from sediment exceed the allowed limit values both in September 2013 and as well as during 2009–2013 monitoring period. As shown in **Table 8**, the *A. cygnea* were found to have a greater capacity for metal accumulation (especially for Cu, As, Cr, Zn) than *U. pictorum* shells.

The bioaccumulation level varied depending on species, metals type, and sampling sites. No significant differences were observed between bioaccumulation factors of Cu, Zn, and Ni calculated for *U. pictorum* collected in July and September. The BSAFsed values were subunitary maintained. It was observed a difference considering the sampling points, respectively, at Murighiol the bioaccumulative metals impact (Ni and Zn) was greater compared to Uzlina. This aspect may be explained by the dredging works for the canal enlarging/widening to facilitate navigation, allowing a better water circulation from the branch inside the canal.

This preliminary study for the metal bioaccumulation capacity in the shell mollusks from Danube Delta aquatic system showed that essential metals involved in metabolic processes (such as Fe, Mn, Zn, Cu, and Mg) have a greater storage capacity than those toxic (such as Pb and Cd). The statement was also confirmed in other studies [83–85].

All the biota sediment bioaccumulation factors were subunitary, which indicated a slowly bioaccumulation process occurred in the studied aquatic ecosystems.

3.4. Preliminary risk assessment

Risk characterization is required for all chemicals as an estimation of their exposure and adverse effects on the environmental compartment. Generally, this is based on Predicted Environmental Concentration (PEC) and Predicted No Effect Concentration (PNEC) calculation, in terms of exposure and assessment of effects [86].

In order to estimate the current contamination of Danube surface water and sediment with metals, we use the average of the measured environmental concentrations (MEC) as PEC values, for the period 2009–2013 at Murighiol and Uzlina. The PNEC value calculation was made using an assessment factor (AF) of 1000 applied for acute toxicity values—LC50 (96 h) or 10 applied for chronic toxicity values—MATC for *C. carpio* (our laboratory tests), which expresses the degree of uncertainty in the actual environmental extrapolation [87]. The risk quotients (RQs) between MEC values and acute or chronic PNECs were calculated, and the level of risk was expressed as: insignificant risk (RQs <0.1); low risk (RQs <1); moderate risk (RQs <10), and high risk (RQs >10). The estimated RQs for the most detected metals in Danube water and sediment (Ni, Cd, Cr, Cu, Pb, and Zn) were summarized in **Table 9**.

Metal	MEC (µg/L)*		PNEC (µg/L)		RQs acute		Risk level	RQs chronic		Risk level
	S7	S8	Acute (AF = 1000)	Chronic (AF = 10)	S7	S8		S7	S8	
Surface water										
Ni	12.4	2.60	65.8	10.0	0.18	0.03	L/I	1.24	0.26	L
Cd	0.36	0.37	0.16	0.10	2.25	2.31	M	3.60	3.70	M
Cr	5.38	3.52	120	100	0.04	0.02	I	0.05	0.04	I
Cu	12.9	14.5	2.17	5.00	5.94	6.68	M	2.58	2.90	M
Pb	2.15	2.17	30.1	100	0.07	0.07	I	0.02	0.02	I
As	1.82	1.73	0.40	0.50	4.55	4.32	M	3.64	3.46	M
Zn	9.58	8.15	12.2	60.0	0.78	0.66	L	0.16	0.14	L
Sediment										
Ni	30.8	39.3	65.8	10.0	0.46	0.59	L	3.08	3.93	M
Cd	0.50	0.51	0.16	0.10	3.12	3.18	M	5.00	5.10	M
Cr	31.6	29.2	120	100	0.26	0.24	L	0.32	0.29	L
Cu	35.1	47.0	2.17	5.00	16.2	21.7	H	7.01	9.41	M
Pb	22.3	21.3	30.1	100	0.74	0.70	L	0.22	0.21	L
As	9.61	9.30	0.40	0.50	24.0	23.3	H	19.2	18.6	H
Zn	88.5	96.9	12.23	60.0	7.23	7.92	M	1.48	1.62	M

* Average of concentrations in period of 2009–2013; I—insignificant risk; L—low risk; M—moderate risk; H—high risk.

Table 9. Estimated acute and chronic RQs at Murighiol (S7) and Uzlina (S8) for *Cyprinus carpio*.

The results showed different levels of risk in accordance with detected environmental concentration of metals, the acute and chronic toxicity and the environmental compartment (water or sediment). In water, Cr and Pb showed insignificant risk; Ni and Zn showed a low risk; and Cd, Cu, and As highlight a moderate risk considering both acute and chronic effects on *C. carpio*. Variations of the RQs depending on sampling location are not observed.

As we expected, the risk level increases within the sediment compartment. The sediment contamination revealed low-to-moderate risk, exception for As and Cu. Therefore, Cr and Pb showed low risk; Ni, Cd, Zn and Cu highlighted moderate risk; and As and Cu could express a high risk on fish *C. carpio*. Cu, Zn, and Ni were constantly present in sediment over the set limits of national norms inducing also their accumulation (see the section *"Field tests— bioaccumulation"*). No variation is observed of the RQs depending on sampling location. Using long-term toxicities in PNECs estimation, the RQs increased for Ni, Cd, and Cr and decreased in case of Cu, Pb, As, and Zn, due to the use of a small applied factor (AF = 10) to chronic toxicities.

The results highlighted a pessimistic view concerning the quality of aquatic ecosystem needed to support the carp fish survival. The concern is related to the constantly presence of metal concentrations especially in sediments (the food provider compartment) which could deter‐ minate the bioaccumulation. The same statement was made in a Romanian study named *"Ecotoxicology of heavy metals in Danube meadow"* [55].

4. Conclusions

The topic of this chapter was based on the assessment of aquatic systems quality related to persistent metal pollution. The toxic metals are the most frequently detected pollutants in the aquatic environmental, and their effects identification are essential to protect the ecosystems integrity as well as human health. Metal pollution is a global problem; thus, the international regulations with regard to the water quality demand compliance with the quality standards in surface water, groundwater, and biota. The use of organisms (such as fish, crustacean, and mollusks) as bioindicators of metal pollution allowed us to obtain valuable information about the effects on the Romanian common species and to estimate the quality of their environment. The results from laboratory toxicity tests showed the highest concentration values that are not relevant for the detected metal concentrations into surface water, but the metals accidentally released and long-term accumulation could create similar conditions to the results of applied tests. Cd, As, Cu, Zn, Pb, Ni, Zr, and Ti have a very toxic and toxic effects for *C. carpio* and could raise concerns because of its importance for human as a fishery resource. Benthic invertebrates' analysis of the bioaccumulation level varied between species, metals type, and sampling sites. The metal analysis in mollusks shell showed that the metals involved in the metabolic processes (Fe, Mn, Zn, Cu, and Mg) had greater storage capacity than the toxic one (Pb, Cd). In case of *V. viviparus* shell, the selectivity of the metal concentration was represented as follows: Fe > Mn > Zn > Cu > Pb > Co > Cd, while the shell of *A. cygnea* had a greater accumulation capacity for Cu, As, Cr, Zn compared to *Unio* sp. The bioaccumulation factors of metals in benthic organisms were subunitary, which indicated a slowly bioaccumulation process occurred in the studied aquatic ecosystems. This conclusion highlighted a bioaccumulation process that can increase the persistence of metals in the ecosystem, with a long-term potential risk in trophic chain. The preliminary aquatic risk assessment calculated for *C. carpio* for the most detected metals both in water and in sediment (Ni, Cd, Cr, Cu, Pb, As, and Zn) revealed insignificant to moderate risk considering the metals measured environmental concentrations, acute and long-term effects. The results highlighted a pessimistic view concerning the quality of aquatic ecosystem needed to support the carp survival. The concern is related to the constant presence of metal concentrations especially in sediments which is the principal food provider, leading to bioaccumulation processes and trophic chain transfer. Future studies have been initiated to understand the long-term effects of metals in aquatic biota and to complete the aquatic risk assessment considering the abiotic factors.

Abbreviations:

Cd	cadmium
As	arsenium
Cu	copper
Pb	lead
Ni	nickel
Zr	zirconium
Ti	titanium
Fe	iron
Zn	zinc
Mn	manganese
Mg	magnesium
Cd	cadmium
Co	cobalt
Cr	chromium
Mo	molybdenum
Se	selenium
Na	sodium
P	phosphorus

S	sulfur
Hg	mercury
CN	cyanide
LC (EC) 50	lethal concentrations for 50% of tested organisms after 96 or 48 h
MATC	maximum acceptable toxicant concentration in aquatic systems
NOEC	no observed effect concentration
LOEC	low observed effect concentration
GOT	glutamic oxaloacetic transaminase
GPT	glutamic pyruvic transaminase
OECD	Organization for Economic Co-operation and Development
PNEC	predicted no-effect concentration
PEC	predicted exposure concentration
MEC	measured environmental concentration
RQ	risk quotient

Author details

Stefania Gheorghe*, Catalina Stoica, Gabriela Geanina Vasile, Mihai Nita-Lazar, Elena Stanescu and Irina Eugenia Lucaciu

*Address all correspondence to: biologi@incdecoind.ro

National Research and Development Institute for Industrial Ecology – ECOIND, Bucharest, Romania

References

[1] Sandu C, Farkas A, Musa-Iacob R, Ionica D, Parpala L, Zinevici V, Dobre D, Radu M, Presing M, Casper H, Buruiana V, Wegmann K, Stan G, Bloesch J, Triebskorn R, Köhler H-R. Monitoring pollution in River Mures, Romania, part I: the limitation of traditional methods and community response. Large Rivers 2008; 18(1–2): 91–106. doi:10.1127/lr/18/2008/91.

[2] Zhang W, Zhang Y, Zhang L, Lin Q. Bioaccumulation of metals in tissues of seahorses collected from Coastal China. Bull Environ Contam Toxicol 2016; 96(3): 281–288. doi: 20.1007/s00128-16-1728-4

[3] Stankovic S, Kalaba P, Stankovic RA. Biota as toxic metal indicators. Environ Chem Lett 2014; 12: 63–84. doi:10.1007/s10311-013-0430-6

[4] Velma V, Tchounwou PB. Chromium-induced biochemical, genotoxic and histopathologic effects in liver and kidney of goldfish, *Carassius auratus*. Mutat Res 2010; 698 (1–2): 43–51. doi:10.1016/j.mrgentox.2010.03.014

[5] Conceiçao Vieira M, Torronteras R, Córdoba F, Canalejo A. Acute toxicity of manganese in goldfish *Carassius auratus* is associated with oxidative stress and organ specific antioxidant responses. Ecotoxicol Environ Saf 2012; 78: 212–217. doi:10.1016/j.ecoenv. 2011.11.015

[6] WFW Directive 2000/60/EC of the European Parliament and of the Council of 23 October 2000 establishing a framework for Community action in the field of water policy.

[7] Directive 2008/1/EC of the European Parliament and of the Council of 15 January 2008 concerning integrated pollution prevention and control.

[8] Directive 2008/56/EC of the European Parliament and of the Council of 17 June 2008 establishing a framework for community action in the field of marine environmental policy (Marine Strategy Framework Directive).

[9] Commission Regulation (EC) No 629/2008 of 2 July 2008 amending Regulation (EC) No 1881/2006 setting maximum levels for certain contaminants in foodstuffs (Text with EEA relevance).

[10] El-Moselhy KhM, Othman AI, Abd El-Azem H, El-Metwally MEA. Bioaccumulation of heavy metals in some tissues of fish in the Red Sea, Egypt. Egypt J Basic Appl Sci 2014; 1(2): 97–105. doi:10.1016/j.ejbas.2014.06.001

[11] Moloukhia H, Sleem S. Bioaccumulation, fate and toxicity of two heavy metals common in industrial wastes in two aquatic mollusks. J Am Sci 2011; 7(8): 459–464. ISSN 1545-1003.

[12] Stankovic S, Jovic M. Health risks of heavy metals in the Mediterranean mussels as seafood. Environ Chem Lett 2012; 10(2): 119–130. doi:10.1007/s10311-011-0343-1

[13] Stankovic S, Jovic M. Native and invasive mussels. In: Nowak J, Kozlowski M (eds). Mussels: ecology, life habits and control. Chapter 1. NOVA Publisher: NY; 2013. p. 1–45. ISBN: 978-1-62618-084-0.

[14] Stankovic S, Stankovic RA. Bioindicators of toxic metals. In: Lichtfouse E et al (eds). Green materials for energy, products and depollution, environmental chemistry for a sustainable world, 3rd ed. Chapter 5. Springer: Berlin; 2013, p.151–228. doi: 10.1007/978-94-007-6836-9_5

[15] Hauser-Davis RA, Calixto de Campos R, Ziolli RL. Fish metalloproteins as biomarkers of environmental contamination. Rev Environ Contam Toxicol 2012; 218: 101–123. doi: 10.1007/978-1-4614-3137-4_2

[16] Frémion F, Bordas F, Mourier B, Lenain JF, Kestens T, Courtin-Nomade A. Influence of dams on sediment continuity: a study case of a natural metallic contamination. Sci Total Environ 2016; 547: 282–294. doi:10.1016/j.scitotenv.2016.01.023.

[17] Hasan MR, Khan MZH, Khan M, Aktar S, Rahman M, Hossain F, Hasan ASMM. Heavy metals distribution and contamination in surface water of the Bay of Bengal coast. Environ Sci 2016; 2(1): 1–12. doi:10.1080/23311843.2016.1140001

[18] Varol M, Şen B. Assessment of nutrient and heavy metal contamination in surface water and sediments of the upper Tigris River, Turkey. Catena 2012; 92: 1–10. doi:10.1016/j.catena.2011.11.011

[19] Equeenuddin SM, Tripathy S, Sahoo P, Panigrahi M. Metal behavior in sediment associated with acid mine drainage stream: role of pH. J Geochem Explor 2013; 124: 230–237. doi:10.1016/j.gexplo.2012.10.010

[20] Vasile G, Cruceru L, Petre J, Iancu V. Complex analytical investigations regarding the bio-availability of heavy metals from sediments. Rev Chim (Bucharest) 2005; 56(8): 790–794.

[21] Griscom SB, Fisher NS. Bioavailability of sediment-bound metals to marine bivalve mollusks: an overview. Estuaries 2004; 27(5): 826–838. doi:10.1007/BF02912044

[22] Roosa S, Prygiel E, Lesven L, Wattiez R, Gillan D, Ferrari BJ, Criquet J, Billon G. On the bioavailability of trace metals in surface sediments: a combined geochemical and biological approach. Environ Sci Pollution Res 2016; 23(11): 10679–10692. doi:10.1007/s11356-016-6198-z

[23] Fu J, Zhao C, Luo Y, Liu C, Kyzas GZ, Luo Y, Zhao D, An S, Zhu H. Heavy metals in surface sediments of the Jialu River, China: their relations to environmental factors. J Hazard Mater 2014; 270: 102–109. doi:10.1016/j.jhazmat.2014.01.044

[24] Bonnail E, Sarmiento AM, DelValls TA, Nieto JM, Riba I. Assessment of metal contamination, bioavailability, toxicity and bioaccumulation in extreme metallic environments (Iberian Pyrite Belt) using *Corbicula fluminea*. Sci Total Environ 2016; 544: 1031–1044. doi:10.1016/j.scitotenv.2015.11.131

[25] Bryan GW, Langston WJ, Hummerstone LG and Burt GR. A guide to the assessment of heavy metal contamination in estuaries using biological indicators. Mar Biolo Assoc UK 1985; 4: 91–110.

[26] Roesijadi G, Robinson WE. Metal regulation in aquatic animals: mechanisms of uptake, accumulation, and release. In: Malins DC and Ostrander GK (eds). Aquatic toxicology: molecular, biochemical, and cellular perspectives. Lewis Publishers: Boca Raton; 1994. p. 387–420. ISBN-13: 978-0873715454. ISBN-10: 0873715454.

[27] Wojtkowska M, Bogacki J, Witeska A. Assessment of the hazard posed by metal forms in water and sediments. Sci Total Environ 2016; 551–552: 387–392. doi:10.1016/j.scitotenv.2016.01.073

[28] Wang W.X, Fisher NS. Modeling metal bioavailability in marine mussels. Rev Environ Contam Toxicol 1997; 151: 39–65. doi:10.1007/978-1-4612-1958-3_2

[29] Chapman PM, Wang F, Janssen C, Persoone G, Allen HE. Ecotoxicology of metals in aquatic sediments: binding and release, bioavailability, risk assessment, and remediation. Can J Fish Aquat Sci 1998 (on-line 2011); 55(10): 2221–2243. doi:10.1139/f98-145

[30] Bryan GW. Some aspects of heavy metals tolerance in aquatic organisms. In: Lockwood APM (ed). Effects of pollutants on aquatic organisms. Cambridge University Press: Cambridge; 1976. p. 7–35.

[31] Eggleton J, Thomas KV. A review of factors affecting the release and bioavailability of contaminants during sediment disturbance events. Environ Int 2004; 30(7): 973–980. doi:10.1016/j.envint.2004.03.001

[32] Adams WJ, Chapman PM. Assessing the hazard of metals and inorganic metal substances in aquatic and terrestrial systems. New York, USA; CRC Press; 2007. ISBN 9781420044416.

[33] Rosado D, Usero J, Morillo J. Assessment of heavy metals bioavailability and toxicity toward *Vibrio fischeri* in sediment of the Huelva estuary. Chemosphere 2016; 153: 10–17. doi:10.1016/j.chemosphere.2016.03.040

[34] Atici T, Obali O, Altindag A, Ahiska S, Aydin D. The accumulation of heavy metals (Cd, Pb, Hg, Cr) and their state in phytoplanktonic algae and zooplanktonic organisms in Beysehir Lake and Mogan Lake, Turkey. Afr J Biotechnol 2010; 9(4): 475–487. ISSN: 1684-5315.

[35] Bere T, Chia MA, Tundisi J.G. Effects of Cr III and Pb on the bioaccumulation and toxicity of Cd in tropical periphyton communities: implications of pulsed metal exposures. Environ Pollut 2012; 163: 184–191. doi:10.1016/j.envpol.2011.12.028

[36] Brraich OS, Kaur M. Ultrastructural changes in the gills of a cyprinid fish, *Labeo rohita* (Hamilton, 1822) through scanning electron microscopy after exposure to Lead Nitrate (Teleostei: *Cyprinidae*). Iran J Ichthyol 2015; 2(4): 270–279. P-ISSN: 2383-1561; E-ISSN: 2383-0964.

[37] Sevcikova M, Modra H, Blahov J, Dobsikova R, Plhalova L, Zitka O, Hynek D, Kizek R, Skoric M, Svobodova Z. Biochemical, haematological and oxidative stress responses of common carp (*Cyprinus carpio* L.) after sub-chronic exposure to copper. Veterinarni Medicina 2016; 61(1): 35–50.

[38] Mishra AK, Mohanty B. Acute toxicity impacts of hexavalent chromium on behavior and histopathology of gill, kidney and liver of the freshwater fish, *Channa punctatus* (Bloch). Environ Toxicol Pharmacol 2008; 26(2): 136–141. doi:10.1016/j.etap.2008.02.010

[39] Eroglu A, Dogan Z, Kanak EG, Atli G, Canli M. Effects of heavy metals (Cd, Cu, Cr, Pb, Zn) on fish glutathione metabolism. Environ Sci Pollut Res 2015; 22(5): 3229–3237. doi: 10.1007/s11356-014-2972-y

[40] Jiang H, Kong X, Wang S, Guo H, Effect of copper on growth, digestive and antioxidant enzyme activities of Juvenile Qihe Crucian Carp, *Carassius carassius*, during exposure and recovery. Bull Environ Contam Toxicol 2016; 96(3): 333–340. doi:10.1007/s00128-016-1738-2

[41] El Basuini MF, El-Hais AM, Dawood MAO, El-Sayed Abou-Zeid A, EL-Damrawy SZ, EL-Sayed Khalafalla MM, Koshio S, Ishikawa M, Dossou S. Effect of different levels of dietary copper nanoparticles and copper sulfate on growth performance, blood biochemical profiles, antioxidant status and immune response of red sea bream (*Pagrus major*). Aquaculture 2016; 455: 32–40. doi:10.1016/j.aquaculture.2016.01.007

[42] Köhler H-R, Sandu C, Scheil V, Nagy-Petrica EM, Segner H, Telcean I, Stan G, Triebskorn R. Monitoring pollution in river Mureş, Romania, Part III: biochemical effect markers in fish and integrative reflection. Environ Monit Assess 2007; 127(1): 47–54. doi:10.1007/s10661-006-9257-y

[43] Triebskorn R, Telcean I, Casper H, Farkas A, Sandu C, Stan G, Colărescu O, Dori T, Köhler H-R. Monitoring pollution in River Mureş, Romania, part II: metal accumulation and histopathology in fish. Environ Monit Assess 2008; 141(1): 177–188. doi:10.1007/s10661-007-9886-9

[44] Zhou Q, Zhang J, Fu J, Shi J, Jiang G, Biomonitoring: an appealing tool for assessment of metal pollution in the aquatic ecosystem. Anal Chim Acta 2007; 14(606): 135–150. doi:10.1016/j.aca.2007.11.018

[45] PAN Pesticides Database, http://www.pesticideinfo.org.

[46] Pickering QH, Gast MH. Acute and chronic toxicity of cadmium to the fathead minnow (*Pimephales promelas*). J Fish Res Board Canada 1972 (on-line 2011); 29(8): 1099–1106. doi:10.1139/f72-164

[47] Rehwoldt R, Menapace LW, Nerrie B, Allessandrello D. The effect of increased temperature upon the acute toxicity of some heavy metal ions. Bull Environ Contam Toxicol 1972; 8(2): 91–96.

[48] Schäfer S, Buchmeier G, Claus E, Duester L, Heininger P, Körner A, Mayer P, Paschke A, Rauert C, Reifferscheid G, Rüdel H, Schlechtriem C, Schröter-Kermani C, Schudoma D, Smedes F, Steffen D, and Vietoris F. Bioaccumulation in aquatic systems: methodological approaches, monitoring and assessment. Environ Sci Eur 2015; 27: 5. doi:10.1186/s12302-014-0036-z

[49] Kominkova D, Nabelkova J. Effect of urban drainage on bioavailability of heavy metals in recipient. Water Sci Technol 2007; 56(9): 43–50. doi:10.2166/wst.2007.736

[50] Schneider L, Belgerb L, Burgerc J, Vogta RC. Mercury bioaccumulation in four tissues of *Podocnemis erythrocephala* (Podocnemididae: Testudines) as a function of water parameters. Sci Total Environ 2009; 407: 1048–1054. doi:10.1016/j.scitotenv.2008.09.049

[51] Salazar MJ, Rodriguez J H, Nieto GL, Pignata ML. Effects of heavy metal concentrations (Cd, Zn and Pb) in agricultural soils near different emission sources on quality, accumulation and food safety in soybean [*Glycine max* (L.) Merrill]. J Hazard Mater 2012; 233–234: 244–225. doi:10.1016/j.jhazmat.2012.07.026

[52] Zhu X, Chang Y, Chen Y., Toxicity and bioaccumulation of TiO$_2$ nanoparticle aggregates in *Daphnia magna*. Chemosphere 2010; 78: 209–215. doi:10.1016/j.chemosphere.2009

[53] Gobas FAPC. Assessing bioaccumulation factors of persistent organic pollutants in aquatic food-chains. In: Stuart H (ed) Persistent organic pollutants. Springer: US; 2001. p. 145–165. doi:10.1007/978-1-4615-1571-5_6

[54] Vădineamu A. "Considerations on the significance of holistic approach of the heavy metals and radioactive pollution problems. Nature Protection". Ocrotirea Naturii 1990; 1(1): 51–54.

[55] Iordache V. Ecotoxicology of heavy metals in Danube meadow. Bucharest, Romania; Ars Docenti; 2009. ISBN 978-973-558-233-3.

[56] Jitar O, Teodosiu C, Oros A, Plavan G, Nicoara M. Bioaccumulation of heavy metals in marine organisms from the Romanian sector of the Black Sea. N Biotechnol 2015; 25; 32(3): 369–378. doi:10.1016/j.nbt.2014.11.004

[57] Vinodhini R, Narayanan R. Bioaccumulation of heavy metals in organs of fresh water fish *Cyprinus carpio* (Common carp). Int J Environ Sci Tech 2008; 5(2): 179–182. doi: 10.1007/BF03326011

[58] Baby J, Raj SJ, Biby ET, Sankarganesh P, Jeevitha MV, Ajisha Su and RAJAN SS. Toxic effect of heavy metals on aquatic environment. Int J Biol Chem Sci 2010; 4(4): 939–952. ISSN 1991-8631.

[59] Saha N, Zaman MR. Evaluation of possible health risks of heavy metals by consumption of foodstuffs available in the central market of Rajshahi City, Bangladesh. Environ Monit Assess 2013; 185(5): 3867–3878. doi:10.1007/s10661-012-2835-2

[60] Ashish T, Amitabh CD. Assessment of heavy metals bioaccumulation in alien fish species, *Cyprinus carpio* from the Gomti River, India. Eur J Exp Biol 2014; 4(6): 112–117. ISSN: 2248-9215.

[61] David IG, Matache ML, Tudorache A, Chisamera Gabriel, Rozylowicz L, Radu GL. Food chain biomagnification of heavy metals in samples from the lower Prut floodplain natural park. Environ Eng Manag J 2012; 11(1): 69–73.

[62] Rainbow PS, Poirier L, Smith BD, Brix KV, Luoma SN. Trophic transfer of trace metals from the polychaete worm Nereis diversicolor to the polychaete *N. virens* and the decapod crustacean *Palaemonetes varians*. Mar Ecol Prog Ser 2006; 321: 167–181.

[63] Stanescu E, Stoica C, Vasile G, Petre J, Gheorghe S, Paun I, Lucaciu I, Nicolau M, Vosniakos F, Vosniakos K, Golumbeanu M. Structural changes of biological compartments in Danube Delta systems due to persistent organic pollutants and toxic metals. In: L.I. Simeonov et al. (eds). Environmental security assessment and management of obsolete pesticides in Southeast Europe, NATO Science for Peace and Security Series C: Environment Security. Springer Science+ Business Media: Dordrecht 2013, p. 229–248. doi:10.1007/978-94-007-6461-3_21

[64] Stoica C, Stanescu E, Lucaciu I, Gheorghe S, Nicolau M. Influence of global change on biological assemblages in the Danube Delta. J Environ Prot Ecol 2013; 14(2): 468–479.

[65] Stoica C, Gheorghe S, Paun I, Stanescu E, Dinu C, Petre J, Lucaciu I. Long term biological changes along Danube Delta system after industrialization period. Revista Rom Aqua 2014; 1: 14–20.

[66] Stoica C, Gheorghe S, Lucaciu I, Stanescu E, Paun I, Niculescu D. The impact of chemical compounds on benthic invertebrates from Danube–Danube Delta systems. Soil Sediment Contam Int J 2014; 23(7): 763–778. doi:10.1080/15320383.2014.870529

[67] Cordos E, Rautiu R, Roman C, Ponta M, Frentiu T, Sarkany A, Fodorpataki L, Macalik K, McCormick C, Weiss D. Characterization of the rivers system in the mining and industrial area of Baia Mare, Romania. Eur J Mineral Process Environ Prot 2003; 3(3) 1303–0868: 324–335.

[68] Dumitrel GA, Glevitzky M, Popa M, Vica ML. Studies regarding the heavy metals pollution of streams and rivers in Rosia Montana area, Romania. J Environ Prot Ecol 2015; 16(3): 850–860.

[69] Iordache M, Popescu LR, Pascu LF, Iordache I. Environmental risk assessment in sediments from Jiu River, Romania. Rev Chim (Bucharest) 2015; 66(8): 1247–1252.

[70] Alam MK, Maughan OE. The effect of malathion, diazinon, and various concentrations of zinc, copper, nickel, lead, iron, and mercury on fish. Biol Trace Elem Res 1992; 34(3): 225–236.

[71] EaSI-Pro® View 14.0 Data base for dangerous chemicals, 2005 Haskoning B.V. Available from: http://www.ekotox.eu/component/content/article/118-EASI-pro-view.

[72] Rand MG. Fundamentals of aquatic toxicology: effects, environmental fate and risk assessment, 2nd ed. North Palm Beach, Florida. CRC Press 1995; p. 943. ISBN1-56032-090-7.

[73] Besser JM, Leib KJ. Toxicity of metals in water and sediment to aquatic biota. In: Stanley E. Church, Paul von Guerard, and Susan E. Finger (eds). Integrated investigations of

environmental effects of historical mining in the Animas River watershed. San Juan County. Colorado; 2007. p. 839–849.

[74] Stoica C, Gheorghe S, Petre J, Lucaciu I, Nita-Lazar M, Tools for assessing Danube Delta systems with macro invertebrates. Environ Eng Manag J 2014; 13(9): 2243–2252.

[75] Kesavan K, Murugan A, Venkatesan V, Vijay Kumar B.S. Heavy metal accumulation in molluscs and sediment from uppanar estuary, Southeast coast of India. Thalassas. Int J Mar Sci 2013; 29(2): 15–21.

[76] Findlater G, Shelton A, Rolin T, Andrews J. Sodium and strontium in mollusc shells: preservation, palaeosalinity and palaeotemperature of the Middle Pleistocene of eastern England. Proc Geololists Assoc 2014; 3125(1): 14–19. doi:10.1016/j.pgeola.2013.10.005

[77] Kadar E, Costa V. First report on the micro-essential metal concentrations in bivalve shells from deep-sea hydrothermal vents. J Sea Res 2006; 56(1): 37–44. doi:10.1016/j.seares.2006.01.001

[78] Eisler R. Molluscs. In: Compendium of trace metals and marine biota. 1st ed. Elsevier: Amsterdam; 2010, p. 143–397 ISBN: 978-0-444-53439-2.

[79] Contia ME, Finoia MG. Metals in molluscs and algae: a north–south Tyrrhenian Sea baseline. J Hazard Mater 2010; 181(1–3): 388–392. doi:10.1016/j.jhazmat.2010.05.022

[80] Beek B. Bioaccumulation: new aspects and developments. In: Otto Hutzinger, editor. Handbook of environmental chemistry. 2nd ed. Reactions and processes, Part J. Springer-Verlag: New York; 2000. p. 284. doi:10.1007/10503050

[81] Zver'kova YS. Use of freshwater Mollusk Shells for monitoring heavy metal pollution of the dnieper ecosystem on the territory of Smolensk Oblast. Russ J Ecol 2009; 40(6): 443–447. doi:10.1134/S1067413609060113

[82] Gupta S.K, Singh J. Evaluation of mollusc as sensitive indicator of heavy metal pollution in aquatic systems: a review. IIOAB J Special Issue Environ Manag Sustain Dev 2011; 2(1):49-57.

[83] Shaari H, Raven B, Sultan K, Mohammad Y, Yunus K. Status of heavy metals concentrations in oysters (*Crassostrea* sp.) from Setiu Wetlands, Terengganu, Malaysia. Sains Malaysiana 2016; 45(3): 417–424.

[84] Oros A, Gomoiu M-T. Comparative data on the accumulation of five heavy metals cadmium, chromium, copper, nickel, lead) in some marine species (mollusks, fish) from the Romanian of the Black Sea 2010 (http://www.rmri.ro/Home/Downloads/Publications.RecherchesMarines/2010/paper03.pdf).

[85] Yusoff N.A.M, Long S.M. Comparative bioaccumulation of heavy metals (Fe, Zn, Cu, Cd, Cr, and Pb) in different edible mollusk collected from the estuary area of Sarawak

River 2011; UMTAS Empowering Science, Technology and Innovation Towards a Better Tomorrow. p. 806–811.

[86] TDG (2003). Technical Guidance Document on Risk Assessment, Commission Directive 93/67/EEC on Risk Assessment for new notified substances, European Commission.

[87] Gheorghe S, Lucaciu I, Paun I, Stoica C, Stanescu E. Environmental exposure and effects of some micropollutants found in Romanian surface waters. J Environ Prot Ecol 2014; 5(3): 2014.

Chapter 5

Biofloc Technology (BFT): A Tool for Water Quality Management in Aquaculture

Maurício Gustavo Coelho Emerenciano,
Luis Rafael Martínez-Córdova,
Marcel Martínez-Porchas and
Anselmo Miranda-Baeza

Additional information is available at the end of the chapter

http://dx.doi.org/10.5772/66416

Abstract

Biofloc technology (BFT) is considered the new "blue revolution" in aquaculture. Such technique is based on in situ microorganism production which plays three major roles: (i) maintenance of water quality, by the uptake of nitrogen compounds generating in situ microbial protein; (ii) nutrition, increasing culture feasibility by reducing feed conversion ratio (FCR) and a decrease of feed costs; and (iii) competition with pathogens. The aggregates (bioflocs) are a rich protein-lipid natural source of food available in situ 24 hours per day due to a complex interaction between organic matter, physical substrate, and large range of microorganisms. This natural productivity plays an important role recycling nutrients and maintaining the water quality. The present chapter will discuss some insights of the role of microorganisms in BFT, main water quality parameters, the importance of the correct carbon-to-nitrogen ratio in the culture media, its calculations, and different types, as well as metagenomics of microorganisms and future perspectives.

Keywords: microbial floc, shrimp, fish, microorganisms, nitrogen compounds, metagenomics

1. Aquaculture: state of the art and challenges

In a world where more than 800 million people continue suffering from chronic malnourishment and where the global population is expected to grow by another 2 billion to reach 9.6 billion people by 2050, it is important to meet the huge challenge of feeding our planet while safeguarding its natural resources for future generations [1]. In this context, aquaculture plays a key role in eliminating hunger, promoting health, reducing poverty, as well as generating

jobs and economic opportunities. According to FAO [1], the world food fish aquaculture production expanded at an average annual rate of 6.2% in the period 2000–2012 from 32.4 million to 66.6 million tons, in which Africa grew 11.7%, Latin America and the Caribbean 10%, Asia (excluding China) 8.2, and China 5.5. Employment in the sector has grown faster than the world's population. The sector provides jobs to tens of millions and supports the livelihoods of hundreds of millions. Fish continues to be one of the most traded food commodities worldwide. It is especially important for developing countries, sometimes worth half the total value of their traded commodities.

On the other hand, global aquaculture has yet to face some serious challenges. For instance, aquaculture has been accused of being an unsustainable activity, because of the effluents discharged to the environment which contain excess of organic matter, nitrogenous compounds, toxic metabolites, and elevated rates of chemical and biochemical oxygen demands [2]. Other serious accusations include the competition for land and water, the introduction of exotic species around the globe, the overexploitation of ocean fish stocks to obtain fishmeal and fish oil, the dispersion of pathogens, the development of antibiotic resistance genes, etc. [3, 4].

Furthermore, aquaculture has to constantly deal with other problems, such as the shortage of ingredients and their price volatility. Thus, strategies aimed to overcome these challenges are required. In this regard, the modification of physicochemical variables of the culture system to favor the proliferation of particular biotic communities has been adopted not only to improve the recirculation of nutrients (and the consequent detoxification of the system) but also to use the biomass of such biotic communities as direct food source for the cultured organisms [5]. These kinds of systems, also known as biofloc (BFT) technology systems, promise to solve some of the above challenges and revolutionize aquaculture [6].

2. Definition and applications of biofloc technology (BFT) in aquaculture

Biofloc technology (BFT) is as an environmentally friendly aquaculture technique based on in situ microorganism production. Fish and shrimp are grown in an intensive way (minimum of 300 g of biomass per square meter [7]) with zero or minimum water exchange. In addition, continuously water movement in the entirely water column is required to induce the macroaggregate (biofloc) formation. Nutrients in water (in accordance with a known carbon-to-nitrogen ratio of 12–20:1) will contribute naturally to a heterotrophic microbial community formation and stabilization. These microorganisms play three major roles: (i) maintenance of water quality, by the uptake of nitrogen compounds generating in situ microbial protein; (ii) nutrition, increasing culture feasibility by reducing feed conversion ratio (FCR) and a decrease of feed costs; and (iii) competition with pathogens.

BFT is considered the new "blue revolution" since nutrients can be continuously recycled and reused in the culture medium, benefited by the minimum or zero-water exchange. Also, the sustainable approach of such system is based on the high production of fish/shrimp in small areas. In addition, the bioflocs is a rich protein-lipid natural source of food available in situ

24 hours per day due to a complex interaction between organic matter, physical substrate, and large range of microorganisms. This natural productivity plays an important role recycling nutrients and maintaining the water quality. The consumption of biofloc by shrimp or fish has demonstrated innumerous benefits such as improvement of growth rate, decrease of FCR, and associated costs in feed [8].

Regarding the applications, in the past years, BFT has been used in grow-out phase for tilapia [9, 10] and marine shrimp [11, 12], nursery phase [13–15], freshwater prawn culture [16, 17], broodstock formation and maturation in fish [18] and shrimp [7–19], and as aquafeed ingredient also called as "biofloc meal" [20–22]. In addition, recently BFT also has been applied in carp culture [23], catfish culture [24], and cachama culture [25].

3. Microorganisms as a tool for water quality management

3.1. Main water quality parameters in BFT

Water quality maintenance and monitoring in aquaculture are the essential practices aiming at the success of the growing cycles. Temperature, dissolved oxygen (DO), pH, salinity, solids [total suspended solids (TSS) and settling solids], alkalinity, and orthophosphate are some examples of parameters that should be continuously monitored, especially in BFT. The comprehension and understanding of water quality parameters and its interactions in BFT are crucial to the correct development and maintenance of the production cycle. For example, safety ranges of pH, DO, total ammonia nitrogen (TAN), solids, and alkalinity will lead a health growth and avoid mortalities. N:P ratio (normally using nitrate and orthophosphate values) will influence the autotrophic community that will occur in the system (e.g., chlorophytes versus cyanophytes). The same recommended water quality parameters ranges and/ or normal ranges observed for tropical species (e.g., marine shrimp *Litopenaeus vannamei* and tilapia) in BFT are presented in **Table 1**.

3.2. The role of microorganisms in BFT aquaculture systems

Microorganisms play a key role in BFT systems. The maintenance of water quality, mainly by the control of bacterial community over autotrophic microorganisms, is achieved using a high carbon-to-nitrogen (C:N) ratio, since nitrogenous by-products can be easily taken up by heterotrophic bacteria. High carbon-to-nitrogen ratio is required to guarantee optimum heterotrophic bacteria growth, using this energy for maintenance (respiration, feeding, movement, digestion, etc.) but also for growth and to produce new bacterial cells.

The stability of zero or minimal water exchange depends on the dynamic interaction among communities of bacteria, microalgae, fungi, protozoans, nematode, rotifer, etc. that will occur naturally. Such consortia of microorganism will help on the water quality maintenance and recycling wastes to produce a high-value food. In a study with stable isotopes, Burford et al. [12] estimated a daily nitrogen retention of 18–29% into the shrimp obtained from biofloc biota, while Avnimelech and Kochba [26] found about 25% of assimilation for tilapia, using the same technique.

Parameter	Ideal and/or normal observed ranges	Observations
Dissolved oxygen (DO)	Above of 4.0 mg L^{-1} (ideal) and at least 60% of saturation	For correct fish, shrimp, microbiota respiration, and growth
Temperature	28–30° (ideal for tropical species)	Besides fish/shrimp, low temperatures (~20° C) could affect microbial development
pH	6.8–8.0	Values less than 7.0 is normal in BFT but could affect the nitrification process
Salinity	Depends on the cultured species	It is possible to generate BFT, e.g., from 0 to 50 ppt
TAN	Less than 1 mg L^{-1} (ideal)	Toxicity values are pH dependent
Nitrite	Less than 1 mg L^{-1} (ideal)	Critical parameter (difficult to control). Special attention should be done, e.g., on protein level of feed, salinity, and alkalinity
Nitrate	0.5–20 mg L^{-1}	In these ranges, generally not toxic to the cultured animals
Orthophosphate	0.5–20 mg L^{-1}	In these ranges, generally not toxic to the cultured animals
Alkalinity	More than 100 mg L^{-1}	Higher values of alkalinity will help the nitrogen assimilation by heterotrophic bacteria and nitrification process by chemoautotrophic bacteria
Settling solids (SS)	Ideal: 5–15 mL L^{-1} (shrimp), 5–20 (tilapia fingerlings) and 20–50 mL L^{-1} (juveniles and adult tilapia)	High levels of SS (measured in Imhoff cones) will contribute to the DO consumption by heterotrophic community and gill occlusion
Total suspended solids (TSS)	Less than 500 mg L^{-1}	Idem to SS

Table 1. Main water quality parameters monitored in BFT systems and its ideal and/or normal observed ranges.

Organic matter and nitrogen wastes are a huge problem in aquaculture. Phytoplankton, heterotrophic, and nitrifying bacteria have the most important role in the nitrogen and OM reutilization. Fungi, ciliate, protozoa, rotifer, copepod, and nematode complement the biofloc community, participating in the recycling of organic matter as a part of complex food webs which include the cultured species.

Mutualism and commensalism relationships occur among some group of microorganisms in BFT, e.g., bacteria-bacteria or bacteria-microalgae. In low water exchange cultures, complex biofilms are generated in which coexist heterotrophic and nitrifying bacteria. Inorganic ions are attracted to the surface of these biofilms and the solid surfaces of the substrate, promoting greater nitrification processes [27]. Some bacterial strains have a positive effect on microalgae growth not only for planktonic species but also on attached (benthic) species [28]. The extracellular polysaccharides of benthic diatoms may be used by heterotrophic organisms as carbon source [29].

One current practice in BFT is the use of commercial bacteria consortia (probiotics). The main reasons of probiotics used in BFT are (i) help to stabilize the heterotrophic community and to compete with autotrophic microorganisms (mainly in the initial phases), (ii) help to recycling the organic matter, and (iii) control solids and TAN levels.

3.2.1. Bacteria

The heterotrophic bacteria use the organic compounds as a carbon source. This community can minimize ammonia accumulation in the water column through incorporation as bacterial biomass. Under suitable conditions (temperature, carbon:nitrogen ratio, pH, etc.), bacteria have a fast growth. Leonard et al. [30] estimated that the generation time for the free viable heterotrophic populations was around 2.5 h in laboratory conditions.

Heterotrophic bacteria utilize sugar, alcohol, and organic acids as energy source but exist in specialized species capable of decomposing cellulose, lignin, chitin, keratin, hydrocarbons, phenol, and other substances [31]. Heterotrophic bacteria are able to colonize a high diversity of environments; they are common in soil, freshwater, and saltwater. Aquatic environments are responsible to recycle high amounts of dissolved and particulate organic matter, playing one of the most important roles in the food webs [32]. In biofloc system, the heterotrophic bacteria colonize the feces, molts, dead organisms, and unconsumed food to produce bacterial biomass, which is consumed by detritivores [8]. Brown et al. [33] evaluated the biochemical compositions of seven strains of marine bacteria and reported protein content (dry weight) of 29–49%, carbohydrates 2.5–11.2, lipids 4–6%, and, additionally, the presence of all essential amino acids.

Chemoautotrophic bacterial community (i.e., nitrifying bacteria) obtains energy through oxidation of toxic nitrogen compounds. The nitrifying bacteria are naturally promoted by the presence of ammonia and nitrite as well as the accumulation of flocculated matter (used as substrate). The alkalinity consumed by these microorganisms must be replaced by different sources (i.e., sodium bicarbonate, calcium carbonate, or calcium hydroxide [34]). In laboratory conditions, the generation time of ammonia oxidizer bacteria was estimated to 25 h and nitrite oxidizer to 60 h [30].

The nitrifying bacteria thrive in a wide diversity of environments [35]. Besides the oxygen, toxic nitrogen compounds are the major concern into the biofloc systems. The main sources of ammonia are excretion of cultured organism and the decomposition of nonliving matter (dissolved and particulate). In BFT, three nitrogen conversion pathways occur for the removal of ammonia nitrogen: (a) photoautotrophic removal by algae, (b) autotrophic bacterial conversion from ammonia to nitrate, and (c) heterotrophic bacterial conversion of ammonia nitrogen directly to microbial biomass [36]. In long term, the most efficient process is the autotrophic, in which two bacterial groups are involved: (a) the ammonia-oxidizing bacteria, which obtain their energy by catabolizing unionized ammonia to nitrite, including the genera *Nitrosomonas*, *Nitrosococcus*, *Nitrosospira*, *Nitrosolobus*, and *Nitrosovibrio* and (b) the nitrite-oxidizing bacteria, which metabolize nitrite to nitrate, including the genera *Nitrobacter*, *Nitrococcus*, *Nitrospira*, and *Nitrospina* [37].

3.2.2. Fungi

Fungi are a group of eukaryotic organisms that include unicellular microorganisms, such as yeasts and molds, as well as multicellular fungi. Most yeasts reproduce asexually by mitosis and many others by the asymmetric division [38]. Typically measuring 3–4 μm in diameter, they are widely distributed in freshwater and saltwater (*Candida, Cryptococcus, Rhodotorula*, and *Debaryomyces*). Some marine species live at temperatures as low as −13° and at deeps of 4000 m; some others can nearly saturate brine solutions. Seawater normally contains 10–100 yeast L⁻¹, but in estuarine environments, the number significantly increases [39]. Brown et al. [33] evaluated seven species of yeast to determine their nutritional value and found 25–37% of protein, 21–39% of carbohydrate, and 4–6% of lipid, as well as complete profile of essential amino acids. Yeasts are strictly chemoorganotrophic and require organic forms of carbon which are quite diverse and include sugars, polyols, organic and fatty acids, aliphatic alcohols, and various heterocyclic and polymeric compounds [39].

Fungi, especially yeast (chemoorganotrophic microorganisms), are also reported in biofloc. They use organic compounds as a source of energy. Carbon is obtained mostly from hexose sugars, such as glucose and fructose. In a biofloc culture of tilapia, Monroy-Dosta et al. [40] reported the presence of the yeast *Rhodotorula* sp. during the fifth week, which increases its biomass by the end of the culture period.

3.2.3. Microalgae

Photoautotrophic community (microalgae) also play an important role in the biofloc system. Microalgae assimilate mainly ammonia and nitrate to produce biomass, additionally consume carbon dioxide, and produce oxygen. The divisions of microalgae reported in biofloc cultures are Chlorophyta, Chrysophyta, and Cyanophyta. These microorganisms catch the solar energy, to produce chemical energy (carbohydrates), which is used in their metabolic process.

In biofloc cultures the microalgae can live as free cell into the water column or could form aggregates. In some cases the aggregations of chrysophytes and cyanophytes can measure up to 2 mm in diameter [41]. Their sizes are highly variable, with cells of less than 10 μm to more 50 μm [42].

Chlorophytes are green microalgae that are the most numerous and diverse in the freshwater; they can reproduce massively forming blooms, but, at difference of the cyanophytes, are non-toxic. This division presents a high plasticity and is able to colonize diverse habitats; they are spherical or oblong and may have flagella or not [43].

Chrysophytes are the most representative organisms that correspond into the Bacillariophyceae class (diatoms), which is divided in centric and pennate. The planktonic species are mainly centric; meanwhile, the pennate are commonly benthic. All centric species are marine, while most of the pennate live in freshwater [43]. In aquaculture, diatoms are considered as beneficial algae because they are a source of food and nutrients for most aquatic animals [44].

Cyanophytes are known as the most ancient photosynthetic organisms; they possess a high morphologic and structural variability. During its evolution they have developed various ecophysiological adaptation strategies to survive in extreme environmental conditions [45]. Given its abundance in different environments, the division Cyanophyta is important for nutrient circulation, incorporating nitrogen into the food chain, which makes them primary producers or decomposers.

In aquaculture ponds, excessive concentrations of major nutrients (nitrogen and phosphorus) can lead to uncontrolled microalgae blooms, sometimes are cyanobacteria dominated which is known to produce some toxic compounds to aquatic animals, and can cause unpleasant flavors in cultured species [46]. Several authors have reported the presence of cyanobacteria in biofloc, with concentrations varying according to the biofloc type. Becerra-Dórame et al. [47] reported 2.1×104 cells mL^{-1} in heterotrophic biofloc, while in autotrophic, they found 3.3×106 cells mL^{-1}. Although cyanobacteria can become toxic or problematic, Lezama-Cervantes et al. [48] found several species of Cyanobacteria (*Nostoc* sp., *Anabaena* sp., *Phormidium* sp., *Chroococcus* sp., *Oscillatoria* sp., and *Lyngbya* sp.) in a microbial mats used to culture *L. vannamei* postlarvae and fund evidenced of active grazing by the shrimp.

Aquaculture microalgae are widely used; their nutritional characteristic has permitted to produce crustaceans, fishes, and mollusk in laboratory. Several factors can contribute to the nutritional value of microalgae, including its size and shape, digestibility, biochemical composition, and bioactive compounds as enzymes, vitamins, antioxidants, etc. Microalgae grown to late-logarithmic growth phase typically contain 30–40% protein, 10–20% lipid, and 5–15% carbohydrate; PUFAs derived from microalgae, i.e., docosahexaenoic acid (DHA), eicosapentaenoic acid (EPA), and arachidonic acid (AA), are known to be essential for various farmed species [49].

In general, microalgae are common inhabitants in biofloc, even in a bacteria dominated (heterotrophic). Monroy-Dosta et al. [40] evaluated the microorganism composition in biofloc tilapia culture, indicated that the first microalgae appeared were chlorophytes, followed by diatoms and finally cyanobacteria, and also mentioned that diatoms achieved the highest concentration and cyanobacteria the lowest. Ray et al. [50] cultured *L. vannamei* in zero-water exchange, and differing from the previous authors, in their study the chlorophytes were dominant over the diatoms. Biofloc systems are highly dynamic; Kuang et al. [51] indicate that in nature, certain species of ciliates and rotifers have a selective consumption of microalgae, and therefore may influence their diversity. The physicochemical parameters also affect the microalgae dominance. Maicá et al. [52] observed in a *L. vannamei* biofloc culture greater abundance of chlorophytes in salinity of 2 %, while in 25 % diatoms were the most abundant.

3.2.4. Zooplankton

Protozoa is one of the most relevant microorganism groups in BFT system. They play an essential role (together with bacteria) recycling the organic matter in the system. Both groups are the "basis" of the trophic transfer of energy to the next levels. Protozoa have different body shapes (spherical, oval, and elongated) and often have one or more whip appendages called flagella or many short hair-like structures called cilia. Protozoa are abundant in many types of environments and often are found on the surfaces of submerged rocks, free living into the water column or colonizing the sediment. Ciliates are the largest group of protozoa in nature; they eat bacteria (including cyanobacteria) and small phytoplankton. Some are carnivorous and feed on zooplankton [53].

In nature, ciliates have importance as a live food source for juvenile stages of aquatic animals including small invertebrates. Pandey et al. [54] carried out the analysis of the ciliate *Fabrea salina* to evaluate proximate and biochemical composition. The moisture, protein, fat, carbohydrate, and ash content from natural sources were 86.66%, 56.66%, 36.66%, 1%, and 4%, respectively. Gas chromatographic analysis revealed the presence of fatty acids such as oleic, palmitic, palmitoleic, linoleic, and stearic.

Ballester et al. [14] registered concentrations from 39 to 169 ciliates/mL, in postlarvae *Farfantepenaeus paulensis* biofloc culture. Maicá et al. [52] found an average concentration of 164, 64, and 29 ciliates mL^{-1} in water salinities of 2%, 4%, and 25%, respectively. Furthermore, Monroy-Dosta et al. [40] observed minimum and maximum concentrations of 13 and 39 and also noted a variation in species according to the culture age.

Rotifer belongs to the smaller group of metazoans. Most rotifers are 0.1–0.5 mm long. Their body shape varies widely between groups: they can be spherical, cylindrical, or elongated. The body can be soft or may have a firm covering called lorica. The cilia surrounding a rotifer's mouth form a circle, called a corona or wheel organ. The rapid movements of the cilia create water currents for swimming and feeding [53]. Their diets consist on microalgae, bacteria, yeast, and protozoa [55].

The rotifers are the group of organisms that probably have been largely used to replace *Artemia* as exogenous natural food in larval culture of crustaceans and fish. Campaña-Torres et al. [56] evaluated the proximal composition of the rotifer *Brachionus rotundiformis* cultured in laboratory and reported a dry content of carbohydrate of 15.9–22.7%, lipid 21.4–24.12%, protein 45.7–61.3%, and ash 4.5–4.6%.

Loureiro et al. [57] indicate that rotifers can fragment the flocs and consume bacteria. The mucilage produced by their excretions contributes to new flocs formation [58]. Ballester et al. [14] registered concentrations form 4.6 to 151 org/mL in seawater; besides, Monroy-Dosta et al. [40] reported concentrations between 28 and 96 org/mL in freshwater.

Copepods comprise two main groups: calanoids and cyclopoids. Calanoid copepods have an elongated body and the first pair of antennae is long, whereas cyclopoid have a robust body and a first pair of antennae is short. In general, both use the appendices near the head to create streams to filter or collect food. They feed on bacteria, phytoplankton, detritus, or any other organic material [53].

Farhadian et al. [59] evaluated the proximate composition of copepod *Apocyclops dengizicus* and reported protein levels between 39 and 42% and lipid between 16 and 19%, indicating that nutritional properties varied according to the microalgae used as feed.

Cladocerans posses a body covered by a transparent shell, although it may be yellowish or brown. A pair of appendages called thoracic members are inside the shell, and are important for the capture and transfer of food particles in the mouth. In general, cladocerans eat a wide variety of phytoplankton and suspended matter. They can greatly reduce the abundance of phytoplankton in the water column [53].

As the other zooplankton groups, the cladocerans play an important role into the natural food webs. They could supply a high amount of protein into the biofloc cultures. Berberovic [60] evaluated the elemental composition (CHN) of two *Daphnia* species, reporting the following: C, 46.1%; H, 6.5%; N, 9.7%; and ash, 23.8%, which permit to estimate a protein content of 60.6%. This group of organisms was reported in biofloc system by Emerenciano et al. [61]. Moreover, in a postlarvae culture of *L. vannamei* reared in zero-water exchange, Ferreira-Marinho et al. [62] reported *Cladocera* abundance from 0.89 to 1.16 ind mL^{-1} represented by the genus *Bosmina* (0.39–0.53 ind mL^{-1}) and *Daphnia* (0.50–0.69 ind mL^{-1}).

Nematoda is the other essential group in BFT. The body of these organisms is perfectly cylindrical, coated by a relatively thick noncellular cuticles secreted by the underlying epidermis. The cuticle is composed primarily of collagen [63]. They continuously ingest bacteria and other microbenthic organisms; almost all particles which fit into the buccal cavity are ingested, hinting at a selection mechanism based primarily on particle size. Moens and Vincx [64] proposed six major nematode feeding strategies: (a) microvores; (b) ciliate feeders, (c) deposit feeders sensu stricto, (d) epigrowth feeders, (e) facultative predators, and (f) predators.

Ray et al. [50] mentioned that nematodes are an important group in the biofloc systems, whose abundance is determined by the presence of various ciliates that serve as food. In other studies, Monroy-Dosta et al. [40] observed the appearance of nematodes around the fourth week with average of 25 org mL^{-1} with a maximum of 125 org mL^{-1}, and their abundance were correlated with the ciliates' presence. Loureiro et al. [57] reported the presence of nematodes in the stomach contents of fish grown in the biofloc and suggest that they are a rich source of live food in situ.

4. Carbon:nitrogen (C:N) ratio and its application

The management of the carbon-to-nitrogen ratio (C:N) in BFT is normally divided in two phases: (i) initial and formation phase, utilizing a carbon-to-nitrogen ratio of 12–20:1, and (ii) maintenance phase, utilizing a carbon-to-nitrogen ratio of 6:1, according to the total ammonia nitrogen (TAN) values.

In the beginning of culture period, high carbon-to-nitrogen ratio (12–20:1) in water is a key factor to promote and stabilize the heterotrophic community in BFT [8]. High carbon concentration will induce the nitrogenous by-product assimilation by heterotrophic bacteria and also will supersede the carbon assimilatory capacity of algae, contributing to bacteria growth. Aerobic microorganisms are efficient in converting feed to new cell material (40–60% of conversion efficiency), rather than higher organisms (e.g., micro-herbivores, micro-carnivores, and deposit feeders) that spend about 10–15% to rise in weight. The system is considered "mature" (~30 to 50 days) when SS reaches at least 5 mL/L (measured using Imhoff cones) and TAN and nitrite peaks already occurred. To accelerate the water "maturation" (biofloc equilibrium), an inoculum of a previous BFT culture can be used once sanitary conditions are satisfactory.

It is important to note that as long as the production cycles advance, nitrifying (chemoautotrophic) bacteria play a major role in N-compound control. In addition, suspended particles or solids (bioflocs) also will be increasing over time. With this information in mind, carbon addition could be reduced (or even stopped), preventing the excess of solids (bioflocs) in the cultured system that will lead an excessive DO consumption [65] and shrimp/fish gill occlusions [66].

For the maintenance phase, the monitoring of TAN values is an important tool for water quality maintenance. When values of TAN are higher than 1.0 mg L^{-1}, external carbon source application is recommended with a C:N ratio of 6:1 [36]. In such phase, the use of monosaccharide and oligosaccharide carbohydrate-rich types (e.g., molasses and other sugars) is recommended due to the faster bacteria assimilation and consequently TAN reduction. Same examples of C:N calculations for the phase I and phase II are presented as followed. For both examples, the carbon content of the feed will be considered 50% (based on dry matter). For the carbon source, molasses was chosen and its content in such case is also 50%. It is important to note that the carbon content will change according to the dry matter composition and type of carbon source. In a practical way, dry matter of the feed will be 90%. Fish and shrimp assimilation will be considered 35 and 20%, respectively.

Example 1 (initial and formation phase using a C:N ratio of 20:1) in a tilapia culture tank that receives 4 kg of feed (30% of crude protein) per day.

Calculation 1 (C:N content in the feed)

C: 4 kg of feed × 0.9 (90% dry matter) × 0.7 (30% of fish assimilation or 70% of waste that remains in water)/2 (carbon content of the feed is ~50% based on dry matter) = 1260 g of C

N: 4 kg of feed × 0.9 (90% dry matter) × 0.7 (30% of fish assimilation or 70% of waste that remains in water) × 0.3 (30% crude protein content of feed)/6.25 (constant) = 121 g of N. **The results indicated a ~10:1 C:N ratio of feed.**

Calculation 2 (adjusting the C:N ratio)

If I want a C:N ratio of 20:1, 121 g of N in feed × 20 = I need 2420 g of C. But I already have 1260 g of C (calculated in feed). So 2420 g–1260 g of C = I really need 1160 g of C.

If the molasses has 50% of carbon content (based on dry matter), 1 kg of molasses represents 500 g of carbon. So, 1160 g of carbon requirement will represent **2320 g (or 2.3 kg) of molasses (applied daily until biofloc maturation).**

Example 2 (maintenance phase and C:N ratio of 6:1) in a *L. vannamei* culture tank (30 m^3) that indicates 2.0 mg L^{-1} TAN values.

Calculation 1 (TAN in water)

For 2.0 mg L^{-1} of TAN in a 30 m^3 tank = 0.002 g × 30,000 L = 60 g of TAN

Calculation 2 (adjusting the C:N ratio)

If I want a C:N ratio of 6:1, 60 g of TAN in water × 6 = I need 360 g of C. If my molasses has 50% of carbon content (based on dry matter), 1 kg of molasses represents 500 g of carbon. So, 360 g of carbon requirement will represent 720 g (or 0.72 kg) of molasses (one application and checked after 2–3 days).

5. Economics and types of carbon sources

The carbon sources applied in BFT are often by-products derived from human and/or animal food industry, preferentially cheap and local available. Cheap sources of carbohydrates such as molasses, glycerol, and plant meals (i.e., wheat, corn, rice, tapioca, etc.) will be applied before the fry/postlarvae stocking (fertilization protocols) and during grow-out phase, aiming to (i) provide food for the first stages of growth and (ii) to maintain a high C:N ratio and to control N-compound peaks in the culture tanks, respectively [67].

Depending of the carbon source chosen, organic fertilization could be considered as an important item of The production costs. Local available sources should be tested, but bacteria assimilation's characteristics will certainly need to take into account. Monosaccharide and oligosaccharide simple carbohydrate-rich types (e.g., glucose, sucrose-rich sugars, etc.) versus polysaccharide complex-rich types (e.g., starch and cellulose) will lead different bacteria assimilations, nutritional value, and growth. Crab et al. [16] evaluated the effect of different carbon sources for *Macrobrachium rosenbergii* postlarvae. Besides the price, different sources will lead diverse nutritional value of the flocs. The authors observed that when using glycerol as compared to glucose and acetate, higher values of n-6 PUFA were observed.

For each phase (initial and formation phase or maintenance phase), different sources should be chosen according to the price and purpose. For example, dextrose (high purified sugar) versus molasses; refined sugar versus grains by-products, etc. Grains and tubercles contain high levels of carbon (carbohydrates), as polysaccharides. Some grains used as carbon sources additionally contain protein and lipids. García-Ríos [68] compared three carbon sources into the BFT tilapia fry culture and found that corn meal contains 11.79% of protein and 2.8% lipid; meanwhile, wheat has 15.5% of protein and 3.73% lipid. The unrefined sugar (monosaccharide) without protein and lipid promoted the best growth and the highest protein content into the tilapia tissue. It is possible that the chemical structure of sugar presented a high bioavailability to heterotrophic bacteria, hence, fast increase of bacterial biomass.

6. Groups and metagenomics

Phytoplankton, free and attached bacteria, aggregates of particulate organic matter and grazers, such as rotifers, ciliates and flagellates, and protozoa and copepods are common groups of microorganisms in BFT. As the use, identification and study of microbes in aquaculture have become a usual practice in the last decade [69]. For a long time, techniques based on culture media were used as the main strategy to know the microbial composition of biotic communities, including biofilm and BFT; however, this was a very superficial approach considering that >80% of the bacteria thriving in any environment are readily culturable or unculturable at all [70]. The overcome of culture-independent techniques such as denaturing gradient gel electrophoresis (DGGE) but particularly high-throughput sequencing (next-generation sequencing) increased the depth and coverage of studies aiming to study the microbial diversity of these kinds of conglomerates [71, 72].

Metagenomics is therefore a relative recent genomics subdiscipline that has emerged as a promising scientific tool to analyze the complex genomes contained within microbial communities. However, its use is not yet common in some agro-industrial disciplines such as aquaculture. The reason of this relies in the high cost of this technology; however, prices have significantly decreased during the last decade, and now it is possible for individual laboratories to perform metagenomics studies using high-throughput sequencing.

The study of microbial diversity can be studied with the highest resolution so far, for instance, ribosomal genes such as 16S and 23S have been used as a targeted loci approach for diversity studies of prokaryotic communities. Herein, universal genes are used for the amplification of particular hypervariable regions of these genes [4]; these hypervariable regions contain elements to differentiate organisms. In this regard, this technology offers the possibility to reveal most of the bacteria thriving in any biofloc biomass. However, current sequencing technologies can cover only a fraction (~600 bp) of the ribosomal genes used for taxonomic classification, which means that only 2 or 3 of the 10 hypervariable regions of the 16S gene can be used for this classification. Researchers have made efforts to elucidate whose sequences are the most information richness [73]; however, there is still a loss of information contained on the regions that cannot be covered by these sequencing platforms.

Novel technologies such as single-molecule real-time (SMRT) sequencing have been developed [74]. This particular technology has advantages such as the generation of long reads and high accuracy. Long reads could be useful for sequencing not only ribosomal genes but also larger fragments of DNA, which would serve for multilocus classification. Whether the price for doing metagenomics of bacteria is still high, it is expected to decrease along the following years.

In spite of the incomplete coverage of ribosomal genes by most of the current high-throughput sequencing platforms, the amplification of two or three hypervariable regions of these genes is still a very useful tool to know most of the bacteria contained in biofilms and bioflocs [72] and inclusively to detect novel species, study dynamic population patterns, probiotic activity, etc. Furthermore, metagenomics based on single-gene surveys and random shotgun studies of all accessible genes in any environment could be two useful approaches to study the biological activity and communications of these complex bacterial networks. Whether the use of BFT systems is now a reality and promises to be revolutionary strategy, their biology studied through novel genomic tools is still to provide mass information of these biotic communities.

7. Conclusions and perspectives

Biofloc technology will enable aquaculture grow toward an environmentally friendly approach and biosecurity. Consumption of microorganisms in BFT reduces FCR and consequently costs in feed. Also, microbial community is able to rapidly utilize dissolved nitrogen leached from shrimp/fish feces and uneaten food and convert it into microbial protein, maintaining the water quality. The physical, chemical, and biological interactions that occur into the biofloc systems are complex; further studies can elucidate specific phenomena and their possible applications to other biotechnological fields.

Acknowledgments

The authors like to thank the National Council for Science and Technology (CONACyT), México, and Brazilian National Council for Scientific and Technological Development (CNPq), Santa Catarina Research Foundation (FAPESC) for the research support. The authors also like to thank Sara Mello Pinho for the technical revision.

Author details

Maurício Gustavo Coelho Emerenciano[1]*, Luis Rafael Martínez-Córdova[2], Marcel Martínez-Porchas[3] and Anselmo Miranda-Baeza[4]

*Address all correspondence to: mauricioemerenciano@hotmail.com

1 Santa Catarina State University (UDESC), Aquaculture Laboratory (LAQ), Laguna, SC, Brazil and Animal Science Postgraduate Program (PPGZOO/UDESC), Chapecó, SC, Brazil

2 Universidad de Sonora (UNISON), Hermosillo, Sonora, Mexico

3 Centro de Investigaciónen Alimentación y Desarrollo (CIAD), Hermosillo, Sonora, Mexico

4 Universidad Estatal de Sonora (UES), Navojoa, Sonora, Mexico

References

[1] FAO (Food and Agriculture Organization of the United Nations). The State of World Fisheries and Aquaculture 2014. Opportunities and Challenges. 2014; p. 223. Available at http://www.fao.org

[2] Martínez-Porchas M, Martinez-Córdova LR. World aquaculture: environmental impacts and troubleshooting alternatives. The Scientific World Journal. 2012;389623:1–9.

[3] Naylor RL, Williams SL, Strong DR. Aquaculture – A gateway for exotic species. Science. 2001;294(5547):1655–1656.

[4] Martínez-Porchas M, Vargas-Albores F. Microbial metagenomics in aquaculture: a potential tool for a deeper insight into the activity. Reviews in Aquaculture. 2015 (online published first 6 July 2015).

[5] Martínez-Córdova LR, Emerenciano M, Miranda-Baeza A, Martínez-Porchas M. Microbial-based systems for aquaculture of fish and shrimp: an updated review. Reviews in Aquaculture. 2015;7(2):131–148.

[6] Martínez Córdova LR, Martínez Porchas M, Emerenciano M, Miranda-Baeza A, Gollas-Galván T. From microbes to fish the next revolution in food production. Critical Reviews in Biotechnology. 2016; Early Online: 1–9.

[7] Emerenciano M, Cuzon G, Goguenheim J, Gaxiola G, Aquacop. Floc contribution on spawning performance of blue shrimp *Litopenaeus stylirostris*. Aquaculture Research. 2012a;44:75–85.

[8] Avnimelech Y. Biofloc Technology – A Practical Guide Book. 3rd ed. The World Aquaculture Society, Baton Rouge, Louisiana, United States. 2015.

[9] Avnimelech Y. Feeding with microbial flocs by tilapia in minimal discharge bioflocs technology ponds. Aquaculture. 2007;264:140–147.

[10] Crab R, Kochva M, Verstraete W, Avnimelech Y. Bio-flocs technology application in over-wintering of tilapia. Aquacultural Engineering. 2009;40:105–112.

[11] Burford MA, Thompson PJ, McIntosh RP, Bauman RH, Pearson DC. Nutrient and microbial dynamics in high-intensity, zero-exchange shrimp ponds in Belize. Aquaculture. 2003;219:393–411.

[12] Burford MA, Thompson PJ, McIntosh RP, Bauman RH, Pearson DC. The contribution of flocculated material to shrimp (*Litopenaeus vannamei*) nutrition in a high-intensity, zero-exchange system. Aquaculture. 2004;232:525–537.

[13] Samocha TM, Patnaik S, Speed M, Ali AM, Burger JM, Almeida RV, Ayub Z, Harisanto M, Horowitz A, Brock DL. Use of molasses as carbon source in limited discharge nursery and grow-out systems for *Litopenaeus vannamei*. Aquacultural Engineering. 2007;36:184–191.

[14] Ballester ELC, Abreu PC, Cavalli RO, Emerenciano M, Abreu L, Wasielesky W. Effect of practical diets with different protein levels on the performance of Farfantepenaeus paulensis juveniles nursed in a zero exchange suspended microbial flocs intensive system. Aquaculture Nutrition. 2010;16:163–172.

[15] Emerenciano M, Ballester ELC, Cavalli RO, Wasielesky W. Biofloc technology application as a food source in a limited water exchange nursery system for pink shrimp *Farfantepenaeus brasiliensis* (Latreille, 1817). Aquaculture Research. 2012b;43:447–457.

[16] Crab R, Chielens B, Wille M, Bossier P, Verstraete W. The effect of different carbon sources on the nutritional value of bioflocs, a feed for *Macrobrachium rosenbergii* postlarvae. Aquaculture Research. 2010;41:559–567.

[17] Pérez-Fuentes JA, Pérez-Rostro CI, Hernández-Vergara MP. Pond-reared Malaysian prawn *Macrobrachium rosenbergii* with the biofloc system. Aquaculture. 2013;400–401:105–110.

[18] Ekasari J, Zairin M, Putri DU, Sari NP, Surawidjaja EH, Bossier P. Biofloc-based reproductive performance of Nile tilapia *Oreochromis niloticus* L. broodstock. Aquaculture Research. 2015;46:509–512.

[19] Emerenciano M, Cuzon G, Arevalo M, Mascaró M, Gaxiola G. Effect of short-term fresh food supplementation on reproductive performance, biochemical composition, and fatty acid profile of *Litopenaeus vannamei* (Boone) reared under biofloc conditions. Aquaculture International. 2013a;21:987–1007.

[20] Kuhn D, Boardman G, Lawrence A, Marsh L, Flick G. Microbial floc meal as a replacement ingredient for fish meal and soybean protein in shrimp feed. Aquaculture. 2009;296:51–57.

[21] Bauer W, Prentice-Hernandez C, Tesser MB, Wasielesky W, Poersch LHS. Substitution of fishmeal with microbial floc meal and soy protein concentrate in diets for the Pacific white shrimp *Litopenaeus vannamei*. Aquaculture. 2012;342–343:112–116.

[22] Neto HS, Santaella ST, Nunes AJP. Bioavailability of crude protein and lipid from biofloc meals produced in an activated sludge system for white shrimp, *Litopenaeus vannamei*. Brazilian Journal of Animal Science. 2015;44(8):269–275.

[23] Najdegerami EH, Bakhshi F, Lakani FB. Effects of biofloc on growth performance, digestive enzyme activities and liver histology of common carp (*Cyprinus carpio* L.) fingerlings in zero-water exchange system. Fish Physiology Biochemistry. 2016;42(2):457–465.

[24] Yusuf MW, Utomo NBP, Yuhana M, Widanarni. Growth performance of catfish (*Clarias gariepinus*) in biofloc-based super intensive culture added with *Bacillus* sp. Journal of Fisheries and Aquatic Science. 2015;10(6):523–532.

[25] Poleo G, Aranbarrio JV, Mendoza L, Romero O. Cultivo de cachama blanca en altas dens- idades y en dos sistemas cerrados. Brazilian Journal of Agricultural Research (PAB). 2011;46(4):429–437.

[26] Avnimelech Y, Kochba M. Evaluation of nitrogen uptake and excretion by tilapia in bio floc tanks, using 15 N tracing. Aquaculture. 2009;287(1):163–168.

[27] Timmons, M.B., J.M. Ebeling, F.W. Wheaton, S. T. Summerfelt, and B.J. Vinci. 2002. Recirculating Aquaculture Systems, 2nd edition. Northeastern Regional Aquaculture Center. Publication No. 01-002. Cayuga Aqua Ventures. Ithaca, NY. 769 pp.

[28] Fukami K, Nishijima T, Ishida Y. Stimulative and inhibitory effects of bacteria on the growth of microalgae. Hydrobiologia. 1997;358:185–191.

[29] Bruckner CG, Bahulikar R, Rahalkar M, Schink B, Kroth PG. Bacteria associated with benthic diatoms from Lake Constance: phylogeny and influences on diatom growth

and secretion of extracellular polymeric substances. Applied and Environmental Microbiology. 2008;74(24):7740–7749.

[30] Leonard N, Blancheton JP, Guiraud JP. Populations of heterotrophic bacteria in an experimental recirculating aquaculture system. Aquacultural Engineering. 2000;22(1):109–120.

[31] Glazer AN, Nikaido H. Microbial Biotechnology: fundamentals of Applied Microbiology. Cambridge University Press, Cambridge, New York, USA, 2007.

[32] Van FB, Meyer-Reil LA. Biomass and metabolic activity of heterotrophic marine bacteria. In: Advances in Microbial Ecology. Plenum Press, New York, USA. 1982; pp. 111–170.

[33] Brown MR, Barrett SM, Volkman JK, Nearhos SP, Nell JA, Allan GL. Biochemical composition of new yeasts and bacteria evaluated as food for bivalve aquaculture. Aquaculture. 1996;143(3):341–360.

[34] Furtado PS, Poersch LH, Wasielesky W. Effect of calcium hydroxide, carbonate and sodium bicarbonate on water quality and zootechnical performance of shrimp *Litopenaeus vannamei* reared in bio-flocs technology (BFT) systems. Aquaculture. 2011;321(1):130–135.

[35] Koops HP, Pommerening-Röser A. Distribution and ecophysiology of the nitrifying bacteria emphasizing cultured species. FEMS Microbiology Ecology. 2001;37(1):1–9.

[36] Ebeling JM, Timmons MB, Bisogni JJ. Engineering analysis of the stoichiometry of photoautotrophic, autotrophic, and heterotrophic removal of ammonia–nitrogen in aquaculture systems. Aquaculture. 2006;257(1):346–358.

[37] Hagopian DS, Riley JG. A closer look at the bacteriology of nitrification. Aquacultural Engineering. 1998;18:223–244.

[38] Hartwell LH, Culotti J, Pringle JR, Reid BJ. Genetic control of the cell division cycle in yeast. Science. 1974;183:46–51.

[39] Walker GM. Yeast physiology and biotechnology. John Wiley & Sons. Chichester, UK,1998.

[40] Monroy-Dosta MDC, Lara-Andrade D, Castro-Mejía J, Castro-Mejía G, Emerenciano M. Composición y abundancia de comunidades microbianas asociadas al biofloc en un cultivo de tilapia. Journal of Marine Biology and Oceanography (RBMO) 2013;48(3):511–520.

[41] Bowling L. Freshwater phytoplankton: diversity and biology, In: Suthers IM, Rissik D (Eds.). Plankton: A Guide to Their Ecology and Monitoring for Water Quality. CSIRO Publishing. Melbourne, Australia, 2009; pp. 115–139.

[42] Decamp O, Conquest L, Cody J, Forster I, Tacon AG. Effect of shrimp stocking density on size-fractionated phytoplankton and ecological groups of ciliated protozoa within zero-water exchange shrimp culture systems. Journal of the World Aquaculture Society. 2007;38(3):395–406.

[43] Graham LE, Wilcox LW. Algae. New Jersey: Prentice-Hall. 2000.

[44] García N, López-Elías JA, Miranda A, Huerta MMPN, García A. Effect of salinity on growth and chemical composition of the diatom *Thalassiosira weissflogii* at three culture phases. Latin American Journal of Aquatic Research. 2012;40(2):435.

[45] López-Rodas V, Maneiro E, Costas E. Adaptation of cyanobacteria and microalgae to extreme environmental changes derived from anthropogenic pollution. Limnetica. 2006;25(1–2):403–410.

[46] Pearl HW, Tucker CS. Ecology of blue-green-algae in aquaculture ponds. Journal of the World Aquaculture Society. 1995;26:109–131.

[47] Becerra-Dórame MJ, Martínez-Córdova LR, Martínez-Porchas M, Lopez-Elías JA. Evaluation of autotrophic and heterotrophic microcosm-based systems on the production response of *Litopenaeus vannamei* intensively nursed without Artemia and with zero water exchange. Israeli Journal of Aquaculture – Bamidgeh. 2011;63:1–7.

[48] Lezama-Cervantes C, Paniagua-Michel J. Effects of constructed microbial mats on water quality and performance of *Litopenaeus vannamei* post-larvae. Aquacultural Engineering. 2010;42(2):75–81.

[49] Brown MR. Nutritional value and use of microalgae in aquaculture. Avances en Nutrición Acuícola VI. Abstract of VI Internacional Simposium of Aquaculture Nutrition. 2002;3:281–292.

[50] Ray JA, Seaborn G, Leffler WJ, Wilde BS, Lawson A, Browdy LC. Characterization of microbial communities in minimal-exchange, intensive aquaculture systems and the effects of suspended solids management. Aquaculture. 2010;310:130–138.

[51] Kuang Q, Bi Y, Xia Y, Hu Z. Phytoplankton community and algal growth potential in Taipinghu reservoir, Anhui Province, China. Lakes & Reservoirs: Research & Management. 2004;9:119–124.

[52] Maicá PF, Borba MR, Wasielesky W. Effect of low salinity on microbial floc composition and performance of *Litopenaeus vannamei* (Boone) juveniles reared in a zerowater-exchange super-intensive system. Aquaculture Research. 2012;43:361–370.

[53] Kobayashi T, Shiel R, King JA, Miskiewicz GA. Freshwater zooplankton: diversity and biology In: Suthers, IM, & Rissik, D (Eds.). Plankton: A Guide to Their Ecology and Monitoring for Water Quality. CSIRO Publishing. Melbourne, Australia, 2009; pp.157–179.

[54] Pandey BD, Yeragi SG, Pal AK. Nutritional value of a heterotrichous ciliate, Fabrea salina with emphasis on its fatty acid profile. Asian Australasian Journal of Animal Sciences. 2004;17(7):995–999.

[55] Ben-Amotz AD, Fishler R, Schneller A. Chemical composition of dietary species of marine unicellular algae and rotifers with emphasis on fatty acids. Marine Biology. 1987;95(1):31–36.

[56] Campaña-Torres A, Martínez-Córdova LR, Martínez-Porchas M, López-Elías JA, Porchas-Cornejo MA. Productive response of *Nannochloropsis oculata*, cultured in dif-

ferent media and their efficiency as food for the rotifer *Brachionus rotundiformis*. Phyton. 2012;81:45–50.

[57] Loureiro KC, Wilson WJ, Abreu PC. Utilização de protozoários, rotíferos e nematódeos como alimento vivo para camarões cultivados no sistema BFT. Atlantica. 2012;34(1):5–12.

[58] Pérez AJD. Aplicación y evaluación de un reactor de contactores biológicos rotativos (RBC o biodiscos), a escala de laboratorio como tratamiento de los lixiviados generados en el relleno sanitario de la Pradera. Master's degree thesis (Urban Engineering) Engineering Faculty. University of Medellín. 2010; pp. 259.

[59] Farhadian O, MdYuso F, Mohamed S. Nutritional values of *Apocyclops dengizicus* (Copepoda: Cyclopoida) fed *Chaetocerous calcitrans* and *Tetraselmis tetrathele*. Aquaculture Research. 2009;40:74–82.

[60] Berberovic R. Elemental composition of two coexisting Daphnia species during the seasonal course of population development in Lake Constance. Oecologia. 1990;84(3):340–350.

[61] Emerenciano M, Gaxiola G, Cuzon G. Biofloc Technology (BFT): A Review for Aquaculture Application and Animal Food Industry. INTECH open science_ open minds. 2013b;Cap 12:301–327. http://dx.doi.org/10.5772/53902.

[62] Ferreira-Marinho Y, Otavio-Brito L, Silva CVF, Santos IGS, Olivera-Gàlvez A. Effect of addition of *Navicula* sp. on plankton composition and postlarvae growth of *Litopenaeus vannamei* reared in culture tanks with zero water exchange. Latin American Journal of Aquatic Research. 2014;42(3):427.

[63] Johnstone IL. The cuticle of the nematode *Caenorhabditis elegans*: a complex collagen structure. Bioessays. 1994;16(3):171–178.

[64] Moens T, Vincx M. Observations on the feeding ecology of estuarine nematodes. Journal of the Marine Biological Association of the United Kingdom. 1997;77(01):211–227.

[65] Vinatea L, Gálvez AO, Browdy CL, Stokes A, Venero J, Haveman J, Lewis BL, Lawson A, Shuler A, Leffler JW. Photosynthesis, water respiration and growth performance of *Litopenaeus vannamei* in a super-intensive raceway culture with zero water exchange: interaction of water quality variables. Aquacultural Engineering. 2010;42:17–24.

[66] Schveitzer R, Arantes R, Costódio PF, Espírito Santo CM, Arana LV, Seiffert WQ, Andreatta ER. Effect of different biofloc levels on microbial activity, water quality and performance of *Litopenaeus vannamei* in a tank system operated with no water exchange. Aquacultural Engineering. 2013;56:59–70.

[67] Avnimelech Y. Carbon nitrogen ratio as a control element in aquaculture systems. Aquaculture. 1999;176:227–235.

[68] García-Ríos L. Crecimiento, sobrevivencia y calidad de crías de Tilapia del Nilo (Oreochromis niloticus) cultivadas en biofloc con diferentes fuentes de carbono. Master's degree thesis. Sonora State University, Navojoa, Sonora, México. 2015;42 p.

[69] Moriarty DJW. The role of microorganisms in aquaculture ponds. Aquaculture. 1997;151:333–349.

[70] Streit WR, Schmitz RA. Metagenomics the key to the uncultured microbes. Current Opinion in Microbiology. 2004;7:492–498.

[71] Xia S, Wang F, Fu Y, Yang D, Ma X. Biodiversity analysis of microbial community in the chem-bioflocculation treatment process. Biotechnology and Bioengineering. 2005;89(6):656–659.

[72] Martínez-Córdova LR, Martínez-Porchas M, Porchas-Cornejo MA, Gollas-Galván T, Scheuren-Acevedo S, Arvayo MA, López-Torres MA. Bacterial diversity studied by next-generation sequencing in a mature phototrophic Navicula sp-based biofilm promoted into a shrimp culture system. Aquaculture Research. 2016 (online published first 16 April 2016).

[73] Soergel DA, Dey N, Knight R, Brenner SE. Selection of primers for optimal taxonomic classification of environmental 16S rRNA gene sequences. The ISME Journal. 2012;6(7):1440–1444.

[74] Roberts RJ, Carneiro MO, Schatz MC. The advantages of SMRT sequencing. Genome Biology. 2013;14(7):405.

Modelling

Modelling Impact of Adjusted Agricultural Practices on Nitrogen Leaching to Groundwater

Matjaž Glavan, Andrej Jamšek and Marina Pintar

Additional information is available at the end of the chapter

http://dx.doi.org/10.5772/66324

Abstract

The aim of the research was to determine how changes in the management of agricultural land (cultivation techniques, fertilisation, type of crop and crop rotation) influence on the leaching of nitrogen from the soil profile. Research was conducted in the Drava River plain in Slovenia. The impact of 31 different scenarios of potential change in agricultural land management was evaluated using the Soil and Water Assessment Tool (SWAT) model. The research was located on the shallow aquifer with alluvial bedrock composite from carbonate and silicate layers, which is the main source of drinking water in the area. The results of the SWAT model version 2009 showed that with the constant climate and land management technology, the magnitude of nitrogen leaching from the soil profile is mainly influenced by soil properties. The most drastic effect on the increase of nitrogen leaching showed vegetable production technology, followed by cereals (corn, wheat and barley). Vegetable production even in ecological production by Slovenian standards can result in similar leaching potential as conventional farming, due to unfavourable conditions originating from soil properties (shallow soil profile). Effects of grassland production may lead to 76–98% reduction in nitrogen loss from soil profile in comparison to current practices.

Keywords: nitrogen balance, leaching, agriculture, SWAT, environment

1. Introduction

The purpose of this paper is to present the scientifically based starting point in the development of sustainable farming in water protection areas. The issue of proper agricultural management on water protection areas is very complex since two, in management of the environment, quite different ecosystem services (water and food providing) have to coexist [1].

After the 1950s, the area of arable land and the quantity of used mineral fertilisers worldwide and in Europe increased sharply [2]. Intensive agricultural production and greater density of animals have influenced on increased input of nitrogen on land and leaching into water bodies causing deterioration of groundwater and surface water resources quality. From the first serious attempt to change the impact of agriculture on water quality in Europe, with the adoption of the Nitrate Directive (91/676/EEC), it has been 25 years. Therefore, the main objectives of the expert community are to determine the transport and balance of nitrogen from agricultural land and its impact on water bodies and changing agricultural practices towards sustainable agriculture. The results of higher environmental awareness accompanied with measures adopted in agriculture policy can be seen in substantial gradual drop in consumption of mineral fertilisers in European Union (EU) member states [3]. These results were achieved through many different actions such as political decision of EU to act, designation of nitrate-vulnerable zones (NVZs) and establishment of Codes of Good Agricultural Practice for farmers on voluntary basis, establishment of action programmes to be implemented by farmers within NVZs, and establishment of national monitoring and reporting system every 4 years for each member state [3]. To be more precise, there are some measures within cover action programmes which are crucial for the success, such as regular education of farmers, subsidy payments, cross-compliance in agriculture, implementation of new crop varieties, organic and no-till farming, promoting a 3-year rotational scheme, promoting nitrogen fixation plants, green manure plants and nitrogen catch crops, and so on.

In Slovenia, groundwater accounts for 98% of all sources of drinking water supply, so the effective protection of groundwater quality is of the utmost importance for the health of the population [4, 5]. But unfortunately, main areas of groundwater resources such as Drava Plain in Slovenia spatially coincide with the most intensive agricultural areas. Therefore, is nitrogen in these areas together with the plant protection products the main groundwater pollutant? Coincidence of natural geological and climate conditions, development in agriculture production management and past inappropriate decisions by authorities caused that many of drinking groundwater sources are at a high risk or even not suitable for use [6]. While Slovenia assigns the whole country as nitrate-vulnerable zone by Nitrate Directive and almost all farmers implemented Codes of Good Agricultural Practice, areas of additional special protection of drinking water groundwater resources are defined as water protection areas (WPAs). The basic function of WPA is conservation of drinking water quality of all water resources, which are intended for the supply of the population. Each of the EU member states committed themselves to the Water Framework Directive (2000/60/EC) with aim to implement a variety of environmental measures and maintain or improve good quantitative and chemical status of all groundwater and surface water bodies [7]. On this basis, each member state had to prepare river basin management plans and define water bodies' quality status and actions to achieve ultimate WFD goal of good water quality. All actions and quality status are carefully monitored and reported to European commission. In the case that member state in not fulfilling its own plan European Commission begins process of determining liability which could lead to the imposition of a fine to member state. One of the reporting activates of each member state is also annual report on gross nitrogen budget (GNB) and net nitrogen budget (NNG) which is prepared on the basis of Eurostat/OECD methodology [8]. The GNB is calculated as the balance between inputs (consumption of fertilisers, manure input, atmospheric deposition, biological fixation, seed-sand planting materials and crop residues) and outputs (crop harvest, harvest and grazing of fodder, crop residues removal and stock changes of N in soil) of nutrients to the agricultural soil [9]. The GNB serves as a measure of the total potential threat of nitrogen surplus or deficit in soils to the environment. Long-term deficit means loss of agriculture land productivity and excess means higher potential for pollution and eutrophication of water resources.

In the EU, WFD is proposed to use different modelling strategies to define the most cost-effective and especially environmentally effective actions with a purpose of finding balance between preserving water quality and sustaining food production. The European Commission has been in pursuit of the best model suitable for modelling nutrient losses from agricultural systems in European condition funded by the EUROHARP project [10]. Among the large ensemble of models was Soil and Water Assessment Tool (SWAT) together with NL-CAT, TRK and EveNFlow proved to be one of the best for hydrology and water quality modelling. However, researchers emphasise that there is no single model which could be used in all conditions and produce reliable results. Soil and Water Assessment Tool model is one of the open source models capable of fast and effective evaluation of agricultural practices impact on water bodies [7, 11, 12]. In the first place, it was developed to model best management practices (BMPs) in agriculture. Although these seem to be an easy task, the model requires large amount of data on management practices and water quality measurements to produce reliable results. BMPs in agriculture are one of the most often modelled scenarios; however, a combination of local hydrology, terrain, soil, land use, climate and management practices makes them constantly attractive [12–15]. BMPs in agriculture can also be described as agri-environmental measures (AEMs) introduced by farmers due to practical and cost-effective reasons. They influence on erosion processes and sediment transport, fertilisers and plant protection products transport and leaching. Based on their efficiency, they prevent pollutants to enter water bodies and conserve drinking water supply and water habitats while maintaining agricultural production [13, 14].

The aim of the paper is to investigate the impact of different adjustments in the management of agricultural land (cultivation techniques, fertilisation, type of crop and crop rotation) on the nitrogen leaching from the soil profile. For this, 31 different BPM scenarios of potential changes in agricultural land management were evaluated using Soil and Water Assessment Tool model.

2. Materials and methods

2.1. Study area

The River Drava Plain (Dravsko polje) aquifer study area (293.2 km^2) is located in the north-eastern part of Slovenia (**Figure 1**). The Drava Plain altitude is relatively small and ranges between 200 and 250 m without any distinct slopes. The plain is divided on four alluvial terraces. The agricultural land lies above an intergranular aquifer with specific soil characteristics which are the result of the deposition of river sediments. The river Drava deposited sediments of Quaternary sand and gravel in the area which forms extensive alluvial aquifer. The aquifer is very well permeable with the permeability coefficient of about 5×10^{-3} m/s. The aquifer is unconfined and exposed to the intake of pollutants from the surface. The area is suitable for intensive agriculture (grain production) due to the favourable terrain and structure of land ownership. According to data on land use prevails arable (44%) followed by the forest (20%), urban (19%) and grassland (9%). Other land-use classes are represented by 1% or less. Soils are shallow and contain many sand particles and larger rocks. Due to continental climate with spring rainfall and hot and relatively dry summers, drought often occurs on these soils.

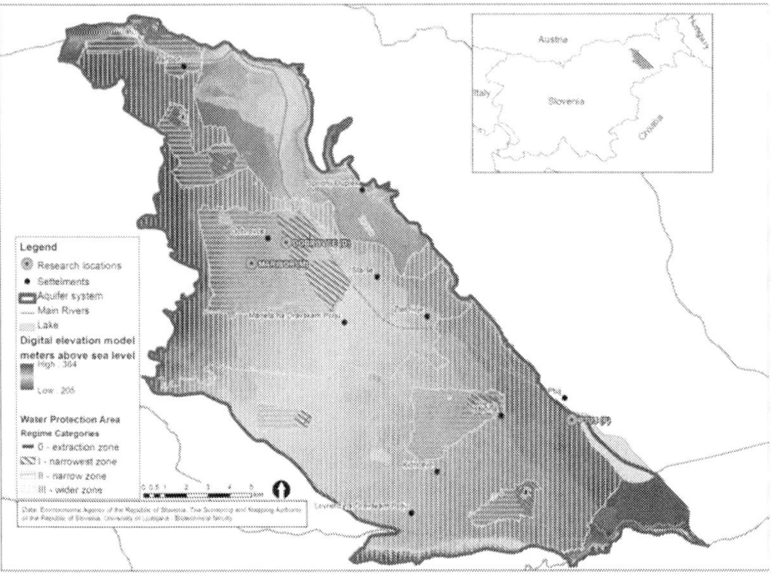

Figure 1. The river Drava Plain study area land use, elevation and water protection area.

Geographically speaking, the Drava Plain area is located in sub-Pannonian Slovenia, which is characterised by continental to sub-continental climate, with lowest rainfall quality in winter and spring months (January to April) and the highest in the summer months (June to September) due to typical stormy rainfall events. The average annual rainfall amounts (1981–2010) measured at the Maribor Airport and Ptuj were 935 and 959 mm, respectively. The average annual minimum temperature measured at the Maribor Airport was 5.3°C and maximum 15.3°C. The average minimum temperature for the meteorological winter (December to February) was −3.8°C and meteorological summer (June-August) 25.4°C.

In the area of the river Drava Plain, two regulations on water protection areas are in force, which protect the aquifer as the primary source of drinking water in the area. The measured concentration of nitrate (NO_3^-) in groundwater is at many monitoring points, in excess of the WFD-recommended concentrations for drinking water (50 mg NO_3^-/l) (**Figure 2**).

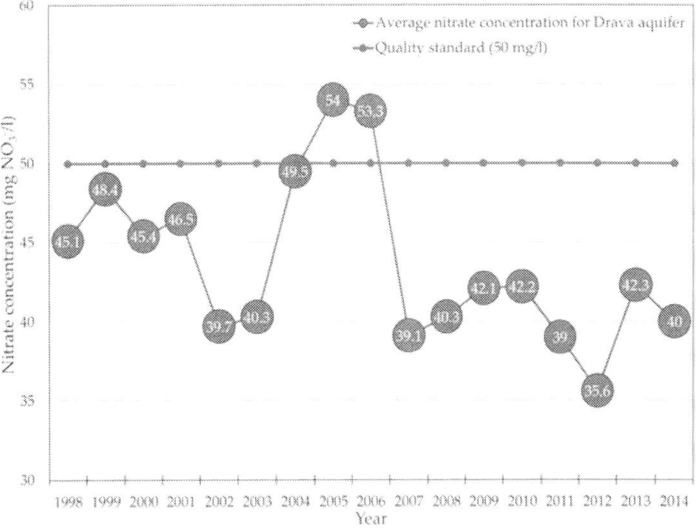

Figure 2. Average annual nitrate concentration (mg NO_3^-/l) for Drava aquifer between 1998 and 2014 calculated by Slovenian Environmental Agency on the basis of several monitoring points.

Three research locations were included in the study (**Figure 1**): Dobrovce (253 m.a.s.l.) mostly vegetable horticultural production, narrowest WPZ zone I (1.53 ha); Maribor (262 m.a.s.l.) with a grain crop rotation, narrow WPZ zone II (15.12 ha); Ptuj (218 m.a.s.l.) mixed crop rotation for seed production, wider WPZ zone III (2.25 ha) (**Figure 1**).

2.2. SWAT model description

Processes in the unsaturated zone were modelled with the Soil and Water Assessment Tool model ArcSWAT version 2009.10.1 [16]. The model was developed to assist water managers in evaluating the impact of agricultural activities in the river basins. The core of the model was developed in the early 1990s under the United States Department of Agriculture (USDA). The model was at the beginning called SWRRB and was created by joining three existing models CREAMS, EPIC and GLEAMS [12].

The SWAT model has the capability to predict the impact of land use and land management on the water quality and quantity and transport of sediment and soluble materials from agriculture in large river basins with the complex heterogeneous topography, soils, land use and land management conditions over long periods of time [12]. It is capable of modelling numerous agricultural management practises, agri-environmental measures, climate changes, scenarios of future land-use development, plant growth and biomass development. It operates on annual, monthly, daily and even on hourly time scale. Its open source code enables scientists to connect the model with others such as APEX, ALMANAC and MODFLOW and upgrade it for better performance such as SWIM and SWAT-G [12, 17, 18].

Diffuse sources of nutrients and their transport routes are in the SWAT model strongly linked to the water cycle, which is influenced by water and solar energy. When precipitation falls on the soil, it may follow different preferential pathways such as surface runoff or shallow subsurface runoff (transfer of N and P) and vertical leaching into the shallow aquifer (transfer of N). The nitrogen balance in the soil and groundwater depends on many factors (biological, climatic and physico-chemical properties of the soil). Detailed explanation of the SWAT model strengths, weaknesses, opportunities and threats in catchment modelling is given in preceding book chapter [19].

2.3. Database

For the preparation of the model, data were collected ranging from spatial data (digital elevation model (DEM) map, land-use map and classification and soil map and its properties), time series (weather such as daily precipitation, temperature, solar radiation, wind speed, relative humidity; crop rotations such as type of farmed culture, crop rotation sowing and harvesting dates; tillage such as the type of tillage and machines, the time of basic soil preparation, cultivation during growth and after; fertilisation such as the type, time of use, quantity and nutrient content), attribute data (soil parameters such as thickness of horizons, hydrological group, bulk density, texture, colour, rocks, organic matter, hydraulic conductivity, plant available water and soil erosivity—MUSLE; plant growth such as leaf area index, the development of dry biomass and average yield) to data for model calibration (soil water content (SWC)) (**Table 1**). Information about the type, quantities and dates of the use of fertiliser was obtained from Agricultural Extension Service (Chamber of Agriculture and Forestry of Slovenia—Unit Maribor) and farmers. Data acquisition began in early July 2011, and lasted over the entire period of the survey until 2013. When the data were collected, they were during the research gradually entered in the SWAT model to prepare a base scenario of the base current agricultural practices in all three research locations. This was a base for further development of scenarios of potential crop rotations including good agricultural practices for protecting water protection zones (WPZs) in the Drava Plain area.

2.4. Agricultural land management scenarios

The scenarios were developed with the aim to determine how changes in agricultural practices (crop rotation, fertilisation rate and type of plant varieties) influence the leaching of nitrogen below the plant roots from the soil profile. In designing the scenarios, we relied on the guidelines for the scientifically grounded fertilisation, issued by the Ministry of Agriculture, Forestry and Food [20], own expertise and information from agricultural producers (farmers) in the study area.

Depending on the availability of data, we prepared a total of seven sets of scenarios with 31 possible combinations of alternative rotations and managements (A. Basic rotations with modified fertilising norms, B. Basic rotations with introduced new crops, C. Grassland use, D. The most common rotations in the research area, E. Adapted the most common rotations, F. Organic rotations, G. Water protection zones regime rotations) (**Table 2**). Scenarios can serve only as indicative information as actual future development of agricultural land management is impossible to predict. EU Common Agricultural Policy (CAP) changes at least once in 7 years with possible mid-term changes in each of the member states after evaluation of the national Rural Development Programmes (RDPs). Agricultural policy can throughout financial stimulants dictates simultaneous sustainable agriculture and protection of water resources. In designing the scenarios, we relied on the guidelines for professionally justified fertilisation, issued by the Ministry of Agriculture, Forestry and Food [20], own expertise and on information from farmers.

Climate	Data source
Precipitation, avg., min. and max. temperature, relative humidity, energy of global solar radiation, average wind speed	Slovenian Environmental Agency (ARSO), Meteorological station Maribor Airport, Precipitation station Ptuj
Soil	
Digital soil map, the number and depth of horizons, rooting depth, soil-bulk density, field capacity, wilting point, saturated hydraulic conductivity, soil colour, organic matter, texture (clay, silt, sand), erosivity soil—MUSLE	Ministry of agriculture, forestry and food (MKGP), field measurements, pedotransfer functions, laboratory measurements, calculations
Land-use management practices	
number of rotation years, crop type, time of sowing, planting and harvesting, time of fertilisation, type of fertiliser, the method and depth of applications, time of tillage, type of machines and tillage depth, rooting depth, height of plants, plant nutrient content, the biomass development, the potential yield, harvest index, LAI	Farmers, Chamber of Agriculture and Forestry of Slovenia (KGZS)—Unit Maribor, SWAT database
Terrain, land use	
Digital elevation model (DEM), land-use map, land-use classification	Ministry of agriculture, forestry and food (MKGP), The Surveying and Mapping Authority of the Republic of Slovenia (GURS)

Table 1. Model input database.

Scenario/location		Crops in rotation	Elemental fertiliser (kgN/ha per year)		
Base	Dobrovce	ca-tn/on/po/pe/po	120-0/27/100(organic)/75/174		
	Maribor	or-cl/co/ww-cl/co/wb-cl/co/ww	150-0/111/109-0/111/109-0/111/109		
	Ptuj	ww-bw/ww-cl/fp/ww-cl/cas	161-0/161-0/27/161-0/122		
Scenarios of adjusted agricultural practices			Average fertiliser application (kg elemental N/ha per year)		
			Dobrovce	Maribor	Ptuj
Base scenarios			127	135	130
A.	1	Medium-stocked soil for a high yield	–	171	152
	2	Medium-stocked soil for average yield	–	114	90
	3	Medium-stocked soil for a low yield	–	69	60
	4	Without livestock manure	103	–	–
	5	With livestock manure	–	180	150

Scenarios of adjusted agricultural practices			Average fertiliser application (kg elemental N/ha per year)		
			Dobrovce	Maribor	Ptuj
B.	6	without corn (and with soya)	–	76	–
	11	Integrated vegetable crop rotation 1	162	–	–
	12	Integrated vegetable crop rotation 2	213	–	–
	13	Integrated vegetable crop rotation 3	122	–	–
C.	7	Grassland — four-cut-BMP			182
	8	Grassland — three-cut-BMP			122
	9	Grassland — two-cut-BMP			47
	10	Grassland — one-cut-BMP			0
D.	14	Average cattle rotation			183
	15	Average pig rotation			170
	16	Average arable crop rotation			160
	17	Average permanent grassland			434
E.	18	Cattle rotation no livestock manure			147
	19	Cattle rotation no maize (with soya)			87
	20	Pig rotation no livestock manure			153
	21	Pig rotation no maize (with soya)			37
F.	22	Organic vegetable rotation			47
	23	Organic field crop rotation			60
G.	24	WPZ I — cattle			134
	25	WPZ II/WPZ III — cattle			177
	26	WPZ I — pig			144
	27	WPZ II/WPZ III — pig			170
	28	WPZ I — arable crop			142
	29	WPZ II/WPZ III — arable crop			160
	30	WPZ I — permanent grassland			191
	31	WPZ II/WPZ II — permanent grassland			415

Key: ca, cabbage; tn, turnip; on, onion; po, potatoes; pe, peppers; or, oilseed rape; cl, clover; co, corn; ww, winter wheat; wb, winter barley; bw, buckwheat; fp, field peas; cas, cabbage for seeds; BMP, best management practices according to guidelines for the scientifically grounded fertilisation [20]; WPZ, water protection zone; I, narrowest WPZ zone (stricter regime); II, narrow WPZ zone; III, wider WPZ zone.

Table 2. Agricultural land management scenarios.

In the first set (A.) are three scenarios for Maribor and Ptuj, where fertilisation of basic rotation changed depending on the quantity of yield (A. Scenarios 1–3) and one with organic fertiliser (cattle slurry) introduced in to rotation with strictly mineral fertilisers (5). For the location of Dobrovce, organic fertilisers are replaced by mineral (Scenario 4). In the second set (B.) is one scenario for the Maribor rotation, where soya replaced corn (Scenario 6) and three scenarios for Dobrovce with alternative vegetable rotations, with one legume as a main crop and winter greening (Scenarios 11–13). In the third set (C.) are four scenarios including four-cut, three-cut, two-cut and extensive one-cut (no fertilisers) grassland (Scenarios 7–10). In the fourth set (D.) are average cattle/dairy, pig, arable rotation and for the research area typical permanent grassland management (Scenarios 14–17). In the fifth set (E.) are two variations of modified cattle/dairy and pig rotations, one without organic fertilisers (Scenarios 18 and 20) and the other with soya replacing corn (Scenarios 19 and 21). In the sixth set (F.) are average horticultural (vegetable) and organic rotation (Scenarios 22 and 23). In the seventh set (G.) are average cattle/dairy, pig, arable rotation and permanent grassland management with included fertilisation rates required in WPA (Scenarios 24–31). The results of the alternative scenarios were compared with the baseline scenarios (business as usual).

2.5. Calibration process and data analysis

Model-testing procedures were carried out on daily level for all three research locations. Simulation period was split on warm-up, calibration and validation period. Warm-up period was excluded from comparison due to model setting up the water and nutrient cycle balance. Model calibration and validation were performed with comparison of measured and simulated soil water content data at research location. These were the only available data for testing whether water cycle in the soil profile is functioning adequately. Calibration and validation periods are as follows: Ptuj December 2011-March 2012 and April-May 2012, respectively, Maribor November 2011 and December 2011, respectively, and Dobrovce July-August 2011 and August to September, respectively. Parameters for automated and manual calibration were selected based on sensitivity analysis tool in ArcSWAT [21] and expert knowledge of the research area. Ten parameters were selected, including CN2, ESCO, GW_REVAP, REVAPMN, CANMX, FFBC, SOIL_BD, SOL_AWC, SOL_K and SOL_ALB. Sensitivity analysis, calibration and validation procedures for these three locations are in-depth explained in previous publication [22].

Model performance was determined with comparison of measured and simulated time series via graphical or visual comparisons and objective function called percent bias ($PBIAS$) [23]. It measures the average tendency (higher or lower) of simulated values to be different than observed ones. Negative $PBIAS$ values mean excess water in simulation and positive values mean lack of water in simulation. Visual comparison was used due to important share of rocks in the soil which impact probes measurements of soil water content at the research locations. Simulated values were acceptable if they fall within minimum- and maximum-measured values.

To obtain useful and informative results, simulation of base and alternative management scenarios was run for a period of 12 years (2000–2011) for all three locations. The first three warm-up years (2000–2002) were excluded from result analysis. Results were analysed on the basis of hydrological response unit (HRU) obtained from SWAT OUTPUT.HRU data file on daily, monthly and annual level for the period of 9 years (2003–2011). Results include analysis of nitrogen balance for base and alternative agricultural land management scenarios. The main model output variables for nitrogen balance are nitrogen fertiliser applied (N_APP), N added to soil profile by rain (NRAIN), N fixation (NFIX), fresh organic to mineral N (F-MN), active organic to mineral N (A-MN), active organic to stable organic N (A-SN), denitrification (DNIT), plant uptake (NUP) and N leached from the soil profile (NO3L). All variables are expressed as kilograms of N per hectare (kg N ha^{-1}).

Wilcoxon rank-sum non-parametric test was used for the detection of significant differences between base and alternative scenarios. We compared the average annual values of two independent samples of equal size ($n1 = n2 = 9$). The results of alternative agricultural land management scenarios are statistically significantly different from base situation, if the Wilcoxon test value exceeds 62 at $\alpha = 0.05$ or 70 at $\alpha = 0.20$.

3. Results and discussion

3.1. Calibration

Water that enters the soil profile can move by several possible routes. Soil water can be removed from the soil by plant uptake or evaporation (evapotranspiration) or may percolate vertically through the soil horizons below the bottom of the soil profile, or laterally as surface runoff and interflow. The majority of the soil water is removed through evapotranspiration. Correct preparation of soil parameters is verified by soil water content and plant growth rate results.

PBIAS statistical test shows that simulated SWC at all three research locations is within a reasonable range and in good agreement with measured values (**Table 3** and **Figure 3**). The results fall within the very good category [23]. From **Figure 3**, SWC is well seen declining during prolonged periods of drought and the rise of SWC after precipitation events.

Parameter	Default	Range	Ptuj		Maribor		Dobrovce	
			SAR	FCV	SAR	FCV	SAR	FCV
Cn2	D*	±25%	4	+5	3	+10	6	+25
Esco	0.95	0.5–1	3	0.87	5	0.85	4	0.80
Gw_Revap	0.02	0.02–0.20	10	0.10	9	0.09	10	0.06
Revapmn	750	0–1000	8	760	10	634	9	530
Canmx[a]	0	0–20	7	2.3	4	2.5	5	2
FFBC	0	0–1	6	0.93	8	0.94	7	0.95
Sol_Bd	D*	±25%	5	+2	6	+10	3	+21
Sol_Awc	D*	±25%	1	+8	1	+12	1	+18
Sol_K	D*	±25%	2	0	2	−5	2	−10
Sol_Alb	D*	±25%	9	+1	7	−2	8	−3
Calibration	0 = optimal		−7.59		4.33		9.61	
Validation	+ values (%) = underestimate − values (%) = overestimate		4.76		−8.73		−7.64	

SAR, Sensitivity Analysis Rank; FCV, Final Calibrated Value; [a], forest, permanent crops, grassland + arable; D*, depends on soil type, land use and modeller set-up; Cn2, SCS runoff curve number for moisture condition II; Esco, soil evaporation compensation factor; Gw_Revap, groundwater "revap" coefficient; Revapmn, threshold depth of water in the shallow aquifer for "revap" to occur; Canmx, maximum canopy index; FFBC, Initial soil water storage expressed as a fraction of field capacity water content; Sol_Bd, moist bulk density; Sol_Awc, available water capacity of the soil layer; Sol_K, saturated hydraulic conductivity; Sol_Alb, moist soil albedo.

Table 3. Sensitivity analysis and daily time-step soil water content (SW) calibration and validation performance statistics for the Ptuj, Maribor and Dobrovce research locations.

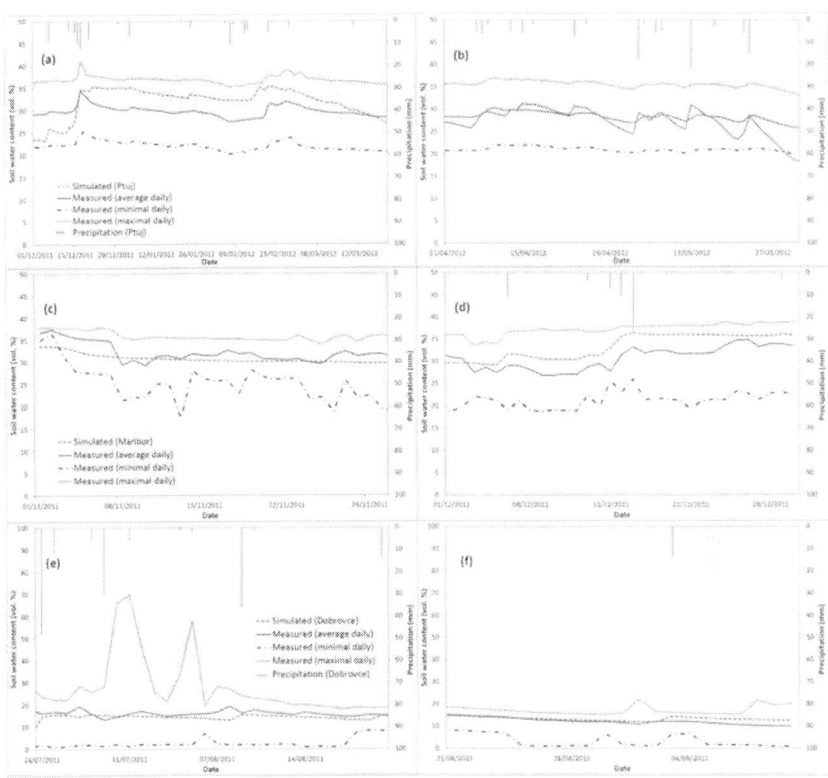

Figure 3. Visual comparison of soil water content calibration and validation at various periods for Ptuj (a and b), Maribor (c and d) and Dobrovce (e and f) research locations.

Based on the results of PBIAS test and visual comparison of the simulated and measured values of SWC, we can argue that the SWAT model is well enough calibrated to be suitable for carrying out simulations of SWC and nitrate leaching from the soil profile. It is necessary to be cautious in interpreting the results because the period of SWC measurement was short and calibration and validation periods do not cover all possible weather and land management events.

3.2. Nitrogen balance

3.2.1. Base scenarios

The base scenarios show a high average annual variability in nitrogen leaching from the soil profile (**Table 4**). Comparison of base rotations from practice between themselves showed that production technologies with higher N intake have negative impact on the balance of N causing higher leaching. Results of the model show that the same technology (rotation) is not suitable for all soil types (**Table 4**). As shown in the example from Maribor with rotation suitable for relatively deep soils, can this rotation cause from two to three times greater N leaching if used on shallow soils of Dobrovce. Measures for controlling nitrogen fertilisers' application are not defined on the basis of soil properties, according to the current regulation for the WPA of the Drava Plain. Areas of regimes I, II and III have been determined in order to prevent microbiological contamination of drinking water wells. The results show that for the purpose of preventing the negative N balance, the WPA zones and regime should be designed according to the soil properties. This is even more important because Water Framework Directive obliges member states to improve the water quality status of the entire aquifer and not only that part in the vicinity of wells.

Research location	Nitrogen leached from the soil profile(kg N ha^{-1} year)											
	Rotation											
	Ptuj				Maribor				Dobrovce			
↓Soil↓	avg	StDv	min	max	avg	StDv	min	max	avg	StDv	min	max
Ptuj	51.3	43.4	1.2	109.1	32.4	17.5	0.7	56.2	71.4	50.1	13.2	152.8
Maribor	71.1	76.0	4.0	180.9	59.9	27.5	22.2	103.9	85.5	49.7	8.8	159.0
Dobrovce	91.5	105.5	6.0	267.6	97.5	62.3	12.3	208.4	91.1	56.1	4.9	188.3

avg, average; StDv, standard deviation; min, minimal; max, maximal; shaded cells, results of rotation management and soils type from the same research location.

Table 4. Average annual nitrogen leaching (kg N ha^{-1}) from the soil profile (model SWAT) for all three base rotations of the three research locations Ptuj (P), Maribor (M) and Dobrovce (D) for the research period 2003–2011.

3.2.2. Agricultural land management scenarios

Current base fertilisation rates at the Maribor research locations are higher than the rates for the average yield are (Scenario 2) (**Table 5** and **Figure 4**) [20]. On replacing part of the organic fertiliser with the mineral (Scenarios 4 and 5) and vice versa, organic animal fertilisers were shown to cause higher excess N in the balance (**Table 5** and **Figure 4**). It is necessary to invest in the education of producers and to strengthen the control of fertilisation plans. Analysis of soil properties is required to check how much fertiliser can soil hold and how much can be applied at given soil conditions to achieve optimum yields and to avoid excessive N leaching.

Comparison of Dobrovce base rotation similar to organic and conventional integrated horticultural rotation (Scenarios 11–13) with fertilising norms for optimal production of vegetables has shown that outdoor horticultural production in Dobrovce shallow and sandy soils with gravel parent material is probably not the optimal use of agricultural land from the water protection point of view (**Table 5** and **Figure 4**). Much better results for N leaching were archived by organic field crop rotation (Scenario 23) with N-leaching yields lower from base rotations (**Figure 5**). Organic farming in WPZ is, beside water quality, also in pursuit of other goals, such as increased biodiversity, animal welfare and ban of synthetic plant protection products.

Nitrogen (kg N ha⁻¹ per year)						
	Scenario					
	Base*	1	2	3	4	5	6
Applied	130:135:127	170	112	67	101	157	88
P – leached	51	50	43	41		66	
M – leached	60	85	44	24		83	24
D – leached	91				87		

	Base	7	8	9	10	11	12	13
Applied	130:135:127	172	122	43	0	144	220	144
P – leached	51	1	2	5	4			
M – leached	60	5	8	9	9			
D – leached	91	26	26	18	14	68	146	89

	Base	Cattle			Pig			Grassland
		14	18	19	15	20	21	17
Applied	130:135:127	178	146	39	180	152	40	433
P – leached	51	38	33	38	56	53	50	12
M – leached	60	62	53	58	79	74	69	21
D – leached	91	96	79	76	110	101	88	74

	Base	16	22	23
Applied	130:135:127	166	47	13
P – leached	51	52	65	47
M – leached	60	63	79	59
D – leached	91	79	91	69

	Cattle			Pig			Arable			Grassland		
	14	24	25	15	26	27	16	28	29	17	30	31
Applied	178	139	173	180	125	180	166	127	166	433	190	415
P – leached	38	5	37	56	6	56	52	38	52	12	1	8
M – leached	62	16	61	79	19	79	63	47	63	21	4	19
D – leached	96	44	94	110	41	110	79	58	79	74	21	68

*Applied amount of nitrogen fertilisers: Ptuj (P) – 130 kg N ha⁻¹; Maribor (M) – 135 kg N ha⁻¹; Dobrovce (D) – 127 kg N ha⁻¹. Leached – nitrogen leached past the bottom of the soil profile
1 – fertilising for large yield; 2 – fertilising for average yield ; 3 – fertilising for small yield; 4 – base rotation/no organic fertilisers; 5 – base rotation/only organic fertilisers: 6 – soya replaces corn; 7 – grass/4 cuts; 8 – grass/3cuts; 9 – grass/1 cut only ; 10 – grass/1 cut no fertilisers; 14 –cattle/dairy cows rotation; 15 – pigs rotations; 16 –arable rotation; 17 – average grassland; 18 – cattle rotation/no organic fertilisers; 19 – cattle rotation/soya replaces corn; 20 – pigs rotations/ no organic fertilisers; 21 – pig rotation/soya replaces corn; 22 – organic vegetable rotation; 23 – organic field crops rotation; 24 – 14 with WPZ I regulations; 25 – 14 with WPZ II and III regulations; 26 – 15 with WPZ I regulations; 27 – 15 with WPZ II and III regulations; 28 – 16 with WPZ I regulations, 29 – 16 with WPZ II and III regulations; 30 – 17 with WPZ I regulations; 31 – 17 with WPZ II and III regulations

Table 5. Comparison of average annual applied fertiliser (mineral and organic) and nitrogen leaching from soil profile (kg N ha⁻¹) between base and alternative scenarios for the research locations Ptuj (P), Maribor (M) and Dobrovce (D) in the research periods 2003–2011.

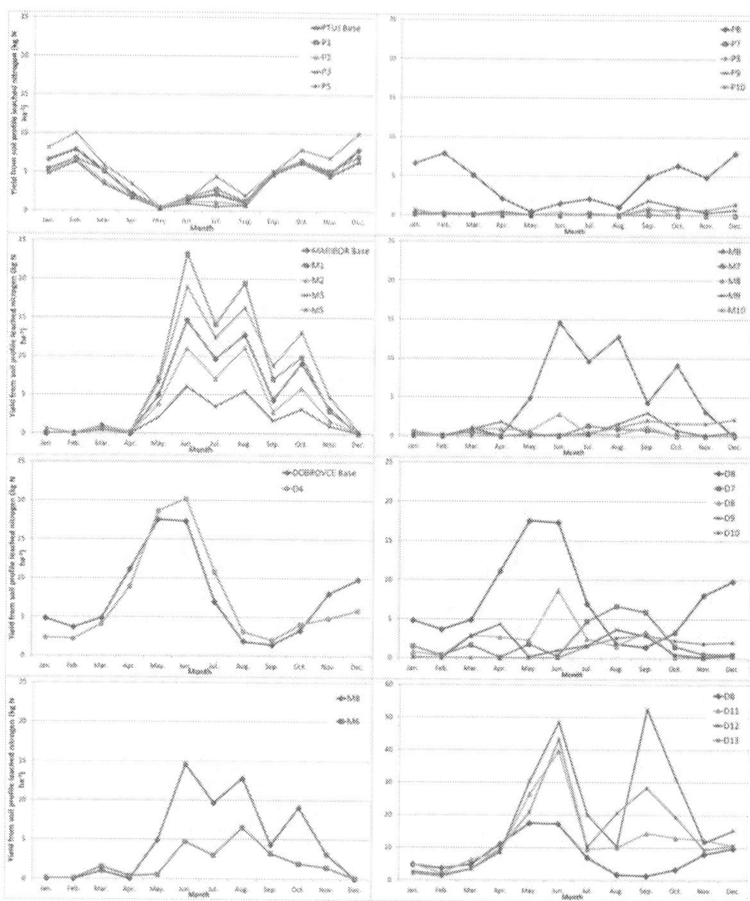

Figure 4. Comparison in simulated average monthly nitrogen leaching (kg ha⁻¹) between base and alternative agricultural management (Scenarios 1–13) for research locations Ptuj (P), Maribor (M) and Dobrovce (D) in the period between 2003 and 2011 (Scenarios key in **Tables 2** and **5**).

Grassland land use and management proved to be an extremely beneficial for soil N balance (Scenarios 7–10). Interestingly, the four-cut intensive grassland without excessive use of animal manure contributes to a drastic reduction in N leaching (**Table 5** and **Figure 4**). Through the process of modelling permanent pasture with average production technology, it was found that farmers on average spread slurry three times per year (in some cases even more) in addition to that they spread mineral fertilisers (Scenario 17). This practice causes on shallow soils such as in Dobrovce heavy losses of N, which are comparable to those in the arable fields. This shows that regulation on banning the organic fertilisers especially liquid animal manure (slurry) in the WPA I is appropriate and eligible measure. Awareness of this is even more important as currently a major part of slurry is applied on arable land as part of corn field fertilisation and not on the grassland areas.

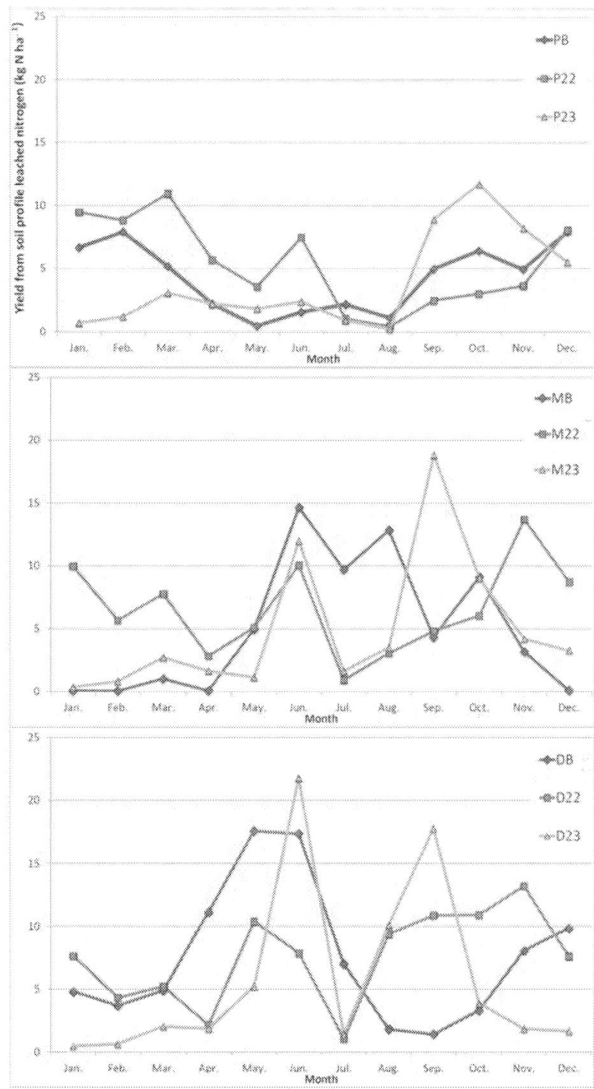

Figure 5. Comparison in simulated average monthly nitrogen leaching (kg ha⁻¹) between base and alternative organic vegetable (22) and field crops (23) agricultural management scenarios for research locations Ptuj (P), Maribor (M) and Dobrovce (D) in the period between 2003 and 2011 (Scenarios key in **Tables 2** and **5**).

One of the options for the reduction in N leaching could be expanding ban on organic fertilisers with exclusion of cattle and pig slurry from the practice also on WPZ II and III. This could lead in farmers' revolt and dramatic socio-economic changes on short term and restructuring the farm production on long term. This was investigated in Scenarios 18 (cattle farms) and 20 (pig farms) (**Table 5** and **Figure 6**). However, results did not show dramatic changes in the reduction of N leaching as organic fertilisers were substituted with mineral ones. In addition to that, a new problem would emerge as surplus N would need to be properly treated. The same was simulated when we replaced corn with soya beans (Scenarios 19 and 21) (**Table 5** and **Figure 6**). Although applied amount of fertilisers (organic and mineral) were reduced in the rotation, the nitrogen from symbiotic fixation was still released in the environmental and subject of mineralisation. On annual level, legumes fixate 150–250 kg N per hectare [20].

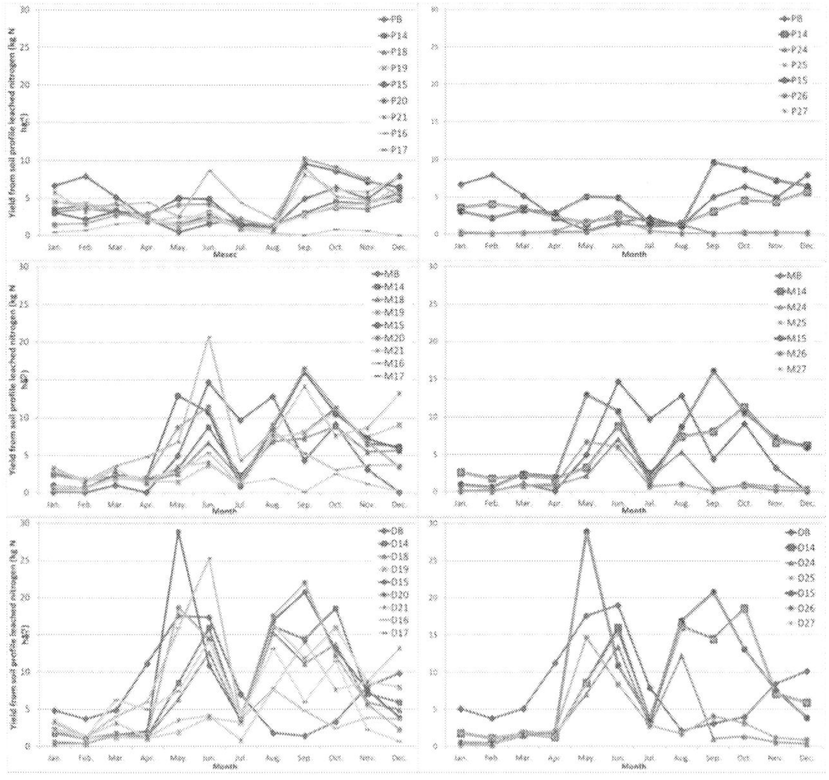

Figure 6. Comparison in simulated average monthly nitrogen leaching (kg ha⁻¹) between base and alternative agricultural management (Scenarios 14–21 and 24–27) for research locations Ptuj (P), Maribor (M) and Dobrovce (D) in the period between 2003 and 2011 (Scenarios key in **Tables 2** and **5**).

Rotations adapted to WPA zone I regime (Scenarios 24, 26, 28 and 30) reduce losses of N while the development of biomass and yield is not affected (**Table 5** and **Figure 6**). The main reason for this is the ban on the use of liquid animal manure and strictly controlled application of mineral N during the growing season. The effects of the measures are not equally effective in all areas. Efficiency is strongly related to the soil properties. This type of scenarios is a very attractive option for regulators (State), with few very relevant side effects on agriculture, for which the regulator will have to provide answers and solutions. The first effect is the surplus of livestock manure, the second is the cost for mineral fertilisers and the third, the control of measures implementation if the zone I regime would be extended over greater area.

Measures of WPA zone II and III regime have minimal effects on arable land which means that farmers can practically farm without any serious limitations (Scenarios 25, 27, 29 and 31) (**Table 5** and **Figure 6**). It is also possible that farmers adapted production technologies according to the requirements of the regulations for water bodies in the area of the Drava Plain. Results show stable N balance, which is similar to the average situation outside of WPA. Given the fact that WPA zone I regime covers only a small part of the Drava Plain (2.3%), the effect of these measures on the quality of groundwater is minimal (**Figure 2**). In addition to that in the large central part of the Drava Plain with shallow soils and under the WPA zone II and zone III regimes, a normal agricultural practice is taking place. The results of the SWAT model show that it is possible to reduce the quantity of the applied and thus also leached N, without any important effect on biomass or yield production.

4. Conclusions

The results show that the soil-type properties have the greatest impact on the nitrogen balance, with the same technology of production and weather conditions. Comparison of base- and adapted-farming practices with each other showed that the same agricultural practice is not suitable for all soil types. According to current regulation of WPA measures, restricting the intake of nitrogen fertilisers is not defined in terms of soil type. The results show that for the purpose of preventing the negative nitrogen balance, it is necessary to design WPZ regimes according to the soil types.

It is also important to increase control over the implementation of the measures prescribed by the regulation for the aquifer water bodies in the Drava Plain and the Rural Development Programme of the Republic of Slovenia, especially through cross-compliance under EU Common Agricultural Policy.

The comparison also showed that in locations Ptuj and Maribor fertilise more than is recommended for the average yield, according to national guidelines. Replacing part of organic fertilisers for the mineral and vice versa showed that organic fertilisers cause excess nitrogen that is available for leaching. For each type of soil, it is necessary to check the nutrient's holding capacity. It is necessary to ensure that we know what quantity of nutrients is required at a given soil properties to achieve optimal or even maximal yields for preventing leaching of excessive nitrogen. We recommend more efforts in introducing crops that require less inputs of nitrogen for growth (e.g. soya beans), and also in raising awareness of the need to reduce the fertilisation rates by their number and quantity.

In the organic farming, it is necessary to introduce fertilisation management based on the soil properties. Results show that outdoor vegetables production on shallow and sandy soil is not optimal agricultural practice from the drinking water protection perspective and is in certain situations even comparable to conventional production.

Grassland use is a good alternative to arable. Different methods of farming practices on grassland land use, also intensive ones, have according to national guidelines proved to be extremely beneficial from the nitrogen balance perspective. However, it is necessary, with the help of professional services, to clearly specify the amount and type of N fertiliser rate, which has to be dependent on the soil type and properties.

Agricultural practices adjusted to the stricter WPA I regime considerably reduce the loss of nitrogen from the soil and do not impact the yield which remains stable. The current state represents a good balance between the benefits for good-quality status of drinking water and economic situation of agricultural holdings.

On the other side, we have less strict WPA II and III regime minimal effects on N leaching from arable land which could mean that producers cultivate land without any serious restriction or they adjusted agricultural practices according to the requirements of the WPA regulations.

Measures of WPZ II and III regimes are virtually no different than the average conventional production practices outside the WPZ. Since these two areas occupy the vast majority of WPZ and because we have, in regard to the commitments adopted from Water Framework Directive, achieved good-quality status of groundwater throughout the aquifer, it is necessary to change the current approach of forming WPZ regimes. The current system for determining the WPZ regimes is positioned so that it fully ignores the characteristics of the soil.

Assessing the impact of the scenarios was done with the knowledge of the uncertainties of the model. Uncertainties were associated with the establishment of production technologies, rotations, dates of harvest, grass-cut dates, dates of mechanical tasks and dates of fertilisers' application. All these data are just an average estimate, as each farmer has their own time schedule and technology of production, which varies according to the type of crops, crop and livestock species, intensity of agricultural production and changing weather conditions. Additional uncertainty originates in soil maps, firstly because of spatial resolution, and secondly because the model requires information on wilting point, field capacity and hydraulic conductivity which are not part of standard soil map.

According to the results, we suggest that future measures or WPA regimes are formed according to the type and properties of the soil and not only on the groundwater flow direction and proximity of drinking water wells. Constant communication with the land owners and cultivators in the area and their regular education is vital for successful transformation of the area. We recommend more efforts of professional agricultural services in the introduction of agricultural crops that require less inputs of nitrogen for growth such as soya and in raising awareness on the need for reducing fertiliser norms by its number and quantity. Grassland is a good alternative to the arable use, but help of professional services is needed to specify the type of fertiliser N and norms depending on the soil type and properties.

The results of this study are a product of a one-computer model for catchment modelling (SWAT) and understanding of one catchment modeller. Therefore, the final assessment of the scenarios should never be regarded as definitive, but only as a possible response of the system to changes. Model results and their interpretation by the modeller must lead to constructive discussions with the aim of achieving and maintaining good water quality in the research area of the Drava Plain, which is also the aim of the Water Framework Directive.

Acknowledgements

Financial support for this study was provided by the Slovenian Research Agency founded by the Government of the Republic of Slovenia, Ministry for Agriculture, Forestry and Food, Contract number 1000-10-281058. The report is also available from e-site: http://www.dlib.si/details/URN:NBN:SI:DOC-HWYIJSVV.

Author details

Matjaž Glavan[1*], Andrej Jamšek[2] and Marina Pintar[1]

*Address all correspondence to: matjaz.glavan@bf.uni-lj.si

1 Department for Agronomy, Biotechnical Faculty, University of Ljubljana, Ljubljana, Slovenia

2 Chamber of Agriculture and Forestry of Slovenia (KGZS) — Unit Maribor, Maribor, Slovenia

References

[1] Glavan M, Ceglar A, Pintar M. Assessing the impacts of climate change on water quantity and quality modelling in small Slovenian Mediterranean catchment—lesson for policy and decision makers. Hydrol Process. 2015;29(14):3124–44.

[2] Brown LR. Full planet, empty plates: the new geopolitics of food scarcity. New York, NY, USA: WW Norton & Company; 2012.

[3] Eurostat. Agri-environmental indicator—mineral fertiliser consumption. Luxembourg: Eurostat, European Commission; 2012; Available from: http://ec.europa.eu/eurostat/statistics-explained/index.php/Agri-environmental_indicator_-_mineral_fertiliser_consumption#Further_Eurostat_information [cited 2016 11th May 2016].

[4] Poje M, Dobnikar-Tehovnik M, Krajnc M, Trisic N, Krsnik P, Mihorko P. Quality of surface sources of drinking water in Slovenia. Ljubljana, Slovenia: Slovenian Environmental Agency; 2008.

[5] Koroša A, Mali N. Review of emerging organic pollutants in groundwater in Slovenia. Geologija. 2012;55(2):243–62.

[6] Andelov M, Kunkel R, Uhan J, Wendland F. Determination of nitrogen reduction levels necessary to reach groundwater quality targets in Slovenia. J Environ Sci. [Article]. 2014;26(9):1806–17.

[7] Volk M, Liersch S, Schmidt G. Towards the implementation of the European Water Framework Directive? Lessons learned from water quality simulations in an agricultural watershed. Land Use Policy. 2009;26(3):580–8.

[8] Wick K, Heumesser C, Schmid E. Groundwater nitrate contamination: Factors and indicators. J Environ Manage. [Article]. 2012;111:178–86.

[9] Eurostat. Nutrient Budgets – Methodology and Handbook. Version 1.02. Luxembourg: Eurostat and OECD; 2013.

[10] Kronvang B, Behrendt H, Andersen HE, Arheimer B, Barr A, Borgvang SA, et al. Ensemble modelling of nutrient loads and nutrient load partitioning in 17 European catchments. J Environ Monitor. 2009;11(3):572–83.

[11] Arnold JG, Moriasi DN, Gassman PW, Abbaspour KC, White MJ, Srinivasan R, et al. SWAT: Model use, calibration, and validation. T Asabe. 2012;55(4):1491–508.

[12] Gassman PW, Reyes MR, Green CH, Arnold JG. The soil and water assessment tool: Historical development, applications, and future research directions. T Asabe. 2007;50(4):1211–50.

[13] Pearce NJT, Yates AG. Agricultural best management practice abundance and location does not influence stream ecosystem function or water quality in the summer season. Water 2015;7(12):6861–76.

[14] Holmes R, Armanini DG, Yates AG. Effects of best management practice on ecological condition: does location matter? Environ Manage. 2016;57(5):1062–76.

[15] Liu RM, Zhang PP, Wang XJ, Wang JW, Yu WW, Shen ZY. Cost-effectiveness and cost-benefit analysis of BMPs in controlling agricultural nonpoint source pollution in China based on the SWAT model. Environ Monit Assess. 2014;186(12):9011–22.

[16] Arnold JG, Srinivasan R, Muttiah RS, Williams JR. Large area hydrologic modeling and assessment - Part 1: Model development. J Am Water Resour Assoc. 1998;34(1):73–89.

[17] Lenhart T, Fohrer N, Frede HG. Effects of land use changes on the nutrient balance in mesoscale catchments. Phys Chem Earth. 2003;28(33–36):1301–9.

[18] Krysanova V, Arnold JG. Advances in ecohydrological modelling with SWAT—a review. Hydrolog Sci J. 2008;53(5):939–47.

[19] Glavan M, Pintar M. Strengths, weaknesses, opportunities and threats of catchment modelling with Soil and Water Assessment Tool (SWAT) Model. In: Nayak P, editor. Water Resources Management and Modeling. Rijeka, Croatia: InTech; 2012. p. 310.

[20] Mihelič R, Čop J, Jakše M, Štampar F, Majer D, Tojnko S, et al. Guidelines for professionally justified fertilisation (Smernice za strokovno utemeljeno gnojenje). Ljubljana: Ministry of Agriculture, Forestry and Food of Republic of Slovenia; 2010.

[21] Green CH, van Griensven A. Autocalibration in hydrologic modeling: using SWAT2005 in small-scale watersheds. Environ Modell Softw. 2008;23(4):422–34.

[22] Glavan M, Pintar M, Urbanc J. Spatial variation of crop rotations and their impacts on provisioning ecosystem services on the river Drava alluvial plain. Sustain Water Qual Ecol. 2015;5:31–48.

[23] Moriasi DN, Arnold JG, Van Liew MW, Bingner RL, Harmel RD, Veith TL. Model evaluation guidelines for systematic quantification of accuracy in watershed simulations. T Asabe. 2007;50(3):885–900.

Chapter 7

Modeling Agricultural Land Management to Improve Understanding of Nitrogen Leaching in an Irrigated Mediterranean Area in Southern Turkey

Ebru Karnez, Hande Sagir, Matjaž Gavan,

Muhammed Said Golpinar, Mahmut Cetin,

Mehmet Ali Akgul, Hayriye Ibrikci and Marina Pintar

Additional information is available at the end of the chapter

http://dx.doi.org/10.5772/65809

Abstract

Nitrogen (N) cycle dynamics and its transport in the ecosystem were always an attracting subject for the researchers. Calculation of N budget in agricultural systems with use of different empirical statistical methods is common practice in OECD and EU countries. However, these methodologies do not include climate and water cycle as part of the process. On the other hand, big scale studies are labor and work intensive. As a solution, various computer modeling approaches have been used to predict N budget and related N parameters. One of them is internationally established Soil and Water Assessment (SWAT) model, which was developed especially for modeling agricultural catchments. The aim of this study was to improve understanding of N leaching with simulation of agricultural land management (fertilization, irrigation, and plant species) in hydrological heavily modified watershed with irrigation-depended agriculture under Mediterranean climate. The study was conducted in Lower Seyhan River Plain Irrigation District (Akarsu) of 9495 ha in Cukurova region of southern Turkey. Intensive and extensive water and nitrogen monitoring data (2008–2014), soil properties, cropping pattern, and crop rotation were used for the SWAT model build, calibration, and validation of the model.

Keywords: crop management, irrigation, nitrogen balance, SWAT, modeling

1. Introduction

In arid and semiarid regions, freshwater resources are under the ever increasing pressure of many current issues such as population increase, economic development, climate change, and pollution [1]. Water quality is a major concern and expressed by its biological, chemical, physical, and aesthetic properties [2]. The water quality is determined by a number of factors such as electrical conductivity, pH, amount of salts, dissolved oxygen, levels of microorganisms, nutrients, heavy metals, quantities of pesticides, and herbicides [3]. These factors can lead to the problems (salinity, infiltration, toxicity, and nutrients), which are extensively present in many watersheds with irrigated agriculture [4–7].

Nitrogen leaching from agricultural land is a main pollutant in many countries in the world [7, 8]. In agricultural areas of the European Union (EU), fertilizer contribution as nonpoint source pollution to the surface water is estimated to be 55% [9]. The European Union Water Framework Directive (WFD) has issued important regulations in order to reduce the environmental impact of nitrogen due to agriculture and to keep water bodies in good quality state; based on the EU Drinking Water Directive (80/778/EEC), the accepted maximum admissible concentration for the nitrate was set as 50 mg l^{-1} [10].

On the other hand, nitrogen is an essential nutrient for adequate plant growth, and mostly used as type of fertilizer [11]. During the N cycle, it undergoes many processes in soil, water, and atmosphere level [12–14]. Nitrogen cannot be used directly by the plants and animals until it is converted into its available compounds and forms. Nitrate ions in soil are usually in dissolved form in the soil solution, and it can easily be lost to leaching as water moves through the soil profile due to the rapid dynamism [15, 16].

Understanding of nitrogen dynamics in the nature, nitrogen balance or nitrogen budget becomes more of an issue about prevention of environmental pollution and economic losses on a country basis. Nitrogen balance studies have been continued for over 170 years [17]. There are different ways of defining nitrogen budgets in empirical statistical methods, depending on the measurements and modeling. Calculation of N budget in agricultural systems by this way is a common practice in OECD and EU countries. This method does not include explaining the processes of nutrient cycle in the soil-plant-atmosphere system but follows statistical methodology at national and regional levels to determine nitrogen budget [18–20].

Measured nitrogen budgets in soil-plant-atmosphere level are based on the conservation of mass of nitrogen in the system. A previous study carried out [21, 22] aimed at evaluating nitrogen fluxes by measuring agronomic system in Akarsu Study Area in southern Turkey. As part of the findings, it was found that considerable amounts of nitrate are lost to drainage and shallow groundwater. During the study years, nitrogen budget calculations resulted in unaccounted values ranging from 40 to 60 kg N ha^{-1} [23].

As known, Mediterranean climate is characterized by mild rainy winters and hot dry summers [24]. Annual and interannual changes in dry and wet periods result in change of water balance and water level fluctuations especially in the areas where Mediterranean climate is dominating [25]. Based on the recent years' ongoing drought events and therefore water scarcity, irrigation scheduling and types need to be reevaluated. Recently, best management techniques such as drip irrigation [26] and rain water harvesting techniques [27] have been tried to put into practice in order to save both irrigation water and fertilizers. In the Mediterranean climate, irrigation is inevitable for maximizing the crop yield [28]. To increase crop yield and quality and at the same time to decrease the leaching below the rooting zone, managing nutrient concentrations in irrigation water is necessary, according to crop requirements [29].

Many tools are available to observe impacts of reduced irrigation and fertilization under agriculture best management practices (BMPs) scenario. Among those tools are different hydrological models capable of defining the nitrogen dynamics at the watershed level like AGNPS, AnnAGNPS, ANSWERS, ANSWERS-Continuous, CASC2D, DWSM, HSPF, KINEROS, MIKE SHE, APEX, and SWAT. And these are only a few of watershed modes, which are currently and commonly under the service of scientists and practitioners [30]. Soil and water assessment tool (SWAT) model is one of the tools developed to predict water and nutrient dynamics [31–34].

The aim of this study was to improve understanding of (a) the effects of bypass flows due to irrigation on the calibration of SWAT model, (b) irrigation return flow (IRF) and/or drainage generating processes, and (c) N leaching dynamics with simulation of agricultural land management (fertilization, irrigation, and plant species) under Mediterranean climate conditions.

2. Materials and methodology

2.1. Study area

The Akarsu Irrigation District (AID) study area is located in the Mediterranean coastal region, between 36°51′45″ and 36°57′35″ N latitudes, and 35°24′10″ and 35°36′20″ E longitudes in Turkey. The district covers an area of 9495 ha (irrigation area), and hydrological area is 11,308 ha in the Lower Seyhan Plain (LSP) and has been irrigated for over 60 years under conventional irrigation and drainage infrastructures. Until 1994, the national irrigation agency, i.e., State Hydraulic Works (DSI), was responsible for the management, operation, and maintenance of the district. Management of the irrigation and drainage system in the district was taken over by the water users in 1994. Akarsu Water User Association has been responsible for the irrigation management, operation, and repairing issues in the district since 1994. Irrigation water has been provided from Seyhan Dam (L6, L3, and L7 in **Figure 1**), in case of water shortage in the system during the peak irrigation season or if irrigation water is not diverted to the main irrigation canal through L6, then pumping station is activated and some water is diverted from Ceyhan River (Abdioglu Pumping Station, L9 in **Figure 1**). The irrigation water in Seyhan Dam has excellent water quality ($0.33 \leq EC \leq 0.50$ dS m^{-1}, EC = 0.43 dS m^{-1}). However, electrical conductivity (EC) of Ceyhan River is slightly higher than Seyhan ($0.41 \leq EC \leq 0.80$ dS m^{-1}, EC = 0.58 dS m^{-1}). The drainage water flows through open ditches along the downstream areas and finally discharges into the Mediterranean Sea.

In the study area, the Mediterranean climate is dominant, summers are hot and dry winters are mild and rainy. Precipitation is mostly in the form of rain (average of 659 mm) that usually falls during winter and spring [35]. Temperature in June, July, and August is very high (average 33.3°C); winter months are cool with reasonable temperatures (average 10.5°C) [36]. While the long-term (1929–2014) mean temperature is 27.4°C, the long-term mean total evaporation is about 1559 mm annually (coefficient of variation <27%). According to the long-term data, soil moisture and soil temperature regimes are defined as xeric and thermic by Ref. [37].

In the area, 1st April–30th September is defined as irrigation season (IS), while 1st October–1st April is defined as nonirrigation season (NIS). However, these dates may change a little by precipitation and climatic conditions.

The soils of Akarsu consist of 11 different soil series (Incirlik, Arikli, Yenice, Innapli, Arpaci, Canakci, Mursel, Ismailiye, Golyaka, Gemisure, and Misis). The model-related physical and chemical characteristics of these soil series are recorded from Ref. [37] and verified to be used in the SWAT model. As an example, only the data of six common soil series are given in **Table 1**. Arikli (29.5%), Incirlik (25.3%), and Yenice (12.2%) series cover 67% of the entire study area. Innapli (1.03%) and Mursel (0.7%) have got the minimum distributions.

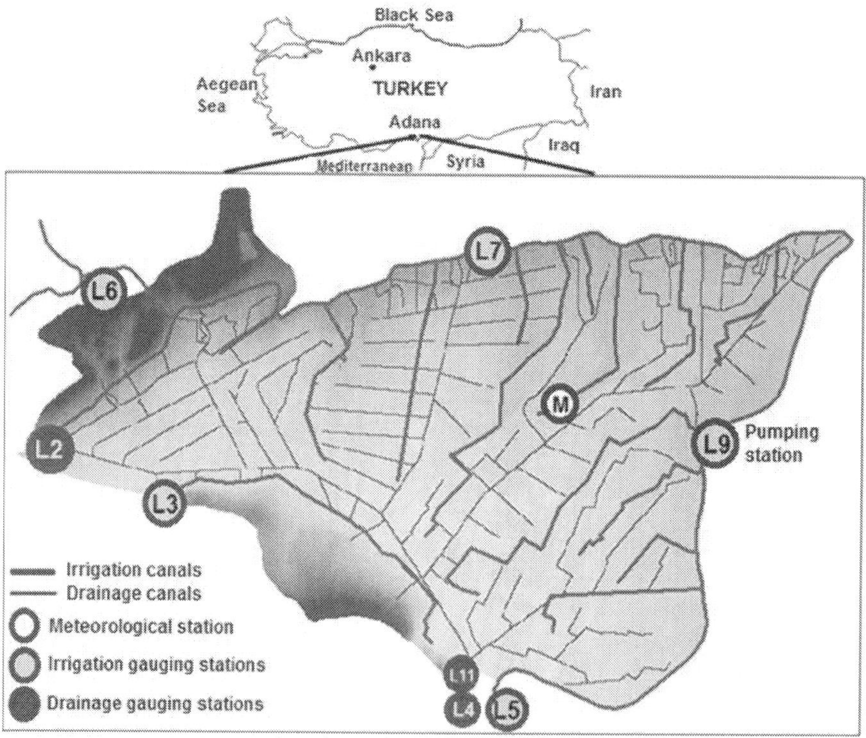

Figure 1. The Akarsu study area.

Soil series	Depth (cm)	Texture class[1]	Sand	Silt	Clay	Rock	BD[2]	OM[3]	AWC[4]	K_{sat}[5]
Incirlik	0–13	C	12	26	62	2.5	1.4	1.25	0.226	2.8
	13–78	C	16	24	60	2.5	1.6	0.62	0.233	0.55
	78–150	C	14	26	60	1.0	1.6	0.33	0.301	2.9
Arikli	0–13	SiC	8	29	63	3.5	1.3	1.25	0.268	0.63
	13–30	SiC	8	30	62	2.5	1.4	0.62	0.268	0.16
	30–57	SiC	4	29	67	1.5	1.4	0.33	0.287	0.4
	57–100		7	31	62			1.24		
	110–114		6	31	63			0.91		
	114–150		3	42	55			0.84		
Yenice	0–14	C	14	32	54	2.5	1.57	1.61	0.218	191.5
	14–32	C	12	30	58	1.8	1.59	1.21	0.259	328
	32–92	CL	14	30	56	1.0	1.48	0.80	0.219	729
	92–118		18	31	51			0.54		
Misis	0–24	C	25	23	52	1.5	1.63	1.61	0.212	3.33
	24–45	C	25	21	54	1.5	1.53	1.21	0.232	1.7
	45–64	SCL	23	21	56	1.7	1.49	0.93	0.247	1.7
	64–86		22	19	59		1.51	0.80		
	86–120		22	18	60			0.67		
	120–140		51	23	26			0.13		
Canakci	0–10	SL	25	47	28	1.5	1.51	1.37	0.208	23.9
	10–39	CL	21	55	24	1.8	1.34	1.17	0.171	16.8
	39–60	CL	29	39	32	1.6	1.58	1.50	0.157	11.5
	60–73		35	43	22			0.39		
	73–94		28	49	23			0.46		
	94–112		13	52	35			0.63		
	112–150		22	48	30			0.36		
Gemisure	0–21	C	2	26	72	1.5	1.45	1.53	0.15	2.4
	21–36	C	3	24	73	1.5	1.35	1.47	0.15	2.4
	36–78	C	3	22	75	1.8	1.39	1.34	0.15	2.4
	78–120		4	19	77			1.07		

[1] L, loam; C, clay; S, sand; Si, silt.
[2] Bulk density (g cm^{-3}).
[3] Organic matter (%).
[4] Plant available water capacity (mm H_2O mm soil depth^{-1}).
[5] Saturated hydraulic conductivity (mm h^{-1}).

Table 1. Soil properties for the Akarsu study area.

2.2. Database

The SWAT model input data, which is used in the project, is listed in **Table 2**. The 25 m resolution digital elevation model was derived by Akgul [38]. The chemical and physical properties of soils were gathered from Ref. [37], and these data were checked and verified with various measurements and laboratory analysis. Soil albedos and values of USLE were calculated by using the equations given in Ref. [39]. Soil series characteristics were interpreted and soil hydrologic group codes were assigned to each soil series based on the run-off generating characteristics. Daily irrigation return flow rates were determined by the data observed at the Inlet (L2, L11) and Outlet (L4) drainage monitoring stations. Nitrate concentrations were determined in water samples collected via automatic sampler located in L4 gauging site.

Data type	Resolution	Source	Description/properties
Topography (DEM)	25 m × 25 m	[38]	Elevation, slope, channel slopes, overland
Land cover/land use	10 m × 10 m	[35]	Land cover, land use classification
Soils	10 m × 10 m	[37]	Spatial soil variability, soil types, soil physical properties; bulk density, texture, saturated hydraulic conductivity classes, etc.
Drainage network		[35]	Drain spacing, length of cannels, drainage divides, etc.
Climate data		Adana State meteorological station and meteorological monitoring gage (L8)	Daily precipitation, temperature (max., min.), solar radiation, wind speed, relative humidity
Agricultural management practices		Farmer questionnaires in Akarsu and field surveys (face to face)	Planting, fertilizer application rates and timing, tillage, harvesting dates, irrigation water management and amount, etc.
Daily irrigation return flow rate (outlet)		1 monitoring and sampling station (L4 in **Figure 1**)	Daily flow (m^3 day^{-1})
Daily irrigation return flow rate (inlet)		2 monitoring and sampling stations (L2, L11)	Daily flow (m^3 day^{-1})
Daily irrigation return flow nitrate load (outlet)		1 monitoring and sampling station	Daily NO_3-N load (kg day^{-1})
Daily irrigation return flow nitrate load (inlet)		Two monitoring and sampling stations (L2, L11)	Daily NO_3-N load (kg day^{-1})

Table 2. Model input data and the sources.

2.3. Agricultural land management

The SWAT model has eight main components: hydrology, weather, sedimentation, soil temperature, crop growth, nutrients, pesticides, and agricultural management [30]. Watershed hydrology is affected by vegetation types, soil properties, geology, terrain, climate, land use practices, and spatial patterns of interactions among these factors [40].

The area is suitable for various agricultural productions with its favorable climatic and productive land conditions. Cropping pattern data have been assessed since 2006, and the likely crop rotation has been decided for the modeling practices. According to the data, land use and cropping pattern varied from year to year depending on the market and cultivation conditions. Based on the assessments, we have set five different crop rotations plus fruit orchards and citrus plantations (**Table 3**), which have been well adopted by the farmers in the region. Based on the recent years' evaluation, the main crops in the area were wheat, corn, citrus, cotton, and vegetables (**Table 3**). Agricultural management practices were determined based on the current surveys carried out at the local field and farmers' level.

Year	Soil tillage and crop growing period	Crops	Inorganic nitrogen fertilizer (kg elemental N ha^{-1})	Irrigation water (mm)
Rotation 1				
1	16th Mar.–16th Sep.	C1^1	385	1168
1/2	20th Nov.–07th June	WW^2	230	383
2/3	15th June–10th Oct.	S2^3	120	870
3	16th Mar.–16th Sep.	C1^1	385	1168
3/4	20th Nov.–1st June	WW^2	230	383
4	15th June–10th Oct.	S2^3	120	870
Rotation 2				
1	15th June–10th Oct.	S2^3	120	870
2	16th Mar.–16th Sep.	C1^1	385	1168
2/3	20th Nov–07th June	WW^2	230	383
3	15th June–10th Oct.	S2^3	120	870
4	16th Mar.–16th Sep.	C1^1	385	1168
Rotation 3				
1	15th Mar.–15th Oct.	Co^4	290	1535
2	15th Apr.–10th Sep.	P1^5	210	1068.33
3	15th Mar.–15th Oct.	Co^4	290	1535
4	16th Mar.–16th Sep.	C1^1	385	1168.33
Rotation 4				
1	15th June–25th Oct.	P2^6	210	800
2	16th Mar.–16th Sep.	C1^1	385	1168.33
2/3	20th Nov.–07th June	WW^2	230	383.33
3	15th June–25th Oct.	P2^6	210	800
4	15th Mar.–15th Oct.	Co^4	290	1535
4/1	20th Nov.– 07th June	WW^2	230	383.33

Year	Soil tillage and crop growing period	Crops	Inorganic nitrogen fertilizer (kg elemental N ha^{-1})	Irrigation water (mm)
Rotation 5				
1	20th June–30th Oct.	C2[7]	330	858.33
2	16th Mar.–16th Sep.	C1[1]	385	1168.33
2/3	20th Nov.–07th June	WW[2]	230	383.33
3	20th June–30th Oct	C2[7]	330	858.33
4	15th Mar.–15th Oct.	Co[4]	290	1535
4/1	20th Nov.–07th June	WW[2]	230	383.33
Orchards and citrus[+]				
Perennial	15th Mar.–8th Oct.	Orchards	250	1238.33
Perennial	1st Oct.–27th Sep.	Citrus	335	1040

[1] C1, first crop corn.
[2] WW, winter wheat.
[3] S2, second crop soybean.
[4] Co, cotton.
[5] P1, first crop peanut.
[6] P2, second crop peanut.
[7] C2, second crop corn.
[+] All kinds of operations done to orchards and citrus between these dates.

Table 3. Agricultural land management crop rotations used in the model.

The proportion of this land use type in the hydrological model area (11,308 ha) is: AGRL (Agricultural Area) (64.56%), ORAN (Citrus) (21.49%), ORCD (Orchards) (1.74%), WPAS (Winter Pastures) (9.20%), URMD (Settlement area (Medium Density)) (1.64%), and URLD (Settlement area (Low Density) (1.36%)). The agricultural areas in the study area contain various annual crops such as first crop corn, second crop corn, winter wheat, first crop soybean, second crop soybean, peanuts, and cotton.

2.4. SWAT model description

The soil and water assessment tool is one of the recent models, known as a catchment area or watershed scale model, developed by Arnold et al. [31] and improved in the last 30 years [41]. It is a semidistributed hydrological model, which is a physically based, long period of simulation, lumped parameter, and derived from agriculture management systems models such as CREAMS, EPIC, and GLEAMS [41, 42]. The model separates selected basin to subbasins and hydrologic response units (HRU) comprised of identical hydrological properties such as land use, soil, and slope [43]. SWAT is an efficient tool to predict the impact of nitrogen cycle and land management practices on water, sediment, nutrient, and pesticide with the ArcSWAT module [44]. The nitrogen cycle can be represented by the SWAT model in the soil profile and shallow aquifer. SWAT comprises two pools that are inorganic forms of nitrogen (NH_4^+ and NO_3^-) and three pools that are organic forms of nitrogen in the soil [45–47]. Nitrate and organic N into the nitrogen cycle, N removal from soil to water sources, and amounts of NO_3-N included in lateral flow, runoff, and percolation can also be represented by the SWAT model [45]. The SWAT model could sufficiently predict sediment and nutrient statuses as well as tile drainage NO_3-N losses [48, 49].

The prediction of land management practices is important as well as nitrogen cycle to provide the progress of future socioeconomic stability and sustainable use of natural resources and to search the impact of human activities on a given basin [50, 39]. SWAT has a capability to estimate the effects of land management practices on sediment, water, and agricultural chemical yields in large complex watersheds with varying soils, land use, and management conditions over a long-term time [43, 51–54].

2.5. Calibration process of the base model

Calibration and validation are key processes in reducing the uncertainty and increasing user experience in its predictive results, making the software a practical model and leading to user competence.

The adjustment of model parameters is described as calibration. These parameters are associated with checking results toward observations to assure the same response in time [55]. A number of calibration techniques, comprising manual calibration method and automated method, improved for the SWAT model [39]. The model calibration is done manually and finalized by SWAT-CUP (Calibration and Uncertainty Programs). SWAT-CUP is an interface known as "automated model calibration method" that was improved for SWAT to connect with a link between the input and output of a calibration program and the model [56]. The SUFI-2 algorithm was used for sensitivity analysis, model calibration, and validation process. The warm up period was set for 1 year.

The calibration of SWAT is completed in three phases [39]. The first phase is the determination of most sensitive parameters (such as Alpha_Bf, Canmx, Ch_K2, Ch_N, Cn2, Esco, Gw_Delay, Gw_Revap, Gwqmn Surlag for flow, and Nperco, Al1 CMN, Hlife_Ngw for nitrate-nitrogen) [57]. The second phase is model calibration with use of statistical methods such as Pearson coefficient of correlation (R^2), Nash-Sutcliffe efficiency (ENS), and percent bias (PBIAS). The final phase is validation process for hydrological calibration and nitrogen calibration of the model.

Validation, known as the part of simulation, can be done without modifying any parameter values adjusted during calibration for a different time series to input data and also for the same time period at a different spatial location [58]. In this study, daily measured values of irrigation and irrigation return flows, and also nitrate loads for the year of 2008 were used for the warm up period. SWAT was calibrated with daily values over a 4-year period from 2009 to 2012 for hydrological years and used daily values for nitrogen. The 2-year time period from 2013 to 2014 was used for validation of hydrology and nitrogen.

3. Results and discussion

3.1. Calibration of drainage flows

Calibration process of the model used in this specific research was first completed with hydrologic calibration and followed by the drainage nitrogen. In general, calibration and validation of water quality models are typically performed with data collected at the outlet of a watershed to be able to assess possible pollution risks. In Akarsu, daily measured data were used during the model processes. The most sensitive parameters for hydrologic calibration process were SURLAG, GW_Delay, Revapmn, GW_Revap, and Esco (**Table 4**), while Nperco, Cmn, Hlife, and Ngw are the sensitive ones for nitrogen calibration.

Parameter	File extensions	Explanation
Alpha_BF	.gw	Base flow recession factor, days
GW_DELAY	.gw	Groundwater delay, days
SURLAG	.bsn	Surface runoff lag coefficient, days
ESCO	.bsn	Soil evaporation compensation factor
GWQMN	.gw	Threshold depth for ground water flow to occur, mm
GW_REVAP	.gw	Groundwater "revap" coefficient
Ch_N2	.rte	Manning's n
SOL_AWC	.sol	Available water capacity
CN2	.mgt	SCS curve number, antecedent moisture condition II, for crop land use
REVAPMN	.gw	Threshold water depth in shallow aquifer for percolation to deep aquifer to occur
CH_K2	.rte	Effective hydraulic conductivity in main channel alluvium
NPERCO	.bsn	Nitrate percolation coefficient
HLIFE_NGW	.gw	Half-life of nitrate in shallow aquifer (days)
CMN	.bsn	Rate factor for humus mineralization of active organic nutrients (N).
AI1	.wwq	Fraction of algal biomass that is nitrogen (mg N mg alg^{-1}).

Table 4. SWAT input parameters for river flow and nitrogen calibrations.

Based on the model outputs, the SWAT model is reliable enough to be used in nonnatural catchments such as Akarsu Irrigation District where drainage network is not topography-driven but man-made. Additionally, hydrologic water dynamics such as inflows, outflows, and the whole water balance are well defined since 2006. The area is affected by routine agricultural management activities, i.e., irrigation and fertilization in specific.

Three recommended quantitative statistics, determination (R^2), Nash-Sutcliffe efficiency (NSE), and PBIAS, in addition to the graphical techniques for visual examination have been used to assess the hydrologic model performance [59], i.e., model calibration and validation. These performance indicators of the model (R^2, NSE, and PBIAS) during calibration period of 2009–2012 have been found as 0.62, 0.57, and 6.3, respectively (**Table 5**). Typically, values of R^2

greater than 0.50, while values of NSE between 0.0 and 1.0, and values of PBIAS ±25% for streamflow calibration are generally considered as acceptable levels [59]. In addition, model validation was made by utilizing the daily data for 2013 and 2014 period. The performance statistics for the validation period were 0.67, 0.59, and –10.04 for R^2, NSE, and PBIAS, respectively (**Table 5**).

Variable	R^2	NSE	PBIAS
Calibration (2009–2012)			
Daily drainage flow	0.62	0.57	6.3
Daily nitrogen loss	0.47	–0.63	88.1
Validation (2013–2014)			
Daily drainage flow	0.67	0.59	–10.04
Daily nitrogen loss	0.50	–0.20	72.9

Table 5. Objective function statistics for drainage flow and nitrogen in drainage.

Descriptive statistics for observed and simulated (calibration and validation) were resented in **Table 6**, indicating that model performance was satisfactory with the mean values of 3.51, 2.98 m^3 s^{-1} for calibration period and 2.71 and 2.98 for validation period. Similarly, other descriptive statistics for observed and simulated flow values were in good agreement.

The visual examination of observed versus predicted drainage flows for the calibration (**Figure 2**) and validation periods (**Figure 3**) indicated adequate calibration and validation. Therefore, SWAT simulations and observed data were in good agreement visually and statistically. SWAT-CUP automatic calibration results for the sensitive parameters were presented in **Table 7**. These parameters are reasonable enough to accept performance of the model [56] in a well-defined agricultural catchment of Akarsu where anthropogenic factors affecting hydrological processes are very preponderant.

Because the study area is under irrigation in dry periods of the year, it was necessary to consider irrigation amounts of field and horticultural crops grown in the region. Therefore, during the calibration period, irrigation requirements of the crops were estimated by using universal reference evapotranspiration method of Penman-Monteith. Then, using the crop coefficients of FAO [60], net irrigation requirements of irrigated crops were obtained and used in management files as a model input. For the calibration, the created base model with net irrigation amounts and routine fertilizer rates were saved in crop rotations. The actual irrigation bypass flows were determined through running different simulations by adapting calibrated SWAT parameters given in **Table 7**. Finally, it was determined that 40% of the total diverted irrigation water to the district at any time was directly draining into the drainage system as bypass flow.

3.2. Calibration of nitrogen in drainage water

After calibrating the hydrologic part of the model with a successful performance, nitrate simulation was confidently applied with the appropriate water parameters. All the nitrogen inputs were incorporated in the management files as fertilizer, water, and soil point sources.

Average daily NO_3-N loads (kg day^{-1}) were selected as water quality parameter and calculated based on daily discharge data (m^3 day^{-1}) at L4 gauging station (outlet).

	Calibration period (2009–2012)		Validation period (2013–2014)	
	Observed	Simulated	Observed	Simulated
	Drainage flows (m^3 s^{-1})			
Mean	3.51	2.98	2.71	2.98
Median	3.48	2.69	2.74	2.62
Mode	1.04	1.86	1.23	2.35
Standard dev.	2.02	1.93	1.45	1.46
Kurtosis	5.54	10.43	−1.01	−0.62
Skewness	1.70	1.96	0.33	0.38
Minimum	0.73	0.09	0.58	0.31
Maximum	14.06	18.65	6.05	7.27
CV%	58	65	54	49
	Nitrogen in drainage (kg d^{-1})			
Mean	1810.2	443.1	938.7	552.6
Median	1376.4	243.9	774.0	457.6
Mode	1775.5	229.3	–	1135.0
Standard dev.	1659.0	639.9	608.9	533.2
Kurtosis	15.8	21.9	5.8	57.8
Skewness	3.4	4.0	2.0	5.8
Minimum	143.4	10.4	64.9	39.7
Maximum	14024.7	6403.0	4826.0	7599.0
CV%	92	144	65	96
Sample size (n)	1461		730	

Table 6. Descriptive statistics of drainage flows and nitrogen loads in drainage for observed and simulated during calibration and validation periods.

Objective function statistics, R^2, NSE, and PBIAS in specific, for nitrogen in drainage were defined as 0.47, −0.63, and 88.1% for the calibration and 0.50, −0.20, and 72.9% for validation, respectively (**Table 5**). As indicated by Moriasi et al. [59], the PBIAS ±70% for N is accepted as a performance criteria.

Average daily NO_3-N loads (kg day^{-1}) of the selected water quality parameter was also calculated based on daily discharge data (m^3 day^{-1}) at L4 gauging station at the outlet of the district. Based on the graphical presentation in **Figures 4** and **5**, overlapping of the both measured and calibrated lines for N cannot be considered as perfect because nature and dynamics of N in the whole system, even though the statistics are reasonably acceptable. Similar underestimation with the data of only 2009 and 2010 was also recorded in the same location [38]. It is important to point out that calibration and validation of the model are sensitive to time periods, instead of using daily data, monthly data were more suitable to modeling purpose of N [61].

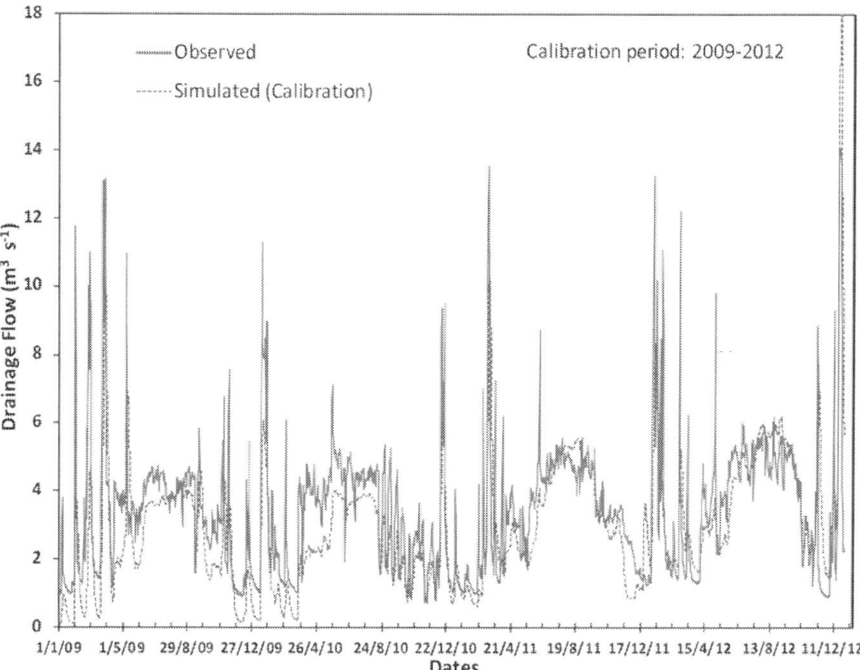

Figure 2. Daily drainage discharge (m³ s⁻¹) calibration for the Akarsu catchment outlet L4.

This basin is not natural instead it is a man-made hydrologically well-defined area in a semiarid Mediterranean region where it is subjected to intensive irrigation and fertilizer applications by anthropogenic activities. Imported N loads by irrigation water, rainfall, and inorganic fertilizer inputs make the calibration and validation difficult and relatively weak. There are three district-specific conditions in the area to be pointed out for nitrogen and nitrogen balance: the canals being open despite high rates of ET, irrigation also taking place outside of irrigation season, and the possible loses of irrigation water to drainage.

In terms of management practices, there are two planting seasons in a year; the crop rotations used in the model include all the planting and harvesting dates. Except for perennial crops, the crop pattern (land use) varies from year to year. The model permits use of only one land use map in HRU delineation; for this reason, rotation calendars were made to be utilized within the model. Farmer behavior and knowledge are diverse, and the use of nitrogen fertilizers and irrigation is intense.

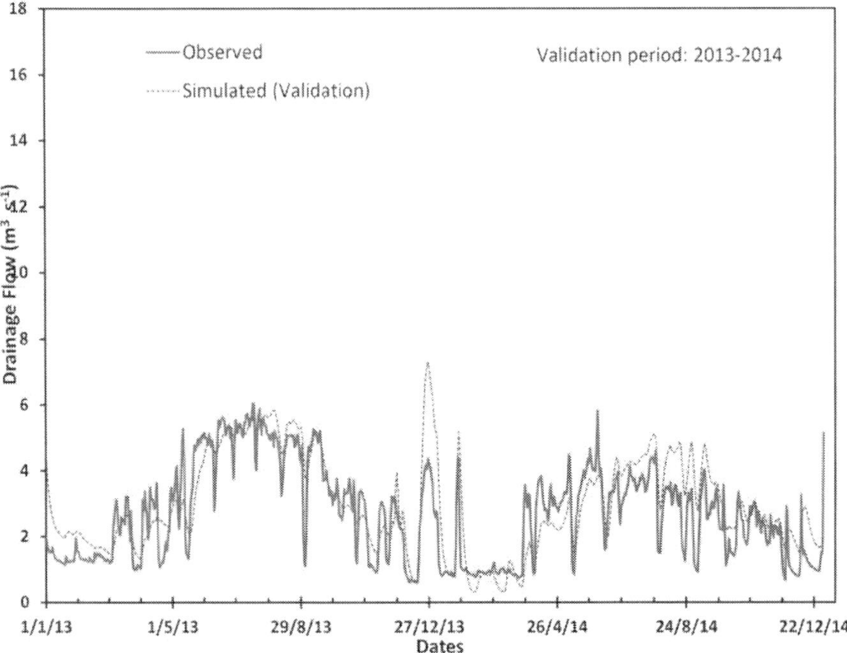

Figure 3. Daily drainage discharge ($m^3\ s^{-1}$) validation for the Akarsu catchment outlet at L4.

Parameter	Default	Range	Calibrated values
CN2	83	35–98	73.9
Alpha_BF	0.048	0–1	0.55
GW_Delay	31	0–500	36.08
Gwqmn	1000	0–5000	4187.5
Surlag	4	1–24	0.42
Esco	0.95	0–1	0.837
Revapmn	750	0–1000	488.75
Ch_K2	0	−0.01 to 500	378.75
Gw_Revap	0.02	0.02–0.2	0.089
Ch_n2	0.014	−0.01 to 0.3	0.266

Table 7. Sensitive hydrologic model parameters for SWAT.

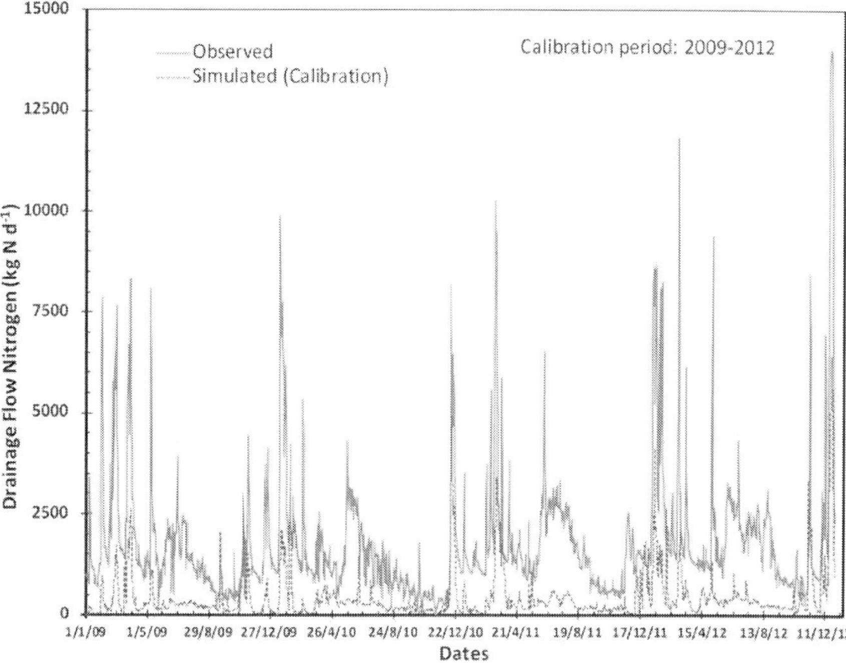

Figure 4. Nitrogen load (kg day^{-1}) at L4 (outlet) calibration and validation period on monthly level.

3.3. Nitrogen balance

Nitrogen calibration was carried out on daily basis. Average daily NO_3-N loads (kg day^{-1}) of selected water quality parameter were calculated based on daily discharge data (m^3 day^{-1}) at L4 gauging station. **Table 5** and **Figures 4** and **5** created for nitrogen did not show a strong relationship between measured and simulated values. One of the main reasons is that for hydrologic reasons inclusion of the two hilly pasture areas (**Figure 1**) into the 9495 ha hydrologically well-defined Akarsu irrigation district by extending the area to 11,308 ha. Therefore, when the actual N inputs were distributed in a larger area the prediction became lower. Also, since the soils are climatically suitable to nitrification, greater amount of nitrogen especially from the inorganic fertilizers may be quickly transformed to nitrate in a very short time period and leached to the drainage [62]. As also discussed by Abbaspour et al. [56], amount of nitrogen fertilizer leached below the root zone, which is 0–90 cm in the study, is under-estimated. In addition, fertilizer application level may be higher than that of the recorded from our three consecutive survey data. Therefore, it may cause higher measured NO_3 concentrations in drainage. Overall, since the irrigated area is under very intensive agricultural management practices including irrigation and very dynamic fertilization, it is quite possible to underestimate the N leaching to the drainage. For example, SWAT model prediction was very successful for calibration (and validation) of rivers accounting the dynamics of nitrate transport [56].

Nitrogen balance variables are given in **Table 8**. The sums of nitrate nitrogen leached from the soil profile in kg NO_3–N(NO3L) and N uptake by plants (NUP) from 2009 to 2014 are reasonably in agreement with the amount of applied nitrogen (N APP). The remaining inputs in the so-called man-made research area are coming from the N content of irrigation water, rainfall, mineralization of soil organic matter, and transforms of N forms into readily available NH_4^- and NO_3. Based on the climatic conditions, amount of rainfall, thus leaching to drainage, and groundwater, varies year to year. For example, in 2013, total rainfall was 349 mm, which was the lowest figure among the other years of the study (ranged 349–951 mm). The reflection of this unusual rainfall was clearly performed in **Figure 5**, which is for the simulation period. **Figure 4** clearly indicates that impacts of rainfall in winter and irrigation applications in

Year	Nitrogen balance variables (kg N ha⁻¹ year⁻¹)		
	N_APP[1*]	NO3L	NUP
2009	329.2	196.8	270.0
2010	368.1	212.8	228.3
2011	310.9	234.7	181.3
2012	368.1	256.3	175.1
2013	329.2	159.2	277.7
2014	368.1	249.3	254.6

[*] N_APP, NO3L, and NUP stand for applied, leached, and taken-up nitrogen at the catchment level.

Table 8. Temporal variability of nitrogen balance by SWAT modeling for the Akarsu region (2009–2014).

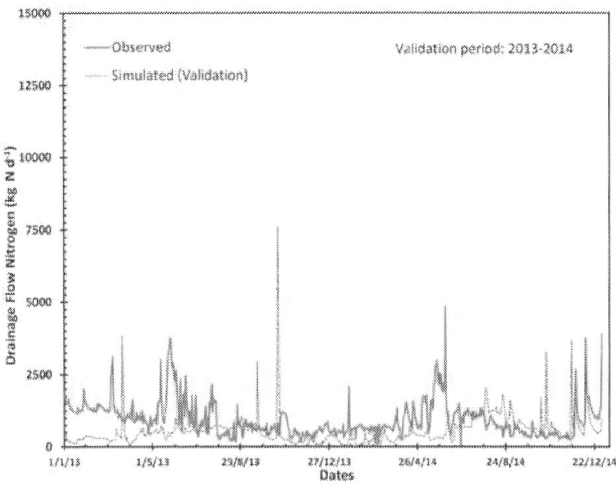

Figure 5. Nitrogen load (kg day⁻¹) at L4 (outlet) calibration and validation period on monthly level.

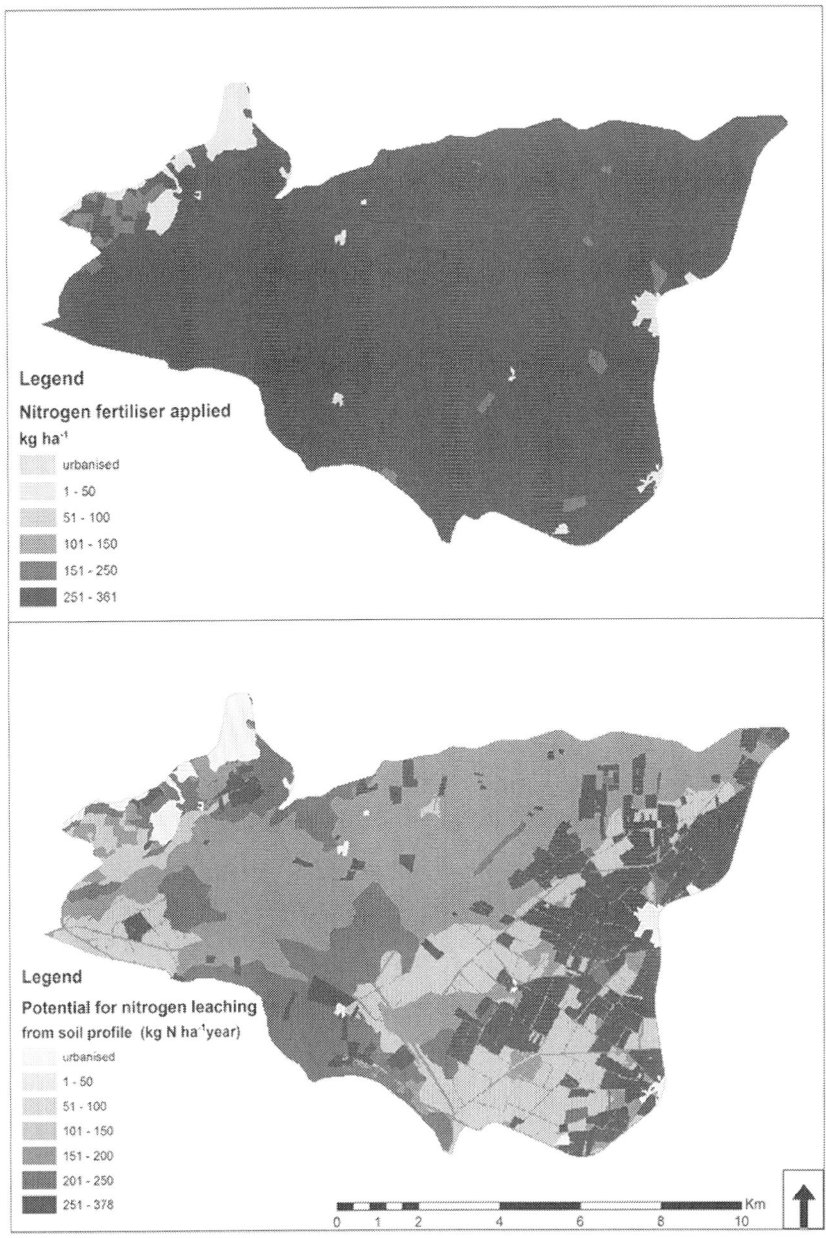

Figure 6. Comparison between average nitrogen fertilizers applied (kg ha^{-1}) and potential for nitrogen leaching (kg ha^{-1}) below the bottom of the soil profile in Akarsu study area in the period between 2009 and 2014.

summer are the most important drivers of the N leaching. Conflicting performance ratings of N calibration seen in **Figures 4** and **5** might be attributed to above mentioned two drivers. In addition, routine fertilizer applications are exceedingly high than the recommended levels, i.e., 380 kg N ha^{-1} is applied to corn while only 240 kg N ha^{-1} is the expert recommendation for corn in the region [63]. This results in high potential for nitrogen leaching (**Figure 6**).

4. Conclusions

Distributed watershed models are known as the very powerful tools both for scenario development and for simulating the effects of watershed dynamics management on soil and water resources. This study was aimed to improve understanding of (a) the effects of bypass flows due to irrigation on the calibration of the SWAT model, (b) irrigation return flow and/or drainage generating processes, and (c) N leaching dynamics with simulation of agricultural land management (fertilization, irrigation, and plant species) under the Mediterranean climate conditions. To this aim, the research was conducted in an irrigated agricultural catchment of Akarsu irrigation district. Visual examination of data used in modeling has indicated that drainage flows and nitrogen-leaching processes are not governed by the natural processes in the catchment but mostly by anthropogenic activities.

Model calibration and validation were carried out to determine the most sensitive and appropriate parameter values for the drainage flows generated by the agricultural catchment. Although daily flow data were used in modeling, quantitative model performance evaluation statistics (R^2, NSE, and PBIAS) revealed clearly that the calibrated SWAT model produced rather satisfactory simulation results at the catchment outlet in wet, average, and dry years. In the irrigated catchment, irrigation water losses directly from irrigation channels to drainage ditches, i.e., bypass flows, has direct influence on calibrating hydrologic part of the SWAT model. In this case, the SWAT model findings helped us to highlight that almost 40% of diverted irrigation water has been recklessly squandered in the irrigation scheme. It is almost impossible to quantify bypass flow magnitudes in such irrigation system without using any modeling tools.

Furthermore, modeling exercises showed that the SWAT model run results were sensitive on crop rotations due to the fact that runoff by precipitation and irrigation applications are affected by the land use and land cover types. Contrary to the expectations, daily nitrate modeling results were not able to yield rather satisfactory model performance statistics, indicating that simulated daily nitrogen loads data in drainage were not sufficiently matched with the measured ones. Visual evaluation of measured and simulated nitrogen graphs showed implicit signals that measured nitrogen data might involve some inherent uncertainties and irregularities at the catchment level. Based on the findings, as highlighted in the literature [59], we concluded that model performance can be improved to some extent by increasing the time step from daily to monthly or yearly level for the nitrogen data with involves inherent uncertainties. These uncertainties should be considered when calibrating, validating, and evaluating watershed models because of differences in inherent uncertainty between measured flow, sediment, and nutrient data.

Improved fertilization practices are not only necessary for farmer's economy but also crucial for preserving soil and water resources. In recent years, especial soil analysis in the study area became a very useful tool for fertilizer subsidizes and expert recommendations. However, recommendations can not only be related to and designed by the soil analysis, it should be comprehensively evaluated in a broader environment. At this stage, a suitable model performance enables modeling more sensitive management practices like the fertilizer rates.

Acknowledgements

This work is a results of cooperation between researchers included in Slovenia-Turkey bilateral project financed by Slovenian Research Agency (ARRS) and the Scientific and Technological Research Council of Turkey (TUBITAK, Project No: 213O057).

Author details

Ebru Karnez[1], Hande Sagir[2], Matjaž Gavan[3]*, Muhammed Said Golpinar[4], Mahmut Cetin[4], Mehmet Ali Akgul[5], Hayriye Ibrikci[2] and Marina Pintar[3]

*Address all correspondence to: matjaz.glavan@bf.uni-lj.si

1 Faculty of Agriculture, Cukurova University, Adana, Turkey

2 Soil Science and Plant Nutrition Department, Cukurova University, Adana, Turkey

3 Department for Agronomy, Biotechnical Faculty, University of Ljubljana, Ljubljana, Slovenia

4 Department of Irrigation Engineering, Faculty of Agriculture, Cukurova University, Adana, Turkey

5 The Sixth Regional Directorate of State Hydraulic Works (DSI), Adana, Turkey

References

[1] Wheater HS, Mathias SA, Li X, editors. Groundwater modelling in arid and semi-arid areas. International Hydrology Series, Cambridge University Press; New York, USA; 2010.

[2] Office of Environment and Heritage, 2015. http://www.environment.nsw.gov.au/water/waterqual.htm

[3] National Marine Sanctuaries, 2011. floridakeys.noaa.gov/ocean/waterquality.html

[4]   FAO Corporate Document Repository, 1985. http://www.fao.org/docrep/003/t0234e/T0234E01.htm#ch1.3

[5]   Koc C. A study on the pollution and water quality modeling of the river Buyuk Menderes, Turkey. Clean-Soil, Air, Water. 2010; 38 (12): 1169–1176.

[6]   Volk M, Liersch S, Schmidt G. Towards the implementation of the European Water Framework Directive? Lessons learned from water quality simulations in an agricultural watershed. Land Use Policy. 2009; 26: 580–588.

[7]   Du B, Saleh A, Jaynes DB, Arnold JG. Evaluation of swat in simulating nitrate nitrogen and atrazine fates in a watershed with tiles and potholes. American Society of Agricultural and Biological Engineers. 2006; 49 (4): 949–959.

[8]   O'Shea L, Wade A. Controlling nitrate pollution: an integrated approach. Land Use Policy. 2009; 26:799–808.

[9]   Kersebaum KC, Steidl J, Bauer O, Piorr H-P. Modelling scenarios to assess the effects of different agricultural management and land use options to reduce diffuse nitrogen pollution into the river Elbe. Physics and Chemistry of the Earth, Parts A/B/C. 2003; 28: 537–545. DOI: 10.1016/S1474-7065(03)00090-1

[10]   Croll BT, Hayes CR. Nitrate and water supplies in the United Kingdom. Environmental Pollution. 1988; 50: 163–187.

[11]   Marschner H.Marschner's Mineral Nutrition of Higher Plants (Third Edition Book). Academic Press, Elsevier; Oxford, UK; 2002. ISBN: 978-0-12-384905-2.

[12]   Zhu ZL, Chen DL. Nitrogen fertilizer use in China-Contributions to food production, impacts on the environment and best management strategies. Nutrient Cycling in Agroecosystems. 2002; 63 (2): 117–127.

[13]   Raun WR, Schepers JS. Nitrogen management for improved use efficiency. In JS Schepers and WR Raun, editors. Nitrogen in agricultural systems. American Society of Agronomy Monograph no. 49. American Society of Agronomy, Madison, WI, USA; 2008. pp. 675–695.

[14]   Smil V. Nitrogen cycle and world food production. World Agriculture. 2011; 2: 9–13.

[15]   Kladivko EJ, Scoyoc GEV, Monke EJ, Oates KM, Pask W. Pesticide and nutrient movement into subsurface tile drains on a silt loam soil in Indiana. Journal of Environmental Quality. 1991; 20: 264–270.

[16]   Ng HYF, Tan CS, Drury CF, Gaynor JD. Controlled drainage and subirrigation influences tile nitrate loss and corn yields in a sandy loam soil in Southwestern Ontario. Agriculture Ecosystems and Environment. 2002; 90: 81–88. DOI: 10.1016//S0167-8809(01)00172-4

[17]   Meisinger JJ, Schepers JS, Raun WR. Crop nitrogen requirement and fertilization. In JS Schepers, et al. editors. Nitrogen in agricultural systems. Agronomy Monograph. 49. ASA, CSSA, and SSSA, Madison, WI; 2008. pp. 563–612.

[18] Kopinski J, Tujaka A, Igras J. Nitrogen and phosphorus budgets in Poland as a tool for sustainable nutrients management. Acta Agriculturae Slovenica. 2006; 87 (1): 173–181.

[19] Ozbek FS, Leip A. Estimating the gross nitrogen budget under soil nitrogen stock changes: a case study for Turkey. Agriculture Ecosystems & Environment. 2015; 205: 48–56.

[20] EC. Development of agri-environmental indicators for monitoring the integration of environmental concerns into the common agricultural policy. Communication from the Commission to the Council and the European Parliament. COM, 508 final. Commission of the European Communities, Brussels. 2006. Available from: http://eur-lex.europa.eu/LexUriServ/LexUriServ.do?uri=COM2006.0508:FIN:EN:PDF

[21] Cetin M. Nitrogen input-output mass balance study results in the selected Mediterranean countries. International Seminar: diagnosis and control of diffuse pollution in mediterranean irrigated agrıculture, Zaragoza, Spain, 20–21 October 2010 (http://www.iamz.ciheam.org/qualiwater/ZaragozaSeminar/INDEX%20OF%20PRESENTATIONS)

[22] Karnez E, Evaluation of nitrogen budget inputs and outputs in wheat and corn growing areas in lower Seyhan plain (Ph.D. Dissertation). Adana, Turkey: Çukurova University (in Turkish); 2010.

[23] Ibrikci H, Cetin M, Karnez E, Topcu S, Kirda C, Ryan J, Oguz H, Oztekin E, Dingil M. Plant contribution to the nitrogen budget under irrigation in the Cukurova region of Southern Turkey. In: World Soils Congress Australia; 1–6 August 2010.

[24] Lionello P, Malanotte-Rizzoli P, Boscolo R, Alpert P, Artale V, Li L, Luterbacher J, May W, Trigo R, Tsimplis M, Ulbrich U, Xoplaki E. The Mediterranean climate: an overview of the main characteristics and issues. Developments in Earth and Environmental Sciences. 2006; 4: 1–26.

[25] Beklioğlu M, Çakıroğlu A, Tavşanoğlu N, Levi E, Özen A, Coppens J, Balaman SB, Jeppesen E. Impact of climate and nutrient enrichment on ecology of shallow lakes of Turkey using multiple approaches. In: Ecology and evolutionary biology symposium (EEBST 2015); Ankara, Turkey, 05–07 August 2015.

[26] FAO-ISRIC. 1990. Guidelines for profile description. 3rd Edition. Rome.

[27] Ramblinrobert. Irrigating with rainwater in a Mediterranean climate. Urban Agroecology. 2012. Available from: urbanagroecology.org/2012/12/05/irrigating-with-rainwater-in-a-mediterranean-climate

[28] Cavero J, Barros R, Sellam F, Topcu S, Isidoro D, Lounis A, Ibrikci H, Cetin M, Williams JR, Aragüés R. APEX simulation of best irrigation and n management strategies for off-site n pollution control in three Mediterranean irrigated watersheds. Agricultural Water Management. 2012; 103: 88–99.

[29] Hagin J, Lowengart A. Fertigation for minimizing environmental pollution by fertilizers. Fertilizer research. 1996; 43 (1): 5–7.

[30] Borah DK. Water resources models. In: K.C. Ting, D.H. Fleisher, and L.F. Rodriguez, editors. System analysis and modeling in food and agriculture, Encyclopedia of Life Support Systems; EOLSS Publisher, UNESCO; Oxford, UK; 2009.

[31] Arnold JG, Srinivasan R, Muttiah RS, Williams JR. Large area hydrologic modeling and assessment: Part I. Model development. Journal of American Water Resources Association. 1998; 34(1): 73–89.

[32] Panagopoulos Y, Makropoulos C, Mimikou M. A multi-objective decision support tool for rural basin management. In: http://www.iemss.org/sites/iemss2012/proceedings.html. International Environmental Modelling and Software Society (iEMSs). International Congress on Environmental Modelling and Software. Managing Resources of a Limited Planet, Pathways and Visions under Uncertainty, Sixth Biennial Meeting; Leipzig, Germany; 2012.

[33] Ertürk A, Ekdal A, Gürel M, Karakaya N, Guzel C, Gönenç E. Evaluating the impact of climate change on groundwater resources in a small Mediterranean watershed. Science of the Total Environment. 2014; 499: 437–447. DOI: 10.1016/j.scitotenv.2014.07.001

[34] Savé R, Herralde FD, Aranda X, Pla E, Pascual D, Funes I, Biel C. Potential changes in irrigation requirements and phenology of maize, apple trees and alfalfa under global change conditions in Fluvià watershed during XXIst century: results from a modeling approximation to watershed-level water balance. Agricultural Water Management. 2012; 114: 78–87.

[35] Çetin M, İbrikçi H, Berberoğlu, S, Gültekin U, Karnez E, Selek B. Analysis and optimization of irrigation efficiencies in order to reduce salinization impacts in intensively used agricultural landscapes of the semiarid Mediterranean Turkey (MedSalin). Project Final Report, Funded by TÜBİTAK-BMBF, Program Code: 2527, Project #: TUBITAK 108O582. 2012; 186.

[36] Kaman, H. Yield response of maize genotypes to conventional deficit irrigation and partial root drying (Ph.D. Dissertation). Adana, Turkey: Çukurova University (in Turkish); 2007.

[37] Dinc U, Sarı M, Senol S, Kapur S, Sayın M, Derici MR, Cavusgil V, Gök M, Aydın M, Ekinci H, Ağca N, Schlinchting E. Soils of Cukurova Region, Cukurova Üniversitesi Ziraat Fakültesi Yardımcı Ders Kitabı, No: 26, 2. Baskı. C¸ukurova University Publishing, Adana, Turkey (in Turkish); 1988.

[38] Akgul MA. Modelling water and nitrate budget in left bank irrigation of lower Seyhan plain. Adana, Turkey: Cukurova University, Institute of Natural and Applied Sciences, (in Turkish with English abstract); 2015.

[39] Arnold JG, Kiniry JR, Srinivasan R, Williams JR, Haney EB, Neitsch SL. SWAT and Water Assessment Tool, Input/Output Documentation, Chapter 20 SWAT Input Data: MGT, 243 p. http://swat.tamu.edu/documentation/2012-io. 2012

[40] Tomer MD, Dosskey MG, Burkart MR, James DE, Helmers MJ, Eisenhauer DE. Methods to prioritize placement of riparian buffers for improved water quality. Agroforestry Systems. 2009; 75: 17–25. DOI: 10.1007/s10457-008-9134-5

[41] Gassman PW, Reyes MR, Green CH, Arnold JG. The soil and water assessment tool: historical development, applications, and future directions. Transactions of the ASABE. 2007; 50 (4): 1211–1250.

[42] Jain KS, Tyagi J, Singh V. Simulation of runoff and sediment yield for a Himalayan watershed using SWAT model. Journal of Water Resource and Protection. 2010; 2: 267–281. DOI:10.4236/jwarp.2010.23031

[43] Maharjan M, Babel MS, Maskey S. Reducing the basin vulnerability by land management practices under past and future climate: a case study of the Nam Ou River Basin, Lao PDR. Hydrology and Earth System Sciences Discuss. 2014; 11: 9863–9905. DOI:10.5194/hessd-11-9863-2014.

[44] Li L, Jiang D, Hou X, Li J. Simulated runoff responses to land use in the middle and upstream reaches of Taoerhe River basin, Northeast China, in wet, average and dry years. Hydrological Processes. 2013; 27: 3484–3494. DOI: 10.1002/hyp.9481.

[45] Neitsch SL, Arnold JG, Kiniry JR, Williams JR. SWAT and Water Assessment Tool Theoretical Documentation Version. http://swat.tamu.edu/documentation 2009.

[46] Clayton S, Muleta M. Approaches to control nitrate pollution in the San Joaquin watershed. In: World Environmental and Water Resources Congress Proceedings. 2012, 3: 2232–2235. DOI: 10.1061/9780784412312.224

[47] Akhavan S, Abedi-Koupai J, Mousavi SF, Afyuni M, Eslamian SS, Abbaspour KC. Application of SWAT model to investigate nitrate leaching in Hamadan–Bahar Watershed, Iran. Agriculture, Ecosystems and Environment. 2010; 139: 675–688.

[48] Santhi C, Arnold JG, Williams JR, Dugas WA, Srinivasan R, Hauck LM. Validation of the SWAT model on a large river basin with point and nonpoint sources. Journal of American Water Resources Association. 2001; 37: 1169–1188.

[49] Moriasi DN, Gowda PH, Arnold JG, Mulla DJ, Ale S, Steiner JL. Modeling the impact of nitrogen fertilizer application and tile drain configuration on nitrate leaching using SWAT. Agriculture Water Management. 2013; 130: 36–43. DOI: 10.1016/j.agwat.2013.08.003

[50] Pintar M, Kompare B, Uršič M, Bremec U, Gabrijelčič E, Sluga G, Globevnik L. The impact of land use on nutrient concentration in upper streams of waters in Slovenia. Integrated Watershed Management: Perspectives and Problems. Springer, Netherlands; 2010. pp. 190–199. DOI: 10.1007/978-90-481-3769-5_16

[51] Neitsch SL, Arnold J, Kiniry JR, Williams JR. Soil and Water Assessment Tool (SWAT)–Theoretical Documentation, USDA Agricultural Research Service, Temple, Texas, 2001.

[52] Cau P, Paniconi C. Assessment of alternative land management practices using hydrological simulation and a decision support tool: Arborea agricultural region, Sardinia. Hydrology and Earth System Sciences Discussion. 2007; 11: 1811–1823.

[53] Pongpetch N, Suwanwaree P, Yossapol C, Dasananda S, Kongjun T. Using SWAT to assess the critical areas and nonpoint source pollution reduction best management practices in Lam Takong River Basin, Thailand. Environment Asia 8.1. 2015; 41–52. DOI: 10.14456/ea.2015.6

[54] Ullrich A, Volk M. Application of the soil and water assessment tool (SWAT) to predict the impact of alternative management practices on water quality and quantity. Agricultural Water Management 2009; 96 (8): 1207–1217.

[55] Abraham LZ, Roehrig J, Chekol DA. Calibration and validation of SWAT hydrologic model for meki watershed, Ethiopia. In: Conference on International Agricultural Research for Development, Tropentag 2007 University of Kassel-Witzenhausen and University of Göttingen; 9–11 October 2007.

[56] Abbaspour KC, Yang J, Maximov I, Siber R, Bogner K, Mieleitner J, Zobrist J, Srinivasan R. Spatially-distributed modelling of hydrology and water quality in the prealpine/alpine Thur watershed using SWAT. Journal of Hydrology. 2007; 333: 413–430.

[57] Glavan M, Pintar M, Urbanc J. Spatial variation of crop rotations and their impacts on provisioning ecosystem services on the river Drava alluvial plain. Sustainability of Water Quality and Ecology. 2015; 5: 31–48.

[58] Benaman J, Shoemaker CA, Haith DA. Calibration and validation of Soil and water assessment tool on an agricultural watershed in upstate New York. Journal of Hydrology Engineering. 2005; 10 (5): 363–374.

[59] Moriasi DN, Arnold JG, Van Liew MW, Bingner RL, Harmel RD, Veith TL. Model evaluation guidelines for systematic quantification of accuracy in watershed simulations, Transactions of the ASABE. 2007; 50 (3): 885–900.

[60] Allen RG, Pereira LS, Raes D, Smith M. Crop evapotranspiration: guidelines for computing crop water requirements. Rome: FAO, (Irrigation and Drainage Paper, 56); 1998. 300 p.

[61] Narasimhan B, Srinivasan R, Arnold JG, Di Luzio M. Estimation of long-term soil moisture using a distributed parameter hydrologic model and verification using remotely sensed data. Transactions of the ASAE. 2005; 48(3): 1101–1113.

[62] Ibrikci H, Cetin M, Karnez E, Wolfgang AF, Tilkici B, Bulbul Y, Ryan J. Irrigation-induced nitrate losses assessed in a Mediterranean irrigation district. Agricultural Water Management. 2015; 148: 223–231.

[63] Ulger A, Ibrikci H, Cakir B, Guzel N. Influence of nitrogen rates and row spacing on corn yield, protein content, and other plant parameters. Journal of Plant Nutrition. 1997; 12: 1697–1709.

Chapter 8

Identification of Seawater Quality by Multivariate Statistical Analysis in Xisha Islands, South China Sea

Mei-Lin Wu, Jun-De Dong and You-Shao Wang

Additional information is available at the end of the chapter

http://dx.doi.org/10.5772/66227

Abstract

Xisha waters are considered to be in pristine condition, while facing the fast increasing stress under anthropogenic activities. Water quality around Yongxing Island (YX) has been measured in May, 2012. The results show that the water quality is of the first class standards as compared to the water quality of China, with insignificant difference among the monitoring stations. Robust principal component analysis (PCA) was used to identify the spatial pattern of water quality. YX is characterized by high DO, salinity, and Chl-a with low nutrients, indicating phytoplankton photosynthesis is stronger in YX island waters than the rest of the study areas. Beidao (BD) is characterized by high NH_4-N and COD, and low pH, implying that these areas may have higher organic matter decomposition than rest of the areas. The water quality monitoring stations should cover spatially and temporally around Xisha waters for protecting the marine environment.

Keywords: water quality, Xisha, principal component analysis, marine environment protection, nutrients

1. Introduction

Isolated oceanic islands with limited area are surrounded by seawater, without sitting on the continental shelves. They may be a natural experimental area to examine the ecological changes and impacts that accompany human arrival [1]. The island's ecosystems have pristine environmental conditions, but the interference of humans poses serious threats to the delicate and vulnerable ecological processes of the island [2]. Anthropogenic loading of

polluted materials from urbanized and industrialized land to the isolated islands cannot be directly performed, but it may be brought to these open waters by ocean currents and atmospheric deposits. Human activities in the island including tourism, overfishing, and aquaculture farms have determined influence on the water quality around the islands. To some extent, natural progress including hydrodynamics, typhoons, and other weather conditions also have significant influence on the water quality.

Xisha Islands located in the northern South China Sea (nSCS), consist of more than 20 islands and atolls. The Chinese government established Sansha city on Yongxing Island of the Xisha Islands in 2012. Sansha is the smallest prefecture-level city, by both population and land area in China. The residents in Sansha are about 1443, while the floating population was up to 2000 by the end of December, 2013 (http://www.sansha.gov.cn/). The total land area of Sansha is less than 13 km^2. Because Sansha government pays attention to environmental protection, about 2.92 million dollars will be spend to build desalination systems and grow trees on Xisha in the hope of turning the island into a new oasis (http://www.news.xinhuanet.com). In view of this, it is necessary to conduct the environmental and ecological monitoring in Xisha waters, in order to suggest the better management activities to protect environment around the islands. The ecological conditions in the islands were intensively attracted worldwide. The decline in number of seabirds and the remaining birds are caused by increasing human disturbance [3]. The coverage of living hermatypic corals have sharply reduced, while the dead coral coverage had sharply increased from 2005 to 2009 [4]. Coral species dramatically decreased in the past several decades in Yongxing Island [5]. Fish resources are abundant in Xisha waters, which can be exploited to a certain extent [6]. Even though the coral reefs of Xisha islands are considered to be the healthiest and most resilient in the northern South China Sea, it is facing living environmental problems including coral bleaching events, diseases and natural disasters, especially due to anthropogenic activities. Because coral reefs thrive in oligotrophic conditions, pristine water quality is a crucial contributor for the growth of coral reefs. However, water quality characteristics in these islands have not been reported so far. Consequently, it is lack of related information on understanding variation influence on physicochemical properties and phytoplankton under the human activities.

The purpose of this study is to present data on the water quality in Xisha waters. The spatial pattern of water quality was assessed by multivariate statistical analysis. Meanwhile, the key driving factors that control the water quality have been identified. From all these information not only do the people recognize the water quality status, but also give suggestion to establish an effective way for further environmental and ecological assessment of Xisha waters.

2. Materials and methods

This study was conducted in May 2012, around Yongxing Islands (YX), connected with the Beidao (BD), Shidao (SD) and Zhaoshu (ZS) islands with rich coral reef community.

Twelve sampling stations (**Table 1**) were selected along the coast of the above-mentioned studyarea (**Figure 1**). Physical parameters such as salinity and pH were measured *in situ* using a quanta water quality monitoring system (Hydrolab Corporation, USA). Discrete water samples were taken at 0.5 m below the surface, and 1 m above the sea bottom using 5-l GO FLO bottles. The surface water was only taken when the water depth was less than 5 m. Water from the surface and bottom layers were taken when the depth was more than 5 m. Water quality parameters including nutrients (nitrite, nitrate, ammonia, phosphate, and total phosphate), chemical oxygen demand (COD), and chlorophyll (Chl-a) were estimated using standard methods from "The specialties for oceanography survey" (GB17378.4-1998, GB17378.4-2007, China). Dissolved oxygen (DO) (mg·L^{-1}) was determined using Winkler titrations.

Stations	Sampling depth (m)	Longitude	Latitude
ZS1	1/8	112.2609	16.9722
ZS2	1/10	112.2641	16.9694
ZS3	1/6	112.2719	16.9716
BD1	1/8	112.3003	16.9659
BD2	1/6	112.3101	16.9601
BD3	1/7	112.3149	16.9571
SD1	1/8	112.3409	16.8493
SD2	1/4	112.3465	16.8494
SD3	1/4.6	112.3515	16.8473
YX1	1/4	112.3592	16.8364
YX2	1/6.5	112.3560	16.8335
YX3	1/4.5	112.3503	16.8264

Symbol "1/number" in the column of sampling depth displays the sampling layer. For example, string "1/8" exhibits sampling was conducted in surface layer-1m below the surface and bottom layer-8 m, respectively.

Table 1. Sampling stations and their locations.

Descriptive statistics and multivariate statistical analysis were carried out for coastal water samples using MATLAB 2014. Since the ratio of samples to the variables is 2:1, the classical principal component analysis might fail. Robust principal component analysis (RPCA) is still effective even if there are a few anomalous observations and even observation samples are less than number of variables [7–9]. Thus, RPCA is employed to understand the spatial pattern of water quality.

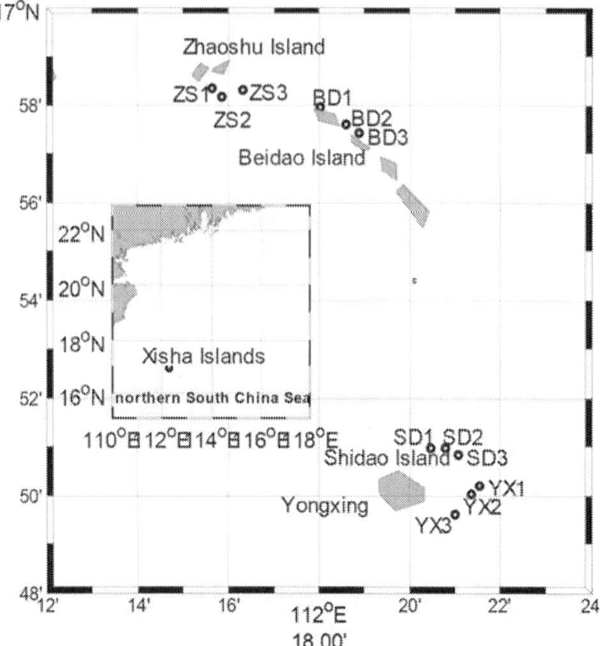

Figure 1. Monitoring stations in the studying area. Four sampling areas are Yongxing south (YX), Shidao (SD), Beidao (BD) and Zhaoshu (ZS), respectively.

3. Results and discussion

Descriptive statistics of nine water quality parameters is shown in **Table 2**. The NO_2-N level at most of sampling stations is below the detected limit (data not shown). pH ranged from 8.00 to 8.06, with the narrow range. Chl-a has a relatively low value in the range from 0.02 to 0.06 μg L^{-1}. COD varied from 0.16 to 0.31 mg L^{-1}, with the mean value of 0.24 mg L^{-1}. Nutrients display a big change with higher standard deviation than pH and COD. According to the seawater quality standard of China (GB 3097–1997), all these water quality (pH, DO, COD, and nutrients) ranges from the first class quality standard scale. Thus, Xisha waters belong to the first class water quality, which is namely, the pristine water.

	pH	salinity	DO (mg L^{-1})	COD	PO$_4$-P (μmol L^{-1})	NO$_3$-N	NH$_4$-N	Chl-a (μg L^{-1})	TP (μmol L^{-1})
Mean	8.04	33.35	6.61	0.24	0.07	0.68	0.97	0.03	0.20
Min	8.00	32.99	6.02	0.16	0.03	0.36	0.43	0.02	0.13
Max	8.06	33.50	7.81	0.31	0.29	2.25	1.79	0.06	0.45
Std	0.02	0.12	0.45	0.04	0.06	0.47	0.39	0.01	0.08

Table 2. Descriptive statistics of nine water quality parameters.

In this study, different water quality parameters display different spatial characteristics (**Figure 2**). From the graph, it is observed that pH and Chl-a are higher in YX than the rest of the study stations. Salinity, COD, NH_4-N, and TP are maximum in SD. The maximum of NO_3-N is observed in ZS, with the minimum of DO also in ZS. However, results of one-way analysis of variance show the insignificant difference among the four sampling areas, suggesting the water quality mainly affected by the natural progress, with the least disturbed anthropogenic activities.

pH is significantly and positively correlated with DO ($r = 0.71$, p-value $= 0.00$). It is possible that in an environment with low levels of organic matter and low levels of respiration, any hydrogen ions released are being absorbed by the alkalinity of the surrounding seawater [10]. COD is lower than the threshold limit of the first class water quality standard and is insignificantly related with other water quality parameters, implying the low organic matter in this environment. On the other hand, the chemoautotrophic process of nitrification is also associated with DO and pH. The process uses up oxygen and also releases hydrogen ions which can cause a fall in pH. However, NO_3-N is insignificantly related with NH_4-N and DO. It implies that nitrification is not a main controlling factor on dissolved nitrogen for transformation. That is to say, other physical-biochemical processes play an important role in dissolved nitrogen cycle. NH_4-N is significantly negatively related with Chl-a ($r = -0.50$, p –value $= 0.02$), while NO_3-N is insignificantly related with Chl-a. NH_4-N may be the preferred N source for phytoplankton growth in the study area. NH_4-N is assimilated primarily by phytoplankton in Pearl River Estuary, nSCS [11].

Robust principal component analysis rendered the first three significant factors (eigenvalue >1.0) that explained 67.02% of the total variance of data set. The results of the RPCA were visualized in the form of ordination diagrams based on PC1 and PC2. The parameter lines were obtained from the factor loadings of the original variable (**Figure 3a**). The closer the two-parameter lines lie together, the stronger the mutual positive correlation is [12]. pH and DO have the highest positive correlation coefficients, so the two lines are very close with acute angle between them. Similarly, an obtuse angle between two lines represents a negative correlation, etc., NH_4-N and Chl-a. The lengths of the parameters represent the relative explanatory power of sampling stations within the ordination in PC1 and PC2 (**Figure 3b**). In the fourth quadrant, sampling stations in YX are characterized by high DO, salinity and Chl-a, and low nutrients, as these parameter lines are located in this quadrant. It suggests that this area may have a high photosynthesis rate and high primary production. According to Redfield ratio, the average ocean photosynthesis and aerobic respiration can be expressed as follows:

$$106\,CO_2 + 16\,NH_3 + H_3\,PO_4 + 122\,H_2\,O \;\rightarrow\; \left(CH_2O\right)_{106}(NH)_{16}(H_3\,PO_4) + 106\,O_2 \qquad (1)$$

Due to photosynthetic activities, nutrients (NH_4-N and PO_4-P) are quickly taken up by the algae, which result in a shift of the equilibrium [Eq. (1)]. That is to say, phytoplankton photosynthesis is stronger in YX island waters than rest of the areas, suggesting phytoplankton consumes more nutrients from the water, and then produces more oxygen. Photosynthetic organisms release O_2 and assimilate nutrients, thereby community respiration attenuating the pH and DO decline. It suggests that photosynthesis keeps balance with respiration, therefore leaving the system without extra organic matter in these areas. The loading of COD is negative in PC1, indicating less organic matter in these areas than rest of the areas (**Figure 3**).

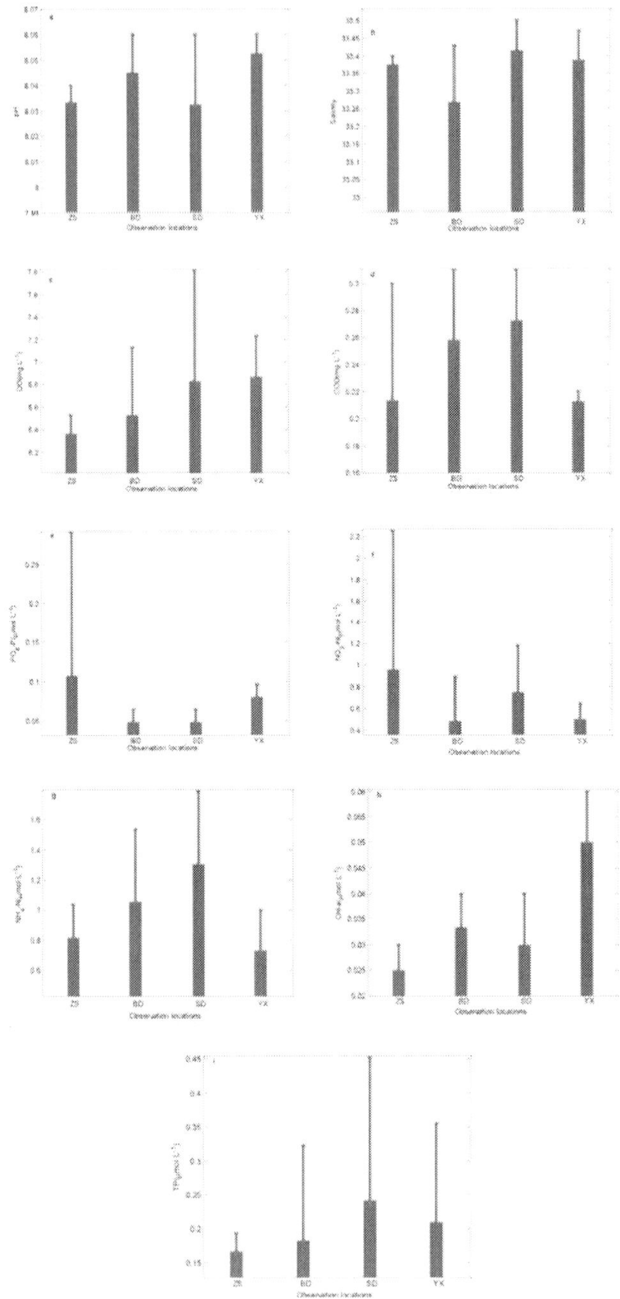

Figure 2. Spatial distributions of water quality, (a) pH, (b) salinity, (c) DO, (d) COD, (e) phosphate, (f) nitrate, (g) ammonia, (h) chlorophyll and (i) total phosphate. ZS, BD, SD, and YX denote Zhaoshu, Beidao, Shidao, and Yongxing, respectively.

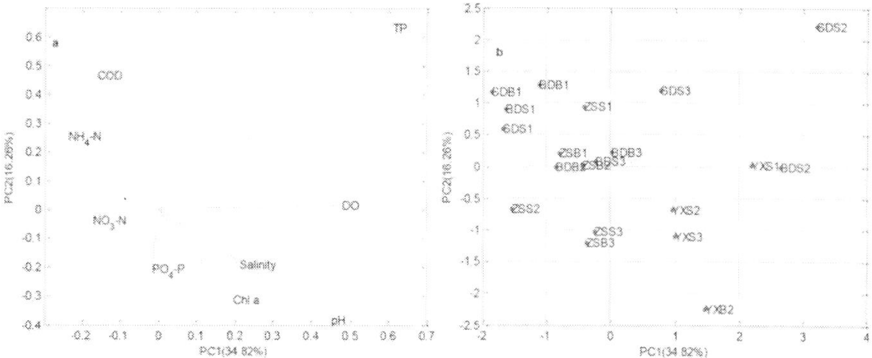

Figure 3. Results of robust principal component analysis: (a) loadings of water quality parameters and (b) scores of monitoring stations. The first two letters denote the monitoring area. "S" and "B" in the third letter denote the surface and bottom layers, respectively. The number denotes the monitoring station.

On the other hand, sampling stations in ZS1, SD1 and most of sampling stations in BD are characterized by high NH_4-N and COD, and low pH (**Figure 3**). From this phenomenon, it implies that these areas may have higher organic matter decomposition than rest of the areas. Algal debris and other organic matters can be converted into carbon dioxide and NH_4-N [Eqs. (2) and (3)] under biochemical oxidation, resulting in the increase of NH_4-N and a decrease in pH.

$$\left(CH_2O\right)_{106}(NH)_{16}(H_3PO_4)+106\,O_2 \;\rightarrow\; 106\,CO_2 + 16\,NH_3 + H_3PO_4 + 122\,H_2O \qquad (2)$$

$$NH_3 + H^+ \;\rightarrow\; NH_4^+ \qquad (3)$$

NH_4-N in these stations is insignificantly related with NO_3-N ($r = -0.31$, p-value > 0.10), while ammonia concentration is higher than NO_3-N except in ZSB1, ZSS2, and SDS1. NH_4-N production has excess ammonia reduction during nitrification process, resulting in that nitrification may be secondary biochemical progress. This is in disagreement with the study [13]. In this study, the presence of nitrate is mainly due to processes such as nitrification. Water quality may be associated with coral community in Xisha waters. Different dominant coral species reside in the different areas around Yongxing Island, with different species index and cover [5].

As the abovementioned, Xisha waters have spatial variations due to different ecological characteristics. Nutrient levels are below the threshold of the first class water quality of China, Xisha waters are considered in pristine environment so far. However, Sansha city on central Yongxing Island of the Xisha islands was built up by China in 2012. With the development of Sansha, more and more infrastructure has been built in Xisha islands. Special marine features have a great deal of potential for attracting various types of tourists. The elevated nutrient level may promote negative responses such as an increase in bleaching susceptibility of coral community [14]. Consequently, Xisha islands is intensively facing the stress of anthropogenic activities.

4. Conclusion

In this study, only summer water samples were taken at the several limited islands around Yongxing Island. Results from multivariate statistical analysis shows that Xisha waters are considered to be in pristine condition, while facing the increasing stress under anthropogenic activities. Thus, in the next environment and ecological monitoring plan of Xisha waters, more islands around Yongxing will be integrated into one unit. Xisha Deep Sea Marine Environment Observation Station, South China Sea Institute of Oceanology, Chinese Academy of Sciences have been established in 2007, which is a good platform to conduct the ecosystem observation focusing on Xisha waters. This organization will help us to obtain the information on the variability where water quality will be easily polluted. The present collected data can present a baseline for future studies in Xisha waters, and even in other isolated island waters.

Acknowledgements

This research was supported by the National Natural Science Foundation of China (Nos. 31270528, 41430966 and 41206082), Natural Science Foundation of Guangdong (No. S2013020012823), Scientific Research Project of Guangzhou (Nos. 201504010005 and 201504010006), the project of Guangdong Provincial Department of Science and Technology (No. 2012A032100004) and the projects of knowledge innovation program of State Key Laboratory of Tropical Oceanography (No. LTOZZ1604).

Author details

Mei-Lin Wu[1,2,4,*], Jun-De Dong[1,2,3] and You-Shao Wang[1,2]

*Address all correspondence to: mlwu@scsio.ac.cn

1 State Key Laboratory of Tropical Oceanography, South China Sea Institute of Oceanology, Chinese Academy of Sciences, Guangzhou, China

2 Key Laboratory of Marine Bio-resources Sustainable Utilization, South China Sea Institute of Oceanology, Chinese Academy of Sciences, Guangzhou, China

3 Tropical Marine Biological Research Station in Hainan, Chinese Academy of Sciences, Sanya, China

4 Xisha Deep Sea Marine Environment Observation Station, Chinese Academy of Sciences, Sansha, China

References

[1]   Connor, S.E., van Leeuwen, J.F.N., Rittenour, T.M., van der Knaap, W.O., Ammann, B., and Bjorck, S., The ecological impact of oceanic island colonization—a palaeoecological perspective from the Azores. Journal of Biogeography, 2012, **39**(6): 1007–1023 DOI: 10.1111/j.1365-2699.2011.02671.x.

[2] VishnuRadhan, R., Thresyamma, D.D., Sarma, K., George, G., Shirodkar, P., and Vethamony, P., Influence of natural and anthropogenic factors on the water quality of the coastal waters around the South Andaman in the Bay of Bengal. Natural Hazards, 2015, **78**(1): 309–331 DOI: 10.1007/s11069-015-1715-9.

[3] Cao, L., Pan, Y.L., and Liu, N.F., Waterbirds of the Xisha archipelago, south China sea. Waterbirds, 2007, **30**(2): 296–300 DOI: 10.1675/1524-4695(2007)30[296:wotxas]2.0.co;2.

[4] Wang, D.-R., Wu, Z.-J., Li, Y.-C., Chen, J.-R., and Chen, M., Analysis on variation trend of coral reef in Xisha. Acta Ecologica Sinica, 2011, **31**(5): 254–258.

[5] Huang, H., Lian, J.S., Huang, X.P., Huang, L.M., Zou, R.L., and Wang, D.R., Coral cover as a proxy of disturbance: A case study of the biodiversity of the hermatypic corals in Yongxing Island, Xisha Islands in the South China Sea. Chinese Science Bulletin, 2006, **51**: 129–135 DOI: 10.1007/s11434-006-9129-4.

[6] Sun, D.-R., Lin, Z.-J., and Qiu, Y.-S., Survey of coral reef fish resources of the Xisha Islands. Zhongguo Haiyang Daxue Xuebao, 2005, **35**(2): 225–231.

[7] Croux, C. and Ruiz-Gazen, A., High breakdown estimators for principal components: the projection-pursuit approach revisited. Journal of Multivariate Analysis, 2005, **95**(1): 206–226.

[8] Wu, M.L., Wang, Y.S., Sun, C.C., Wang, H.L., Lou, Z.P., and Dong, J.D., Using chemometrics to identify water quality in Daya Bay, China. Oceanologia, 2009, **51**(2): 217–232.

[9] Wu, M.L. and Wang, Y.S., Using chemometrics to evaluate anthropogenic effects in Daya Bay, China. Estuarine Coastal and Shelf Science, 2007, **72**(4): 732–742.

[10] O'Boyle, S., McDermott, G., Noklegaard, T., and Wilkes, R., A simple index of trophic status in estuaries and coastal bays based on measurements of pH and dissolved oxygen. Estuaries and Coasts, 2013, **36**(1): 158–173 DOI: 10.1007/s12237-012-9553-4.

[11] Xu, J., Glibert, P.M., Liu, H.B., Yin, K.D., Yuan, X.C., Chen, M.R., and Harrison, P.J., Nitrogen sources and rates of phytoplankton uptake in different regions of Hong Kong waters in summer. Estuaries and Coasts, 2012, **35**(2): 559–571 DOI: 10.1007/s12237-011-9456-9.

[12] Wu, M.L., Wang, Y.S., Dong, J.D., Sun, C.C., Wang, Y.T., Sun, F.L., and Cheng, H., Investigation of spatial and temporal trends in water quality in Daya Bay, South China Sea. International Journal of Environmental Research and Public Health, 2011, **8**(6): 2352–2365 DOI: 10.3390/ijerph8062352.

[13] Praveena, S.M. and Aris, A.Z., A baseline study of tropical coastal water quality in Port Dickson, Strait of Malacca, Malaysia. Marine Pollution Bulletin, 2013, **67**(1–2): 196–199 DOI: 10.1016/j.marpolbul.2012.11.037.

[14] Sawall, Y., Teichberg, M.C., Seemann, J., Litaay, M., Jompa, J., and Richter, C., Nutritional status and metabolism of the coral Stylophora subseriata along a eutrophication gradient in Spermonde Archipelago (Indonesia). Coral Reefs, 2011, **30**(3): 841–853 DOI: 10.1007/s00338-011-0764-0.

Relationship between Land Use and Water Quality and its Assessment Using Hyperspectral Remote Sensing in Mid-Atlantic Estuaries

Gulnihal Ozbay, Chunlei Fan and Zhiming Yang

Additional information is available at the end of the chapter

http://dx.doi.org/10.5772/66620

Abstract

Mid-Atlantic coastal waters are under increasing pressures from anthropogenic disturbances at various temporal and spatial scales exacerbated by the climate change. According to the National Oceanic Atmospheric Association (NOAA), 10 of the 22 estuaries in the Mid-Atlantic, including the Chesapeake Bay, exhibit high levels of eutrophic conditions while seven, including Delaware Bay, exhibit low conditions. Chesapeake Bay is the largest estuarine system in the United States and undergoes frequent eutrophication and low dissolved oxygen events. Although substantially lower in nutrients compared to other Mid-Atlantic Estuaries, the biological, chemical, and ecological status of the Delaware Bay has changed in the past few decades due to high coastal tourism, increased local resident populations, and agricultural activities which have increased nutrient inputs into this shallow coastal bay. As stated by the Academy of Natural Sciences, although the nutrient load has reduced since the Clean Water Act, years of nutrient accumulation, contaminations, and sedimentation have impacted estuarine systems substantially, long-term monitoring is lacking, and ecological responses are not well quantified. Eutrophication within the Bays has degraded water quality conditions advanced by sedimentation. Understanding the quality of the water in any aquatic ecosystem is a critical first step in order to identify characteristics of that ecosystem and draw conclusions about how well adapted the system is in terms of anthropogenic activity and climate change. Determining water quality in intertidal creeks along the Chesapeake and Delaware coastlines is important because land cover is constantly changing. Many of these tidal creeks are lined with forested riparian buffers that may be intercepting nutrients from running off into the waterways. Identifying water conditions, coupled with the marsh land cover, provides a strong foundation to see if the buffer systems are providing the ecosystem services they are designed to provide. Our primary goal in this chapter is to provide research findings on the application of the hyperspectral remote sensing to monitor specific land-use activities and water quality. Along with hyperspectral remote sensing, our monitoring was coupled with the integration of remotely sensed data, global

positioning system (GPS), and geographic information system (GIS) technologies that provide a valuable tool for monitoring and assessing waterways in the Mid-Atlantic Estuaries.

Keywords: remote sensing, hyperspectral image, water quality, total suspended solids, chlorophyll-*a*, turbidity

1. Introduction

More than half of the United States population inhabits coastal and estuarine areas [1]. These regions are under pressure as rapid population growth, overutilization of natural resources, and removal of such resources as agriculture and urban development become more predominant. Although this growth and development provide economic opportunity, they also alter local ecosystems by changing land use and land cover; causing changes in soil and groundwater chemistry, watershed-level hydrology, and dissolved nutrients in waterways, particularly in the form of nitrogen and phosphorous species [2]. According to Nehrling [3], throughout the United States, commercial fertilizer use has experienced nearly a threefold increase, from approximately 6.8 million metric tons in 1960 to about 20.0 million metric tons in 2011 in order to accommodate rapid population growth and need for increased food quantities.

According to the National Oceanic and Atmospheric Administration (NOAA) National Ocean Service [4], estuaries have been seriously impacted by various anthropogenic activities and disturbances, and many are seriously degraded by pollution. Toxic substances such as chemicals and heavy metals, nutrient pollution (resulting in eutrophication), and pathogenic bacteria and viruses are the pollutants having the greatest impact on the health of estuarine waters [4]. Estuaries by their nature are transitional areas between the land and the sea including both freshwater and saltwater environments and have the accumulative impacts of both land and water activities. By far, large-scale changes from draining, filling, damming, or dredging are the greatest threats to estuaries [4]. These activities result in immediate destruction, loss of estuarine habitats, or irreversible changes in the environment.

Given the increasing number of water quality issues such as harmful algal blooms (HABs) in economically significant waters of the world, methods are needed for early detection of these issues and their sources, which when used with other environmental and historic data can alert authorities a threatening condition. Remote sensing (RS) has the potential to provide accurate synoptic views of water quality conditions over a large spatial extent and can be used to estimate turbidity and algal biomass, and have the potential to identify algal taxonomic groups. Detection of water quality issues such as dominant algal taxonomic groups can provide coastal managers with information regarding those blooms that are potentially composed of harmful algae. As such, remote sensing is an important component of overall water quality and HAB-monitoring strategy. However, coastal and estuarine waters are optically complex and the monitoring of HABs is limited in these environments by remote sensing [5].

Major issues facing wetland habitats include long-term changes in land cover and land use (generally including agricultural activities, habitat destruction, encroachment, and historic diking of estuarine habitat) [6], shoreline hardening, shoreline erosion, and shoreline alterations anthropogenically induced to change the structure and function of the actual marsh surface such as hydrology, marsh topography, plant community, nutrient retention, tidal flooding, detritus accumulation, and availability to secondary producers [7–11].

Land-use activities by humans cause considerable changes in the magnitude and nature of many biological, chemical, and geological (-physical) constituents that are delivered and mixed in coastal water in the Mid-Atlantic Estuaries. Consequently, monitoring intensities of changes in the estuaries and bays' water properties is important to understand the propagation and modification of terrestrial components coming from the land and also their effect on coastal environments in response to catastrophic events and climate change. However, the distributions of the biogeochemical constituents of water remain poorly understood due to limited spatial and temporal sampling. Hyperspectral remote sensing provides a continuous spectrum of radiance (or reflectance) values associated with each pixel of an acquired image to capture a unique spectral signature of water quality indicators including salinity, chlorophyll-*a* (Chl-*a*), turbidity (Secchi depth), total suspended solids (TSSs), and dissolved organic materials ((DOMs)…etc.) which act as proxies for biogeochemical exchange in various coastal waters. This type of monitoring can help to understand and quantify relationships in the spatial and temporal structure of coupled physical-biogeochemical processes for the Mid-Atlantic Estuaries ecosystems.

Remote-sensing technology can also be used as a tool to help monitor the impact of invasive species in coastal areas although this is a difficult and complex task. Ozbay et al. [12] discussed how *Phragmites australis*, the common reed, has been widespread in tidal marshes of the eastern coast of the United States in the last 50 years [13, 14] and how their abilities to supplant other wetland vegetation and decrease wetland biodiversity has attracted considerable research attention in the United States. As stated by the authors, about 10–15% of Delaware's coastal wetlands are now estimated to be invaded with tall, dense stands of *P. australis* [15, 16].

1.1. Blackbird Creek Watershed

The Delaware Bay is where the Delaware terrestrial system meets the Delaware coastal waters, leading to complex and nonlinear-mixing processes of terrestrial and marine properties. However, scientists have little knowledge of many details of the biogeochemical cycles in the bay. Furthermore, biogeochemical variability in the Delaware Bay is affected by a number of processes occurring at a wide range of temporal and spatial scales. Considering the complex and dynamic nature of the Delaware Bay water, developing a sustainable monitoring system suitable for the varying scales of the processes occurring in this region is a difficult task, especially in response to catastrophic events and climate change.

As Stone [2] stated, between 2010 and 2014, the coastal state of Delaware has grown faster than the United States as a whole—4.2–3.3%, growth, respectively [17]. The Blackbird Creek Watershed is located in southern New Castle County, Delaware, and drains roughly 80 km^2 and flows into the Delaware River just upstream from the Delaware Bay [18].

This largely semi-pristine forested watershed provides many recreational activities, and therefore varying degrees of anthropogenic effects disturb the ecosystem. Monitoring the Blackbird Creek Reserve has been on-going since mid-1970 for the detection of ecosystem changes and established management practices. Blackbird Creek is a part of the Delaware National Estuarine Research Reserve (DNERR) which is associated with the NOAA for maintaining the estuarine system. Just over a decade ago, the Delaware Department of Natural Resources and Environmental Control (DNREC) identified that roughly 36.1% of the watershed is designated for agricultural use, and an additional 13.2% for urban use [19]; **Figure 1**. However, based on more current data from the Delaware Geospatial Exchange, approximately 44% of the watershed is designated to agriculture and only about 4% to urban use.

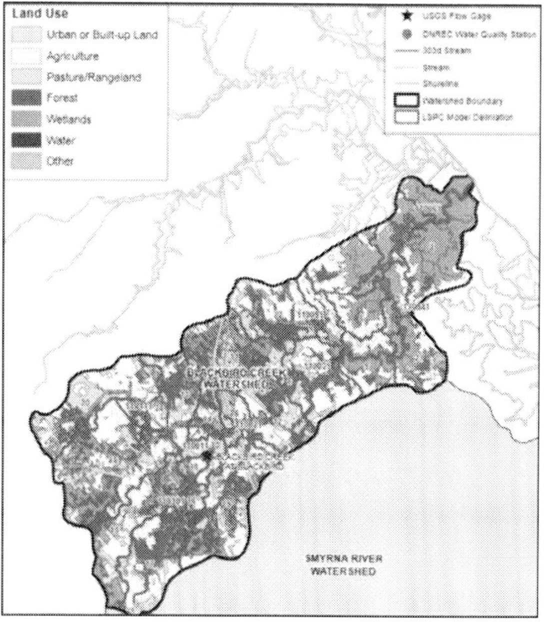

Figure 1. Land use practices in Blackbird Creek, Delaware, as described by LSPC Model Segmentation (map copied from Stone 2016, map by DNREC [19].

The Blackbird Creek tidal marsh ecosystem is dominated by various salt marsh grasses and macro fauna. Similar to many estuaries in the United States, this ecosystem is threatened by the surrounding land-use activities especially agricultural and residential activities [12].

Stone [2] reported how population growth and land-use changes put pressure on ecosystems. As the population increases, more land is cleared for development and agricultural purposes. Increased impervious surfaces from roads, parking lots, and other infrastructure can prohibit rain from entering the soil system. This can result in the conversion of the rainfall to runoff, flowing into storm drains, rivers, and streams, and collecting sediments, harmful pollutants, and nutrients along the way. The monitoring of nutrient runoff into waterways and how nutrients are cycled through estuarine ecosystems are of particular importance in the Blackbird Creek Watershed. Rabalais et al. [20] and Boesch et al. [21] discussed how nutrient loading impacts streams, especially in large watersheds. Nutrients entering estuarine systems such as Blackbird Creek or the Choptank River often do so through both surface water and groundwater inputs. As Stone [2] reported, some agricultural practices, including over-tilling and unnecessary loosening of soil, over-irrigating cropland, and over-applying fertilizers, can send more soil particles and nutrients into the creek and, subsequently, into Mid-Atlantic Estuaries.

Delaware's agricultural land use has greatly impacted Delaware estuaries, in particular, poultry farming. Based on the report by the United States Department of Agriculture (USDA) cited by the News Journal [22], over 550 million broiler chickens were produced on the Delmarva Peninsula in 2012 alone, an increase of almost 14,000% since the early 1900s. According to the USDA National Agricultural Statistics Service [23], per year, Delaware state growers produced about 216 million birds alone. Poultry litter is overabundant as a by-product of the large chicken industry in the state of Delaware. Poultry litter is high in nutrients, particularly in phosphorous, and is often used as a fertilizer for pasture and hayfields throughout Delaware, including in the Blackbird Creek Watershed [24]. This litter contains high concentrations of water-soluble phosphorus and is readily transported as farm runoff. Excessive or improper fertilizer application can result in phosphorous buildup considering that the N:P ratio in plants is 8:1 while only 3:1 in poultry litter [25]. The United States Environmental Protection Agency (USEPA) identified phosphorous concentrations of 0.31 mg/L or greater to be detrimental to aquatic organisms, and is therefore the maximum value that should ever be in bioavailable form in waterways [26]. Studies have shown the detrimental impacts that phosphorous has on aquatic ecosystems, including eutrophication [27–29], mortalities of fish and invertebrates [30], and stream community shifts from a heterotrophic to an autotrophic state [31].

Major habitat change in the Blackbird Creek ecosystem is attributed to the abundance of invasive *P. australis*. Although invasive species are a less widely discussed human-caused disturbance in the watershed, it is by far the most noticeable one in the creek. Ozbay et al. [32] monitored potential impacts of *P. australis* invasion on aquatic species such as the blue crab population in Blackbird Creek Watershed to help understand the changing ecosystems in Delaware's coastal environment. Numerous studies have shown that *P. australis* invasions negatively impact essential fish habitats [33, 34]. Currently, it is not clear how invertebrates have responded to *P. australis* invasions. Research by Jivoff and Able [35] suggests that marsh surface vegetation influences the way blue crabs use marsh surface habitats. In the state of Delaware, *P. australis* is extensively studied, but most of these studies focus on the mechanism and prevention of *Phragmites'* invasion [36, 37]. Very few studies involved the detection and mapping of *P. australis* and its impact on fishery habitat in Delaware.

1.2. St. Jones River Watershed

The St. Jones River Watershed, located in central Kent County, covers 23, 327 ha (57,643 ac). This 16-km long river extends from Silver Lake Dam to the Delaware Bay. The St. Jones

Reserve, part of DNERR land, covers 1517 ha (3750 ac) and is located in the lower part of this watershed and features tidal brackish-water and salt marshes dominated by salt marsh salt hay (*Spartina patens*), cordgrass (*S. alterniflora*), and open water of creek, river, and bay areas, buffered by farmlands, freshwater-wooded fringe, and meadows but urbanized at upstream non-tidal areas [38]. It consists of three of the four hydro-geomorphic regions: poorly drained uplands, well-drained uplands, and beaches/tidal marshes/lagoons/barrier islands [39]. In the St. Jones Watershed, 39% of wetlands are tidal salt or brackish marshes, 35% are freshwater flats and 24% are riverine wetlands, and less than 2% are freshwater tidal or depressions [40]. There are several unique and threatened wetland features including bald cypress, Atlantic white cedar, and coastal plain ponds [40].

Agriculture and developed land in the St. Jones Watershed comprise almost 70% of the entire watershed's acreage. **Figure 2** shows the distribution of land use across the watershed from the National Land Cover Dataset (NLCD). Agricultural land and non-tidal wetlands are spread throughout the watershed. While forested patches are mainly located in the northern portion, large areas of tidal wetlands occur in the southeastern portion of the watershed. Since 2002, this watershed has experienced significant changes in land use and cover. The conversion of farmed land into residential housing led to a 6.1% increase in developed land and a 6.5% decrease in agricultural land. There were also losses in freshwater wetlands, particularly in non-tidal forested and scrub-shrub wetland habitats [41].

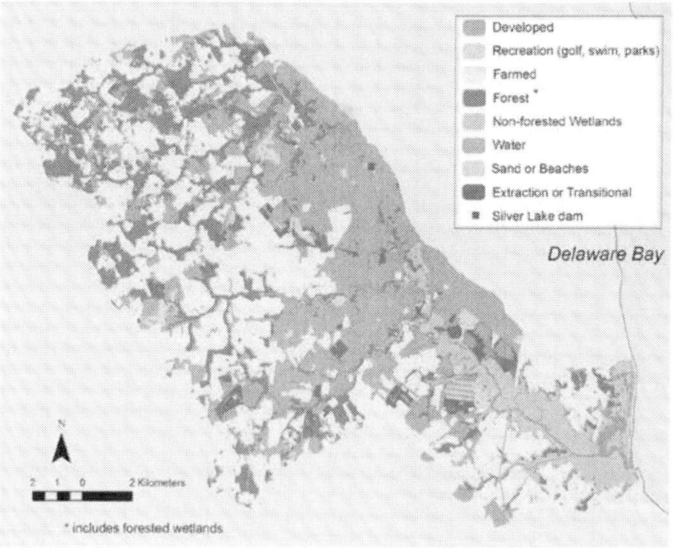

Figure 2. Land cover for the St. Jones River Watershed in 2007 based on National Land Cover Dataset (NLCD) land use categories [42].

Water quality problems in the St. Jones Watershed are attributed to sediment, nutrient, and chemical runoff [39]. Non-point pollution sources including runoff are mainly from roads (potentially containing oil, antifreeze, salt, and/or gas), lawn fertilizers and pesticides, and so on, dominate the St. Jones Watershed. Wastewater from septic systems is a source of nutrients in groundwater [39]. Moreover, channelization and ditching for agriculture and mosquito control led to altered natural wetland functions [39]. In addition, the invasive species such as *P. australis* have spread throughout fresh to brackish wetlands [39].

Rogerson et al. [42] reported that tidal sites with higher wetland condition scores resulted in lower avian species richness composed of primarily wetland-specific species, based on the intensive surveys of the avian community and vegetative biomass. Unfortunately, large areas of headwater flats in St. Jones River Watershed have been lost due to mainly agricultural production and development. According to Rogerson et al. [42] "Thirty-five percent of wetlands across the watershed were flats and had an average condition of 81, ranging from 57 to 94. Using condition categories, 37% of flat wetlands were minimally stressed, 47% were moderately stressed, and 16% were severely stressed." Over the last 50 years, over half (53%) of the 1385 ha of flats had not been forested (e.g., clear-cut, selective cut). As stated by Rogerson et al. [42], "Flats in higher condition had minimal garbage or dumping, low coverage by invasive plants, minimally altered micro-topography, and had a low occurrence of wetland ditching." Based on their assessment, forestry activity including cutting and harvesting within wetlands, and development and agriculture in buffers seem to be the major source of stressors in the watershed. Per Rogerson et al. [42], "Riverine wetlands, found adjacent to streams and rivers, accounted for 24% of the watershed's wetland acreage and had an average condition score of 72. Over half (55%) were considered minimally stressed, with low occurrences of invasive species, fill, and ditching, but frequently had dumping in addition to development and roads in the buffer. The severely stressed portion (10%) had condition scores as low as 27, related to the high prevalence of forestry activity, dumping, fill, storm water inputs and development within the buffer." They noted that wetlands with greater condition scores result in greater amounts of total-below ground biomass and a greater ratio of total-above-to-total-below ground biomass. Developmental activities such as residential, commercial, and/or transportation in the 100-m assessment site buffer were strongly related to the occurrence and frequency of wetland stressors such as storm water inputs, garbage, dumping, and invasive plants.

The watershed wetland condition report by Rogerson et al. [42] provided site-specific recommendations to improve wetland management, maximize natural benefits of tidal and non-tidal wetlands, and encourage informed decisions concerning the future of wetlands in Delaware. Previous research by Tiner [41] confirmed that the Delaware Bay basin was converted significantly to agriculture and residential development, particularly in non-tidal forested and scrub-shrub wetland habitats in St. Jones River Watershed. Emergent tidal wetland losses resulted in the excavation and filling of hydric sediments and the creation of impoundments [41]. As stated by Rogerson et al. [42], changes in land-cover and land-use patterns, especially near wetlands, are important to consider when evaluating wetland condition and its health.

1.3. Chesapeake Bay Watershed

The Chesapeake Bay—the largest estuary in the United States—is an incredibly complex ecosystem that includes important habitats and food webs. The Chesapeake Bay watershed spans more than 16,575,900 ha and encompasses parts of six states—Delaware, Maryland, New York, Pennsylvania, Virginia, and West Virginia—and the entire District of Columbia. Almost 18 million people live in the Chesapeake Bay watershed. The Chesapeake Bay's land-to-water ratio is 14:1: the largest of any coastal water body in the world. It is a very productive ecosystem that supports important recreational and commercial fisheries [43]. As shown in **Figure 3**, anthropogenic activities on land have a large impact on the Bay's health, most importantly; the water quality in Chesapeake Bay has declined significantly in recent decades, driven by the human population growth and changes in land use that together increased the nutrient loads [44].

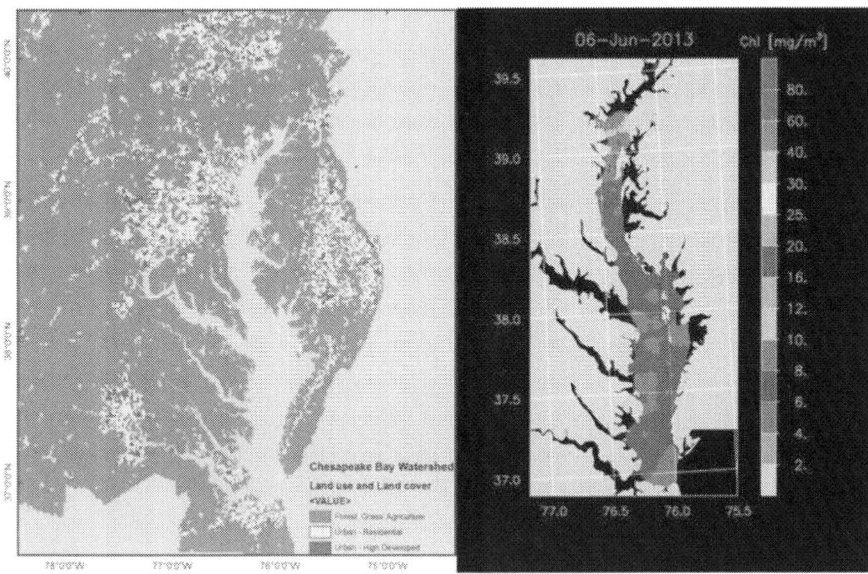

Figure 3. Chesapeake Bay Watershed land use and land-cover map showing majority land uses for forest, grass and agricultural activities and residential development (data from MD DRN LULC 2010 census), and the surface Chlorophyll-*a* concentration in the Chesapeake Bay by aircraft remote sensing in 2013 [45].

Soil and water quality protection are major challenges that must be addressed in modern day agricultural systems. Fifty billion liters of water flow into the Chesapeake Bay from over 150 streams and rivers. It supports more than 2700 species of plants and animals and produces 227 million kg of seafood each year. However, today, various factors have negatively impacted water quality and the deterioration of the overall health of the bay ecosystem. The Chesapeake basin water is highly impacted by agricultural development, residential development, effluent discharge from sewage treatment plants, and residential development by Pennsylvania Association of Conservation Districts, Inc. [46]. Point and non-point source (NPS) pollutants have caused ecosystem eutrophication which stimulates hypoxia, anoxia, frequent fish kills, increased turbidity, loss of submerged aquatic vegetation, and changes in food web structure [21]. Excess nutrients and sediment pollution originated by the agricultural activities in Delmarva's tributaries are a major concern.

According to Chesapeake Bay Foundation Report [47], the three major contributors to the poor health of the Chesapeake Bay include nitrogen, phosphorus, and sediment. As stated in the report, excess nitrogen and phosphorus fuel unnaturally high levels of algae growth in the water, blocking sunlight from reaching submerged aquatic vegetation such as grass bed that serve as food and habitat. In addition to hindering the growth of aquatic vegetation, algal die-off results in an increase in bacterial population that decomposes the dead algae and aquatic vegetation, as well as consumption of dissolved oxygen in the water. The Chesapeake Bay Foundation Report [47] states that fertilizers, wastewater, septic tank discharges, air pollution, and runoff from farms, cities, and suburbs result in excessive nutrients.

Erosion, coastal alteration, and construction-related development result in excessive amounts of sediment. Excessive sedimentation from tiny particles of dirt, sand, and clay floating in the water clouds the water, blocking sunlight from reaching submerged aquatic vegetation. Aquatic species such as oysters, a keystone species, and other bottom-dwelling organisms can be smothered when that sediment finally settles to the bottom. Major sources of pollution in the Chesapeake Bay include agricultural runoff by 41%, air pollution by 25%, wastewater treatment and factories by 16%, urban and suburban storm water runoff by 15%, and septic tank failures and leaks by 3%. As the major contributor to the pollution, agricultural runoff includes animal waste and fertilizers which wash off agricultural land or contaminate groundwater, which in turn pollute rivers and streams, and the Bay. **Figure 4** illustrates these pollutants and their origins [47].

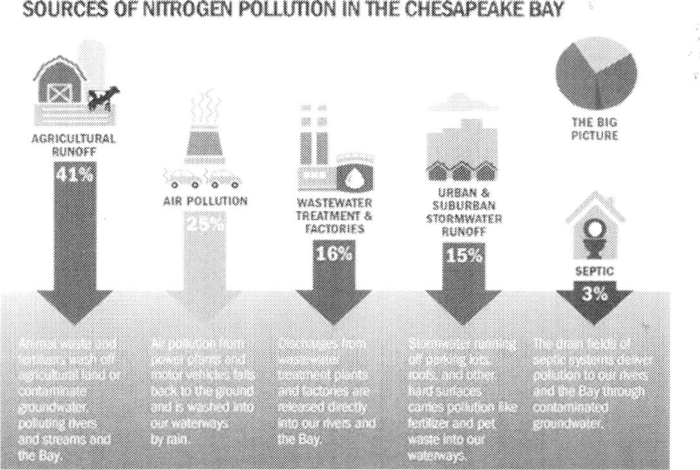

Figure 4. Illustration shows the sources of nitrogen pollution in the Chesapeake Bay [47].

Specifically, the waste from poultry production has raised serious concerns about treatment and disposal of the chicken manure along the shores of the Chesapeake Bay. With over 12 million cubic meters of chicken waste produced by over 523 million chickens produced each year in just Maryland and Delaware alone, serious risks in the coastal states have been raised. Diminishing crop lands have led to an increased use in manure load limits for cropland soil enrichment, resulting in excess nutrients from the poultry farms to flow into the ditches, streams, and rivers feeding into the Chesapeake Bay [48]. According to Ogejo [49], regardless of the size or type of farm, animal, livestock, and poultry producers need to manage manure for better economic returns and environmental protection. Similar to other basins in the mid-Atlantic, the Chesapeake water basin is excessively impacted by agricultural development, residential development, habitat loss, and effluent discharge from sewage treatment plants [50].

1.4. Remote sensing: multispectral and hyperspectral analyses

During the last few decades, remote sensing has allowed continuous monitoring opportunities of areas and objects over longer period and larger scale on Earth [51]. According to NOAA [52], remote sensing basically is the science of obtaining information about the areas or objects from a distance such as a satellite or an aircraft. As they stated, remote sensors, which can be on satellites or mounted on aircraft, collect data by detecting the energy that is reflected from Earth. Those sensors are either passive or active sensors. The most common source of radiation detected by passive sensors is reflected sunlight, while active sensors such as a laser-beam remote-sensing system use internal stimuli to collect data about Earth [52]. **Figure 5** shows an example of the Chesapeake Bay: Land Remote Sensing Image by using A Landsat 8 Surface Reflectance Mosaic by United States Geological Survey [53].

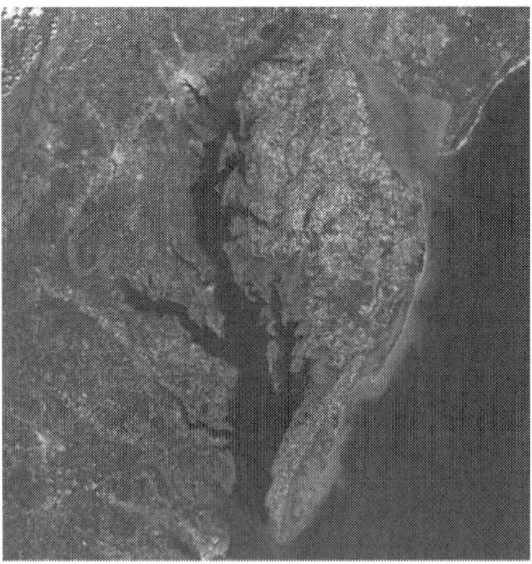

Figure 5. Chesapeake Bay land remote-sensing image by using A Landsat 8 Surface Reflectance Mosaic on April 4, 2016 [53].

Coastal waters are under increasing pressures from anthropogenic disturbances at various temporal and spatial scales [54, 55]. Water quality monitoring is vital for assessing such impacts, and further provides important information for sustainable water resource managements. Water quality analysis is an extremely important tool used to determine the overall health of an aquatic ecosystem, as a healthy environment will impact aquatic life forms and affect how they interact with the environment and each other. We are identifying how land use affects water quality, especially as it pertains to direct human interaction and indirect runoff from agricultural fields. Estuaries receive continuous inputs of biogeochemical constituents from their fresh water sources. For example, discharge of nutrient-rich water from urban, forest, salt marshes, agriculture, and ranching areas increases organic and inorganic sediments accumulation in Blackbird Creek. Nutrients can be measured by monitoring water quality and estimating distributions of matters in water bodies [56]. Although a wealth of new knowledge generated over the last several decades about these ecosystems, the spatial and temporal patterns of biologic and physical processes as well as anthropogenic influences are not fully understood.

While traditional monitoring programs can provide valuable water quality information, these programs are time and labor intensive, and lack the necessary temporal or spatial information needed for better decision making. Remote sensing offers the most effective means for frequent, synoptic water quality measurements [57–59]. Remote sensing offers a unique perspective because of the synoptic view and those quantitative algorithms can be used to extract geophysical and biophysical information [60]. Optically complex estuarine environment such as Blackbird Creek can be accurately mapped, measured, and characterized using remote-sensing techniques in order to develop models to map water quality characteristics of estuaries and freshwater tidal wetlands.

Ozbay et al. [12] discussed how recording the presence of *P. australis* within a tidal marsh zone presents challenges in the physical execution of a vegetation survey in an environment with restricted mobility, an overall study region with an area measurement on the order of tens to hundreds or even thousands of square kilometers, as well as the accuracy of species identification via sensor capabilities and various classification methodologies. Remote sensing utilizing satellite or aerial imagery enables one to more to easily collect data from afar without having to be physically present in harsh environments. Likewise, remotely sensed imagery can be collected at various resolutions, providing detail on a multitude of scales. Lastly, various algorithms and sensor capabilities provide varying degrees of accuracy for species identification.

According to GIS Geography [61], multispectral and hyperspectral remote sensing are being used for many applications in the fields of agriculture, ecology, oil and gas, oceanography, and atmospheric studies to better understand the world we live in [61]. However, their applications are depending on the information needed. Having a finer spectral detail in hyperspectral images gives better capability to see the unseen in finer details such as specific fish population in water. Technically, the main difference between multispectral and hyperspectral is the number of bands and how narrow the bands are to one another. Per GIS Geography [61] "multispectral imagery generally refers to 3 to 10 bands that are represented in pixels.

Each band is acquired using a remote sensing radiometer. Hyperspectral imagery consists of much narrower bands (10-20 nm). A hyperspectral image could have hundreds of thousands of bands. This uses an imaging spectrometer." According to Fisher et al. [62], "a collection of several monochrome images of the same scene taken by a different sensor makes a multi-spectral image that we referred them as a band. Multi-band image also called as multi spectral is a RGB color image that consists of a red, a green and a blue image taken with a sensor sensitive to a different wavelength". Example is provided as hyperspectral remote sensing distinguishes between three minerals because of its high spectral resolution while the multispectral Landsat Thematics Mapper could not.

An example given by Ozbay et al. [12] provided a comparison of the application of multispectral versus hyperspectral imaging with an important factor in the classification or monitoring of a wetland species, such as *P. australis*, where utilizing remotely sensed data involved the concept of scale. The extent of an intended study area has a great impact on the necessary available data to analyze. As discussed by Ozbay et al. [12], a sensor with high spatial resolution provides finer detail compared to a medium-resolution sensor over the same geographic area. If a specific area has a large geographic extent, the use of satellite imagery with a coarse resolution (10–30 m), such as *Landsat* Thematic Mapper (TM) or Le Systeme Pour l'Observation de la Terre (SPOT), may be sufficient. However, if the primary interest is on the order of several hundred meters or a few kilometers in lineal extent, aerial photography or the use of high-resolution (1–4-m) imagery, such as IKONOS or *QuickBird*, may be more appropriate or beneficial. The use of imagery with high spatial resolution may not necessarily result in improved detection or classification accuracy.

In general, multispectral data can be useful to determine broad vegetation classes and hyperspectral data can be effective to differentiate vegetation classes at the species level [12]. Authors noted that high-resolution imagery or hyperspectral data can lose effectiveness because of different vegetation species with similar biochemical and biophysical properties [63]. They also discussed that results may suffer due to spectral variations in the same species as a result of age differences, soil and water background, or stresses most notably in the near- to mid-infrared bands issues.

Classification of various wetland vegetation species including *P. australis* poses a complex challenge utilizing remote-sensing techniques. As stated by Shippert [64], standard multi-spectral image classification techniques are generally developed to classify multispectral images into broad categories while hyperspectral imagery provides an opportunity for more detailed image analysis for specific narrow range. Hyperspectral data-derived vegetation classifications generally are possible due to the ability to distinguish species using known reference spectra or spectral libraries.

As stated by Ozbay et al. [12], the cost of field validation of multi- or hyperspectral imagery is relatively high, limiting the use of remote-sensing data due to the uncertainty of the classifications required to manage natural resources [65]. A small wetland site may be adequately and reasonably covered with high-resolution satellite imaging. On the other hand, if the size of a study site becomes larger, it will become more cost-effective to utilize medium-resolution satellite imagery over high-resolution satellite imagery. As stated by the authors "Whereas aerial photography may be suitable for smaller wetlands or non-frequent observations, mapping or monitoring on a regional scale and increased frequency with such imagery would be costly and time-consuming to process."

Hyperspectral remote sensing provides a new class of Earth observation data for improved Earth surface characterization. For example, it has been used accurately to characterize complex coastal environments [66, 67] and has the potential for mapping and detecting harmful algal blooms [68], major sediment pigments of benthic substrates [69], and suspended particulate matter [70]. The high spectral resolution provides the opportunity to develop and evaluate advanced methods of spectral shape analysis, such as derivative spectroscopy that can accurately distinguish subtle features in spectra and may be critical for discriminating optically significant water constituents [71, 72].

Various empirical, semi-analytical, and analytical ocean color models have been developed last few decades to derive the water quality parameters of interest such as concentrations of total suspended solids (TSS), chlorophyll-a (Chl-a), and the absorption of colored dissolved organic matter (CDOM)) [73]. However, coastal and estuarine waters are optically complex and the signal that a remote-sensing detector collects is a mixed signal composed of various water optically active constituents from different sources [59]. For the case of the mixing of two waters, attempts to apply the ocean color models could result in poor predictive ability in retrieval of various water quality proprieties [74].

Jo et al. (unpublished data) [75] discussed the advantage of using high spectral and spatial resolutions as they illustrated in **Figure 6** where **Figure 6A** shows chlorophyll-a concentration in the Delaware Bay, calculated from MODIS data (500-m spatial resolutions), **Figure 6B** shows the Enhanced Thematic Mapper Plus (ETM+) (30-m spatial resolutions) on the same day, and **Figure 6C** shows an aerial photograph from the Helikite (Helium Kite, http://www.allsopp.co.uk/) at around 30-cm spatial resolution to monitor specific locations such as oyster beds continuously [76].

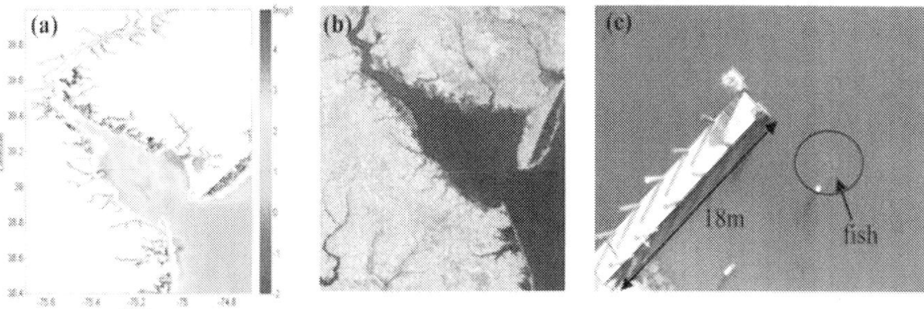

Figure 6. (A) Chlorophyll-a concentration from MODIS at 500-m spatial resolution; (B) combined three bands from ETM+ of Landsat 7 at 30-m spatial resolution; and (C) aerial photo of Helikite at 30-cm spatial resolution in August 31, 2011 (Jo et al. unpublished data) [75].

1.5. Water quality: suspended solids, turbidity, phosphorus, and chlorophyll-*a*

The water quality conditions in many fresh and coastal marine systems have progressively declined as the human population increases last century. This is often in response to increased runoff of nutrients from encroaching agricultural and urban areas due to housing development [77–79].

Over half of the Blackbird Creek Watershed is composed of agricultural lands. These lands are fertilized heavily to accommodate the timely growth of corn, soybeans, and sorghum grass. This fertilization is a concern for marsh habitat and local nekton of commercial and ecological importance. The excess nutrient loading tends to favor the growth of aggressive, invasive, often weedy, plant species which displace native plants. In extreme cases, nitrogen and phosphorous loading cause algae growth so great that the blooms block out sunlight to submerged aquatic vegetation, a crucial habitat for larval and juvenile fish and crabs [78, 79].

Stone [2] discussed that the overabundance of both reactive nutrients and non-reactive constituents can be detrimental to the quality of waterways, while they may not be acutely toxic. Turbidity is a measure of how clear or cloudy a water sample is and, by proxy, a suitable measurement to determine suspended sediments and solids which could settle out onto the riverbeds and banks [80]. Davies-Colley and Smith [81] reported that sediments in suspension can have the potential to smother benthic biota, irritate fish, crab and shellfish gills, and transport adsorbed contaminants. While sediments do have the capability of settling out, there is also a potential for them to remain in suspension and reduce the visual range of sighted organisms to seek out prey and/or members of the opposite gender for reproductive activity. High turbidity reduces the light penetration into the water and lowers the amount of available light availability. This, in turn, reduces the amount of phytoplankton production, possibly impacting higher trophic levels in a bottom-up model [82–85].

De Robertis et al. [83] and Kirk [86] documented the role of phosphorous in receiving waters. In systems like Blackbird Creek and Choptank River, the contribution of phosphorous to surface water is more likely from crop fields instead of household wastewater. Olli et al. [87] discussed that most of the phosphorous is generally in the form of bound reactive phosphate attached to eroded particles that have run off into the surface water not readily available for uptake by plant. On the other hand, Correll [88] and Reddy et al. [89] reported that phosphorous was retained in water and sediments via adsorption, complex formation, chemical precipitation, and biogeochemical reactions or recycled in the system. The compounds may be enzymatically hydrolyzed to form orthophosphate once phosphorous enters the water. This is the only form of phosphorus that can be taken up by algae, plants, and bacteria.

Blackbird Creek and other Mid-Atlantic watersheds influenced by agricultural land-use practices are susceptible to phosphorous enrichment. Many agricultural crops that are grown adjacent to waterways are treated with fertilizer containing high phosphate concentrations. As most of the phosphorus that enter aquatic environment precipitate by interacting with cations, phosphorous in freshwater can be retained in sediments by interactions with cations such as iron (Fe) and aluminum (Al). In seawater environments, deposited phosphorous is in large part returned to the overlying water through remineralization, making it biologically available for the consumption of phytoplankton and have better cycling of phosphorus [90].

RMB Environmental Laboratories Inc. [91] reported that chlorophyll-*a* is a green pigment in plants, algae, and phytoplankton that allows photosynthesis. The amount of phytoplankton growth in water bodies depends on various factors such as water temperature, water transparency, predation by zooplankton, and the availability of phosphorus and nitrogen (nutrients). Natural seasonal variations changed phytoplankton concentrations, and drastic changes in the abovementioned parameters due to hurricanes, storms, heat waves, and so on have detrimental effects on overall phytoplankton diversity and biomass [92]. Per RMB Environmental Laboratories Inc. [91], spring time, water is more transparent while there are more nutrients available due to the spring turnover; however, the water temperature is still low and this limits algal growth. As the water temperature increases during the summer, algae grow and its concentrations are higher. More nutrients can get washed into the lake if there are heavy rains that trigger more algal blooms. With increase in algal concentration, water becomes less transparent. Ritchie and Cooper [93] reported that eutrophication can be quantified by the concentration of chlorophylls contained in phytoplankton. According to USEPA (2000), monitoring the concentration of chlorophylls in phytoplankton can be used to evaluate and control eutrophication in water bodies.

According to Fondriest Environmental Inc. [94], the concentration of chlorophyll-*a*, as the green pigment used by phytoplankton in photosynthesis, is a good indicator of the amount of phytoplankton biomass in aquatic environments. Chlorophyll-*a* concentrations can be monitored in several forms consisting of chlorophylls a, b, c, d, and e. Concentrations of those forms determine the trophic state of water bodies, and high concentration most often indicates eutrophication [94].

Chlorophyll-*a* concentration estimated from the blue-to-green ratio of water-leaving radiances is used as a proxy for the phytoplankton biomass—for its distribution, variability, and growth rate studies [95]. This approach has been very successful in providing data for global distributions of algal biomass and in formulating estimates of global ocean primary production [96]. Nevertheless, knowledge of chlorophyll-*a* concentration is not sufficient to properly assess the biogeochemical and geophysical properties of the ocean surface layer [97].

Stumpf et al. [98] published results on monitoring *Karenia brevis* blooms in the Gulf of Mexico using SeaWiFS ocean color imagery, and indicated "the need to detect *Karenia brevis* blooms requires additional capability than simply identifying chlorophyll patterns." The capability of determining algal taxonomic composition can yield important information about overall water quality, eutrophication, and the distribution of HABs. A shift in algal species composition can be an important indicator of changes in water quality and may have serious environmental repercussions [99]. Therefore, the capability of determining this composition, as well as an estimation of algal biomass (chlorophyll-*a* concentration), by remote sensing is needed for the detection and monitoring of HABs. These data, when combined with other data, such as in situ data of nutrients and meteorological data, can potentially create models to forecast the development of blooms and HABs.

Stone [2] reported "as climate changes, it is simply not possible to discuss aquatic ecosystems without a detailed description of current and future trends in temperature, salinity, dissolved oxygen, pH, and the interactive effects of each of these variables on each other. This is also true with respect to change in land use. When land is modified to accommodate human needs, surface runoff, evapotranspiration, stream discharge, and sediment transport are all influenced, which in turn could have implications on the aforementioned parameters. If there is greater runoff from a crop field that was converted from a forested wetland, for example, there may be increased turbidity, reduced water temperature, and lower salinity. The potential ramifications in such a scenario would be monumental." The integrated use of remotely sensed data, geographic information system (GIS), and global positioning system (GPS) has enabled natural resource managers, state and federal governments, and consultants to develop management plans and policies for a variety of natural resource management applications for long-term sustainable options [100].

2. Case studies

Following case studies provide research-based discussion on the application of hyperspectral remote sensing and its specific modification for the research needs. In this chapter section, our primary focus is on Mid-Atlantic Estuaries, specifically Blackbird Creek Watershed, St. Jones River Watershed, and Chesapeake Bay Watershed.

2.1. Geographic information system (GIS) and remote sensing (RS) application in land-use classification and its relationship with water quality

Blackbird Creek is characterized by extensive salt marshes and large native populations of the salt marsh cordgrass (*S. alterniflora*). Blue crabs, as well as several other ecologically and economically important aquatic species, prosper in salt marsh environments where the leaves, roots, and stems of native plants provide sources of food and shelter from predators for many juvenile animals. Much of Delaware's other coastal wetlands including Blackbird Creek has been subject to a loss of biodiversity over the past several decades due largely to the common reed (*P. australis*) invasion. As reported by Roeske [101], this may be considered a highly disturbed ecosystem due to the invasion of *P. australis* and the intensive management (i.e., herbicide spraying) that have occurred since the early 1990s. Roeske [101] also reported that glyphosate is one of the most widely used herbicides in the world, especially with the advent of genetically modified organisms (GMOs), giving outcome to glyphosate-resistant crops. This chemical is referred to as a systemic, broad-spectrum herbicide, killing plants only a few days after application by blocking an enzyme that is vital in the production of amino acids and proteins [102]. Sadly, glyphosate is not species specific and generally kills all marsh vegetation. Roeske [101] stated that the long term and most effective management strategies for the control of *P. australis* have utilized a combination of the removal techniques including mowing, burning, herbicide treatment, increased salinity, and so on. These conditions encourage native plant colonization while pushing out non-native *Phragmites* by providing constant stress through these removal techniques. Roeske [101] also discussed that burning alone may actually reinvigorate the populations in subsequent years and thus these methods must be utilized in a strategic manner over the course of several growing seasons if any permanent removal is to be accomplished [103].

Stone [2] reported that ecologically and economically important resident fish and crustacean species in Blackbird Creek can be impacted indirectly by phosphate loading. Increased nutrient concentrations can increase phytoplankton productivity in aquatic ecosystems, which in turn can result in a drawdown of oxygen via increased respiration and decomposition of algae. Lowery and Tate [104] reported that dissolved oxygen reductions to less than 2 mg/L resulted in an increase of blue crab hemolymph lactate activity.

High sediment concentrations in tidal creeks and estuaries are possibly due to transport through the catchment basin via tidal activity and wind-induced sediment resuspension [105], especially when land is modified to accommodate human uses. Wischmeier and Smith [106] reported a strong chance of the erosion of sediment from recently tilled land on any kind of slope due to rain and wind, and subsequent runoff into waterways, which could lead to a variability in light availability on the scale of hours (tides) and days (calm vs. windy days). In systems such as Blackbird Creek, primary production may be compromised due to diminished light penetration which could result in reduced zooplankton production and lead to lower recruitment of estuarine species [107], thus altering the flow of energy through food webs since the autotrophic base would not be efficient enough to sustain higher trophic levels [108].

Agricultural land use in the Blackbird Creek Watershed was identified using ArcGIS layers available in the Delaware Geospatial Exchange. To select agriculture treatments, the watershed was gridded out into 500 × 500-m cells and the percentage of agricultural land cover within each cell was calculated. Water quality analysis was conducted based on these treatments in the Blackbird Creek. Any correlation between orthophosphate concentration and turbidity in relation to the land use based on the percent agricultural use was the main focus of the project (**Figure 7**).

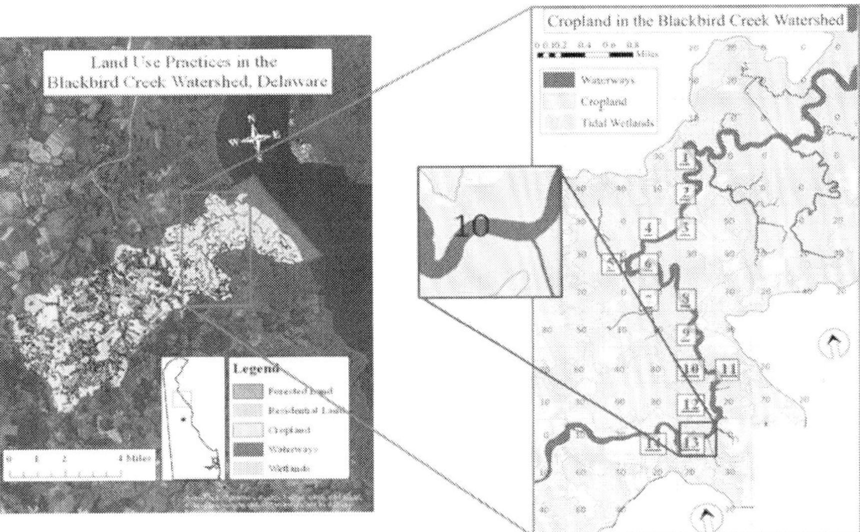

Figure 7. Sampling strategy shows the grid navigable tidal watershed into 500 m × 500 m cells and identifies the percentage of agriculture land cover in each cell. Samples were taken within each cell and differences of dependent variables across the percentages of agriculture were tested [2].

A considerable body of literature is available that explains the role of non-point source phosphorous runoff plays in the eutrophication of receiving water [109]. For agricultural watersheds such as Blackbird Creek, fertilization of soils greater than the total crop phosphorus uptake requirements leads to a buildup of phosphorous in the soil over time [110]. Studies have shown the positive relationship between soil phosphorus and surface water phosphorus, especially during high discharge events [111], indicating an association between soil management practices and surface water phosphorus concentration.

Holliday et al. [112] found 1:1 correlation between turbidity and total suspended solids for silt and clay fractions, but a smaller ratio for bulk-soil samples. Nevertheless, their research suggested that turbidity can be used to estimate sediment concentrations for fine soil fractions, which is, what the mouth of Blackbird Creek is primarily composed of. Other research has shown that reactive phosphorus increases with increased total suspended sediment concentrations [87–89, 113]. Based on these studies, our first goal was to identify trends in orthophosphate in accordance with measurements of turbidity.

Collection of water samples was undertaken over the course of 4 years (2012–2015) at various times of day between March and November. At each site (**Figure 7**), water was collected into dark polyethylene bottles (rinsed three times with sample water), then placed on ice for transport to the laboratory for nutrient analysis where samples were measured for orthophosphate and turbidity concentrations. Because orthophosphate is reactive in nature, a subsample of water was filtered on site into a separate bottle in order to minimize the potential for such reactions. Orthophosphate concentrations were determined using HACH Method No. 8048, the ascorbic acid method. This reaction provided information on soluble reactive phosphate concentrations between 0.02 mg L^{-1} PO_4^{3-} and 2.50 mg L^{-1} PO_4^{3-}. The measurements were performed using a HACH DR 3900 spectrophotometer (HACH, Loveland, CO). Turbidity was measured with an automatic wavelength selection on a YSI 9500 Photometer (YSI Incorporated, Yellow Springs, OH) and presented in Formazin Turbidity Units (FTUs), which is generally equivalent to Nephelometric Turbidity Units and Jackson Turbidity Units (JTUs). This method allowed for measurements between 5 and 400 FTU.

Based on the data by Stone [2], there were no correlations between turbidity and orthophosphate concentration. Linear regression analysis showed that there was a negative trend between the two parameters ($y = -5.0822x + 52.734$), but with an r^2-value of 0.0016, this cannot be considered statistically significant (**Figure 8**). It did appear that the tidal regime was important with respect to the turbidity concentration. On incoming tides, there was a negative trend ($y = -20.486x + 52.776$; $r^2 = 0.0628$) (**Figure 9**), whereas on outgoing tides, there was a positive trend ($y = 20.619x + 46.201$; $r^2 = 0.0175$) (**Figure 10**). Neither of these trends was significant, but it is interesting to note the opposite slopes as they relate to the tidal movement upon collection. When the tide was receding, there was a positive slope in the data. Thus, orthophosphate concentration was greater with greater turbidity, while the opposite was true on incoming tides. It is hypothesized here that this could be due to the source of these parameters as they regard the tide. On incoming tides, there was greater phosphorus concentration (0.62 mg PO_4^{3-}/L ± 0.020 SE) than on outgoing tides (0.55 mg PO_4^{3-}/L ± 0.024 SE). On incoming tides, there was less turbidity (38.72 FTU ± 1.776 SE) than on outgoing tides (47.72 FTU ± 2.965 SE). Other unpublished research [101] has shown that orthophosphate concentrations upstream were considerably lower than those that were found at the sites from this study. Thus, it may be suggested that the positive trend line in the data on outgoing tides could be a function of reduced orthophosphate at the source and/or the size fractions of the sediment particles in the upstream portion of the tidal creek. Further research is necessary to verify this postulate.

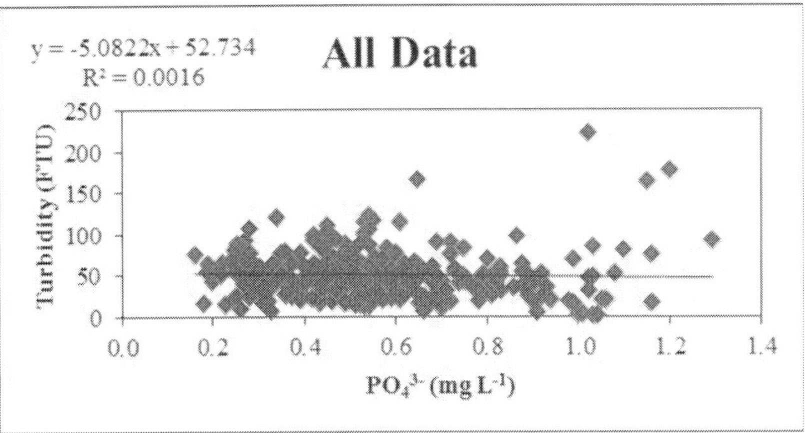

Figure 8. Correlations between turbidity and orthophosphate (PO_4^{3-}) concentration for both incoming and outgoing tides water [2].

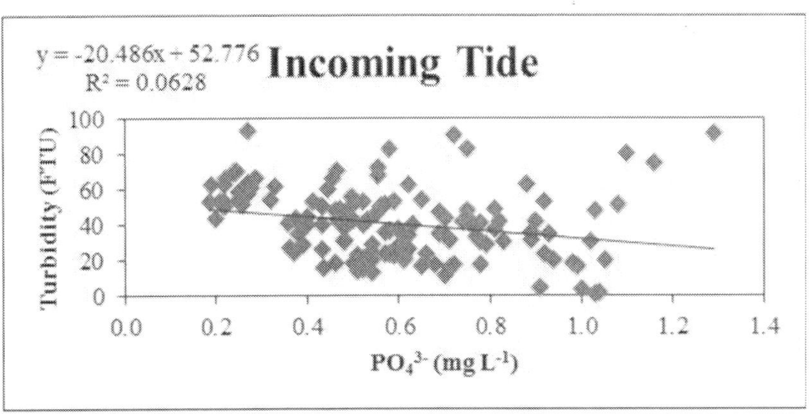

Figure 9. Correlations between turbidity and orthophosphate (PO_4^{3-}) concentration for incoming tide water [2].

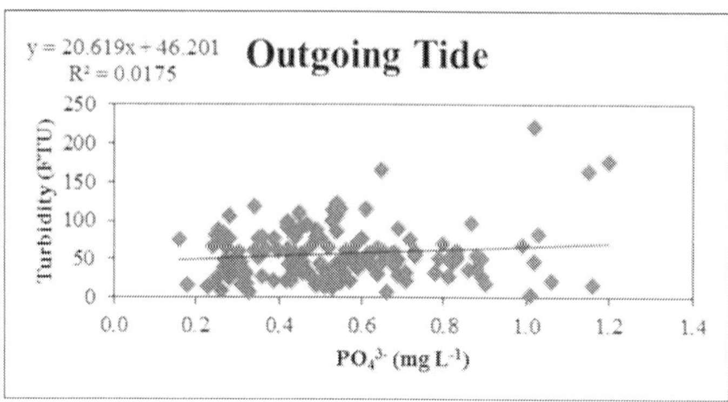

Figure 10. Correlations between turbidity and orthophosphate (PO_4^{3-}) concentration for outgoing tide water [2].

2.2. Application of AISA hyperspectral analysis in investigation of total suspended solids and chlorophyll-*a* in the Blackbird Creek, Delaware

The many biogeochemical processes taking places in estuarine environments maintain ecosystem health. Estuaries have a resilience to change, as long as the disturbance has not reached the threshold of where changes can no longer be reversed. A report by the Department of Interior, in conjunction with the U.S. Fish and Wildlife Service Dahl [114], states that over the past two centuries the continental United States has seen a 53% reduction in total wetland acreage, with Delaware losing approximately 54% of its original wetland habitat. It is generally accepted that this loss of habitat cannot be reclaimed or fully restored to historic structure and function in the near future, and thus represents a constant reminder of the importance of the ongoing management of the remaining wetlands throughout the United States [79].

The amount of direct physical alteration can be estimated by monitoring the distribution of turbidity, suspended solids, each of which is coupled with physical processes, including tidal activity and wind forcing in the Blackbird Creek, where farming and agriculture practices may generate considerable runoff of fertilizers and pesticides [115]. The strength of hyperspectral data analysis techniques lies in their ability to provide both spectral and spatial views of surface water quality parameters that are typically not possible from in situ measurements. Hyperspectral Airborne Imaging Spectroradiometer for Applications (AISA) airborne sensors with high spectral and spatial resolution can be used for monitoring environmental changes in optically complex turbid environments.

Hyperspectral remote sensing has the potential to provide accurate synoptic views of water quality conditions over a large spatial extent, particularly the spatial distributions of total suspended solids (TSS) and phytoplankton that can be retrieved using remotely sensed data. This technique was applied in Blackbird Creek. Our second goal was to develop a procedure for the total suspended solids, turbidity, and total chlorophyll-*a* using method of hyperspectral data (Salem et al. unpublished data) [116]. Those characteristics were used to develop models to map water quality characteristics in optically complex waters. Airborne AISA images with high spatial and spectral capabilities were used in a spectral analysis processes. The spectral angle mapper (SAM) classifier model was applied to AISA data to map spatial distribution of total suspended solids, chlorophyll-*a*, and turbidity levels. Photographs were used for results validation. The spectral analysis of wavelengths provided information on the distribution and concentrations of turbidity and Chl-*a* in Blackbird Creek.

Value increases in water quality parameters such as chlorophyll-*a*, turbidity, total suspended solids, and nutrients are symptomatic of eutrophic conditions. Concentrations of total suspended solids, chlorophyll-*a*, and turbidity parameters can provide insight on the extent of eutrophication and the potential impact on aquatic biota and overall water quality. Suspended solids may serve as a surrogate contaminant in agricultural watersheds since phosphorus, pesticides, and metals adhere to fine sediment particles and can be suspended in tidal estuaries. Estuaries receive continuous inputs of biogeochemical constituents from their fresh water sources and nutrients can be measured by monitoring water quality by estimating distributions of those materials [117]. There is a strong relationship between nutrient input, temperature, and phytoplankton biomass [118] but long-term monitoring of this relationship over a large scale provides information necessary for resource managers. Monitoring changes and impacts of land-use activities using remote-sensing technology provides the amount of data and information resource managers need to prepare management plans and policies for the watersheds.

The relationships between the spectral features of water reflectance and water quality parameters were investigated in Blackbird Creek. The research objective was to establish an empirical model that could be used for retrieval of water quality parameters from airborne hyperspectral AISA data in the Blackbird Creek Watershed. Hyperspectral AISA data provide a rapid and effective water quality-monitoring technique over a large spatial area for effective management of the watershed. Remote-sensing-monitoring results were coupled with the field-testing results. Four years of water quality-monitoring data from Blackbird Creek provided insight in relation to the land-use practices, especially agricultural activities in the watershed, and the presence of invasive *Phragmites*. Flyover photographs of the creek were taken from a small Delaware State University aircraft to help select locations in this research. Areas susceptible to sediment loads and chlorophylls were identified. This also allowed ground truth validation. Previous research by [119] provided the best fit spectral reflectance between 700 and 800 nm for total suspended solids. The scattering peak at ~700 nm was found to be strongly correlated with TSS and turbidity by Härmä et al. [120], Senay et al. [73], and Ammenberg et al. [121]. The image used includes 35 bands from 440 to 865 nm in the visible and infrared wavelengths in this research.

Aircraft-mounted hyperspectral spectrometer, the airborne imaging spectroradiometer for applications, was used to collect landscape images with 35 bands in Blackbird Creek. This type of spectroradiometer mounted in small aircraft can collect landscape images with a high spatial of 3.00 m × 3.00-m pixel size and a spectral resolution of 225 bands. Three segments of AISA images were selected to create a spectral library for phytoplankton (chlorophyll-*a*) in tidal water with high turbidity in the estuaries such as Blackbird Creek (**Figure 11**). Representative spectra for total suspended solids and chlorophyll-*a* in turbid water were selected in the visible and near infrared region (NIR) for areas with similar characteristics. These types of spectral signatures are very useful in either small or large water bodies to monitor chlorophyll concentration (heavy, moderate, or light). AISA data were also used to measure water turbidity in Blackbird Creek and how this related to the primary productivity other than clarity of water. According to Ritchie and Cooper [93], AISA provides a better set of spectral bands in different wavelengths than what Landsat can provide. The spectral angle mapper classification model was used to monitor chlorophylls and total suspended solids based on its chemical composition using spectral signatures for each component (Salem et al. unpublished data) [116].

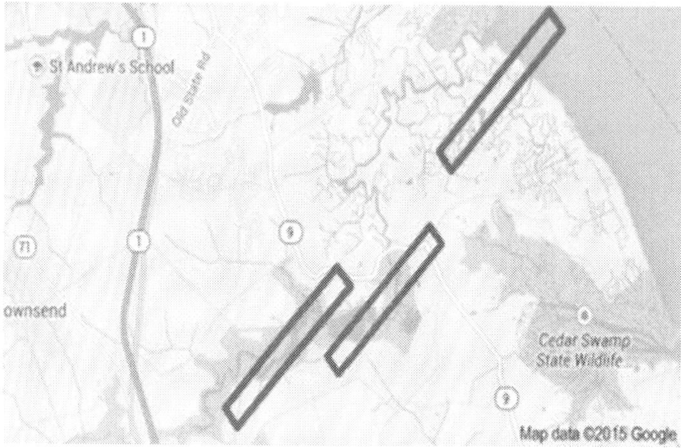

Figure 11. The three segments of AISA data used in the Blackbird Creek and Delaware Bay (Salem et al. unpublished data) [116].

The reflectance spectra of water with high chlorophyll concentration showed the characteristic absorption at the blue (~400 to ~500 nm) and red (~600 to ~680 nm) wavelength regions, which results in low reflectance, especially at ~680 nm wavelength (see **Figure 12**). Also, there is a phytoplankton-scattering peak at ~700 nm; this could be the result from a combination of chlorophyll fluorescence and water absorption at wavelength longer than 700 nm. The TSS reflectance appeared in all wavelengths from ~430 to ~860 nm, but TSS reflectance increases in clear water more than turbid water in the infrared range of the spectrum, and reflectance decreases in the infrared range due to high absorption of light by sediments in turbid water. Turbid water dominated the spectra on the visible (VIS) range of the spectrum, the high reflectance of total suspended solids appeared between (~460 and ~680 nm) and the sediments absorption of light appeared in the infrared portion between 680 and 860 nm which shows that sediments in turbid water absorb light in infrared range and reflect it on the visible range. The algae bloom spectra in turbid water show that chlorophyll absorption of light of phytoplankton bloom has no effect in the signatures taken away from the shoreline due to high turbidity which reduce the absorption effect of chlorophyll.

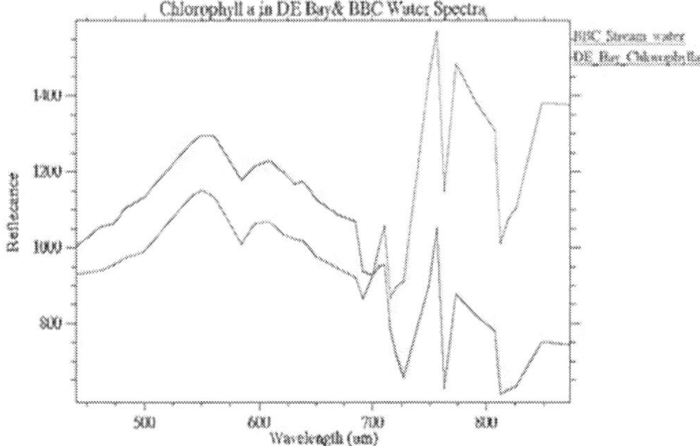

Figure 12. Spectral reflectance of chlorophyll and turbid water on the Blackbird Creek (BBC) and Delaware Bay (Salem et al. unpublished data) [116].

The signatures selected from the image of Blackbird Creek appeared in blue in **Figure 12** show low reflectance in the visible range of the spectra between (430 and 684 nm) due to absorption effect of the blue-green through red light of Chl-*a* and also show high reflectance of TSS reflectance in the near infrared range (727–813 nm) due to less turbid water in the Blackbird Creek than in the Delaware Bay. Low turbid water results in high absorption effect of chlorophylls in the visible range but low reflectance effect of TSS in the infrared range of the turbid water spectrum selected from Blackbird Creek.

Figure 13 shows classified image of different levels of suspended solids in the Blackbird Creek where high concentration of total suspended solids in relatively clear water is shown in the red color and vegetation in green color. The red signature in the spectral collection showed very high water reflectance and low impact on Chl-*a* absorption in the visible range from 430 to 680 nm because low turbid water increases the reflectance in the infrared range between 710 and 860 nm. The reflectance in yellow shows high absorption of Chl-*a* due to relatively clear water. Suspended solids and Chl-*a* show more increase in reflectance and absorption features because of less turbid water.

Results by Salem et al. (unpublished data) [116] indicated that high turbidity influences spectral reflectance scattering at ~700 nm of suspended solids, masking the spectral absorption of Chl-*a* due to the effect of high total suspended solids. These results reveal a positive correlation between water clarity and Chl-*a* concentration with the reflectance troughs caused by chlorophyll absorption at ~680 nm. This case study has demonstrated the feasibility to estimate the relative Chl-*a* levels and total suspended solids concentrations in the Delaware Bay and Blackbird Creek with limited ground-truthing data available (Salem et al. unpublished data). Through the use of remote sensing, hyperspectral imagery analysis combined with aerial photographs of the area and a general understanding of the conditions in the region can be estimated by monitoring relative Chl-*a* levels and total suspended solids concentration in applications of water quality management and policy decision.

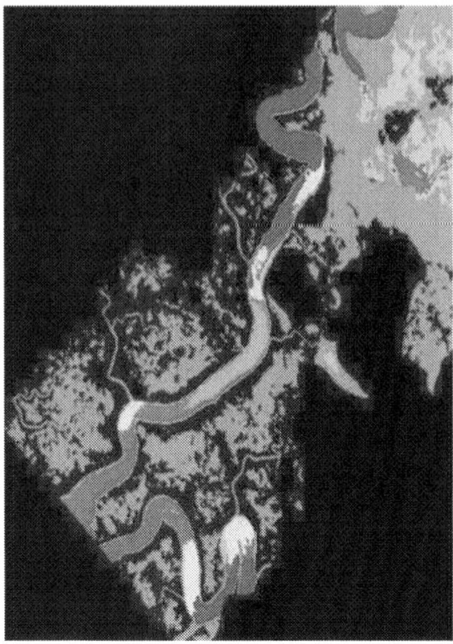

Figure 13. Classified image shows TSSs in red and vegetation in green and different levels of TSSs concentration are shown (Salem et al. unpublished data) [116].

2.3. Hyperspectral remote sensing of phytoplankton taxonomy in estuarine waters

Harmful algal blooms (HABs) have been observed to occur throughout the marine and aquatic environments of Earth during many different seasons, and under widely differing physical/chemical conditions. These blooms involve individual organisms which range in size from *microplankton (>20 μm), large nanoplankton (8–20 μm), small nanoplankton (1–8 μm), and picoplankton (<1 μm)* [122]. HABs are not known to have simple, direct correlations with location, seasonality, temperature, salinity, pH, insulation, nutrient concentrations, or other broadly collected physical oceanographic data [123]. Given the increasing number of occurrences of HABs in economically significant waters of the world, methods to be developed for early detection of the predominant type of bloom organism, which when used with other environmental and historic data, can alert authorities of a potential HAB versus a benign bloom condition.

Normally, the study of HABs is conducted by boats and point sampling, which is time consuming, expensive, and limited in spatial scale. Conversely, remote sensing has the capability of acquiring synoptic, regional-scale information, and is the only available technique that combines the possibility of frequent, large spatial coverage [124, 125] at reasonable cost. Ocean color imagery and products, from space-borne instruments such as Coastal Zone Color Scanner (CZCS) on Nimbus-7, Sea-viewing Wide Field-of-view Sensor (SeaWiFS) on board OrbView2, MODerate Resolution Imaging Spectrometer (MODIS) on EOS Terra and Aqua satellites, and the Medium-Resolution Imaging Spectrometer (MERIS) on ENVISAT, have been used to study ocean surface phenomena for many years [126–129]. Most of these sensors focused on the detection of chlorophyll-a, the pigment common to all algae [130].

The techniques required to identify and quantify algal taxonomic composition are typically by time-consuming cell counts or more recently by high-performance liquid chromatography (HPLC) [131]. Many algae have accessory pigments that are taxonomically specific. Therefore, the detection of specific accessory pigments can often reveal the taxonomy of algae present in the aquatic system [132, 133]. Furthermore, each individual pigment has its own light absorption spectral features, so the detection of these optical features by remote sensing can discriminate the pigments and thus allow determination of the algal taxonomic composition. However, most of the sensors currently in ocean remote sensing cannot detect the algal accessory pigments because of a small number of spectral bands, that is, eight bands for SeaWiFS, six bands for MODIS, and the wide bandwidth associated with these limited numbers of spectral bands. Therefore, increasing the spectral resolution is critical to improve the capability of discriminating different algal groups by remote sensing. An (hyperspectral) imaging spectrometer is often defined as an instrument composed of numerous, contiguous, narrow spectral bands used to identify a material by its unique spectral features, and is the instrument of choice to identify algal groups as an imaging spectrometer.

Another challenge of remote sensing of algal blooms and HABs is the optical complexity in coastal and estuarine waters. Previously, most of the remote-sensing studies were focused on offshore case 1 waters (clear oceanic water, those waters whose optical properties are determined primarily by phytoplankton (Chl-*a*) and related colored dissolved organic matter (CDOM) concentration and detritus degradation products) where optical properties are fully determined by phytoplankton populations (phytoplankton and bacterial plankton) and their retinue (associated detrital material and CDOM), which are well correlated with concentrations of chlorophylls. This is in contrast to coastal Case 2 waters (waters whose optical properties are significantly influenced by other constituents such as mineral particles, CDOM, or microbubbles, whose concentrations do not correlate with phytoplankton concentration), where other substances, for example, sediments and dissolved materials, bottom features, also influence radiation propagation [134, 66]. Studies on the biogeochemistry in the coastal regions using satellite sensors are lacking due to the optical complexity of the coastal environment and sensor characteristics. Darecki et al. [135] studied the optical characteristics of two contrasting case 2 waters and their influence on ocean color remote-sensing algorithms, and concluded that, for accurate determination of chlorophyll from satellite-borne sensors in different coastal waters, a wider choice of spectral bands than currently available was needed.

Hyperspectral remote sensing better defines spectral signatures and provides more independent information to potentially resolve simultaneously phytoplankton, suspended matters, CDOM, and bottom contributions [136]. As discussed previously, multi-spectral remote-sensing instruments that measure chlorophyll-*a* concentration are valuable in determining the bloom distributions and variability of chlorophyll in the case 1 waters. To identify the dominant algal group of a bloom, it is necessary to quantify the absorption features by accessory or marker pigments beyond chlorophyll-*a* [137]. Hyperspectral measurements can resolve spectral features that are relatively narrow, such as those of the accessory pigments, and therefore provide the possibility of identifying phytoplankton by taxonomic groups. Phytoplankton pigments, chlorophyll fluorescence, and in situ absorption spectra have been combined in studies to characterize micro-algal properties [138–142].

A number of research groups have used in situ or laboratory-based instruments to study the spectral properties of specific HAB groups in recent years. Millie et al. [140] demonstrated the utility of using bio-optical parameters for detecting and characterizing HABs by studying the photosynthetic pigments of *Gymnodinium breve*, a red tide dinoflagellate frequently appearing in Florida Bay. They concluded that the hypothetical assemblages of *G. breve* with increased concentration can be discerned from the absorption and the fourth-derivative information; however, the absorption spectra and the fourth-derivative alone may not identify the contribution of a chlorophyll c-containing taxon to the composite spectrum of a mixed assemblage. Richardson and Kruse [143, 144] were also successful in separating major types of algae and other parameters in Florida Bay using an AVIRIS imaging spectrometer.

To characterize the bio-optical properties of the coastal regions, the U.S. government has funded field campaigns to collect large suites of measurements from instruments based on numerous platforms: airborne, shipborne as well as in situ instruments. Those related to hyperspectral technology include HyCODE (Hyperspectral Coupled Ocean Dynamics Experiments) funded by the Office of Naval Research (ONR) [145, 146] and CICORE (California Center for Integrative Coastal Observation, Research and Education) funded by NOAA, NSF, and ONR. These campaigns with their large databases provide scientists with possibilities to study coastal ecosystems as never before. Although these field efforts were not specifically designed to "catch" the HABs' formation, maintenance, and dissipation, they help to build the bases for science and technology advancements in coastal region research including HAB studies.

In order to study the ecology of HABs and to address HAB forecast, detection, monitoring, and mitigation, a systematic effort is needed to develop methodologies that can be used uniformly for all types of HABs. Detecting the taxonomic grouping of phytoplankton is the initial step in identifying those classes that are harmful per geographic region. This information ultimately leads to managers and the public being accurately informed in terms of ecological and health issues. The bio-optical properties of the algal groups and the subsequent development of remote-sensing algorithms help by connecting laboratory and in situ studies with remote-sensing technology. These retrieval algorithms can further be incorporated into future operational satellite-monitoring systems to monitor specific taxonomic characteristics of HABs.

During a hyperspectral flyover mission supported by NOAA Environmental Cooperative Science Center (ECSC) in the Chesapeake Bay region in July, 2005 (**Figure 14**), over 20 GB of airborne hyperspectral imagery data were collected using the AISA (Airborne Imaging Spectrometer Applications)-Eagle hyperspectral scanner on an aircraft (Piper Saratoga, owned and operated by the University of Nebraska-Lincoln) at an altitude from 1525 to 3050 m, which provides 1- to 3-m spatial resolution for the target areas. The spectral range for the AISA-Eagle scanner is 400–1000 nm, and is programmable with regard to the number of individual spectral bands. The AISA-Eagle is a push-broom sensor with minimal smile and keystone. The hyperspectral data consisted of 95 discrete channels with band resolution of 2.5-nm full width at half maximum (FWHM) from the spectral range of 400–1000 nm.

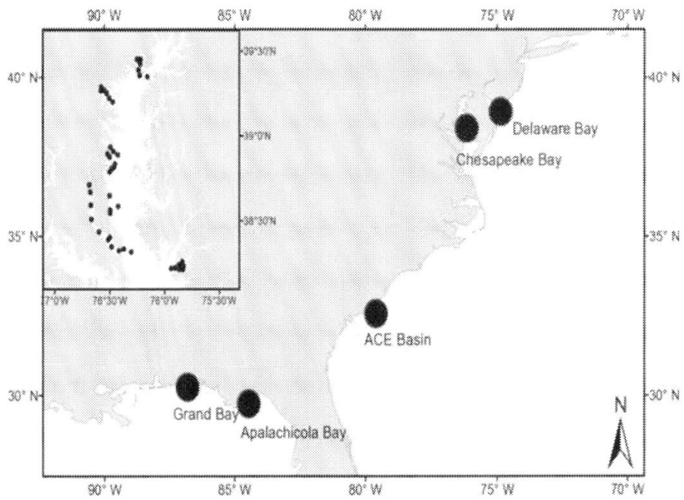

Figure 14. Map of the five estuarine systems located on the east and south coasts of US where the bio-optical data were collected from 2002 to 2008. The locations of each field station in the Chesapeake Bay observation from 2005 to 2008 are also shown in the upper left map [147].

The algorithms developed in the laboratory and field studies were applied to retrospectively analyze the aerial spectrometry imagery. The imagery was first atmospherically corrected using a recognized standard atmospheric correction program, FLAASH. Once the data have been converted into reflectance, further analysis was carried out using ENVI (the Environment for Visualizing Imagery, ITT Visual Information Solutions) software for algal population classification and biomass estimation. ENVI software includes a series of spectral analysis programs that can be used for such operations as locating the target spectra within the imagery by comparison with library of known spectra (end-member spectra) for specific algal groups. The results from imagery analysis were further compared to the ground-truthing data, which verifies the results of imagery analysis and further refines the approach for detecting HABs by hyperspectral remote sensing.

The results by Fan and Warner [147] show that the estuarine environment is optically complex, and chlorophyll, suspended sediments, and CDOM all influence the water reflectance. Chlorophyll-*a* has a prominent absorption trough near 675 nm, and the depth of the trough could be indicative of chlorophyll-*a* concentrations. High sediment concentration increases overall reflectance and amplifies pigment absorption features. The high CDOM concentration suppresses reflectance in the visual spectral range.

The Environment Cooperative Science Center (ECSC) data show no correlation between chlorophyll-*a* concentration and R490/R555 ratio [148]. An empirical algorithm was developed to more accurately predict chlorophyll-*a* concentrations in estuarine waters using hyperspectral remote sensing. The wavelength of chlorophyll-*a* absorption (675-nm band), the wavelengths sensitive to chlorophyll-*a* (650- and 700-nm bands) absorption, as well as the wavelengths reflecting the effects of CDOM (550-nm band) and other suspended sediments (440-nm band) were included in this model. This algorithm can account for 72% of the variance in the chlorophyll-a data across a large area of estuarine waters, a considerable improvement compared to the previous algorithm (**Figure 15B**). By using the NIR region of the spectrum (600–750 nm), our attempts to estimate chlorophyll-*a* concentration, as well as the ability to filter out the effects of suspended sediments and CDOM, are promising.

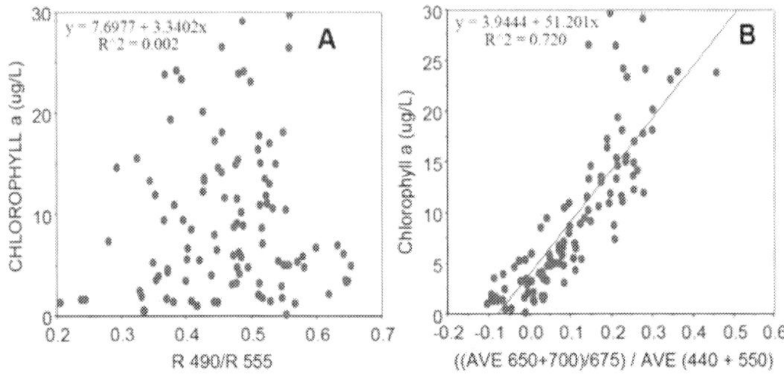

Figure 15. (A) Plot of the relationship between the model (R490/R555) and Chl-*a* concentration (µg/L) data collected across ECSC study sites; (B) plot of new algorithm (Ave (R650+R700)-R 675)/(Ave R440+ R550) versus Chl-*a* concentration (µg/L) collected across ECSC study sites [149].

Overall, our data suggest a great variability of optically active constituents (e.g., Chl-*a*, TSS, and CDOM) among the different estuarine systems, as well as the different field stations within a single estuarine system (**Figure 16**). The correlation analysis further indicates that the optically active water constituents are not related to each other (**Figure 16**) in this study. The determination coefficient ($R2$) of linear relationship between Chl-*a* and TSS is less than 0.01, suggesting that TSS and Chl-*a* did not co-vary together, and the major sources of TSS in the estuarine systems in this study might be from the land runoff. Similarly, no correlations were observed for concentrations of TSS and Chl-*a* versus CDOM absorbance, suggesting that the complex origins of these water optically active constituents, and phytoplankton (e.g., Chl-*a*) are not the only driver controlling the water optical properties.

The water reflectance $R(\lambda)$ (**Figure 17A**) measured in this study also displayed a large variation both in magnitude and in shape over the visible and NIR spectral regions. This variability is the direct result of largely uncorrelated optically active water constituents with various concentrations.

A common spectral pattern as shown in **Figure 17B** suggests that most of the spectra have a reflectance peak around 570 nm. This reflectance peak should be the results of the minimal absorption by Chl-*a* and the backscattering by particulates, such as TSS. A secondary reflectance peak was observed at red/NIR spectral range around 695 nm; this peak could be the result of Chl-*a* absorption at 675 nm and the strong water absorption at wavelength longer than 700 nm. Also, as shown in **Figure 17B**, the standard deviation of reflectance had similar pattern as the mean of reflectance, and it is wavelength dependent with a larger variance at the green spectral region.

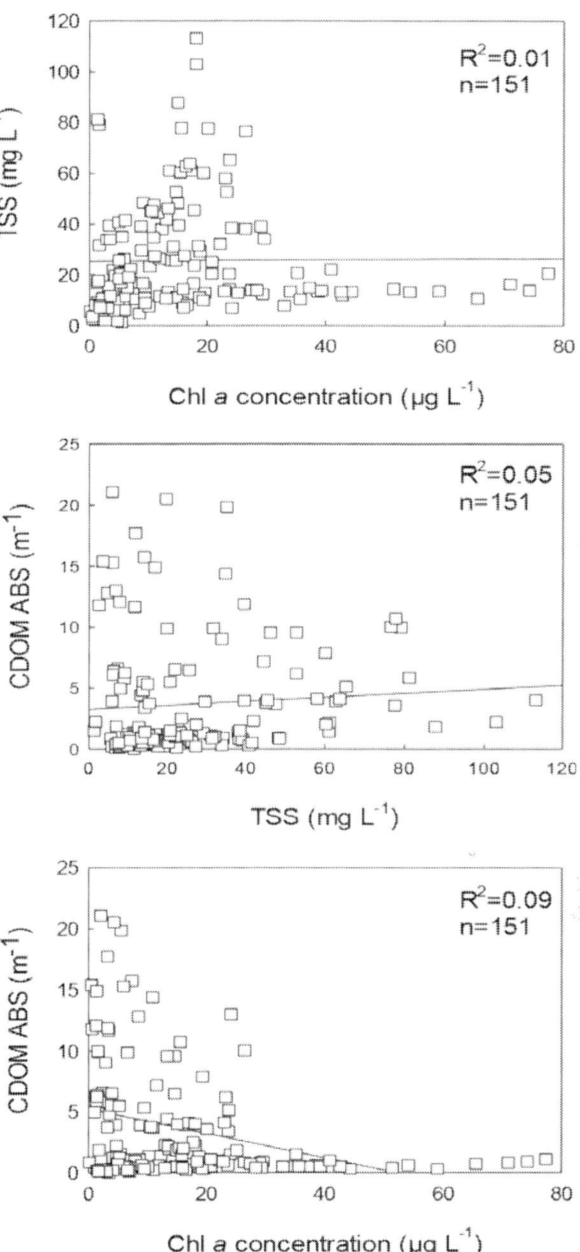

Figure 16. The relationships of bio-optical properties (e.g. Chl-*a* concentration, TSS concentration, and CDOM absorbance) of 151 field stations [147].

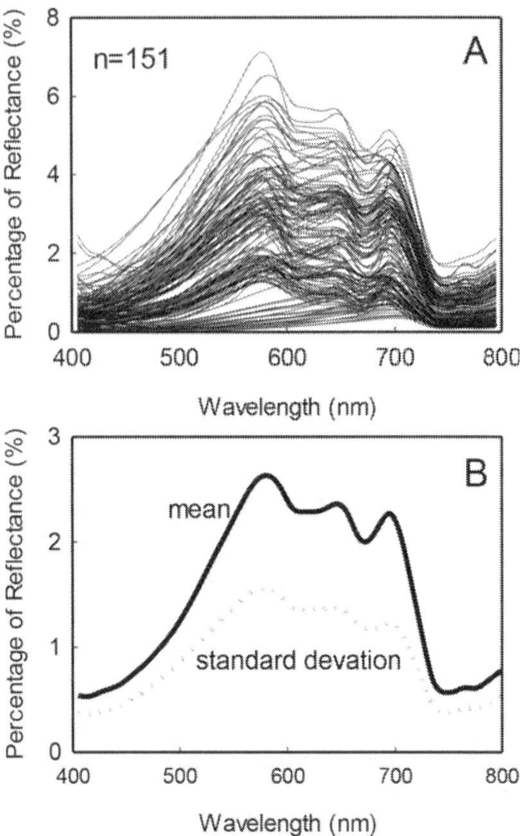

Figure 17. (A) Water reflectance $R(\lambda)$ spectra measured at 151 field stations from five estuarine systems on the east and south coasts of US; (B) the mean and standard deviation of the spectra [147].

A principal component analysis (PCA) of the reflectance dataset yields three dominant principal components which represented more than 97% of the total variance of the in situ water irradiance reflectance $R(\lambda)$ in our study (**Figure 18**). The first principal component accounts for 72.8% of the total variance, and displays the positive loadings across all wavelengths. Its spectral pattern is similar to the mean water reflectance observed in **Figure 17B**. This similar pattern suggests the backscattering by suspended particles controlling the overall water reflectance $R(\lambda)$. The second principal component accounts for 20.4% of total variance in the dataset. Its spectral shape was negative in blue and green spectral regions, with the minimal loadings around 560 nm. This feature corresponds to the minimal absorption by phytoplankton (e.g., Chl-a absorption) at this spectral range. Furthermore, the positive loadings in the red and NIR region could be contributed to the Chl-a absorption at 675 nm as well as the water absorption at this spectral range. So, the second principal component could be described as the effects of Chl-a (phytoplankton populations) absorption on water reflectance.

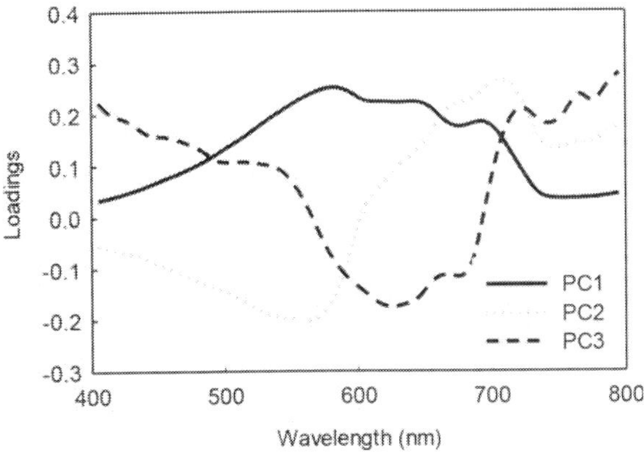

Figure 18. The loadings of the first three principal components of the correlation matrix of the reflectance $R(\lambda)$ dataset. The figure illustrates the weights by which the original spectral bands are weighted to construct the first PCs.

Figure 19 shows the biplot of the first two axes from canonical correspondence analysis (CCA), and it shows how the water reflectance in different in situ stations is influenced by water optical constituents (OACs), for example, Chl-*a*, TSS, and CDOM. The influences of OACs are indicated by vectors whose directions describe the gradients and whose lengths are proportional to their importance. While both TSS and CDOM have significant ($p < 0.01$) effects on first spectral axis, Chl-*a* has significant correlation ($p < 0.01$) with second spectral axis and nearly perpendicular to the TSS-CDOM gradient. So, the first CCA axis could represent the backscattering of the water, and TSS will increase the backscattering, while CDOM will suppress it. The second CCA axis could represent the influence of Chl-*a* (phytoplankton) on water reflectance. Together, these two axes explain most of the variability (96%) in the reflectance dataset.

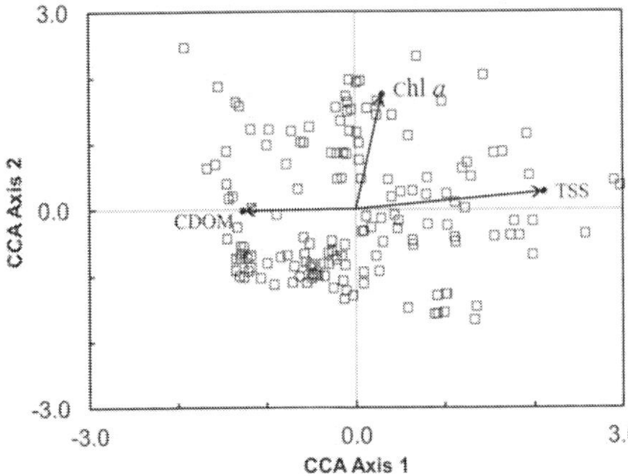

Figure 19. Biplot of the first two spectral axes of canonical correspondence analysis and the field stations. The influence of Chl-*a*, TSS, and CDOM is indicated by the vectors on this biplot [147].

This research advanced the basis of hyperspectral remote sensing under the optically complex coastal waters. The results provided necessary scientific information for using hyperspectral remote sensing in coastal and estuarine waters to monitor algal blooms. It also provides information leading toward the identification of the predominant bloom organism in estuarine waters. Knowing the type of organism as well as the synoptic view is of critical importance for coastal managers in their decisions relating to blooms, as well as being a more timely and economic method for this determination.

2.4. Spectral analysis of water reflectance for hyperspectral remote sensing of water quality in estuarine waters

Although hyperspectral remote sensing offers an effective approach for frequent, synoptic water quality measurements over a large spatial extent, the optical complexity of brackish estuarine water makes water quality monitoring by remote sensing a challenge. The third goal was to develop algorithms for hyperspectral remote sensing of water quality based on in situ spectral measurement of water reflectance [150].

During a hyperspectral remote-sensing study in the Chesapeake Bay, water reflectance spectra $R(\lambda)$ and the discrete water samples were collected at 11 field stations from 2008 to 2011 in the Patuxent River, a tributary of Chesapeake Bay (**Figure 20**). In this study, the relationships between the water reflectance $R(\lambda)$ and water quality parameters were examined to establish empirical algorithms that could be used for retrieval of water quality parameters from airborne hyperspectral data for rapid and effective water quality monitoring.

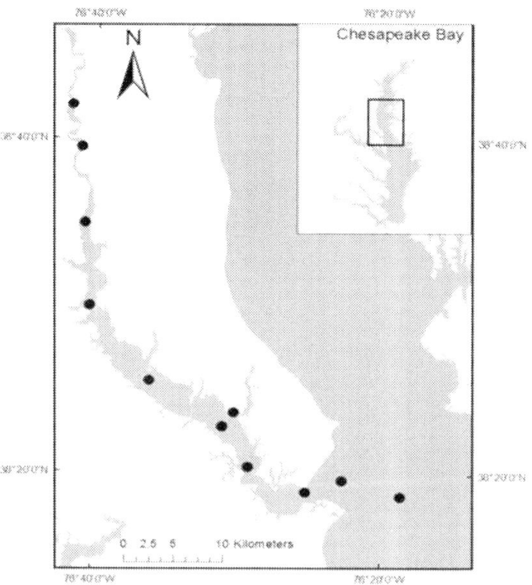

Figure 20. Map of the study area and the locations of the field stations for *in situ* measurement on the Patuxent River. The location of the study area in Chesapeake Bay watershed is shown in the upper right map [150].

Also, the aerial hyperspectral imagery of the Patuxent River with 2-m spatial resolution was acquired in the summer of 2005 using AISA-Eagle VNIR remote hyperspectral sensor (Center for Advance Land Management Information Technologies, University of Nebraska, Lincoln, NE). AISA-Eagle is a push-broom hyperspectral system with 1000-pixel swath width which can collect hyperspectral data at 2-m spatial resolution in 97 contiguous bands (2.5-nm bandwidth from 435 to 730 nm and 10-nm bandwidth from 730 to 950 nm). A total of six segments over a 50-km stretch of the Patuxent River were sampled during the summer of 2005.

The ground-truthing data suggest that water reflectance $R(\lambda)$ measured in this study displayed a high degree of variation because of the largely uncorrelated bio-optical properties. However, the influences on the general shape and magnitude of the reflectance by bio-optical constituents were still observed in this study. As shown in **Figure 21**, high chlorophyll-*a* concentration water shows characteristic absorption at the blue (400–500 nm) and red (600–680 nm) wavelength regions. Also, there is a phytoplankton-scattering peak at ~700 nm, which is the result from a combination of chlorophyll absorption and the strong water absorption at wavelength longer than 700 nm. This general influence on spectral reflectance by phytoplankton pigments (i.e., chlorophylls) is also consistent with our previous studies [147, 151]. Our data also suggest that TSS-dominated water shows high reflectance across a wide spectra range from 560 to 700 nm. This influence of TSS in regulating water reflectance spectra was also observed in the previous studies [152, 153]. For the water with high CDOM concentrations, its reflectance was characterized by low reflectance across the blue and green spectral regions, with very small reflectance peaks near 660 and 700 nm [151].

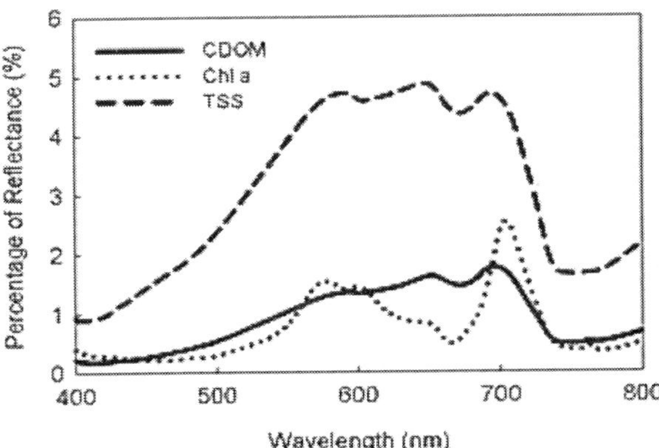

Figure 21. The general spectral features for coastal water that are dominated by high concentrations of Chl-*a*, TSS concentrations, and absorption of CDOM [147].

The general features of reflectance spectra observed in this study further provided insights for wavelength bands selection that are likely to be used for retrieval models to retrieve water quality parameters (i.e., Chl-*a*, TSS, and CDOM) [151]. Based on the in situ measurements in this study, band ration algorithms were developed to establish the relationships between water reflectance and selected water quality parameters, and this approach has been suggested as one of the most appropriate methods elsewhere [154]. Several previous studies showed strong relationships between log-transformed water quality data and reflectance values from hyperspectral data [151].

Chlorophyll-*a* model: As one of the most commonly measured water quality parameters, Chl-*a* has a significant correlation with the reflectance ratio of 700–670 nm in our dataset (**Figure 22A**). This was also found to be the case in several previous studies [155]. This relationship is due to the backscattering by phytoplankton at 700 nm and the strong absorption at 670 nm. The regression model is of the following form:

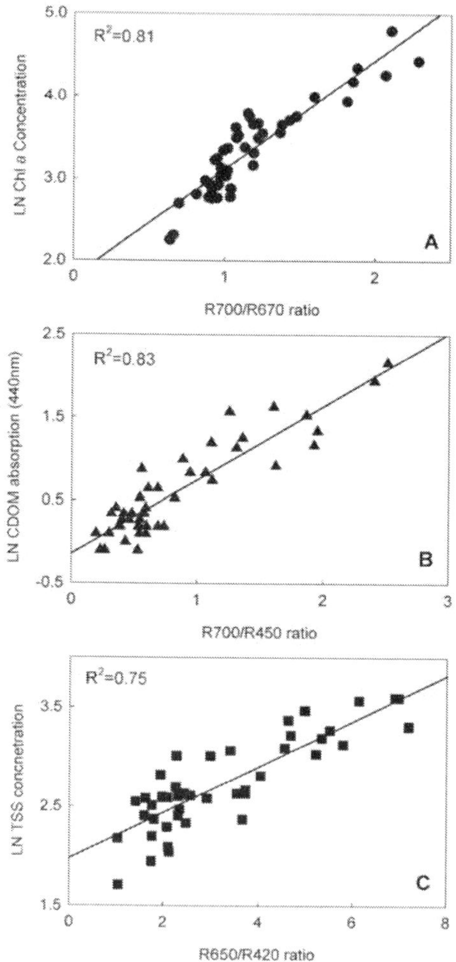

Figure 22. Scatter plots of *in situ* log-transformed water quality parameters versus corresponding band ratio values: (A) Chl-*a*; (B) CDOM; and (C) TSS [150].

$$LN(\text{Chl-}a) = 1.29\left(\frac{R700}{R670}\right) + 1.83, \quad \text{with } R^2 = 0.81 \tag{1}$$

CDOM model: Because of low CDOM concentration in lot oceanic areas, the impacts of CDOM on spectral variability have been largely neglected in some previous studies [156, 157]. This study shows that CDOM concentration strongly suppressed the reflectance at blue-to-green wavelength region, and then this effect declined exponentially toward the longer wavelength (**Figure 21**). Based on this CDOM spectral feature, the regression model (**Figure 22B**) was developed as the following form:

$$LN(\text{CDOM}) = 0.89\left(\frac{R700}{R450}\right) - 0.15, \quad \text{with } R^2 = 0.83 \tag{2}$$

TSS model: TSS represents both living organic solids (mainly phytoplankton) and inorganic suspended solids which mainly contribute to scattering of light. So, as shown in **Figure 21**, water with high TSS concentration tends to have high reflectance, especially at 500–600-nm regions. Remote-sensing algorithms for TSS reported in literature are less consistent and more dependent on the specific bio-optical conditions [158]. In this dataset, the relationships between log-transformed TSS concentration showed good correlation with the reflectance ratio of 650:420 nm. The regression model (**Figure 22C**) is of the following form:

$$LN(\text{TSS}) = 0.23\left(\frac{R650}{R420}\right) + 1.98, \quad \text{with } R^2 = 0.75 \tag{3}$$

Water quality mapping: The ASIA-Eagle hyperspectral imagery acquired in this study was processed and classified by using EXELIS ENVI 4.7 and ESRI ArcGIS 10.1. The ASIA imagery was first corrected geospatially and radiometrically to "at platform reflectance," then atmospheric correction was conducted on the image using the FLAASH in ENVI using the band at 820-nm wavelength to produce the image of "water reflectance." This "water reflectance" imagery is further classified by using a supervised classification in ENVI based on the selected water region of interests (ROIs). The terrestrial features were masked out to create the water-only image for further analysis. Finally, the band ratio models developed in this study were applied to these radiometric data in the image to create the pixel-level water quality maps. **Figure 23** shows the pixel-level maps of water quality (Chl-*a*, TSS, and CDOM) in a small section of the flight segment at the mouth of the Patuxent River. These maps clearly demonstrate the heterogeneity of water quality over a relatively small area. A patch of phytoplankton bloom (dinoflagellate *Prorocentrum minimum*, based on field observation) with high Chl-*a* concentration was observed in the middle of the river, while the Chl-*a* concentration in other area was relatively low. The water quality maps also show that the water with high TSS concentration was mainly located in the small creeks, coves, and river banks, suggesting that the land runoff might be the major source for TSS in water column. The CDOM absorption was also found high in the small creeks and coves around the river, as well as in the area with phytoplankton blooms. This suggests that the origin of CDOM could be both terrestrial or produced during the algal bloom.

Overall, the feasibility of hyperspectral remote sensing that can capture the fine-scale variation of water quality parameters is illustrated in **Figure 23**. In these maps, the algal bloom could be very patchy, and there was a large variation in these water quality parameters over a relatively small body of water. These techniques could provide vast water quality information that would not be seen by conventional monitoring program which probably only involve several sampling stations in the same area. However, the retrieval models in this study were derived from the dataset that is specific to the study area, which do not necessarily represent the coastal waters in other areas. The accuracy of such algorithms is always subject to the location of ground-truthing dataset.

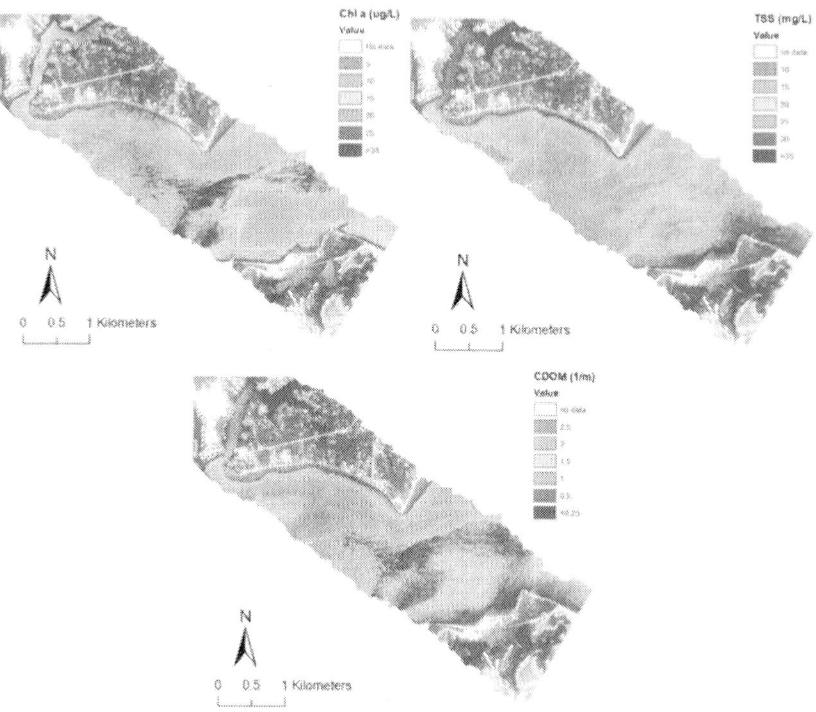

Figure 23. Maps of Chl-*a*, TSS concentrations, and CDOM absorption at the mouth for the Patuxent River, summer 2005 [150].

2.5. Estimating chlorophyll-*a* concentrations in the coastal St. Jones River Watershed using AISA images

The St. Jones River Watershed (**Figure 24**) is located in central Kent County, Delaware, and drains a portion of the coastal plain. The total area of the St. Jones River Watershed is 23,300 ha, including most of the capital city of Dover and a portion of Dover Air Force Base [157]. The upper St. Jones River is impounded by a dam 17 km upstream from the bay to form Silver Lake, a municipal recreation area. Much of the eastern portion of the watershed consists of wetlands, agricultural lands, and forests, including lands and waters managed for waterfowl, wild turkey, deer, and other wildlife.

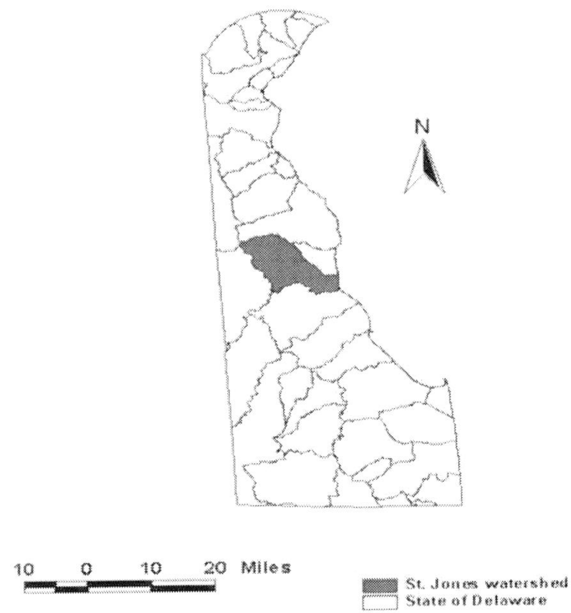

Figure 24. Location of the study watershed within the state of Delaware [158].

The percentage of various land uses in the St. Jones River Watershed in 1992 and 2001 was calculated based on the USGS land use/cover map of Delaware in 1992 and 2001, and there were significant changes in land use in the St. Jones River Watershed during this period (**Table 1**). While more lands were converted to urban and agricultural land, there was significant loss of wetland and open water in the St. Jones River Watershed. However, agriculture is always dominant land use in the St. Jones River Watershed at both time periods. Thus, it is important to examine chlorophyll-*a*, an indicator of eutrophication induced by agriculture runoff, in water bodies.

Land use	%, 1992	%, 2001
Open water	2.3	1.5
Urban	13.2	17.0
Forest	15.4	14.7
Agriculture	53.1	55.3
Wetland	15.7	10.8
Barren	0.3	0.7

Table 1. Percentage of land use in the St. Jones River Watershed in 1992 and 2001 [160].

In this watershed, eight water segments were in the 303 list for nutrients, dissolved oxygen (DO), and bacteria that were contributed from non-point sources (NPS) [19]. Many stations experience low DO levels less than the state minimum WQS of 4 mg/L with elevated chlorophyll-*a* levels. Also, nutrient concentrations at most of stations were high enough to support algal growth [19]. Therefore, it is essential to monitor chlorophyll-*a* concentrations in water bodies in the St. Jones Watershed.

The Airborne Imaging Spectrometer for Applications (AISA) is a commercially produced hyperspectral push-broom-type imaging spectrometer system capable of collecting data within a spectral range of 430–900 nm. The AISA airborne system, manufactured by Spectral Imaging (Finland), offers a high degree of positional accuracy provided by the combination of a differential global positional system (DGPS) with an integrated navigation system (INS) and real-time Kalman filter.

Aligned with the second goal in this chapter, this study determined if hyperspectral data could be used to measure water quality indicators such as chlorophyll-*a*, TSS, and CDOM in Delaware coastal waters. The correlation between a spectral index and in situ chlorophyll-a concentration was examined. Spectral indices were compared to determine the best index used for retrieval of chlorophyll-*a* from Delaware coastal waters with high confidence.

From July 20 to July 23, 2004, Delaware State University and the NOAA-sponsored Environmental Cooperative Science Center (ECSC) conducted a hyperspectral imaging flyover of the St. Jones River Watershed, Delaware. AISA data were acquired by the Center for Advanced Land Management Information Technologies (CALMIT) at the University of Nebraska using the AISA-Classic as the primary on-board sensor flown in a specially modified Piper Saratoga aircraft. The AISA sensor was configured for 35 spectral bands from 441 to 873 nm, and bandwidths varied for different portions of the spectrum. During the flyover, 25 flight lines (strips) of hyperspectral imagery were obtained between 11 AM and 2 PM covering the major parts of the St. Jones River Watershed. A ground resolution of 3.0 m was achieved by flying at an average altitude of 1500 m. One of the goals for this flyover was to determine if hyperspectral data could be used to measure water quality indicators such as chlorophyll-a, TSS, and CDOM in Delaware coastal waters.

Ground-truthing measurements included water sampling for laboratory analysis (chlorophyll-a, nutrients, and total suspended solids) and in situ measurement of water quality parameters including salinity, conductivity, dissolved oxygen, and temperature. Water samples were collected as close to the aircraft overpass as possible, with the exact coordinates of the sampling stations determined using a Garman GPS unit. All samples were taken from the 0- to 0.4-m surface layer. After acquisition, the AISA images were radiometrically and geometrically corrected by CALMIT using calibration files fed to AISA CaliGeo® Software and converted to Environment for Visualizing Images (ENVI) file format.

ENVI software (RSI, Version 4.2) was used to process preprocessed AISA images. AISA images for the study area and one shape file of sampling sites (GPS points) were loaded into ENVI. To reduce noise in the data, one 3 × 3 kernel window was used at each sampling site to extract spectral profiles, which include values for each band. Then, all spectral profiles were exported and input into Excel for further processing.

In Excel, NIR/Red "index" values were calculated based on different combinations of NIR and red bands, using one near infrared band between 670 and 780 nm and one red band between 620 and 670 nm. Fifty different NIR/red ratios were computed in total. Correlations between spectral indices and in situ chlorophyll-a data were examined and the significance of the correlations tested. If there was a significant relationship between an index and measured chlorophyll-a, the index was examined again to see if it had any significant relationship with measured TSS. The residual difference between estimated chlorophyll-a and measured chlorophyll-a was also tested to determine if there was a significant relationship between the residuals and measured TSS for each initially identified spectral index. This procedure was used to determine if TSS was potentially responsible for any estimate bias in the spectral indices. Based on all three procedures, indices that were a good fit for retrieving chlorophyll-a in Delaware coastal waters were finally identified.

Results are shown in **Table 2**. It shows all ratio indices possessing a significant relationship with in situ chlorophyll-a. There were significant relationships between chlorophyll-a and ratios covering near infrared bands only from 692 to 705 nm. Gitelson [159, 160] stated that reflectance near 700 nm was sensitive to chlorophyll-a in inland and coastal waters, and our result confirmed his finding.

Spectral indices	Correlation coefficients	Spectral indices	Correlation coefficients
R692/R621	0.56	R700/R651	0.53
R692/R632	0.55	R700/R670	0.58
R692/R639	0.57	R705/R621	0.50
R692/R651	0.56	R705/R632	0.49
R692/R670	0.68	R705/R639	0.50
R700/R621	0.54	R705/R651	0.49
R700/R632	0.53	R705/R670	0.52
R700/R639	0.54	R710/R670	0.47
SI	0.53		
Critical value (n = 16) = 0.47 at α = 0.05.			

Table 2. Correlation coefficients between spectral indices and chlorophyll-a in the St. Jones River Watershed [160].

There was no significant relationship between the ratios and in situ TSS found. There was also no significant correlation between the residuals of estimated versus in situ chlorophyll-a and in situ TSS. These results demonstrated that TSS did not have significant impact on estimating chlorophyll-a in Waters in St. Jones River Watershed when using the indices identified from this study.

While significant, the correlation coefficients for sensitive spectral indices are not high (maximum R = 0.68). This may be caused by two reasons. First, there was a time gap between the time to collect AISA images and the time to obtain water quality samples. Second, no atmospheric correction has been performed on the AISA images in this study. Despite these limitations, significant correlations obtained from this study indicate that it is possible to monitor chlorophyll-a using AISA images in Delaware coastal water bodies.

Based on this study, some simple ratio algorithms using spectral bands in the red and red-edge ranges were adequate for quantifying chlorophyll-*a* in coastal waters. These algorithms were validated in other environments such as Nebraska with some modification to the coefficients [160]. Therefore, there is a high potential that chlorophyll-*a* in coastal waters could be estimated using hyperspectral remote sensing without the interference from TSS. NIR/red ratios covering red-edge bands from 692 to 705 nm were found to be useful for retrieving chlorophyll-*a* in Delaware coastal waters when using AISA images and they could work in similar situations in other coastal areas. This suggests the advantage of using ratio-based indices in estimating coastal chlorophyll-*a* concentrations by remote sensing.

3. Final remarks

The impacts of land use on aquatic health and marsh habitat have been a constant battle, and the rate of our wetlands lost is far more than they can be replaced. Tidal marsh ecosystems serve as great examples of dynamic ecosystems that can provide numerous lessons in restoration and management strategies. The ubiquitous incidence of aquatic species as both temporal and spatial perspectives indicates the importance of these types of systems to sustain the health of aquatic species and other important species in the face of anthropogenic and environmental stresses. Thus, it would be pertinent for managers to maintain the health of these areas by preventing any further alterations and damage, especially from nutrient runoff and abusive use of our coastal lands, and the invasion of *P. australis*.

The Blackbird Creek Watershed is composed of only 4% urban development. This provided a great opportunity to study the area that has very little anthropogenic impact, relative to other watersheds. How trophic dynamics can be affected as a result of various land uses is one of the concerns. There are several plots of land designated as cropland in Blackbird Creek. An understanding of crop rotation in these plots both within and between seasons can provide insight on whether changes in crops across years can affect trophic interactions and food web dynamics. Additionally, effectiveness of riparian buffers as blockades for fertilizer runoff is a significant impact on water quality. With sites associated with buffers and sites without buffers, different food web characteristics can be identified, especially if there are inconsistencies in the nutrient concentrations. While this is intriguing, it would be more beneficial if a second watershed with more urban areas and anthropogenic effects was used. The St. Jones Watershed is composed of over 21% urban development. If similar work can be done in each of these watersheds, we may be able to discern the effects of different land uses on Delaware's coastal waterways.

The Chesapeake Bay is under increasing pressures from anthropogenic disturbances at various temporal and spatial scales. Water quality monitoring is vital for assessing such impacts, and further provides important information for sustainable water resource managements. The research in this chapter demonstrates the applications of hyperspectral remote sensing in retrieval of the water quality parameters in such an optically complex system. Further development of retrieval algorithms is still needed in order for the remote sensing to be routinely used in the water quality monitoring.

Based on the studies discussed in this chapter, in almost every watershed studied, major land use and cover change are due to the conversion of agriculture land to developed land in the Mid-Atlantic region. Urban runoff has gradually become a dominate pollution sources to the natural environment, including wetlands and open water bodies. More attention needs to be paid to locate the position and size of land conversion in order to effectively manage and control urban runoff. Also, it is necessary to monitor the spread of invasive plants, especially *P. australis* and its impact on vegetation communities in coastal watersheds within the region. Hyperspectral images have advantages in accurate land classification and can keep track of detailed changes in land use and cover. This will help build more effective land management practices in Mid-Atlantic region.

All studies above reveal that it is feasible to measure water quality indicators including chlorophyll-*a*, TSS, and CDOM using hyperspectral images and simple spectral algorithms. These algorithms involve spectral bands in the red and red-edge ranges and they can capture unique spectral signatures of these water quality indicators such as chlorophyll-*a* in coastal waters. All studies suggest the advantage of using hyperspectral images in estimating chlorophyll a-concentrations, TSS, and CDO in various coastal water bodies. Future research focusing on the direct measurements of in situ inherent optical properties (IOPs) such as absorption coefficients and backscattering for different estuarine systems is needed to develop better algorithms for water quality retrievals which include TSS and CDOM by remote sensing in coastal waters. With application of remote-sensing technology, real-time data collection of water quality parameters and plant community changes for the entire watershed will help resource managers to develop better management strategies.

Acknowledgements

We would like to thank the following individuals for their assistance and support for research discussed and in preparation of this chapter: Matthew Stone, Kris Roeske, Laurieann Phalen, Morgan State University Patuxent Environmental and Aquatic Environmental Laboratory (PEARL) and Delaware State University Aquatic Sciences Laboratory team members for field assistance and the early comments and discussions on the research papers. The authors also thank Dr. Shobha Sriharan at the Virginia State University for her support and Delaware State University for providing support and resources to conduct the research in this chapter. Case studies provided in this chapter are funded in the following order:

Case Study 2.1 is funded by NSF EPSCoR Grant Award# EPS-1301765 and the USDA-NIFA CBG Grant Award# 2013-38821-21246 to the Virginia State University.

Case Study 2.2 is funded by USDA-NIFA CBG Grant Award# 2010-38821-21454 to the Delaware State University and USDA Evans-Allen Grant Award# DELXDSUGO2015.

Case Study 2.3 is partly supported by NOAA-ECSC Grant Award# NA11SEC4810001, the NSF Grant Award# GEO0914546 to the Morgan State University, and the NSF Grant Award#

1036586 to the University of Maryland Eastern Shore and the USDA-NIFA CBG Grant Award # 2011-38821-30892 to the Virginia State University.

Case Study 2.4 is partly supported by NOAA-ECSC Grant Award# NA11SEC4810001, the NSF Grant Award# 1036586 to the University of Maryland Eastern Shore, and the USDA-NIFA Grant Award# 2013-38821-21246 to the Virginia State University.

Case Study 2.5 is funded by NOAA-ECSC Grant Award# NA11SEC4810001.

Author details

Gulnihal Ozbay[1]*, Chunlei Fan[2] and Zhiming Yang[3]

*Address all correspondence to: gozbay@desu.edu

1 Department of Agriculture and Natural Resources, Delaware State University, Dover, DE, USA

2 Patuxent Environmental and Aquatic Research Laboratory, Morgan State University, Baltimore, MD, USA

3 Department of Environmental, Earth and Geospatial Sciences, North Carolina Central University, Durham, NC, USA

References

[1] NOAA National Ocean Service. 2016. What Percentage of the American Population Lives near the Coast? http://oceanservice.noaa.gov/facts/population.html (accessed 15 June 2016).

[2] Stone ML. 2016. Land Use Practice and Water Quality, Blue Crab Population Dynamics and Fish Biodiversity in Blackbird Creek, Delaware. Master's Thesis. Natural Resources Program, Department of Agriculture and Natural Resources, Delaware State University, Dover, Delaware. p. 103.

[3] Nehrling R. 2013. Fertilizer Use and Markets: USDA Website. http://www.ers.usda.gov/topics/farm-practices-management/chemical-inputs/fertilizer-use-markets.aspx (accessed 15 June 2016).

[4] NOAA National Ocean Service. 2008. Estuaries: Human Disturbances to Estuaries. http://oceanservice.noaa.gov/education/kits/estuaries/ estuaries09_humandisturb.html (accessed 15 June 2016).

[5] Fan C, Warner R, Lacouture R. 2011. Hyperspectral Remote Sensing of Phytoplankton Taxonomy in Estuarine Waters. Ecology and Oceanography of Harmful Algal Blooms (ECOHAB), EPA-G2008-STAR-A1. Proposal. p. 38.

[6] Simenstad CA, Tanner CD, Thom RM, Conquest LL. 1991. Estuarine Habitat Assessment Protocol. Report to U.S. Environmental Protection Agency, Region 10, Seattle, Washington. EPA 910/9/91/037. p. 201.

[7] Stribling JM, Cornwell JC. Identification of important primary producers in a Chesapeake Bay tidal creek system using stable isotopes of carbon and sulfur. Estuaries and Coast 1997; 20(1): 77–85. doi:10.2307/1352721.

[8] Weinstein MP, Balletto JH. Does the common reed, *Phragmites australis*, affect essential fish habitat? Estuaries and Coasts 1999; 22: 793–802. doi:10.2307/1353112.

[9] Meyerson LA, Saltonstall K, Windham L, Kiviat E, Findlay S. A comparison of *Phragmites australis* in freshwater and brackish marsh environments in North America. Wetlands Ecology and Management 2000; 8: 89–103.

[10] Windham L, Ehrenfeld JG. Net impact of plant invasion on nitrogen-cycling processes within a brackish tidal marsh. Ecological Applications 2003; 13(4): 883–896.

[11] Able KW, Hagan SM, Brown SA. Mechanisms of marsh habitat alteration due to *Phragmites*: response of young-of-the-year mummichog (*Fundulus heteroclitus*) to treatment for *Phragmites* removal. Estuaries and Coasts 2003; 26(2):484–494.

[12] Ozbay G, Augustine A, Fletcher R. Remote sensing of *Phragmites australis* invasion in Delaware tidal marsh zones: issues to consider. Journal of Geophysics and Remote Sensing. 2003. doi: 10.4172/2169-0049.100e107. doi:10.3368/er.28.3.254.

[13] Orson RA. A paleoecological assessment of Phragmites australis in New England tidal marshes: changes in plant community structure during the last few millennia. Biology Invasions 1999; 1: 149–158.

[14] Warren RS, Fell PE, Grimsby JL, Buch EL, Rilling GC. Rates, patterns, and impacts of *Phragmites australis* expansion and effects of Phragmites control on vegetation, macro-invertebrates, and fish within tidal wetlands of the lower Connecticut River. Estuaries 2001; 24: 90–107.

[15] DNREC. 2016. Delaware Private Lands Assistance Program. http://www.dnrec.delaware.gov/fw/dplap/services/LIP/Pages/PhragmitesFacts.aspx (accessed 15 August 2016).

[16] NERRS. 2009. National Estuarine Research Reserve System. St. Jones River Site Description. http://nerrs.noaa.gov (accessed 18 June 2016).

[17] Index Mundi. 2014. United States—Population Growth Rate (2010–2014) by State http://www.indexmundi.com/facts/united-states/quick-facts/all-states/population-growth#map) (accessed 15 June 2016).

[18] Delaware Watersheds, State of Delaware. 2016. Blackbird Creek. http://delawarewatersheds.org/the-delaware-bay-estuary-basin/blackbird-creek/ (accessed 15 June 2016).

[19] DNREC.2006.BlackbirdCreekWatershedproposedTMDLs.ReportDNRE0080.p.54.http://www.dnrec.delaware.gov/swc/wa/Documents/TMDL_TechnicalAnalysisDocuments/5_BlackbirdTMDLAnalysis.pdf (accessed 15 June 2016).

[20] Rabalais NN, Turner RE, Justic E, Dortch Q, Wiseman WJ, Gupta BKS. Nutrient changes in the Mississippi river and system responses on the adjacent continental shelf. Estuaries 1996; 19:386–407.

[21] Boesch DF, Brinsfield RB, Magnien RE. Chesapeake Bay eutrophication: scientific understanding, ecosystem restoration, and challenges for agriculture. Journal of Environmental Quality 2001; 30:303–320.

[22] The News Journal. 2014. Delaware's Broiler Chicken Output Grows. http://www.delawareonline.com/story/money/business/2014/04/29/delawares-broiler-chicken-output-grows/8475187/ (accessed 15 June 2016).

[23] USDA NASS. 2010. Delaware Agriculture. https://www.nass.usda.gov/Statistics_by_State/Delaware/Publications/DE%20Ag%20Brochure_web.pdf (accessed 12 August 2016).

[24] Massey MS, Davis JG, Sheffield RE, Ippolito JA. 2007. Struvite production from dairy wastewater and its potential as a fertilizer for organic production in calcareous soils. In: International Symposium on Air Quality and Waste Management for Agriculture. CD-Rom Proceedings of the 16–19 September 2007, Conference (Broomfield, Colorado), USA. ASABE Publication Number 701P0907cd.

[25] Sharpley AN, Herron S, Daniel T. Overcoming the challenges of phosphorous-based management in poultry farming. Journal of Soil Water Conservation 2007; 62:375–389.

[26] USEPA. 2001. Ambient Water Quality Criteria Recommendations—Information Supporting the Development of State and Tribal Nutrient Criteria—Rivers and Streams in Nutrient Ecoregion XIV: Report EPA 822B-00-022. p. 84.

[27] Schindler DW. Eutrophication recovery in experimental lakes: implications for lake management. Science 1974; 184(4139):897–899. doi: 10.1126/science.184.4139.897.

[28] Schindler DW. Evolution of phosphorus limitation in lakes. Science 1977; 195(4275):260–262. doi: 10.1126/science.195.4275.260.

[29] Havens KE. Cyanobacteria blooms: effects on aquatic ecosystems. Advances in Experimental Medicine and Biology 2008; 619:733–747.

[30] Anderson DM, Glibert PM, Burkholder JM. Harmful algal blooms and eutrophication: nutrient sources, composition, and consequences. Estuaries 2002; 25:704–726.

[31] Peterson BJ, Hobbie JE, Hershey AE, Lock MA, Ford TE, Vestal JR, McKinley VL, Hullar MAJ, Miller MC, Ventullo RM, Volk GS. Transformation of a Tundra river from heterotrophy to autotrophy by addition of phosphorous. Science 1985; 229:1383–1386.

[32] Ozbay G, Roeske K, Chintapenta LK, Kalavacharla V, Stone M, Phalen L. Land use impacts: the effects of non-native grasses on marsh and aquatic ecosystem. Ecosystems and Ecography 2014; 4:1–5. http://dx.doi.org/10.4172/2157-7625.1000151.

[33] Able KW, Hagan SM. Effects of common reed (*Phragmites australis*) invasion on marsh surface macrofauna: response of fishes and decapod crustaceans. Estuaries 2000; 23: 633–646.

[34] Hunter KL, Fox DA, Brown LM, Able KW. Responses of resident marsh fishes to stages of *Phragmites australis* invasion in three mid Atlantic estuaries. Estuaries and Coasts 2006; 29: 487–498.

[35] Jivoff PR, Able KW. English Title: Blue crab, *Callinectes sapidus*, response to the invasive common reed, *Phragmites australis*: abundance, size, sex ratio, and molting frequency. Estuaries 2003; 26: 587–595.

[36] Seliskar DM, Smart KE, Higashikubo BT, Gallagher JL. Seedling sulfide sensitivity among plant species colonizing Phragmites infested wetlands. Wetlands 2004; 24: 426–433.

[37] Seliskar DM. 2008. Phragmites australis: A Closer Look at a Marsh Invader. http://www1.udel.edu/PR/UDaily/2008/oct/PhragmitesBulletin.pdf (accessed 15 June 2016).

[38] DNERR. 2005. Delaware Estuarine Research Reserve Management Plan. https://coast.noaa.gov/data/docs/nerrs/Reserves_DEL_MgmtPlan.pdf. p. 124 (accessed 18 June 2016).

[39] DNREC. 2005. Delaware Bay and Estuary Whole Basin Assessment Report. Delaware Department of Natural Resources and Environmental Control Document Number 40-01-01/05/02/01, Dover, USA.

[40] State of Delaware 1994. Statewide Wetland Mapping Project (SWMP). Prepared for the State of Delaware's Department of Natural Resources and Environmental Control (DNREC) and for the Department of Transportation (DELDOT), Dover, USA.

[41] Tiner RW. 2001. Delaware's Wetlands: Status and Recent Trends. U.S. Fish and Wildlife Service, Cooperative National Wetlands Inventory Publication, Northeast Region, Hadley, Massachusetts, USA.

[42] Rogerson A, Jacobs A, Howard A. 2010. Condition of Wetlands in the St. Jones River Watershed. January 2010. St. Jones River Watershed Report. Final report submitted to U.S. Environmental Protection Agency Region III for Assistance # WL-97329901-0 to the Delaware Department of Natural Resources and Environmental Control, Water Resources Division/Watershed Assessment Section Dover, Delaware 19904. p. 79.

[43] Harding LW Jr, Perry E. Long-term increase of phytoplankton biomass in Chesapeake Bay. Marine Ecology Progress Series 1997; 157:39–52.

[44] Nixon SW, Oviatt CA, Frithsen J, Sullivan B. Nutrients and the productivity of estuarine and coastal marine systems. Journal of the Limnological Society of Southern Africa 1986; 12:43–71.

[45] Harding LW Jr. Mallonee ME, Perry ES. 2013. Long-term trend analysis of chlorophyll in Chesapeake Bay using shipboard and aircraft observations. Marine Ecology Progress Series (in preparation).

[46] PACD. 2009. PA Chesapeake Bay Education Office. http://pacd.org/education/chesapeake-bay-education-office/ (accessed 20 July 2016).

[47] Chesapeake Bay Foundation Report. 2016. What is Killing the Bay? http://www.cbf.org/how-we-save-the-bay/chesapeake-clean-water-blueprint/what-is-killing-the-bay#continued (accessed 20 July 2016).

[48] The PEW Environment Group. 2011. Big Chicken. Pollution and Industrial Poultry Production in America. Washington, DC. 2004. p. 36. http://www.pewtrusts.org/~/media/legacy/uploadedfiles/peg/publications/report/pegbigchickenjuly2011pdf.pdf (accessed 15 June 2016).

[49] Ogejo JA. 2009. Selecting a Treatment Technology for Manure Management. Virginia Cooperative Extension, Virginia Tech. Publication 442–306.

[50] Chesapeake Bay Program. 2007. The Chesapeake Bay Watershed. http://www.chesapeakebay.net/discover/baywatershed (accessed 15 June 2016).

[51] Atallah L, Lo B, Yang G. Can pervasive sensing address current challenges in global healthcare? Journal of Epidemiology and Global Health 2012; 2(1):1–13. doi:10.1016/j.jegh.2011.11.005.

[52] NOAA National Ocean Service. 2015. What is Remote Sensing? http://oceanservice.noaa.gov/facts/remotesensing.html (accessed 15 June 2016).

[53] USGS. 2016. Land Remote Sensing Program. http://remotesensing.usgs.gov/%5C/gallery/gallerynojs.php?id=399&cat=8 (accessed 15 June 2016).

[54] Fan C, Glibert PM, Burkholder JM. Characterization of the affinity for nitrogen, uptake kinetics, and environmental relationships for *Prorocentrum minimum* in natural blooms and laboratory cultures. Harmful Algae 2003; 2(4): 283–299.

[55] Anderson DM, Burkholder JM, Cochlan WP, Glibert PM, Gobler CJ, Heil CA, Kudela RM, Parsons ML, Rensel JEJ, Townsend DW, Trainer VL, Vargo GA. Harmful algal blooms and eutrophication: examining linkages from selected coastal regions of the United States. Harmful Algae 2008; 8(1): 39–53.

[56] Fourqurean JW, Jones RD, Zieman JC. Process influencing water column nutrient characteristics and phosphorus limitation of phytoplankton biomass in Florida Bay, FL, USA: inferences from spatial distributions. Estuarine, Coastal and Shelf Science 1993; 36(3):295–314.

[57] Kendrick P. Remote sensing and water quality. Journal Water Pollution Control Federation 1976; 48(10): 2243–2246.

[58] Brando VE, Dekker AG. 2003. Satellite hyperspectral remote sensing for estimating estuarine and coastal water quality. IEEE Transactions on Geoscience and Remote Sensing 2003; 41(6): 1378–1387.

[59] Campbell G, Phinn SR, Dekker AG, Brando V E. Remote sensing of water quality in an Australian tropical freshwater impoundment using matrix inversion and MERIS images. Remote Sensing of Environment 2011; 115(9): 2402–2414.

[60] Mustard JF, Sunshine JM. 1999. Spectral analysis for Earth science: investigation using remote sensing data. In: Rencz, A.N. (Ed.), Remote Sensing for the Earth Science: Manual of Remote Sensing, Volume 3, 3rd Edition. Wiley, New York, NY, USA.

[61] GIS Geography. 2016. Multispectral vs Hyperspectral Imagery Explained. http://gisgeography.com/multispectral-vs-hyperspectral-imagery-explained/ (accessed 15 June 2016).

[62] Fisher R, Perkins S, Walker A, Wolfart E. 2003. Multi-Spectral Images. http://homepages. inf.ed.ac.uk/rbf/HIPR2/mulimage.htm (accessed 20 May 2016).

[63] Elhadi A, Mutanga O, Rugege D. Multispectral and hyperspectral remote sensing for identification and mapping of wetland vegetation: a review. Wetlands Ecology and Management 2010; 18: 281–296.

[64] Shippert P. 2012. Introduction to Hyperspectral Image Analysis. Research Systems. Ohio Space Journal. p. 13. http://spacejournal.ohio.edu/pdf/shippert.pdf (accessed 20 May 2016).

[65] Artigas F, Pechmann IC. Balloon imagery verification of remotely sensed *Phragmites australis* expansion in an urban estuary of New Jersey, USA. Landscape and Urban Planning 2010; 95: 105–112.

[66] Bukata, RP, Jerome JH, Kondratyev KY, Pozdnyakov DV. 1995. Optical Properties and Remote Sensing of Inland and Coastal Waters. CRC Press, Taylor and Francis Group, Boca Raton, FL, USA. p. 368.

[67] Chang GK, Mahoney A, Briggs-Whitmire DK, Mobley C, Lewis M, Moline M, Boss E, Kim M, Philpot W, Dickey T. The new age of hyperspectral oceanography. Oceanography 2004; 17(2): 16–23.

[68] Nair RR, Blake P, Grigorenko AN, Novoselov KS, Booth TJ, Stauber T, Peres NMR, Geim AK. Fine structure constant defines visual transparency of graphene. Science 2008; 320(5881):1308–1308. ISSN 0036-8075.

[69] Louchard EM, Reid RP, Stephens FC, Davis CO, Leathers RA, Downes TV. Optical remote sensing of benthic habitats and bathymetry in coastal environments at Lee Stocking Island, Bahamas: a comparative spectral classification approach. Journal of Limnology and Oceanography 2003; 48(1–2): 511–521. The American Society of Limnology and Oceanography, Inc.

[70] Althuis IJA. Suspended particulate matter detection in the North Sea by hyperspectral airborne remote sensing. Aquatic Ecology 1998; 32: 93–98.

[71] Tsai F, Philpot W. Derivative analysis of hyperspectral data. Remote Sensing of Environment 1998; 66:41–51.

[72] Ruffin C, King RL, Younan NH. A combined derivative spectroscopy and Savitzky-Golay filtering method for the analysis of hyperspectral data. GIScience and Remote Sensing 2008; 45 (1):1–15.

[73] Senay GB, Shafique NA, Autrey BC, Fulk F, Cormier SM. The selection of narrow wavebands for optimizing water quality monitoring on the Great Miami River, Ohio using hyperspectral remote sensor data. Journal of Spatial Hydrology 2002; 1(1):1–24.

[74] Hunter PD, Tyler AN, Carvalho L, Codd GA, Maberly SC. Hyperspectral remote sensing of cyanobacterial pigments as indicators for cell populations and toxins in eutrophic lakes. Remote Sensing of Environment 2010; 114(11): 2705–2718.

[75] Jo Y-H, Oliver MJ, Ozbay G. Estimating distributions of biogeochemical constituents using hyperspectral remote sensing in Delaware Bay. Delaware EPSCOR Seed Grant Program 2011–2012. Focal Area: Theme 3: Environmental Observation and Sensing. p. 17.

[76] Jo Y-H, Yan XH. 2011. Integrated altitude-adjustable remote sensing systems and methods, Patent Number, UOD-382USP (in pending).

[77] Vitousek PM, Mooney HA, Lubchenco J, Melillo JM. Human domination of Earth's ecosystems. Science 1997; 277(5325):494–499. doi: 10.1126/science.277.5325.494.

[78] Matson PA, Parton WJ, Power AG, Swift MJ. Agricultural intensification and ecosystem properties. Science 1997; 277(5325):504–509. doi: 10.1126/science.277.5325.504.

[79] Lotze HK, Lenihan HS, Bourque BJ, Bradbury RH, Cooke RG, Key MC, Kidwell SM, Kirby MX, Peterson CH, Jackson JB. Depletion, degradation, recovery potential of estuaries and coastal seas. Science 2006; 312 (1806). doi: 10.1126/science.1128035.

[80] Fondriest Environmental Inc. 2016. Turbidity, Total Suspended Solids & Water Clarity. http://www.fondriest.com/environmental-measurements/parameters/water-quality/turbidity-total-suspended-solids-water-clarity/

[81] Davies-Colley RJ, Smith DG. Turbidity, suspended sediment, and water clarity: a review. Journal of the American Water Resources Association 2001; 37:1085–1101.

[82] Cloern J. Turbidity as a control on phytoplankton biomass and productivity in estuaries. Continental Shelf Research 1987; 7:1367–1381.

[83] De Robertis A, Ryer CH, Veloza A, Brodeur RD. Differential effects of turbidity on prey consumption of piscivorous and planktivorous fish. Canadian Journal of Fish and Aquatic Sciences 2003; 60:1517–1526.

[84] Utne-Palm AC. Visual feeding of fish in a turbid environment: physical and behavioural aspects. Marine and Freshwater Behavior and Physiology 2001; 35:111–128.

[85] Pennock JR, Sharp JH. Phytoplankton production in the Delaware estuary: temporal and spatial variability. Marine Ecology Progress Series 1986; 34:143–155.

[86] Kirk JTO. 1994. Light and Photosynthesis in Aquatic Ecosystems: 2nd Edition. Cambridge University Press: New York City, NY, USA. p. 509.

[87] Olli G, Darracq A, Destouni G. Field study of phosphorous transport and retention in drainage reaches. Journal of Hydrology 2009; 365:46–55.

[88] Correll D. The role of phosphorous in the eutrophication of receiving waters: a review. Journal of Environmental Quality 1998; 27:261–267.

[89] Reddy KR, Admas JF, Richardson C. Potential technologies for remediation of Brownfield. Practice Periodical of Hazardous Toxic and Radioactive Waste Management 1999; 3:61–68.

[90] Conley DJ. Biogeochemical nutrient cycles and nutrient management strategies. Hydrobiologia 2000; 410:87–96.

[91] RMB Environmental Laboratories Inc. 2016. Chlorophyll-*a*. http://rmbel.info/chlorophyll-a/ (accessed 15 June 2016).

[92] Gallina N, Anneville O, Beniston M. Impacts of extreme air temperatures on cyanobacteria in five deep peri-Alpine lakes. Journal of Limnology 2011; 70(2): 186–196. DOI: 10.3274/JL11-70-2-04.

[93] Ritchie JC, Cooper CM. 2001. Remote Sensing Techniques for Determining Water Quality: Applications to TMDLS, USDA, Water Environment Federation.

[94] Fondriest Environmental Inc. 2016. Chlorophyll. http://www.fondriest.com/science-library/water-quality/chlorophyll (accessed 15 June 2016).

[95] Behrenfeld M, Boss E, Siegel D, Shea D. Biospheric primary production during an ENSO transition. Science 2001; 291, 2594–2597.

[96] Lillesand TM, Kiefer RW. 1994. Remote Sensing and Image Interpretation. New York, NY: Wiley and Sons.

[97] Clark D. 1981. Phytoplankton algorithms for the Nimbus-7 CZCS; In Gower, R. (Ed.) Oceanography from Space, Plenum Press, New York, NY, p. 227.

[98] Stumpf RP, Culver ME, Tester PA, Tomlinson M, Kirkpatrick GJ, Pederson BA, Truby E, Ransibrahmanakul V, Soracco M. Monitoring *Karenia brevis* blooms in the Gulf of Mexico using satellite ocean color imagery and other data. Harmful Algae 2003; 2:147–160.

[99] Glibert PM, Magnien R, Lomas MW, Alexander J, Fan C, Haramoto E, Trice M, Kana TM. Harmful algal blooms in the Chesapeake and coastal Bays of Maryland, USA: comparison of 1997, 1998, and 1999 events. Estuaries 2001; 24: 875–883.

[100] Ritchie JC, Seyfried MS, Chopping MJ, Pachepsky Y. Airborne laser technology for measuring rangeland conditions, Journal of Range Management 2001; 54(suppl):A8–A21.

[101] Roeske K. 2013. Assessing the impact of *Phragmites australis* subspecies Australis on blue crab (*Callinectes sapidus*) and fish population dynamics in Blackbird Creek, Delaware. Master's Thesis. Natural Resources Program. Department of Agriculture and Natural Resources, Delaware State University, Dover, DE, USA. p 230.

[102] Friends of the Earth Europe. 2013. The Environmental Impacts of Glyphosate. http://www.foeeurope.org/sites/default/files/press_releases /foee_5_environmental_impacts_glyphosate.pdf> (accessed 15 June 2016).

[103] Fish and Wildlife Service. 2004. Comprehensive Conservation Plan. Accessed at <http://www.fws.gov/mountain-prairie/planning /ccp/ut/fhs/documents /fhs_2004_ccpdraft_4consequences.pdf > (accessed 20 May 2016).

[104] Lowery TA, Tate LG. Effect of hypoxia on hemolymph lactate and behavior of the blue crab *Callinectes sapidus* Rathbun in the laboratory and field. Comparative Biochemistry and Physiology, Part A. 1986; 85: 689–692.

[105] Talke SA, Stacey MT. Suspended sediment fluxes at an intertidal flat: the shifting influence of wave, wind, tidal, and freshwater forcing. Continental Shelf Research 2007; 28: 710–725.

[106] Wischmeier WH, Smith DD. 1978. Predicting Rainfall Erosion Losses—A Guide to Conservation Planning. USDA, Science and Education Administration, Hyattsville, MD, USA. p. 62.

[107] Steele JH, Henderson FW. The role of predation in plankton models. Journal of Plankton Research 1992; 14: 157–172.

[108] Perissinotto R, Pillay D, Bate G. Microalgal biomass in the St. Lucia Estuary during the 2004–2007 drought period. Marine Ecology Progress Series 2010; 405: 147–161.

[109] Carpenter SR, Caraco NF, Correll DL, Howarth RW, Sharpley AN, Smith VH. Nonpoint pollution of surface waters with phosphorus and nitrogen. Ecological Applications Journal 1998; 8: 559–568.

[110] Bennett EM, Carpenter SR, Caraco NF. Human impact on erodible phosphorus and eutrophication: a global perspective. Bioscience 2001; 51: 227–234.

[111] Davis RL, Zhang H, Schroder JL, Wang JJ, Payton ME, Zazulak A. Soil characteristics and phosphorus loss in runoff. Journal of Environmental Quality 2005; 34: 1640–1650.

[112] Holliday CP, Rasmussen TC, Miller WP. 2003. Establishing the relationship between turbidity and total suspended sediment concentration. Proceedings of the 2003 Georgia Water Resources Conference.

[113] Wall GJ, Bos AW, Marshall AH. The relationship between phosphorous and suspended sediment loads in Ontario watersheds. Journal of Soil Water Conservation 1996; 51: 504–507.

[114] Dahl TE. 1990. Wetlands Loss Since the Revolution. Wetland Losses in the United States 1780s to 1980s. National Wetlands Inventory. U.S. Fish and Wildlife Service, 9720 Executive Center Drive, Suite 101, Monroe Building, St. Petersburg, FL, USA.

[115] Smith LED, Siciliano G. A comprehensive review of constraints to improved management of fertilizers in China and mitigation of diffuse water pollution from agriculture. Agriculture, Ecosystems and Environment 2015; 209: 15–25. doi:10.1016/j.agee.2015.02.016.

[116] Salem F, Ozbay G, Roper W. Identification of Suspended Sediments and Chlorophyll-a in the Blackbird Creek, Delaware Using AISA Hyperspectral Data and Analysis. Delaware State University, Department of Agriculture and Natural Resources, Dover, DE, USA. Unpublished Report for the USDA-NIFA CBG Grant Award# 2010-38821-21454. p23.

[117] Shafique NA, Fulk F, Autrey BC, Flotemersch J. Hyperspectral Remote Sensing of Water Quality Parameters for Large Rivers in the Ohio River Basin. ICRW Proceedings 2001; pp. 216–221. http://www.tucson.ars.ag.gov/icrw/proceedings/shafique.pdf (accessed 15 June 2016).

[118] Agawin SR, Duarte CM, Agustı S. Nutrient and temperature control of the contribution of picoplankton to phytoplankton biomass and production. By the American Society of Limnology and Oceanography, Inc., Limnology and Oceanography Journal 2000; 45(3): 591–600.

[119] Ritchie JC, Zimba PV, Everitt JH. 1976. Remote sensing techniques to assess water quality. Photogrammetric Engineering and Remote Sensing 03-907. p. 37. ftp://hrsl.ba.ars.usda. gov/pub/incoming/RSinARS/white_papers/03-907a.pdf. Retrieved by June 15, 2016.

[120] Härmä P, Vepsäläinen J, Hannonen T, Pyhälahti T, Kämäri J, Kallio K, Eloheimo K, Koponen S. Detection of water quality using simulated satellite data and semi-empirical algorithms in Finland. The Science of Total Environment 2001; 268(1–3):107–121.

[121] Ammenberg J, Hjelm O, Quotes P. The connection between environmental manage-ment systems and continual environmental performance improvements. Corporate Environmental Strategy 2002; 9:183–192.

[122] Ansotegui A, Sarobe A, Trigueros JM, Urrutxurtu I, Orive E. Size distribution of algal pigments and phytoplankton assemblages in a coastal estuarine environment: contribu-tion of small eukaryotic algae, Journal of Plankton Research 2003; 25(4):341–355.

[123] Turner JT, Tester PA. Toxic marine phytoplankton, zooplankton grazers, and pelagic food webs. Part 2: The Ecology and Oceanography of Harmful Algal Blooms. Limnology and Oceanography. 1997; 42 (5): 1203–1214.

[124] Tester PA, Stumpf RP, Vukovich FM, Fowler PK, Turner JT. An expatriate red tide bloom: transport, distribution, and persistence. Limnology and Oceanography 1991; 36:1953–1961.

[125] Keafer BA, Anderson DM. 1993, Use of remotely-sensed sea surface temperatures in studies of *Alexandrium tamarense* bloom dynamics. In Toxic Phytoplankton in the Sea, Proceedings of 5th International Conference. Smayda TM and Shimizu Y. (Ed.), Elsevier, Amsterdam. pp. 763–768.

[126] Hovis WA. Nimbus-7 Coastal Zone Color Scanner system description and initial imag-ery. Science 1980; 210: 60–63.

[127] Feldman GC, Kuring N, Ng C, Esaias W, McClain C, Elrod J, Maynard N, Endres D, Evans R, Brown J, Walsh S, Carle M, Podesta G. 1989. Ocean color: availability of the global data set. Eos Trans. AGU, 70, 640.

[128] IOCCG. 1999. Status and plans for satellite ocean colour mission: considerations for complementary missions", In Yoder J. (Ed.), Reports of the International Ocean Colour Coordinating Group, Number 2 IOCCG, Dartmouth, N.S., Canada.

[129] Antoine D, Morel A, Gentili B, Gordon H, Banzon V, Evans R, Brown J, Walsh S, Baringer W, Li A. In search of long-term trends in ocean color. Eos 2003; 84(32) 301: 308–309.

[130] Smith RC, Baker KS. Oceanic chlorophyll concentrations as determined by satellite (Nimbus-7 Coastal Zone Color Scanner). Marine Biology 1982; 66:269–279.

[131] Mantoura RFC, Llewellyn CA. The rapid determination of algal chlorophyll and carotenoid pigments and their breakdown products in natural waters by reverse-phase high performance liquid chromatography. Analytica Chimica Acta 1983; 151:297–314.

[132] Gieskes WW. 1991. Algal pigment fingerprints: clue to taxon specific abundance, productivity, and degradation of phytoplankton in seas and oceans. In Demers S (Ed.), Particle Analysis in Oceanography. NATO ASI Series. Vol. G27 Springer-Verlag, Berlin. pp. 61–99.

[133] Millie DF, Paerl HW, Hurley JP, Kirkpatrick GJ. Microalgal pigment assessments using high performance liquid chromatography: a synopsis of organismal and ecological applications. Canadian Journal of Fisheries and Aquatic Science 1993; 50:2513–2527.

[134] Morel A, Prieur L. Analysis of variation in ocean color. Limnology and Oceanography Journal 1997; 22: 709–722.

[135] Darecki M, Weeks A, Sagan S, Kowalczuk P, Kaczmarek S. Optical characteristics of two contrasting case 2 waters and their influence on remote sensing algorithms. Continental Shelf Research 2003; 23(3–4): 237–250.

[136] Davis CO, Bowles J, Leathers RA, Korwan D, Downes T V, Snyder WA, Rhea WJ, Chen W, Fisher J, Bissett WP, Reisse RA. Ocean PHILLS 18 hyperspectral imager: design, characterization, and calibration. Optics Express 2002; 10(4): 210–221.

[137] Chang GC, Mahoney K, Briggs-Whitmire A, Kohler D, Mobley C, Lewis M, Moline M, Boss E, Kim M, Philpot W, Dickey T. The new age of hyperspectral oceanography. Oceanography 2004; 17(2): 22–29.

[138] Johnsen G, Nelson NB, Jovine RVM, Prezelin BB. Chromoprotein and pigment dependent modeling of spectral light absorption in two dinoflagellates, *Prorocentrum minimum* and *Heterocapsa pygmaea*. Marine Ecology Progress Series 1994; 114: 245–258.

[139] Tester PA, Geesey ME, Guo C, Paerl HW, Millie DF. Evaluating photoplankton dynamics in the Newport River estuary (North Carolina) by HPLC derived pigment profiles. Marine Ecology Progress Series 1995; 124: 237–245.

[140] Millie DF, Schofield OM, Kirkpatrick GJ, Johnson G, Tester PA, Vinyard BT. Detection of harmful algal blooms using photopigments and absorption signatures: a case study of the Florida red tide dinoflagellate, *Gymnodium breve*. Limnology and Oceanography 1997; 42(5): 1240–1251.

[141] Schofield O, Arnone R, Bissett W, Dickey T, Davis C, Finkel Z, Oliver M, Moline M. Watercolors in the coastal zone: What can we see? Oceanography 2004; 17(2): 24–31.

[142] Schofield O, Bergmann T, Oliver M, Irwin A, Kirkpatrick G, Bissett W, Moline M, Orrico C. Inversion of spectral absorption in the optically complex coastal waters of the Mid-Atlantic Bight. Journal of Geophysical Research 2004; 109, C12S04. doi:10.1029/2003JC002071.

[143] Richardson LL, Kruse FA. 1999. Identification and classification of mixed phytoplankton assemblages using AVIRIS image-derived spectra. AVIRIS Workshop, Pasadena, CA, USA.

[144] Richardson LL, Kruse FA. 2000. Use of AVIRIS spectral data to discriminate mixed phytoplankton communities, seagrass beds, and benthic algal mats in Florida Bay, USA. AVIRIS Workshop, Pasadena, CA, USA.

[145] Glenn S, Schofield O. Observing the oceans from the COOL room: our history, experience, and opinions. Oceanography 2003; 16(4): 37–52.

[146] Schofield O, Grzymski J, Bissett WP, Kirkpatrick GJ, Millie DF, Moline M, Roesler CS. Optical monitoring and forecasting systems for harmful algal blooms: possibility or pipe dream? Journal of Phycology 1999; 35: 1477–1496.

[147] Fan C, Warner RA. Characterization of water reflectance spectra variability: implications for hyperspectral remote sensing in estuarine waters. Marine Science 2014; 4(1): 1–9.

[148] O'Reilly JE, Maritorena S, Mitchell BG, Siegel DA, Carder KL, Garver SA, Kahru M, McClain C. Ocean color chlorophyll algorithms for SeaWiFS. Journal of Geophysical Research 1998; 103: 24,937–24,953.

[149] Warner RA, Fan C. Optical spectra of phytoplankton cultures for remote sensing applications: focus on harmful algal blooms. International Journal of Environmental Science and Development 2013, 4(2); 94–98.

[150] Fan C. Spectral analysis of water reflectance for hyperspectral remote sensing of water quailty in estuarine water. Journal of Geoscience and Environment Protection 2014; 2(2): 19–27. s. http://www.scirp.org/journal/gep http://dx.doi.org/10.4236/gep.2014.22004.

[151] Olmanson LG, Brezonik PL, Bauer ME. Airborne hyperspectral remote sensing to assess spatial distribution of water quality characteristics in large rivers: the Mississippi River and its tributaries in Minnesota. Remote Sensing of Environment 2013; 130(0): 254–265.

[152] Toole DA, Siegel DA. 2001. Modes and mechanisms of ocean color variability in the Santa Barbara Channel. Journal of Geophysical Research: Oceans (1978–2012); 106(C11): 26985–27000.

[153] Lubac B, Loisel H. Variability and classification of remote sensing reflectance spectra in the eastern English Channel and southern North Sea. Remote Sensing of Environment 2007; 110(1): 45–58.

[154] Legleiter CJ, Roberts DA. Effects of channel morphology and sensor spatial resolution on image-derived depth estimates. Remote Sensing of Environment 2005; 95(2): 231–247.

[155] Schalles JF. 2006. Optical remote sensing techniques to estimate phytoplankton Chlorophyll-a concentrations in coastal. In Richardson LL, LeDrew EF (Eds.), Remote Sensing of Aquatic Coastal Ecosystem Processes, Science and Management Applications. Springer. Volume 9, pp. 27–79.

[156] Doxaran D, Cherukuru R, Lavender S. Use of reflectance band ratios to estimate suspended and dissolved matter concentrations in estuarine waters. International Journal of Remote Sensing 2005; 26(8): 1763–1769.

[157] DNERR. 1999. Delaware National Estuarine Research Reserve (DNERR) Estuarine Profile. p. 158. http://www.dnrec.delaware.gov/coastal/DNERR/Documents/Estuarine%20Profile%201999.pdf (accessed 15 June 2016).

[158] Yang Z, Reiter M. Land use/cover changes within riparian buffers in the State of Delaware. JEMREST 2008; 4: 39–52.

[159] Gitelson A. 1992.The peak near 700 nm on reflectance spectra of algae and water: relationships of its magnitude and position with chlorophyll concentration. International Journal of Remote Sensing 1992; 13: 3367–3373.

[160] Gitelson A. The nature of the peak near 700 nm on the radiance spectra and its application for remote estimation of phytoplankton pigments in inland waters, Optical Engineering and Remote Sensing, SPIE 1993; 1971: 170–179.

Water Treatment

Review of the Impact on Water Quality and Treatment Options of Cyanide Used in Gold Ore Processing

Benias C. Nyamunda

Additional information is available at the end of the chapter

http://dx.doi.org/10.5772/65706

Abstract

Cyanide has been widely used in several industrial applications such as electroplating photography, metal processing, agriculture, food and the production of organic chemicals, plastics, paints and insecticides. The strong affinity of cyanide for metals such as gold and silver makes it suitable for selective leaching of these metals from ores. Cyanide is highly toxic; hence, there is a need to regulate and limit the amount of cyanide that may be discharged into the environment. Technologies focusing on the use of physical, chemical and biological methods have been developed to reduce the concentration of cyanide and cyanide compounds in wastewaters to permissible limits. This chapter reviews the current and emerging technologies for treatment of cyanide from wastewaters generated in gold mining processes.

Keywords: cyanide, gold, leaching, oxidation, wastewater

1. Introduction

Cyanide is an extremely toxic substance that is produced naturally and artificially. Cyanide has been widely used in several industries applications such as textile, plastics, paints, photography, electroplating, agriculture, metal treatment and mining. The high binding affinity of cyanide for metals such as gold, zinc, copper and silver has enabled it to selectively leach these metals from ores.

Southern Africa is a region rich in minerals such as gold. Countries in the southern Africa boost their economies through vast investments in gold mining. Cyanide leaching has become the dominant gold extraction technology since the 1970s replacing previously used less efficient and more toxic mercury. These gold mines discharge effluent containing toxic cyanides into

natural water bodies posing the greatest threat to the quality of water intended for human use. Therefore, it is imperative to develop effective strategies for the removal of cyanide from aqueous industrial wastewater streams.

2. Gold extraction process using cyanide

Cyanidation is the predominant gold extraction technique since the late nineteenth century. The dissolution of gold in aqueous cyanide is commonly described using Elsner's equation [1]:

$$4Au + 8CN^- + O_2 + 2H_2O \rightarrow 4\left[Au(CN)_2\right]^- + 4OH^- \tag{1}$$

Gold dissolution is an electrochemical process in which oxygen is reduced at the cathodic zone, while gold is oxidised at anodic regions. The precise overall dissolution of gold in alkaline, aerated cyanide solutions taking place at cathodic and anodic regions is represented in Eqs. (2) and (3).

$$2Au + 4CN^- + O_2 + 2H_2O \rightarrow 2\left[Au(CN)_2\right]^- + H_2O_2 + 2OH^- \tag{2}$$

$$2Au + 4CN^- + H_2O_2 \rightarrow 2\left[Au(CN)_2\right]^- + 2OH^- \tag{3}$$

The main merits of cyanidation are the high selectivity of free cyanide for gold dissolution compared to other metals and an extremely high stability constant (2×10^{38}) of the gold cyanide complex [2].

Dilute sodium cyanide solutions within concentration ranges of 0.01–0.05% are used in mines for gold leaching [3]. Gold ore is subjected to physical processes such as milling, grinding and gravity separation prior to the addition of aqueous sodium cyanide to form slurry. The pH of the resulting extracting solution is increased by adding slaked lime or sodium hydroxide to prevent generation of toxic hydrogen cyanide [4]. The slurry pH is maintained at not less than 10.5 during cyanidation to prevent excessive loss of cyanide by hydrolysis through volatilisation of hydrogen cyanide. Oxygen an important component during cyanidation is continuously pumped into the slurry resulting in the formation of dicyanoaurate (I) complex.

Several methods are employed for cyanide leaching of gold ore [5]. However, agitation leaching is commonly used for most ores due to its commercial viability [6]. Leaching is typically done in steel vessels, and the solids are maintained in suspension by air or mechanical agitation.

The gold complex $NaAu(CN)_2$ is then extracted from leach solutions by adsorption onto solid adsorbents such as activated carbon or a synthetic ion exchange resin [7–11]. Activated carbon is the most commonly used adsorbent for gold extraction due to several favourable properties such as high adsorption capacity, good reactivation capabilities, low cost, readily available, high mechanical strength and wear resistance [12].

Gold complexes adsorbed onto activated carbon are eluted to produce concentrated high-grade gold solutions suitable for final gold recovery. Eluents such as sodium hydroxide [13] and organic solvents in aqueous solutions [14] have been used for desorption or stripping of gold from activated carbon.

Gold is extracted from solution into a concentrated solid form by a process termed recovery. Zinc precipitation [15] and electrowinning [16] have been used to treat concentrated gold solutions produced from activated carbon stripping. Eq. (4) represents the electrochemical reduction process for gold.

$$2\left[Au\left(CN\right)_2\right]^- + Zn + 4CN^- \rightarrow 2Au + 4CN^- + \left[Zn\left(CN\right)_4\right]^{2-} \tag{4}$$

The gold recovered from crude undergoes refining to produce crude bullion containing between 90 and 99.5% pure gold [17]. Refining involves roasting the crude gold to convert base metals such as iron, lead, copper and zinc to their respective oxides. This process is then followed by smelting, which removes base oxide impurities in form of slag. The bullion produced can be upgraded further to higher purity platinum group metals by processes such as pyrorefining, hydrorefining and electrorefining [18, 19]. These extraction processes leave behind toxic cyanide tailings.

3. Occurrence of cyanide in environment

Cyanide and related compounds are produced at low levels from plants such as sorghum, cassava, potato, broccoli, cashews and apricots [20]. Cyanide is found in certain bacteria, fungi and algae [21]. Anthropogenic sources of cyanide release also include smoke from cigarettes, automobile exhaust fumes and the production of acrylonitrile. Bulk occurrence of cyanide in the environment is attributed to the human operations in industries, metallurgical and mining activities. Cyanide is mainly produced industrially in form of hydrogen cyanide gas or solid sodium cyanide or potassium cyanide [22].

4. Forms of cyanide in aqueous solution

Compounds of cyanides present in water can be generally classified into total cyanide, complex cyanide and free cyanide [23–25]. These aqueous cyanide compounds exist as simple and complex cyanides, cyanates and nitriles. The most toxic form of cyanide is free cyanide, which exists either as cyanide anion or as hydrogen cyanide (HCN) depending on solution pH. HCN is predominant in aqueous systems at pH below 8.5 and can be readily volatilised [26, 27]. At higher pH values, the free cyanide is mainly in form of the cyanide anion. Aqueous cyanides form complexes with metal ions present in industrial wastewaters. These metallo-cyanide complexes exhibit different chemical and biological stabilities. The complexes are classified as weak acid dissociable (WAD) and strong acid dissociable [28, 29] in accordance with the metal-cyanide bond strength. Cadmium, copper, nickel and zinc form weak acid dissociable complexes that readily dissociate under acidic conditions [28]. Complexes of cyanide with cobalt, iron, silver and gold are strong acid dissociable (SAD). Both forms of complexes dissociate and release free cyanide. The stability of these complexes depends on several factors such as pH, light intensity, water temperature and total dissolved solids.

5. Toxicity of cyanide

Cyanide is extremely toxic to humans and aquatic life. Unlike toxic metal ions, the cyanide anion does not accumulate in the body, but instantaneously results in death of aquatic life and human beings in a short time at low dosages through depressing the central nervous system [30]. Cyanide strongly binds cytochromes inhibiting the electron transport chain in mitochondria and energy release in cells [31]. Liquid or gaseous hydrogen cyanide gains entry into the body through inhalation, ingestion or skin absorption. Exposing animals to hydrogen cyanide has several effects such as headaches, dizziness, numbness, tremor and loss of visual sharpness. Other toxic effects of cyanide include an enlarged thyroid gland, cardiovascular and respiratory problems.

6. Acceptable limits for the use of cyanide

There is need for the treatment of wastewater containing cyanide before discharging into the environment to protect water bodies. As a result of this, several countries and environmental bodies have imposed limiting standards for discharging wastewater containing cyanide to main natural water bodies. **Table 1** shows the set acceptable discharge limits of total cyanides by different organisations.

Agent	Cyanide limit	Reference
The U.S. Environmental Protection Agency (USEPA)	50 ppb (aquatic-biota)	[32]
	200 ppb (drinking)	
India Central Pollution Control Board (CPCB)	0.2 mg/L	[25]
Mexico	0.2 mg/L	[33]

Table 1. Permissible cyanide discharge limits in industrial effluents.

In view of the data outlined in **Table 1**, strategies aimed at cyanide recovery and removal need to be adopted to maintain concentrations within regulatory limits.

7. Cyanide removal strategies from wastewaters

Various cyanide attenuation processes have been successfully implemented in the treatment of industrial effluents. Gold mines in southern Africa have adopted various cyanide attenuation techniques aimed at reducing the toxin level in the tailings to internationally acceptable levels. The common methods of treating cyanide are natural, chemical and biological degradations [34].

7.1. Natural degradation of cyanide

Natural attenuation reactions occur in cyanide solutions placed in ponds or tailings resulting in the reduction in the cyanide concentration. The dominant natural degradation mechanism is volatilisation of hydrogen cyanide with subsequent atmospheric transformations to less toxic chemicals [34]. Other reactions such as hydrolysis, photolysis, oxidation, complex formation, oxidation to cyanate, thiocyanate formation and precipitation also take place. This natural process occurs with all cyanide solutions exposed to the atmosphere.

Cyanide forms complexes with metals ions in solution such as zinc, iron and copper. Ferri- and ferrocyanide complexes are extremely stable under most environmental conditions except when exposed to ultraviolet radiation [22]. Zinc and copper cyanide complexes are relatively unstable and can release free cyanide to the environment. Iron cyanide complexes are precipitated by several metals such as Zn, Cu, Ni, Pb, Sn, Cd and Ag over a wide range of pH (2–11). Cyanide and cyanide metal complexes adsorb onto clay, organic matter and oxides of aluminium, manganese and iron. The adsorbed cyanide can be naturally oxidised by oxygen, hydrogen peroxide and ozone into less toxic cyanate. The cyanate is hydrolysed under acidic conditions into ammonium salt and carbon dioxide. Cyanide can be biodegraded to ammonia, which is further oxidised to nitrate [35]. Under anaerobic conditions, HCN is hydrolysed to formic acid or ammonium formate as shown in Eq. (5).

$$HCN \rightarrow HCOO^- + NH_4^+ \tag{5}$$

Elemental sulphur and sulphur containing ores such as chalcopyrite react with cyanide to produce less toxic thiocyanate [36].

7.2. Chemical, physical and biological methods

Natural methods of cyanide attenuation have failed to produce effluents of acceptable quality. This has led to the development of numerous biological, physical and chemical treatment methods [37].

Among the methods used in removing cyanide from wastewater include photocatalysis [38], biotreatment [39], copper-catalysed hydrogen peroxide oxidation [40], ozonation [33], electrolytic decomposition, alkaline chlorination [22], reverse osmosis, thermal hydrolysis and adsorption [41]. Most of these methods have limited applications due to the high cost, production of toxic residues and incomplete degradation of all cyanide complexes [42, 43]. However, biodegradation of aqueous cyanide ions is cheaper than chemical and physical methods [30].

7.2.1. Chemical oxidation methods

7.2.1.1. Sulphur dioxide/air (INCO) process

This process was developed by a Canadian company, Inco metal limited, in 1984 [44]. The process makes use of air and sulphur dioxide in the catalytic oxidation of free and complexed cyanide to cyanate [37, 45, 46] as shown in Eq. (6). The process is catalysed by aqueous copper (II) ions under controlled pH of 8–10. The pH is normally maintained by addition of lime.

$$SO_2 + O_2 + CN^- + H_2O \xrightarrow{Cu^{+2}} SO_4^{-2} + OCN^- + 2H^+ \tag{6}$$

After completion of the oxidation process, previously metal ions complexed with cyanide such as Zn^{+2}, Cu^{+2} and Ni^{+2} are precipitated as metal hydroxides. This process effectively treats cyanide in slurries and solutions.

7.2.1.2. Alkaline chlorination

In this process, cyanide is oxidised by alkaline chlorine. The process converts all acid dissociable cyanide except for iron cyanide complexes and more stable metal-cyanide complexes. The process is a two-stage process. The first stage involves initial oxidation of free cyanide to cyanogens chloride followed by hydrolysis of cyanogens chloride to cyanate (Eqs. (7) and (8)) at pH 11.

$$CN^- + Cl_2 \rightarrow ClCN + Cl \tag{7}$$

$$ClCN + H_2O \rightarrow OCN^- + 2H^+ + Cl^- \tag{8}$$

During the second stage, cyanate is further oxidised to hydrogen carbonate and nitrogen as shown in Eq. (9). The reaction occurs at pH 8.5.

$$3Cl_2 + 2OCN^- + 6OH^- \rightarrow N_2 + 2HCO_3^- + 6Cl^- + 2H_2O \tag{9}$$

The alkaline chlorination process is primarily applied in the treatment of cyanide solutions rather than slurries, which consume a lot of chorine.

7.2.1.3. Hydrogen peroxide oxidation

Oxidation of cyanide tailings by hydrogen peroxide is more suitable for solutions rather than slurries. The oxidation process is maintained at pH of 9–10 to avoid formation of hydrogen cyanide [47]. The oxidation reaction is catalysed by copper (II) sulphate resulting in the production of carbonate and ammonium (Eqs. (10) and (11)).

$$CN^- + H_2O_2 \xrightarrow{\text{Cu}^{-2}} CNO^- + H_2O \tag{10}$$

$$CNO^- + 2H_2O \rightarrow CO_3^{-2} + NH_4^+ \tag{11}$$

7.2.1.4. Ozonation

Ozone is a superior oxidant to oxygen and has been extensively studied in the oxidation of cyanide [48–51]. Two oxidation mechanisms of cyanide to cyanate by ozone have been proposed, namely simple (Eq. (12)) and catalytic (Eq. (13)).

$$O_3 + CN^- \rightarrow OCN^- + O_2 \tag{12}$$

$$O_3 + 3CN^- \rightarrow 3OCN^- \tag{13}$$

Catalytic ozonation rarely occurs and has only been observed under high acidic conditions. Continued addition of ozone results in the formation of hydrogen carbonate and nitrogen (Eq. (14)).

$$2OCN^- + O_3 + H_2O \rightarrow 2N_2 + 2HCO_3^- \tag{14}$$

7.2.1.5. Peroxymonosulphuric acid

Peroxymonosulphuric acid (H_2SO_5) or Caro's acid [52] is used for cyanide treatment in gold tailings. Caro's acid is prepared in situ by the reaction of hydrogen peroxide with sulphuric acid since it easily decomposes. This acid is mostly used in situations where sulphur dioxide/air cannot be used. Caro's acid oxidises cyanide to cyanate as shown in Eq. (15).

$$H_2SO_5 + CN^- \rightarrow OCN^- + SO_4^{2-} + 2H^+ \tag{15}$$

7.2.1.6. Precipitation of cyanide

Stable cyanide complexes can be precipitated by the addition of complexing agents such as iron. Iron cyanide complexes can coprecipitate other compounds containing cyanide in solution producing solids of cyanide salts. Finely divided insoluble iron sulphide is used for adsorbing free and complexed cyanide in solutions. The adsorption occurs at optimum pH of approximately 7.5. The iron sulphide is prepared from the reaction of iron (II) sulphate and sodium cyanide [53]. If hydrated ferrous sulphate is used, iron (II) cyanide precipitate is produced. Precipitation of iron cyanide occurs at pH between 5 and 6.

7.2.2. Physical methods

Cyanide tailings can be treated using physical methods such as dilution, membrane and electrowinning.

7.2.2.1. Dilution

This is a technique that does not destroy toxic cyanide, but dilute it with an eluent that reduces cyanide concentrations below acceptable discharge limits. Dilution is a cheap simple technique, which is often used as a standalone or in conjunction with other methods as a way of ensuring that discharged effluents are within permissible limits [54]. Dilution is normally an unacceptable method since it does not degrade or reduce the quantity of toxic cyanide exposed to the environment.

7.2.2.2. Membrane technology

Reverse osmosis and electrodialysis techniques using membranes have been used in extracting cyanide from wastewater. Both techniques have been effectively applied in the removal of free and complexed cyanide [55–57].

7.2.2.3. Electrowinning

Strong acid dissociable and weak acid dissociable cyanide complexes can be reduced to metals releasing free cyanide (Eq. (16)) by the application of an electric potential across electrodes immersed in complexes solution. The freed cyanide can then be treated by other processes.

$$\left[M(CN)_y \right]^{x-y} + xe^- \rightarrow M + yCN^- \tag{16}$$

Four electrowinning cell designs have been developed for gold processing, namely Zadra, AARL, NIM graphite chip and MINTEK parallel plate cells [54, 58]. Electrowinning performs well in concentrated solutions and has been predominantly utilised for gold processing. This process is termed as Celec or HSA process [59] when it is used for cyanide regeneration.

7.2.2.4. Adsorption

Activated carbon, resins and minerals have been used for cyanide adsorption from solution. Contact vessels such as elutriation columns, agitation cells, packed-bed columns and loops have been used for this purpose. Various separation techniques such as floatation, gravity separation and screening are applied to remove the adsorbed cyanide from solution. The adsorbent is subsequently transferred into another vessel where cyanide is desorbed into low-volume solution, concentrated, reactivated and recycled.

7.2.2.5. Resins

Resins are typically polymeric beads containing a numerous surface functional groups capable of chelating or ion exchanging. Resins that require a substrate are deposited as thin film, while those that do work without a substrate are mostly used in continuous processes. The first column resin for cyanide recover was developed in 1959 [60]. Metal-cyanide complexes have been reported to adsorb more strongly to resins [61, 62]. The extent of adsorption depends on nature of resin used and how the resin and/or solution is pretreated [63]. Resins are cheaper and more effective than activated carbon since they resist organic fouling, have longer life, desorb faster and regenerate more efficiently [63]. Conventional, commercial strong base resins are most suitable for cyanide recovery since most common cyanide species in gold plant tailings are free cyanide anions within 100–500 mg/L range and the tricyano copper complex, both of which can be extracted directly from pulps using anion exchange resins [22].

7.2.2.6. Minerals

Free and metal-complexed cyanides are adsorbed by solid wastes, soils and ores containing minerals such as bauxite [AlO.OH/Al(OH)$_3$], ilmenite (FeTiO$_3$), haematite (Fe$_2$O$_3$) and pyrite (FeS$_2$). Mineral groups such as zeolites, clays and feldspars are also effective adsorbents [64, 65].

7.2.2.7. Activated carbon

Cyanide packed-bed systems can be used to adsorbed cyanide in dilute solutions [30, 66, 67]. Activated carbon has a relatively high affinity for many metal-cyanide complexes, including the soluble cyanide complexes of copper, iron, nickel and zinc [41, 68, 69]. Cyanide is adsorbed at various sites through chelation, ion exchange, solvation and coulombic interactions. This adsorption technique suffers a major drawback of being technically complex and expensive regeneration of activated carbon [70].

7.2.3. Biological oxidation methods

The use of microorganisms in the degradation of cyanide in tailing ponds has often been found to be potentially inexpensive and environmentally friendly compared to conventional chemical and physical processes [23, 71, 72]. Enzymatic activities associated with certain species of bacteria, fungi and algae are known to oxidise cyanide to less toxic cyanate [20, 73, 74]. Aerobic and anaerobic passive biological treatment processes are cost-effective alternatives to conventional cyanide treatment strategies since they do not need external energy, chemicals and routine maintenance. However, they suffer limitations such as the need for warmer climates (>10°C), large space and long retention times. **Figure 1** shows a flowchart for the aerobic and anaerobic oxidation of cyanides and thiocyanides in gold tailing ponds. Common passive biological treatment processes comprise engineered wetlands containing substrate or a mixture of organic and inorganic compounds like manure, straw, saw dust and limestone [22]. In anaerobic wetlands, bacterial oxidation of cyanides and thiocyanides to sulphates, carbonates and ammonia occurs as illustrated in Eqs. (17) and (18).

$$CN^- + 2H_2O + 0.5O_2 \rightarrow NH_3 + OH^- + CO_3^{2-} \tag{17}$$

$$SCN^- + 2H_2O + 2.5O_2 \rightarrow NH_4^+ + HCO_3^- + SO_4^{2-} + H^+ \tag{18}$$

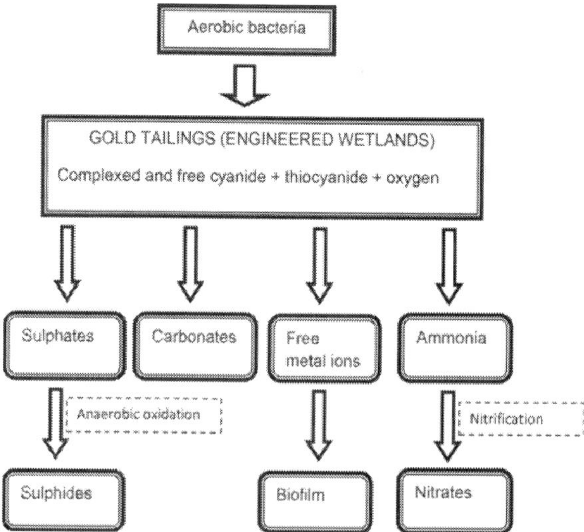

Figure 1. Biological cyanide degradation processes in gold tailings.

The ammonia produced by the aerobic processes provides nutrients for microbial growth and the resultant uptake, sorption, conversion and/or precipitation of cyanides, thiocyanates, sulphates and nitrates by microorganisms [74]. The metals released during the oxidation of cyanide metal complexes are removed from gold tailings by chemical precipitation and/or adsorption on bacterial biofilm. Ammonia also undergoes further oxidation in a two-step nitrification process (Eqs. (19) and (20)).

$$NH_4^+ + 1.5O_2 \rightarrow NO_2^- + 2H^+ + H_2O \tag{19}$$

$$NO_2^- + 0.5O_2 \rightarrow NO_3^- \tag{20}$$

Sulphates undergo anaerobic reduction to sulphides (Eq. (21)). This process is effected by sulphate-reducing bacteria [75]. The sulphide produced is precipitated by metal ions resulting in its removal from aqueous tailings.

$$SO_4^{2-} + 2C_3H_4O_3^- \left(\text{lactate}\right) \rightarrow S^{2-} + 2CO_3^{2-} + 2C_2H_4O_2^- \left(\text{acetate}\right) \tag{21}$$

Several factors influence the biodegradation of cyanide in gold tailings. The most important environmental factors influencing biological treatment include pH, temperature, oxygen levels and nutrient availability. Enzymes that degrade cyanide are generally produced by mesophilic microorganisms, often isolated from soil, with optimum operating temperatures of between 20 and 40°C [34, 43, 76–79]. The availability of nutrient carbon has been found as a limiting factor in the biodegradation of metal-cyanide complexes [75].

Highly acidic and basic conditions have adverse effects on cyanide-degrading microorganisms since bacterial and fungal growth is optimal at pH 6–8 and 4–5, respectively [80]. Cyanide-degrading enzymes have optimum operating pH between 6 and 9. Concentrations of cyanide ions in water or slurries have an impact on the survival and growth of microorganisms. For instance, high cyanide concentrations have been reported to be toxic to *Klebsiella oxytoca* by damaging the nitrile-degrading enzyme, nitrile hydratase [81].

7.3. Emerging technologies on cyanide remediation

Since the 1990s, research has focused on introducing cyanide treatment technologies aimed at reducing costs and producing environmentally friendly products.

Carbon dioxide has been successfully used without a catalyst to replace SO_2 as an inexpensive alternative to the SO_2/air process [82].

Wastewater containing free and complexed cyanides can be oxidised by ultraviolet radiation in the presence of a semiconductor catalyst such as titanium dioxide [33]. When the catalyst mixed in the wastewater is exposed to the sunlight, it generates a highly reactive hydroxyl radical oxidant. These radicals initially convert cyanide to cyanate. Photocatalysis partially dissociates ferricyanide and ferrocyanide complexes to free cyanide and iron hydroxide. Photocatalytic oxidation is effective in relatively clear solutions. In the presence of ozone, ultraviolet oxidation does not produce undesirable by-products such as ammonia.

Solid or liquid cyanide wastes can be thermally decomposed upon treatment at elevated temperatures and pressure in batch or continuous mode [83]. This process is capable of destroying all cyanide complexes. Cyanide hydrolysis occurs in two steps (Eqs. (22) and (23)) producing ammonia and carbonates.

$$NaCN + NaOCl \rightarrow NaCNO + NaCl \tag{22}$$

$$2NaCNO + 3NaOCl \rightarrow H_2O + 3NaCl + N_2 + 2NaHCO_3 \tag{23}$$

This cost-effective process was developed in the early 1990s for the treatment of wastes containing high concentrations of cyanide (100,000 mg/L). Thermal reduction reduces cyanide concentration to approximately 25 mg/L, which can be further oxidised by conventional methods such as ozone or hydrogen peroxide to environmentally permissible levels.

Free cyanide and cyanide complexes containing waste can be treated by electrochemical oxidation. This is an economical and environmentally friendly technique of destroying cyanide. The process results in cyanide ions being destroyed at the anode as metals are deposited at the cathode [27, 84]. During electrolysis, cyanide is initially oxidised at the anode-producing cyanate ions, which are further decomposed to carbon dioxide and nitrogen gas at the cathode (Eqs. (24) and (25)).

$$\text{Anode}: \ CN^- + 2OH^- \rightarrow OCN^- + H_2O + 2e \tag{24}$$

$$\text{Cathode}: \ 2CNO^- + 4OH^- \rightarrow 2CO_2 + N_2 + 2H_2O + 6e^- \tag{25}$$

8. Conclusion

There is a growing use of water as the gold mining activities increase. Water losses should be minimised and recycling adopted as much as possible. Cost-effective and environmentally friendly practices for cyanide treatment need to be implemented. The gold mining industry needs to implement the best practices for cyanide management that are aimed at assisting in protecting human health and reducing environmental impacts through discharge of permissible levels of cyanide in effluents into main water bodies. Such practice will ensure maintenance of good quality of water and sustenance of aquatic life.

Author details

Benias C. Nyamunda

Address all correspondence to: nyamundab@gmail.com

Midlands State University, Manicaland College of Applied Sciences, Department of Chemical and Processing Engineering, Mutare, Zimbabwe

References

[1] Thompson PF. The dissolution of gold in cyanide solutions. Transactions of the Electrochemical Society. 1947; 91(1): 41–71. doi:10.1149/1.3071767

[2] Donato DB, Nichols O, Possingham H, Moore M, Ricci PF, Noller BN. A critical review of the effects of gold cyanide-bearing tailings solutions on wildlife. Environment International. 2007; 33(7): 974–984. doi:10.1016/j.envint.2007.04.007

[3] Rees KL, Van Deventer JSJ. The role of metal-cyanide species in leaching gold from a copper concentrate. Minerals Engineering. 1999; 12(8): 877–892. doi:10.1016/S0892-6875(99)00075-8

[4] Jeffrey MI, Breuer PL. The cyanide leaching of gold in solutions containing sulfide. Minerals Engineering. 2000; 13(10): 1097–1106. doi:10.1016/S0892-6875(00)00093-5

[5] Miller JD, Wan RY, Mooiman MB, Sibrell PL. Selective solvation extraction of gold from alkaline cyanide solution by alkyl phosphorus esters. Separation Science and Technology. 1987; 22(2–3): 487–502. doi:10.1080/01496398708068965

[6] Örgül S, Atalay Ü. Reaction chemistry of gold leaching in thiourea solution for a Turkish gold ore. Hydrometallurgy. 2000; 67(1): 71–77. doi:10.1016/S0304-386X(02)00136-6

[7] McDougall GJ, Hancock RD, Nicol MJ, Wellington OL. Copperthwaite RG. The mechanism of the adsorption of gold cyanide on activated carbon. Journal of South African Institute of Mining and Metallurgy. 1980; 80(9): 344–356.

[8] Yalcin M, Arol AI. Gold cyanide adsorption characteristics of activated carbon of non-coconut shell origin. Hydrometallurgy. 2002; 63(2): 201–206. doi:10.1016/S0304-386X(01)00203-1

[9] Cortina JL, Warshawsky A, Kahana N, Kampel V, Sampaio CH, Kautzmann RM. Kinetics of goldcyanide extraction using ion-exchange resins containing piperazine functionality. Reactive and Functional Polymers. 2003; 54(1): 25–35. doi:10.1016/S1381-5148(02)00180-3

[10] Bachiller D, Torre M, Rendueles M, Díaz M. Cyanide recovery by ion exchange from gold ore waste effluents containing copper. Minerals Engineering. 2004; 17(6): 767–774. doi:10.1016/j.mineng.2004.01.001

[11] Fleming CA, Mezei A, Bourricaudy E, Canizares M, Ashbury M. Factors influencing the rate of gold cyanide leaching and adsorption on activated carbon, and their impact on the design of CIL and CIP circuits. Minerals Engineering. 2011; 24(6): 484–494. doi: 10.1016/j.mineng.2011.03.021

[12] Soleimani M, Kaghazchi T. Adsorption of gold ions from industrial wastewater using activated carbon derived from hard shell of apricot stones—an agricultural waste. Bioresource Technology. 2008; 99(13): 5374–5383 doi:10.1016/j.biortech.2007.11.021

[13]   Espiell F, Roca A, Cruells M, Nunez C. Gold desorption from activated carbon with dilute NaOH/organic solvent mixtures. Hydrometallurgy. 1988; 19(3): 321–333. doi: 10.1016/0304-386X(88)90038-2

[14]   Muir DM, Hinchliffe W, Tsuchida N, Ruane M. Solvent elution of gold from CIP carbon. Hydrometallurgy. 1985; 14(1): 47–65. doi:10.1016/0304-386X(85)90005-2

[15]   Davidson RJ, Veronese V, Nkosi MV. The use of activated carbon for the recovery of gold and silver from gold-plant solutions. Journal of the South African Institute of Mining and Metallurgy. 1979; 281–297.

[16]   Conradie PJ, Johns MW, Fowles RJ. Elution and electrowinning of gold from gold-selective strong-base resins. Hydrometallurgy. 1995; 37(3): 349–366. doi: 10.1016/0304-386X(94)00032-X

[17]   Marsden JO, House CI. The chemistry of gold extraction, Society for Mining, Metallurgy and Exploration. Colorado; Littleton: 2006. p. 147–231.

[18]   Yannopoulos JC. Melting and refining of gold. In the extractive metallurgy of gold. New York, NY: Springer; 1991. p. 241–244. doi:10.1007/978-1-4684-8425-0_11

[19]   Duchao Z, Tianzu Y, Wei L, Weifeng L, Zhaofeng X. Electrorefining of a gold-bearing antimony alloy in alkaline xylitol solution. Hydrometallurgy. 2009; 99(3): 151–156. doi: 10.1016/j.hydromet.2009.07.013

[20]   Haque MR, Bradbury JH. Total cyanide determination of plants and foods using the picrate and acid hydrolysis methods. Food Chemistry. 2002; 77(1): 107–114. doi:10.1016/ S0308-8146(01)00313-2

[21]   Dubey SK, Holmes DS. Biological cyanide destruction mediated by microorganisms. World Journal of Microbiology and Biotechnology. 1995; 11(3): 257–265. doi:10.1007/ BF00367095

[22]   Kuyucak N, Akcil A. Cyanide and removal options from effluents in gold mining and metallurgical processes. Minerals Engineering. 2013; 50: 13–29. doi:10.1016/j.mineng. 2013.05.027

[23]   Zheng DA, Dzombak RG, Luthy B, Sawer W, Lazouska P, Tata, MF, Sebriski JR, Swartling RS, Drop SM, Flaherty JM. Evaluation and testing of analytical methods for cyanide species in municipal and industrial contaminated water. Environmental Science and Technology. 2003; 37: 107–115. doi:10.1021/es0258273

[24]   Ebbs S. Biological degradation of cyanide compounds. Current Opinion in Biotechnology. 2004; 15(3): 231–236. doi:10.1016/j.copbio.2004.03.006

[25]   Desai JD, Ramakrishna C. Microbial degradation of cyanides and its commercial application, Journal of Scientific and Industrial Research. 1998; 57: 441–453.

[26]   Randol International Limited. Water management & treatment for mining & metallurgical operations. Colorado; Golden: 1985. p. 2294–2700.

[27] Flynn CM, Haslem SM. Cyanide Chemistry: Precious Metals Processing and Waste Treatment. US Department of the Interior, Bureau of Mines: 1995.

[28] Young CA, Jordan TS. Cyanide remediation: current and past technologies. In: Erickson LE, Tillison DL, Grant SC, McDonald JP. Proceedings of 10th Conference on Hazardous Waste Research; 23–24 May 1995. Kansas State University, Manhattan, Kansas: p. 104–129.

[29] Botz MM. Overview of cyanide treatment methods, mining environmental management. London: Mining Journal Ltd; 2001. p. 28–30.

[30] Dash RR, Balomajumder C, Kumar A. Removal of cyanide from water and wastewater using granular activated carbon. Chemical Engineering Journal. 2009; 146: 408–413. doi: 10.1016/j.cej.2008.06.021

[31] Ripley EA, Redmann RE, Crowder AA. Environmental effects of mining. Florida: St.-Lucie Press; 1996. p 181–197.

[32] USEPA (Environ. Protection Agency U.S.) Drinking water criteria document for cyanide, Environment Criteria and Assessment Office. Cincinnati. 1985; EPA/600/X-84-192-1.

[33] Parga JR, Shukla SS, Carrillo-Pedroza FR. Destruction of cyanide waste solutions using chlorine dioxide, ozone and titania sol. Waste Management. 2003; 23: 183–191. doi: 10.1016/S0956-053X(02)00064-8

[34] Akcil A. Destruction of cyanide in gold mill effluents: biological versus chemical treatments. Biotechnology Advances. 2003; 21(6): 501–511. doi:10.1016/S0734-9750(03)00099-5

[35] Gurbuz F, Ciftci H, Akcil A. Biodegradation of cyanide containing effluents by Scenedesmus obliquus. Journal of HAZARDOUS materials. 2009; 162(1): 74–79. doi:10.1016/j.jhazmat.2008.05.008

[36] Luthy RG, Bruce Jr SG. Kinetics of reaction of cyanide and reduced sulfur species in aqueous solution. Environmental Science & Technology. 1979; 13(12): 1481–1487. doi: 10.1021/es60160a016

[37] Botz M, Mudder T, Akcil A. Cyanide treatment: physical, chemical and biological processes. Advances in Gold Ore Processing. 2005; 672–700.

[38] Karunakaran C, Abiramasundari G, Gomathisankar P, Manikandan G, Anandi V. Preparation and characterization of ZnO–TiO$_2$ nanocomposite for photocatalytic disinfection of bacteria and detoxification of cyanide under visible light. Materials Research Bulletin. 2011; 46(10): 1586–1592. doi:10.1016/j.materresbull.2011.06.019

[39] Moussavi G, Khosravi R. Removal of cyanide from wastewater by adsorption onto pistachio hull wastes: Parametric experiments, kinetics and equilibrium analysis.

Journal of Hazardous Materials. 2010; 183(1): 724–730. doi:10.1016/j.jhazmat. 2010.07.086

[40]   Kitisa M, Karakayaa E, Yigita NO, Civelekoglua G, Akcilb A. Heterogeneous catalytic degradation of cyanide using copper-impregnated pumice and hydrogen peroxide. Water Research. 2005; 39: 1652–1662. doi:10.1016/j.watres.2005.01.027

[41]   Depci T. Comparison of activated carbon and iron impregnated activated carbon derived from Gölbaşı lignite to remove cyanide from water. Chemical Engineering Journal. 2012; 181: 467–478. doi:10.1016/j.cej.2011.12.003

[42]   Campos MG, Pereira P, Roseiro JC. Packed-bed reactor for the integrated biodegradation of cyanide and formamide by immobilised *Fusarium oxysporum* CCMI 876 and *Methylobacterium sp.* RXM CCMI 908. Enzyme and Microbial Technology. 2006; 38: 848–854. doi:10.1016/j.enzmictec.2005.08.008

[43]   Kao CM, Liu JK, Lou HR, Lin CS, Chen SC. Biotransformation of cyanide to methane and ammonia by *Klebsiella oxytoca*. Chemosphere. 2003; 50: 1055–1061. doi:10.1016/S0045-6535(02)00624-0

[44]   Devuyst EA, Conard BR, Ettel VA. Pilot-plant operation of the Inco SO_2 air cyanide removal process. Canadian Mining Journal. 1982; 103(8): 27–30.

[45]   Mudder TI, Botz M, Smith A. Chemistry and treatment of cyanidation wastes. London: Mining Journal Books; 2001. p. 327–333.

[46]   Nelson GM, Kroeger EB, Arps PJ. Chemical and biological destruction of cyanide: comparative costs in a cold climate. Mineral Processing and Extractive Metallurgy Review. 1998; 19: 217–226. doi:10.1080/08827509608962441

[47]   Kepa U, Stanczyk-Mazanek E, Stepniak L. The use of the advanced oxidation process in the ozone+ hydrogen peroxide system for the removal of cyanide from water. Desalination. 2008; 223(1): 187–193. doi:10.1016/j.desal.2007.01.215

[48]   Carrillo-Pedroza FR, Nava-Alonso F, Uribe-Salas A. Cyanide oxidation by ozone in cyanidation tailings: Reaction kinetics. Minerals Engineering. 2000; 13(5): 541–548. doi: 10.1016/S0892-6875(00)00034-0

[49]   Rowley WJ, Otto FD. Ozonation of cyanide with emphasis on gold mill wastewaters. The Canadian Journal of Chemical Engineering. 1980; 58(5): 646–653. doi:10.1002/cjce. 5450580516

[50]   Soto H, Nava F, Leal J, Jara J. Regeneration of cyanide by ozone oxidation of thiocyanate in cyanidation tailings. Minerals Engineering. 1995; 8(3): 273–281. doi: 10.1016/0892-6875(94)00123-T

[51]   Zeevalkink JA, Visser DC, Arnoldy P, Boelhouwer C. Mechanism and kinetics of cyanide ozonation in water. Water Research. 1980; 14(10): 1375–1385. doi: 10.1016/0043-1354(80)90001-9

[52] Castrantas HM, Manganaro JL, Rautiola CW, Carmichael J. Treatment of cyanides in effluents with Caro's acid. U.S. Patent 5,397,482, issued March 14, 1995.

[53] Adams MD. The removal of cyanide from aqueous solution by the use of ferrous sulphate. Journal of the South African Institute Mining and Metallurgy. 1992; 92: 17–25.

[54] Marsden J, House I. The Chemistry of gold extraction. New York, NY: Ellis Horwood; 1992. p. 160–170.

[55] Kenfield CF, Qin R, Semmens MJ, Cussler EL. Cyanide recovery across hollow fiber gas membranes. Environmental Science & Technology. 1988; 22(10): 1151–1155. doi: 10.1021/es00175a003

[56] Han B, Shen Z, Wickramasinghe SR. Cyanide removal from industrial wastewaters using gas membranes. Journal of Membrane Science. 2005; 257(1): 171–181. doi:10.1016/j.memsci.2004.06.064

[57] Marder L, Sulzbach GO, Bernardes AM, Ferreira JZ. Removal of cadmium and cyanide from aqueous solutions through electrodialysis. Journal of the Brazilian Chemical Society. 2003; 14(4): 610–615. doi:10.1590/S0103-50532003000400018

[58] Nicol MJ, Fleming CA, Paul R.L. The Chemistry of the extraction of Gold, In: G.C. Stanley (Ed.), Extraction metallurgy of gold in South Africa. Pretoria: SAIMM; 1987. p. 834–905.

[59] Kuhn AT. Electrolytic decomposition of cyanides, phenols and thiocyanates in effluent streams—a review. Journal of Chemical Technology and Biotechology. 1971; 21(2): 29–34. doi:10.1002/jctb.5020210201

[60] Goldblatt E. Recovery of cyanide from waste cyanide solution by ion exchange. Industrial and Engineering Chemistry. 1959; 51, 241–246. doi:10.1021/ie51394a022

[61] Avery NL, Fries W. Selective removal of cyanide from industrial waste effluents with ion-exchange. Industrial Engineering Chemistry Product and Research Development. 1975; 14: 102–104. doi:10.1021/i360054a009

[62] Akser M, Wan RY, Miller JD. Gold adsorption from alkaline aurocyanide solution by neutral polymeric adsorbents, Solvent Extraction and Ion Exchange. 1986; 4: 531–546. doi:10.1080/07366298608917880

[63] Fleming CA, Cromberge G. The extraction of gold from cyanide solutions by strong and weak-base anion-exchange resins. Journal of Africa Institute of Mining and Metallurgy. 1984; 84: 125–138.

[64] Strobel G.A. Cyanide utilization in soil. Soil Science. 1967; 103: 299–302.

[65] Saxena S, Prasad M, Amritphale SS, Chandra N. Adsorption of cyanide from aqueous solutions at pyrophyllite surface. Separation and Purification Technology. 2001; 24(1): 263–270. doi:10.1016/S1383-5866(01)00131-9

[66]   Honda S, Kondo G. Treatment of wastewater containing cyanide using activated charcoal. Os. Kogyo Gij. Shik. Koho. 1967; 18: 367.

[67]   Dash RR, Majumder CB, Kumar A. Treatment of metal cyanide bearing wastewater by simultaneous adsorption biodegradation (SAB). Journal of Hazardous Materials. 2008; 152: 387–396. doi:10.1016/j.jhazmat.2007.07.009

[68]   Adhoum N, Monser L. Removal of cyanide from aqueous solution using impregnated activated carbon. Chemical Engineering and Processing: Process Intensification. 2002; 4:117–121. doi:10.1016/S0255-2701(00)00156-2

[69]   Deveci H, Yazıcı EY, Alp I, Uslu T. Removal of cyanide from aqueous solutions by plain and metal-impregnated granular activated carbons. International Journal of Mineral Processing. 2006; 79: 198–208. doi:10.1016/j.minpro.2006.03.002

[70]   Crini G. Non-conventional low-cost adsorbents for dye removal: a review. Bioresource Technology. 2006; 60: 67–75. doi:10.1016/j.biortech.2005.05.001

[71]   Mosher JB, Figueroa L. Biological oxidation of cyanide: a viable treatment option for the minerals processing industry. Minerals Engineering. 1996; 9(5): 573–581. doi: 10.1016/0892-6875(96)00044-1

[72]   Luque-Almagro VM, Huertas MJ, Martínez-Luque M, Moreno-Vivián C, Roldán, MD, García-Gil LJ, Castillo F, Blasco R. Bacterial degradation of cyanide and its metal complexes under alkaline conditions. Applied and Environmental Microbiology. 2005; 71(2): 940–947. doi:10.1128/AEM.71.2.940-947.2005

[73]   Gurbuz F, Ciftci H, Akcil A, Karahan AG. Microbial detoxification of cyanide solutions: a new biotechnological approach using algae. Hydrometallurgy. 2004; 72(1): 167–176. doi:10.1016/j.hydromet.2003.10.004

[74]   Akcil A, Mudder T. Microbial destruction of cyanide wastes in gold mining: process review. Biotechnology Letters. 2003; 25: 445–450. doi:10.1023/A:1022608213814

[75]   Kuyucak N. The Role of microorganisms in mining: generation of acid rock drainage, its mitigation and treatment. European Journal of Mineral Processing and Environmental Protection. 2002; 2(3): 179–196.

[76]   Baxter J, Cummings SP. The current and future applications of microorganism in the bioremediation of cyanide contamination. Antonie van Leeuwenhoek, 2006; 90: 1–17. doi:10.1007/s10482-006-9057-y

[77]   Dumestre A, Chone T, Portal J, Berthelin J. Cyanide degradation under alkaline conditions by a strain of *Fusarium solani* isolated from contaminated soils. Applied and Environmental Microbiology. 1997; 63: 2729–2734.

[78]   Dursun AY, Aksu Z. Biodegradation kinetics of ferrous (II) cyanide complex ions by immobilized *Pseudomonas fluorescens* in a packed bed column reactor. Process Biochemistry. 2000; 35: 615–622. doi:10.1016/S0032-9592(99)00110-7

[79] Babu GRV, Wolfram JH, Chapatwala KD. Conversion of sodium cyanide to carbon dioxide and ammonia by immobilized cells of *Pseudomonas Putida*. Journal of Industrial Microbiology. 1992; 9: 235–238. doi:10.1007/BF01569629

[80] Barclay M, Hart A, Knowles CJ, Meeussen JCL, Tett VA. Biodegradation of metal cyanides by mixed and pure cultures of fungi. Enzyme and Microbial Technology. 1998; 22: 223–231. doi:10.1016/S0141-0229(97)00171-3

[81] Kao CM, Chen KF, Liu JK, Chou SM, Chen SC. Enzymatic degradation of nitriles by *Klebsiella oxytoca*. Applied Microbiology and Biotechnology. 2006; 71: 228–233. doi: 10.1007/s00253-005-0129-0

[82] Randol International Limited. Phase IV – Innovations in gold and silver recovery. Colorado: Golden; 1992. Chapter 54.

[83] Cushnie G. Pollution prevention and control technologies for plating operations, Wastewater Treatment. 2nd ed. London; Amazon: 2009. p. 287–297.

[84] Saarela K, Kuokkanen T. Alternative disposal methods for wastewater containing cyanide: analytical methods for new electrolysis technology developed for total treatment of wastewater containing gold or silver cyanide. In: Pongracz, E. (Ed.), Proceedings of the Waste Minimization and Resource Use Optimization Conference; 2004. p. 107–121 University of Oulu, Oulu. Finland.

Chapter 11

A Comparative Study of Modified and Unmodified Algae (*Pediastrum boryanum*) for Removal of Lead, Cadmium and Copper in Contaminated Water

John Okapes Joseph, Isaac W. Mwangi,

Sauda Swaleh, Ruth N. Wanjau, Manohar Ram and

Jane Catherine Ngila

Additional information is available at the end of the chapter

http://dx.doi.org/10.5772/65745

Abstract

The presence of heavy metals in water is of concern due to the risk toxicity. Thus there is need for their removal for the safety of consumers. Methods applied for removal of heavy metals include adsorption, membrane filtration and co-precipitation. However, studies have revealed adsorption is highly effective technique. Most adsorbents are expensive or require extensive processing before use and hence need to explore for possible sources of inexpensive adsorbents. This research work investigated the use an algal biomass (*pediastrum boryanum*) as an adsorbent for removal of Lead, Cadmium and Copper in waste water in its raw and modified forms. The samples were characterized with FTIR and was confirmed a successful modification with tetramethylethlynediamine (TMEDA). Sorption parameters were optimized and the material was finally applied on real water samples. It was found that the sorption was best at lower pH values (4.2-6.8). Sorption kinetics was very high as more that 90% of the metals were removed from the solution within 30 minutes. The adsorption of copper fitted into the Langmuir adsorption isotherm indicating a monolayer binding mechanism. Cadmium and lead fitted best the Freundlich adsorption mechanism. Sorption of lead and cadmium was of pseudo-second order kinetics, confirming a multisite interaction whereas copper was pseudo-first order indicating a single site adsorption. The adsorption capacity did not improve upon modification but the stability of the material was improved and secondary pollution of leaching colour was alleviated. This implies that the modified material is suitable for application on the removal of metals from water.

Keywords: modified algae, tetramethylethylenediamine, sorption, heavy metals, contaminated water

---

1. Introduction

The release of wastewater into the environment poses a great problem worldwide due to enhancement and mobilization of toxic trace metals due to solubilization [1, 2]. This is enabled by the presence of functional groups capable of forming metal complexes with the metals [3–5]. Unlike organic pollutants which are susceptible to degradation, metal ions remain in the environment available to cause pollution [6]. This makes the presence of heavy metal in the environment a major concern due to their toxicity to various life forms. Even when most metals in the environment are in trace levels or masked by matrices, the presence of wastewater problem exacerbates the toxic nature of heavy metals. The net result is scarcity and insufficient supply of safe water, hence the quality of life. As such, methods for the removal of such contaminants need to be explored to mitigate the effects of metal pollution.

Conventional methods for the removal of metals such as precipitation, coagulation, evaporation and membrane filtration are expensive and not effective when the concentrations are in trace levels ranging from 1 mg l^{-1} to 20 mg l^{-1} [7, 8]. Due o such limitations, a need therefore arises for the development of cost-effective methods to remove heavy metals in waste purification processes. The presence of functional groups within the structure of sorbents from plant origin has received increasing attention for the removal and recovery of heavy metals in aqueous media [9]. However, they have been found to leach organic matter in the water during the treatment process. This has resulted in treating the sorbents first, before applying them for the water treatment activity.

A solution to this was achieved by modification of the raw biomaterial using tetramethylethylenediamine [10]. This work reports on the modification of algae and its subsequent application for the removal of some selected heavy metals through a biosorption process. The effects of modification on secondary pollution and adsorption parameters were also investigated. This was to obtain the information that will contribute to determining adsorption capacity, sorption mechanism and kinetics with a view to apply the material at a point of use. This was intended to contribute to knowledge for social-economic development, to address the availability of clean water to the rural communities who source their water from rivers, dams, boreholes and shallow wells whose water quality is not known. The adsorbent is intended to be applied for purification of water by removal of heavy metals for domestic consumption at a small scale with a view to offer a cheap solution to metal-related toxicity. This is a simple and sustainable water management approach.

Methods applied for separation and preconcentration techniques include adsorption, membrane filtration, cloud-point extraction, solvent extraction and co-precipitation [11]. Studies by Marshall [12] have revealed adsorption by the use of activated carbon to be highly effective for the removal of heavy metals from wastewaters. Despite its extensive use in water and wastewater treatment industries, activated carbon remains an expensive material [13]. In view of this, the need for safe and economical methods for the elimination of heavy metals from polluted and contaminated water has to be explored.

This has necessitated researchers to develop methods to mitigate the effects of heavy metals in water. Low-cost agricultural waste by-products such as sugarcane, bagasse, rice husks, sawdust and coconut husk, oil palm shell, neem bark and maize tassels have been studied and applied for the removal of heavy metals from wastewater. The cost of such materials is an important parameter for comparing the sorbent materials. Such agricultural wastes are abundant, require little processing and therefore have a potential to be applied as low-cost sorbents that are environmentally friendly [14]. But these materials suffered a setback due to leaching of dissolved organic matter in water during the treating process. To overcome such cases of secondary pollution, there is need to explore sorbents for their feasibility for the removal of heavy metals from wastewater. This study investigated the use of algal biomass which is an aquatic plant for the remediation process. Algae does not normally leach colour in the water but only contribute in the water oxidation process and interaction with metals.

Due to those qualities of algae, a solution experienced while using other biosorbents was expected to be solved. The algal biomaterial was modified with tetramethylethylenediamine to form a resin material with suitable functional group to complex with metal ions and remove them from water. The resulting solid material was capable of interacting with metals in water and attracting them to its surface, hence removing them, and was found to be regeneratable and did not leach soluble organic compounds in the treated water.

2. Materials and methods

2.1. Research design

The focus of our study was to synthesize a sorbent by anchoring functional groups capable of interacting with metal ions and removing them from aqueous media. The protocol in preparation was to use non-toxic and environmentally friendly materials. The study was carried out in several parts. This comprised of sampling, synthesis, characterization of the modified material, optimization of removal parameters and then its subsequent application for the removal of fluorides from both synthetic and environmental water samples.

2.2. Chemicals and reagents

All the solutions were prepared in double-distilled water and the reagents were of analytical grade. Metal standard stock solution of 1000 mg/L was prepared by dissolving 1.00 g of the respective metal in one litre of 0.1 M sodium acetate. It was from this solution that subsequent working solutions were prepared from. Separate solutions of 0.01 M nitric acid and 0.01 M sodium hydroxide were prepared and used to adjust the pH of the working solutions to the desired value. The above chemicals and tetramethylethylenediamine (TMEDA) were supplied by Kobian Kenya Ltd. which is Sigma-Aldrich's outlet in Kenya.

2.3. Instrumentation

The modified and unmodified algae were characterized using a Fourier transform infrared (FTIR) spectrometer (FTIR-8400, Shimadzu Tokyo, Japan) to establish the functional groups present. The concentration of metal pollutants in the water samples was determined using atomic absorption spectroscopy (AAS) (Buck, model 210 VGP) set at the optimum operating conditions and wavelength of each respective metal. All pH measurements were done using a calibrated (Jenway 3505) pH metre equipped with a standard calomel electrode (SCE). A constant shaker model CFC 3018 with a water bath was used to shake the samples at the required shaking speeds.

2.4. Sampling and pretreatment of the algae

Samples of the algae *Pediastrum boryanum* were collected in Molo, Nakuru County, Kenya. The algal material was cleaned using tap water, dried, ground into powder and then stored in clean plastic bottles. The dried powdered algae were used as a sorbent and for the modification process.

2.5. Modification of the dried algae

The algae were modified by anchoring tetramethylethylenediamine onto its chemical structure to improve its chelating property [15]. The modification procedure involved chlorination of the biomaterial first and then condensing the resultant with the amino compound. A sample of the algae (30 g) was chlorinated using thionyl chloride ($SOCl_2$) 100 ml, and the mixture heated under reflux at a temperature of 80°C for 4 h with continuous stirring. The chlorinated biomaterial was then washed with 100 ml of distilled water. The solution was filtered using Whatman No. 1 filter paper and dried in an oven set at 60°C. The dry chlorinated sample was then treated with 25.0 ml of tetramethylethylenediamine and refluxed for 3 h to anchor the tetramethylethylenediamine structure into the algal biomass.

2.6. Batch sorption experiments

Sorption studies were carried out on a mechanical reciprocating shaker (SKZ-1 NO. 1007827, India) using plastic screw cap bottles. Batch experiments were conducted to investigate the effects of pH, adsorbent dosage, adsorbate concentration and contact time on the adsorption of Pb^{2+}, Cd^{2+} and Cu^{2+} on the modified and unmodified algae. The pH of the model test solutions containing a known concentration of the metal ion was adjusted to values between 3 and 7. A known weight of the sorbent (0.03 g) was added to each of the solutions and then allowed to equilibrate giving sufficient time for sorption. The resulting mixture was filtered through Whatman No. 42 filter paper, and the metal ions in the filtrate were determined by atomic adsorption spectrophotometry (UNICAM 919).

2.7. Optimization of sorption parameters

2.7.1. Effect of pH on sorption

2.7.1.1. Calibration of the pH metre

The pH metre was calibrated using special buffer tablets for pH 3.0, 5.0, 7.0 and 9.0. Each tablet was dissolved separately in 100 ml of distilled water and then used. The electrode of the pH metre was conditioned with saturated potassium chloride overnight to wet the membrane and make it more sensitive. It was later calibrated with the buffer solutions. This procedure was undertaken before any pH measurements were made [16].

2.7.1.2. Effect of pH of adsorption

The effect of pH of the adsorbate on the adsorption of the metal ions by both modified and unmodified algae was investigated by mixing 0.2 g of the sorbent material with 50 ml of 10 ppm model solution buffered at different pH environments. The pH was brought to the desired values (3–10) by adding drops either 0.01 M nitric acid or 0.01 M sodium hydroxide. The resulting mixture was allowed to equilibrate for 120 min. The resulting mixture was then filtered through Whatman No. 1 and the concentration of the metal ions in the filtrate determined by atomic absorption spectroscopy.

2.7.2. Effect of contact time on sorption

The effect of contact time on sorption of lead (II), cadmium (II) and copper (II) by modified and unmodified algae was done by taking a sample, 0.2 g of the sorbent (modified and unmodified algae) into the plastic bottles and 50 ml of the adsorbate of concentration 10 ppm added. The mixture was buffered to the optimum pH value for each metal and agitated at predetermined time intervals of 2–150 min. The samples were then removed from the shaker and the solutions filtered, and the metal ion concentration in the filtrate was determined.

2.7.3. Effect of initial metal ion concentration on sorption

The extent to which metal ions are adsorbed as a function of the initial ion concentration was investigated by mixing 0.2 g of finely ground modified and unmodified algae separately with 50 ml of varying concentrations of the test solutions, buffered at the optimal pH value for each respective metal. The respective mixtures were allowed to equilibrate for a sufficient duration and then filtered and the concentration of the metal ions in the filtrate was determined.

2.7.4. Effect of sorbent dose on percentage metal removal

The effect of sorbent dose was investigated by agitating 50 ml (10 μg ml^{-1}) of the adsorbate solutions of lead (II), cadmium (II) and copper (II) with various dosages of the sorbents. 0.1, 0.5, 1.0, 1.5 and 2.0 g of modified and unmodified algae were used. The solutions of the adsorbate were buffered to the optimum pH of each respective metal ion under investigation.

The solutions were for 2 h with the temperature set at 25°C. The resulting mixtures were filtered and the concentration of the residual metal ions determined.

2.7.5. Determination of adsorption capacity of modified and unmodified algae

The adsorption capacity was determined by mixing 0.2 g of finely ground sorbent material with 50 ml of varying concentrations of the test metal solution (concentrations 10–250 ppm) buffered at the optimum pH for each respective metal. The mixtures were agitated for 30 min and then filtered, and the concentrations of metal ions were determined.

2.7.6. Adsorption models

The experimental data on metal sorption were also analyzed using adsorption models so as to establish the sorption kinetics and mechanism.

2.7.7. The kinetics of adsorption

To determine the necessary time, different solutions of for adsorption lead (II), cadmium (II) and copper (II) 20 ml containing 10 μg ml^{-1} of the adsorbate were introduced in different sets of plastic bottles containing 0.2 g of the adsorbent, and the pH sets an optimal value for each metal. The mixtures were then introduced in the shaker temperature of 25°C and equilibrated at different time intervals of 2, 5, 10, 20, 30, 60, 90 and 120 min. They were then filtered, and the filtrate was analyzed for adsorbate concentration. The data obtained was treated with Lagergren's [17] pseudo-first-order and Ho et al.'s [18] pseudo-second-order equations to determine molecularity of the adsorption.

The Lagergren first-order and Ho's second-order kinetic models are expressed as shown in equations 5.1 and 5.2, respectively:

$$\ln(C_o - C_t) = Kt + A \qquad (1)$$

$$\frac{1}{q_e} = Kt + A \qquad (2)$$

where C_o is the adsorption per unit mass of adsorbent at equilibrium, K is the adsorption rate constant, A is intercept and C_t is the concentration at time t.

2.7.8. Adsorption isotherms

The experimental data for the effect of metal ion concentration obtained was treated with the Freundlich and Langmuir isotherm models to obtain the adsorption mechanism.

2.7.8.1. *Langmuir isotherm*

For molecules in contact with a solid surface at a fixed temperature, the Langmuir isotherm, developed by Irving Langmuir in 1918, describes the partitioning between gas phase and adsorbed species as a function of applied pressure [19]. Langmuir adsorption isotherm is the widely used isotherm for modeling of adsorption data [20]. Langmuir considered adsorption of an ideal gas on an ideal surface. It is based on the assumption that adsorption can only occur at fixed sites and only hold on one adsorbate molecule (monolayer). All sites are equivalent with no interaction between adsorbed molecules, and the sites are independent as reported by Langmuir [19]. The Langmuir equation was derived from Gibbs approach which takes the form shown in equation 5.3 [19, 21]:

$$q_e = \frac{K_L C_e}{1 + a_L C_e} \tag{3}$$

where C_e is the equilibrium concentration, K_L is the equilibrium constant, q_e is the metal concentration on the sorbent phase at equilibrium in mg g^{-1} and a_L is a Langmuir constant. Eq. (3) can be linearized and often referred to as linearized Langmuir equation as shown in equation 5.4:

$$\frac{C_e}{q_e} = \frac{1}{K_L} + \frac{a_L C_e}{K_L} \tag{4}$$

The experimental data was applied on the equation above, and a plot of $\dfrac{C_e}{q_e}$ against C_e gave a linear regression. This indicates that the adsorption prescribes to the Langmuir model, where the gradient $\left(\dfrac{a_L}{K_L}\right)$ is the theoretical saturation capacity (units in mg g^{-1}) and the intercept is $\dfrac{1}{K_L}$ [19, 22, 23].

2.7.8.2. *Freundlich isotherm*

Freundlich isotherm is an empirical equation based on heterogeneous surface [23]. This is a multi-site adsorption isotherm for heterogeneous surfaces and has a general form as shown in Eq. (5):

$$q_e = K_F C_e^{bF} \tag{5}$$

The equation was linearized by taking logarithms and then applied to determine if the systems are heterogeneous with highly interactive species [24]:

$$\ln q_e = \ln K_F + b_F \ln C_e \qquad (6)$$

where q_e and C_e have the same meaning as in Eq. (3); the numerical value of K_F presents adsorption capacity, and b_F indicates the energetic heterogeneity of adsorption sites [24]. From the data, if a plot of **ln** q_e versus **ln** C_e gave a straight line, it indicates that the adsorption prescribes to the Freundlich model.

3. Results and discussion

3.1. Introduction

The algal material was modified with an amino compound to improve its thermal stability [9]. The resulting product obtained was a water-insoluble solid material. The raw and the modified products were characterized using FTIR to obtain the functional groups present; sorption parameters were established and then used for adsorption experiments in both synthetic solutions and real water samples.

3.2. FTIR analysis of modified and unmodified algae

The modified and unmodified materials were characterized with FTIR, and the resulting spectra are presented in **Figure 1**.

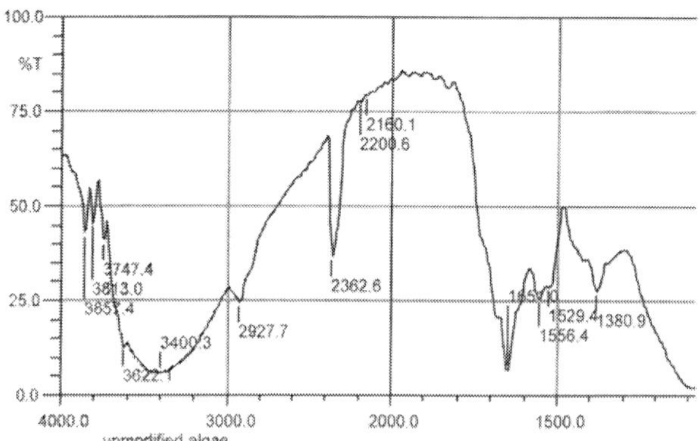

Figure 1. FTIR spectrum of unmodified algae.

The results show the presence of many functional groups capable of metal sorption. The broad and strong band at 3400.3 cm^{-1} could be attributed to either —OH or —NH group [25]. The band at 2927.72 cm^{-1} was assigned to C—H stretches, while the band at 1651.0 cm^{-1} was assigned to stretching —OH, C=O or N=C [25] .The band at 1380.9 cm^{-1} confirms the presence of an amide group of an amide or sulphamide group [26, 27]. The material was modified with tetramethy-lethylenediamine, and results obtained are presented in **Figure 2**.

Upon modification, the band at 3400.3 cm⁻¹ that was attributed to the −NH of an amide shifts to a lower value of 3394.5 cm⁻¹. The intensity of the band also decreases. This can be attributed to suppression influenced by the carbon atoms from anchored tetramethylethylenediamine functional groups. Similar observations were reported by Schluter [28] as he studied the treatment of poly(1.1.1)propellane with lithium organic initiators and then investigated its rigidity. Mwangi and Ngila [29] also recorded the same observation when studying the removal of heavy metals from contaminated water using ethylenediamine-modified green seaweed. New bands are also seen to appear at 1454.2 cm⁻¹ and 2597 cm⁻¹. The band at 1454.2 cm⁻¹ can be attributed to an NO₂ group and that at 2597 cm⁻¹ can be attributed to an additional −OH groups from a phenolic compound. Both the parent and the modified material were applied for sorption experiments.

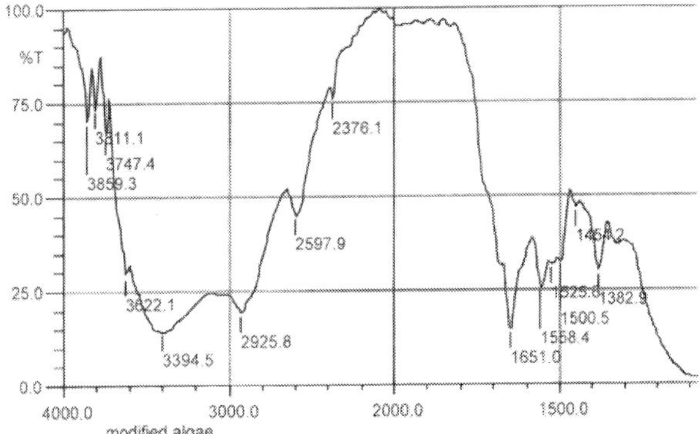

Figure 2. FTIR spectrum of modified algae.

3.3. Effect of pH on sorption

The adsorption of metal ions into the biosorbent is dependent on pH of the solution. pH affects the biosorbent surface charge and degree of ionization. The sorbent also has nitrogen atoms (with a lone pair of electrons) which can be influenced by the pH of the medium. The effect of pH on sorption of lead, cadmium and copper ions are represented in **Figure 3**.

The pH of the solution influences the chemistry of the metal binding sites and the behaviour of the metal itself in solution. The results show that the maximum adsorption for lead was found to occur at a pH of 3.5 by the unmodified sorbent and at a pH of 7.0 for this same metal by the modified adsorbent. There was an increase in the amount adsorbed as the pH increased from 3.5 to 7.0. Beyond this, as the pH increases, the amount adsorbed decreases. Similar results have been reported for other biosorbents [29]. Singh and co-workers (2006) also reported that the highest percentage of lead (II) ions was adsorbed by phosphatic clay at a pH of 5.0 [30]. Similar results were reported by Matheickal et al. [26] when they studied the biosorption of lead by marine alga *Ecklonia radiata*. Benhima et al. [31] observed that there was an increase in lead (II) ion uptake by inert organic matter (IOM) as the pH increased from 2.0 to 6.0. This is in agreement with the observed results for lead (II) ions.

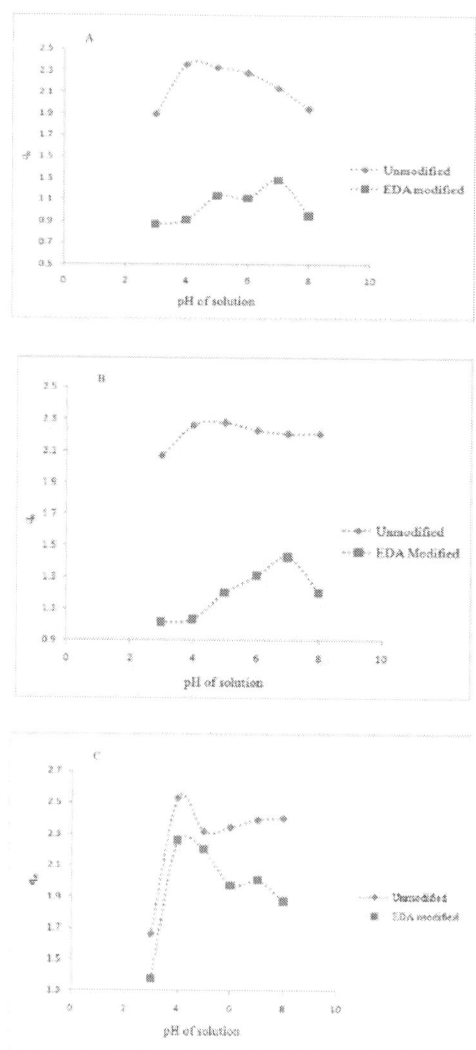

Figure 3. Effect of pH on adsorption of lead, cadmium and copper ions—A, B and C, respectively.

At low pH, the biomass surface would be completely covered with hydrogen ions. H^+ lead (II) ions cannot compete effectively for the adsorption sites. This can be attributed to the fact that protons are strongly competing due to their high concentration. Godhane et al. [32] reported that the minimal sorption obtained at low pH may be due to high mobility of protons and partly due to the fact that the solution pH influences the sorbent surface charge.

The unmodified biosorbent has a maximum adsorption for cadmium at a pH of 5.2, while the modified form was at a pH of 6.7. Similar results were obtained by Singh and co-workers [30] when they investigated the adsorption of cadmium using phosphatic clay. They observed maximum adsorption at a pH of 5.4.

Copper unlike the other metals has a maximum adsorption at a pH of 4.2 for both the modified and unmodified sorbents. This can be explained by the small size of copper giving it a high polarizing power on electrons of the adsorbent [33].

The sorbent has nitrogen atoms (with alone pair of electrons) as well as other functional groups, all of which may be influenced by pH. At low pH, the adsorbent is positively charged because the pH is lower than the isoelectric point or point of zero charge (PZC), i.e. pH < PZC. At such low pH range, adsorption is poor due to the charge on the adsorbent [34]. At high pH (pH > PZC), the adsorbent is negatively charged contributing to the high adsorption [26]. This arises from the fact that when the metal is in solution, it is positively charged and will be attracted to the surface of the negatively charged adsorbent at pH > PZC favouring adsorption. At pH > 7, there is metal hydrolysis leading to precipitation due to formation of hydroxyl metal ions [35].

3.4. Effect of contact time on sorption

The dependence of the adsorption process on the residence time of the adsorbate at the solid solution interphase was studied using batch sorption experiments for both the modified and unmodified algae. The solutions were set at optimal pH values of each metal, and the results of time-dependent adsorption obtained are shown in **Figure 4**.

It was observed that the general uptake rate was fast as over 90% of the adsorption which took place within the first 30 min for all the three metals after which a steady adsorption rate was realized (**Figure 3**). Keskinkan et al. [36] reported similar findings while studying the biosorption of lead (II) ions using aquatic *Ceratophyllum demersum*. Also, Yang and Volesky [37] made similar observations while studying the biosorption of cadmium (II) ions using dead brown *Sargassum fluitans*. The initial rapid uptake may be due to the physical adsorption or ion exchange at the cell surface and the subsequent slower phase due to the other mechanisms such as complexation, micro-precipitation or saturation at the binding sites [38]. Generally, there was a decrease in adsorption for the modified sorbent with the decrease more pronounced for cadmium (II) ions. Different functional groups with different affinities for the metal ions are usually present on the biomass surface; these are significantly altered by modification. New groups are also introduced which affect the binding ability of the sorbent.

An example of a functional group that was affected is the –NH group which appeared at 3400.3 cm^{-1} and after modification shifted to a lower value of 3394.5 cm^{-1}.

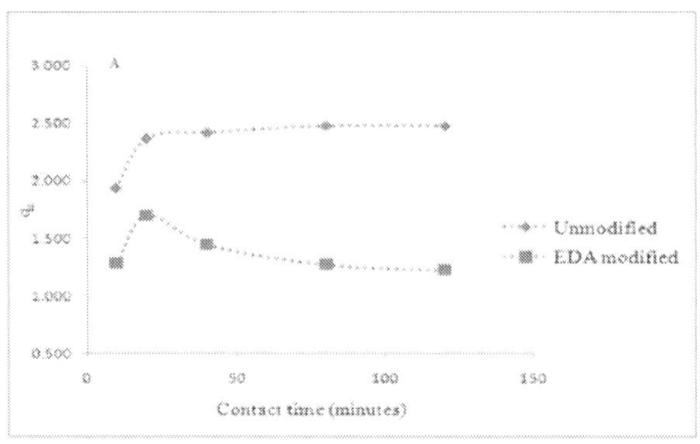

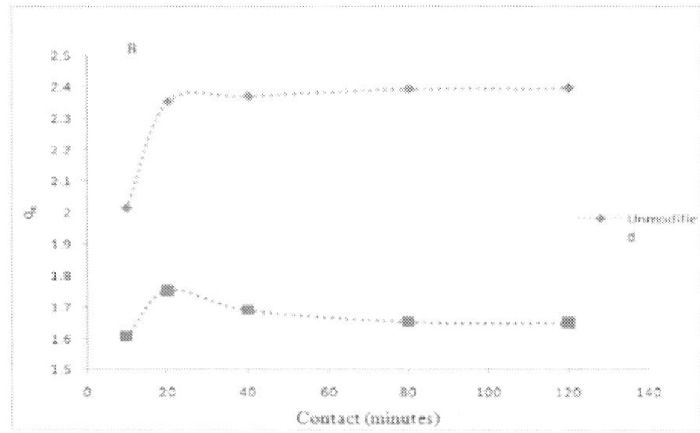

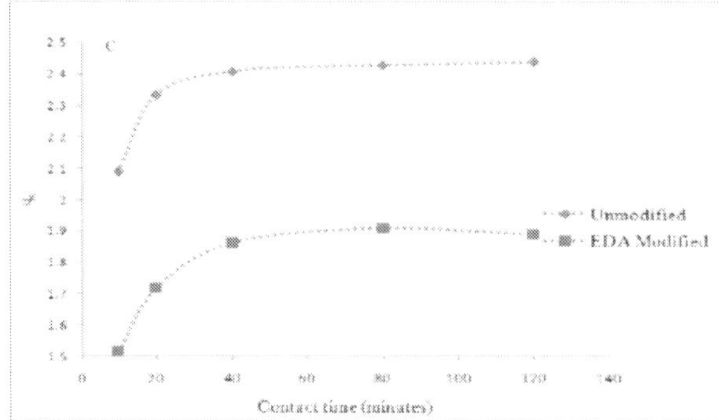

Figure 4. Effect of contact time on sorption of lead, cadmium and copper—A, B and C, respectively.

3.5. Effect of sorbent dosage on sorption

Results presented in **Figure 5** show the effect of varying the mass of the sorbent on the adsorption of the metal ions. The experiment was performed while the solutions were buffered at optimal pH value of each respective metal ion.

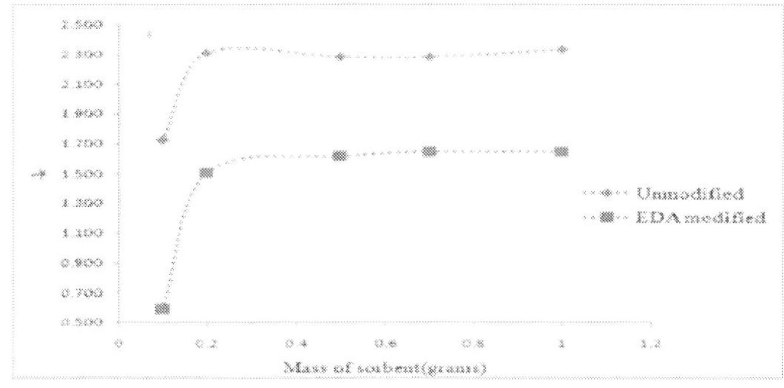

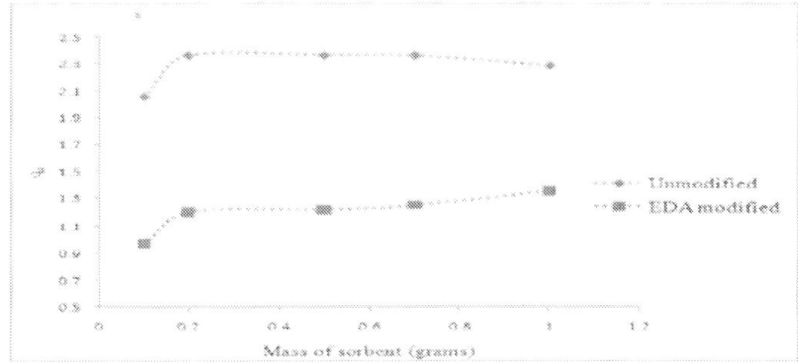

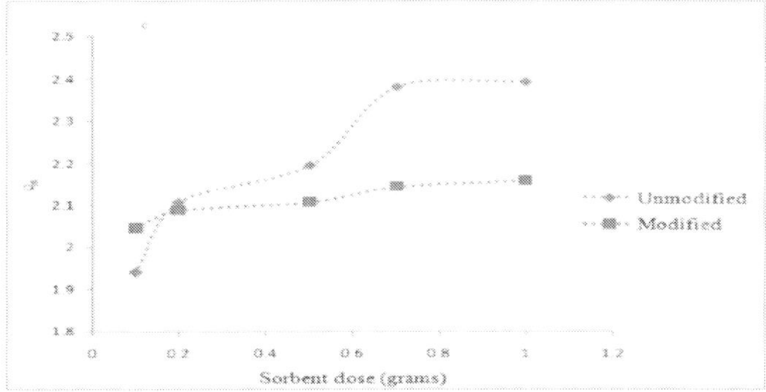

Figure 5. Effect of sorbent dose on sorption of lead, cadmium and copper—A, B and C, respectively.

The general observation is that the amount adsorbed metal ions increased with increase in sorbate dose. This can be attributed to the fact that as the concentration of the adsorbent increases, more adsorption sites are available due to increased surface area. More metal ions can therefore occupy those available active sites [39]. It was also noted that for both the unmodified adsorbates, copper was adsorbed more than both lead and cadmium. This could be attributed to the fact that the affinity to the binding sites is related to the ionic sizes of the respective metals, hence the polarizing power of copper being responsible for the observation. Maximum adsorption by the unmodified sorbate occurred at a mass of about 0.2 g for both lead and cadmium. Mwangi and Ngila [29] reported a similar observation while studying the biosorption of lead and cadmium using green seaweed, *Caulerpa serrulata*. Higher metal ion uptake at low sorbate mass concentrations has been attributed to an increased metal to biosorbent ratio which decreases upon an increase in sorbate mass concentration [40]. The modified sorbate showed maximum adsorption at a mass of 0.7 g for these two metals. For copper, maximum adsorption occurred at a sorbate mass of about 0.7 g for both modified and unmodified algae *Pediastrum boryanum*. A plateau is reached after a certain mass for all the metal ions. The unmodified sorbent material removed more metal ions than the modified material. This decrease in the adsorption capacity could be attributed by the fact that the functional groups in the parent material responsible for binding the metals had higher complexation ability than the anchored group [41]. A similar observation was reported by Drake and co-workers as they studied the sorption of chromium by *Datura inoxia* biomaterial [42].

3.6. Effect of initial ion concentration for determination of adsorption capacity

In order to investigate the adsorption capacity of both modified and unmodified algae, 50 ml solutions of concentrations ranging from 10 µg l⁻¹ to 100 µg l⁻¹ were added to 0.2 g of the biosorbent. **Figure 5** shows the results that were obtained.

A linear pattern of metal uptake was observed followed by some plateau for all metals. This may be attributed to the saturation of the binding sites as the concentration of the ion increases resulting to a steady state. The saturation is more pronounced at very high concentrations [31]. This is attributed to the fact that concentration is the driving force for metal ions to occupy the available binding sites [38]. It was observed that there was no significant difference in the sorption of copper and cadmium by the parent and modified material (see **Figure 6**).

The experimental data above was fitted into the Langmuir and Freundlich adsorption isotherm represented by equations (3) and (4) in Chapter 3 to determine the adsorption capacity for the metal ions using both modified and unmodified algal materials. **Table 1** is a summary of the results that were obtained.

Table 1 shows that the adsorption for both cadmium and lead fitted well with the Freundlich model, while copper fitted well with the Langmuir model. Similar results were obtained by Ng and co-workers [22] as they studied the adsorption of metals on cross-linked seaweed. This indicates that the adsorption process for lead and cadmium is a multi-site or physical sorption as a result of weak Van der Waals forces.

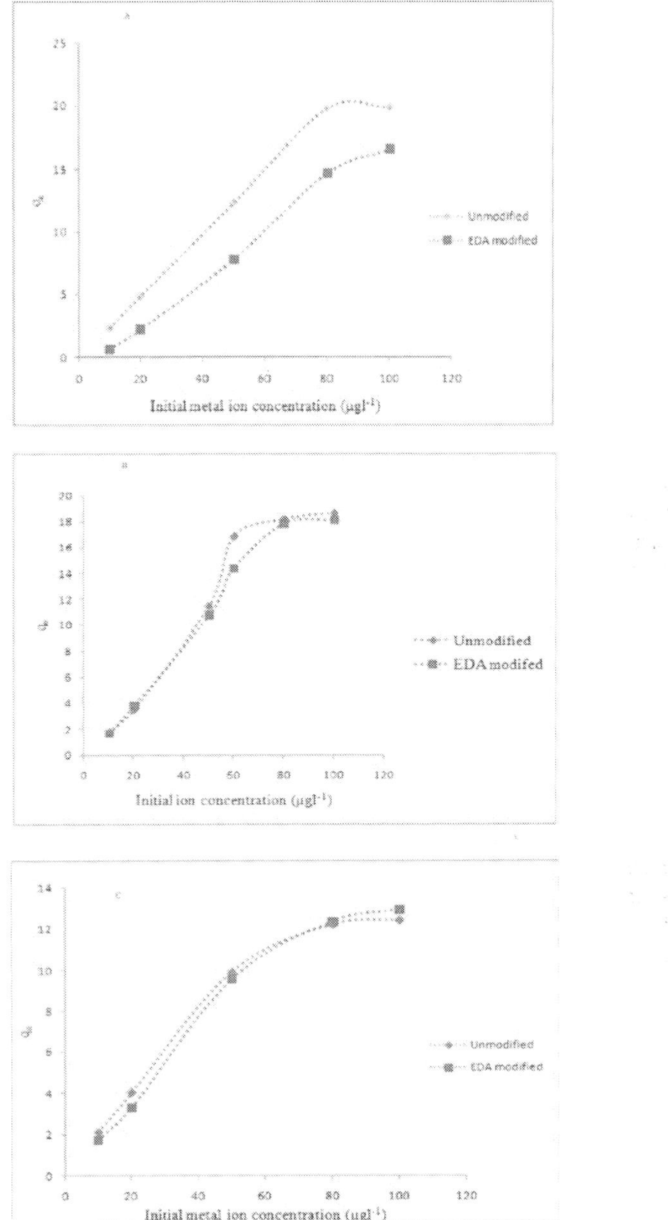

Figure 6. Effect of initial ion concentration on sorption of lead, cadmium and copper ions—A, B and C, respectively.

Metal	Langmuir		Freundlich		Comment
	R^2	Adsorption capacity	R^2	K_f/mg g^{-1}	
Copper					
Unmodified	0.974	0.059	0.605	0.948	Langmuir
Modified	0.512	0.001	0.502	–	Langmuir
Cadmium					
Unmodified	0.043	0.060	0.251	1.137	Freundlich
Modified	0.941	0.276	0.993	2.843	Freundlich
Lead					
Unmodified	0.570	0.021	0.929	0.791	Freundlich
Modified	0.560	0.112	0.906	1.695	Freundlich

Table 1. Results for Langmuir and Freundlich models.

Copper on the other hand fitted well with the Langmuir model with R^2 values of 0.974 for the unmodified and 0.512 for the modified algal material. These values show that the data has a strong correlation and therefore the sorption mechanism can be prescribed to the Langmuir model. The adsorption capacities were found to be 0.059 and 0.001 mg g^{-1} for copper, by the unmodified and modified materials, respectively. Sorption of cadmium and lead prescribed to the Freundlich model with a sorption of 1.137 and 2.843 and 0.791 and 1.695, respectively, by the unmodified and modified adsorbent materials in the same order.

3.7. Kinetics of adsorption of the metal ions

Lagergren's first-order and Ho's second-order kinetics were applied to the data that was obtained [17, 18]. This was used to determine the molecularity of the adsorption and the rate controlling step. A summary of the results that were obtained for all the three metals is shown in **Table 2**.

Metal	Lagergren	Ho	Comment
	R^2	R^2	
Lead			
Modified	0.936	0.954	Pseudo-second order
Unmodified	0.846	0.849	Pseudo-second order
Cadmium			
Modified	0.574	0.584	Pseudo-second order
Unmodified	0.765	0.770	Pseudo-second order
Copper			
Modified	0.512	0.502	Pseudo-first order
Unmodified	0.583	0.573	Pseudo-first order

Table 2. The Lagergren first-order and Ho et al. second-order data.

For all metals the adsorption rates are very fast initially and become almost constant as time increases showing that an equilibrium has been reached.

For cadmium and copper, the adsorption followed pseudo-first-order kinetics showing that only one molecule was involved in the rate determining step [43]. The adsorption of lead followed pseudo-second-order kinetics for both the ethylenediamine modified and the unmodified algal material. Similar results were obtained by Mwangi and Ngila [29] when studying the removal of heavy metals from wastewater using ethylenediamine-modified seaweed.

3.8. Analysis of wastewater samples

Fifty millilitre samples of water from Turi River were placed into plastic bottles and spiked with known concentrations of lead, cadmium and copper. The solution was agitated in a shaker for 30 min and then filtered, and the filtrate was analyzed for lead, cadmium and copper samples using AAS. The percentage of the metal recovered from the water samples was then recorded. Results from this analysis are shown in **Table 3**.

Metal	Concentration added/μgl^{-1}	Concentration recovered/μgl^{-1}	% Recovery
Copper			
Unmodified	00.00	2.050	
	10.00	10.363	86.0
	20.00	18.191	82.5
	30.00	19.871	62
Modified	00.00	2.010	
	10.00	8.191	68.2
	20.00	6.163	28.0
	30.00	7.682	24.0
Cadmium			
Unmodified	00.00	1.137	
	10.00	8.019	72.0
	20.00	14.627	69.2
	30.00	18.370	59.0
Modified	00.00	1.145	
	10.00	8.916	80.0
	20.00	16.070	76.0
	30.00	22.892	73.5
Lead			
Unmodified	00.00	2.185	
	10.00	9.870	81.0
	20.00	17.659	79.6
	30.00	19.633	61.0
Modified	00.00	2.070	
	10.00	9.885	81.9
	20.00	16.729	75.8
	30.00	22.449	70.0

Table 3. The percentage recovery of metals from real water samples.

From the results, the percentage recovery is high at low concentrations but decreased as the concentration of the metal in the water sample increased. The adsorption of copper by the unmodified sample was found to be the best and cadmium the least. For the modified sample, the adsorption of lead was the most and cadmium was the least. This can be explained by the fact that copper ions unlike lead and cadmium have relatively high affinity for ligands containing the nitrogen atom [44]. Such ligands which are smaller than the adsorbent have high affinity for the metal ion due to their high basicity [45]. These results show that the algal biomass has shown good potential to be used in water resource management.

4. Conclusions

The study successfully functionalized the algal material with tetramethylethylenediamine, and the FTIR spectrum provided evidence of its functional groups capable of binding metal ions. Adsorption of the three metals was best at lower pH values (4.2–6.8). Beyond these values, the adsorption decreased considerably. The rate of adsorption was very fast as more that 90% of the metals were removed from the solution within 30 min. The adsorption of copper fitted into the Langmuir adsorption isotherm with R^2 values of 0.974 and 0.512 for the unmodified and modified sorbents, respectively, indicating a monolayer-binding mechanism. Cadmium and lead fitted into the Freundlich adsorption mechanism (R^2 values were Cd modified = 0.993, unmodified = 0.251, Pb modified = 0.906, unmodified = 0.929). The adsorption of lead and cadmium was by Ho's pseudo-second-order kinetics confirming a multi-site interaction, whereas copper followed the pseudo-first-order kinetics, evidence of single-site adsorption. The adsorption by the algal material did not improve upon modification since natural ligands were replaced with ethylenediamine which has a lower stability constant for the metal analytes but minimized leaching of dissolved organic matter. In general the study has shown that the algal material can be used for as an effective sorbent for removal of lead, cadmium and copper from contaminated wastewater at low pH values.

Author details

John Okapes Joseph[1], Isaac W. Mwangi[1*], Sauda Swaleh[1], Ruth N. Wanjau[1], Manohar Ram[1] and Jane Catherine Ngila[2]

*Address all correspondence to: isaacwaweru2000@yahoo.co.uk

1 Department of Chemistry, Kenyatta University, Nairobi, Kenya

2 Department of Chemical Technology, University of Johannesburg, South Africa

References

[1] Love J., Percival E. (1964) The polysaccharides of the green seaweed codium fragile. The water-soluble sulphated polysaccharides. *Journal of Chemical Society* Part II 3338–3345

[2] Matsubara K., Matsuura Y., Bacic A, Liao M.L., Hori K., Miyazawa K. (2001) Anticoagulant properties of a sulfated galactan preparation from a marine green alga, *Codium cylindricum*. *International Journal of Biological Macromolecules* 28:395–399

[3] Selivanovskaya S.Y., Latypova, S., Latypova, V.Z. (2003) The use of bioassays for evaluating the toxicity of sewage sludge and sewage sludge-amended soil. *Journalsof Soils and Sediments* 3 (2): 85–92.

[4] Henry, C.L., Cole, D.W. (1997). Use of bio-solids in the forest: technology, economics and regulations. *Biomass and Bioenergy* 13: 69–277

[5] Losada, M.R., Lopez-Diaz, L., Rodriguez, A. 2001. Sewage sludge fertilisation of a silvopastoral system with pines in northwestern Spain. *Agroforestry Systems* 53: 1–10.

[6] Gupta, V.K., Gupta, M., Sharma, S. (2001) Process development for the removal of lead and chromium from aqueous solution using red mud, an aluminum industry waste. *Water Research* 35: 1125–1134.

[7] Lodeiro, P., Barriada, J.L., Herrero, R., Sastre De Vicente, M.E. (2006) The marine macroalga Cystoseira baccata as biosorbent for cadmium (II) and lead (II) removal: kinetic and equilibrium studies. *Environmental Pollution* 142 (2): 264–273.

[8] Cha, D.K., Song, J.S., Sarr, D. (1997) Treatment technologies. *Water Environment Research* 69: 676–689

[9] Hsien, T.Y., Rorrer, G.L. (1995) Effects of acylation and crosslinking on the material properties and cadmium ion adsorption capacity of porous chitosan beads. *SeparationScience and Technology* 30: 2455–2475.

[10] Chen, J.P., Yang, L. (2005) Chemical modification of Sargassum sp. for prevention of organic leaching and enhancement of uptake during metal biosorption. *Industrial and Engineering Chemistry Research* 44: 9931–9942.

[11] Saracoglu, S., Soylak, M., Elci, L. (2001) Preconcentration of Cu (II), Fe (III), Ni(II), Co(II) and Pb(II) ions in some manganese salts with solid phase extraction method using chromosorb-102 331resin. *Trace Elements and Electrolyte* 18:129–133.

[12] Marshall, W.E., Wartelle, L.H., Boler, D.E., Johns, M.M. and Toles, C.A. (1999) Enhanced metal adsorption by soybean hulls modified with citric acid. *Journal of Bio Resources Technology* 69:263–268.

[13] Guo, Y. J., Qi, S., Yang, K., Yu, Z., Wang, H. (2002) Adsorption of Cr (VI) on micro-and mesoporous rice husk-based activated carbon. *Journal for Materials for Chemistry and Physics* 78:132–137

[14] Bailey, S.E., Olin, T.J., Bricka, R.M., Adrian, D.D. (1999) A review of potentially low-cost sorbents for heavy metals. *Water Resources Journal* 33:2469–2479.

[15] Karsten, E., Erhard, H., Roland, R., Hartmut, H. (2005) Amines, Aliphatic in Ullmann's Encyclopedia of Industrial Chemistry, Wiley-VCH Verlag, Weinheim.

[16] Vogel, A.I. (1978) A Textbook of quantitative Inorganic Analysis Including Elementary Instrumental Analysis. 3rd ed. New impressions, London, p. 910, 1162

[17] Lagergren, S. (1898) The theory of so-called adsorption of soluble substances. Hand-lingar 24 (4): 1–39.

[18] Ho, Y.S., McKay, G., Wase, D.J., Forster, C.F. (2000) Study of the sorption of divalent metal ions on peat. *Adsorption Science Technology* 18: 639–650

[19] Langmuir, I. (1918) The adsorption of gases on plane surfaces of glass, mica and platinum. *Journal of American Chemical Society* 40: 1362–1403.

[20] Tan, C., Xiao, D. (2009) Adsorption of cadmium ion from aqueous solution by ground wheat stem. *Journal of Hazardous Materials* 164: 1329–1363.

[21] Yang, R.T. (1987) Gas separation by adsorption processes. *Journal of the American Chemical Society* 53: 497.

[22] Ng, J.C.Y., Cheung W.H., McKay, G. (2003) Equilibrium studies for the sorption of lead from effluents using chitosan. *Chemosphere* 52 (6): 1021–1030.

[23] Freitas, P.A., Iha, M.K., Felinto, M.C.F.C., Suárez-Iha, M.E.V. (2008) Adsorption of di-2-pyridyl ketone salicyloylhydrazone on Amberlite XAD-2 and XAD-7 resins:characteristics and isotherms. *Journal of Colloid and Interface Science* 323: 1–5.

[24] Freundlich, H.M.F. (1906) Over the adsorption in solution. *Zeitschrift für Physikalische Chemie* 57 (A) 385–470.

[25] Stuart, B. (1996) Modern Infrared Spectroscopy, John Wiley & Sons Inc., New York

[26] Matheickal, J.T., Yu, Q. (1996) Biosorption of lead from aqueous solutions by marine alga, Ecklonia radiate. *Water Science Technology* 34: 1–7.

[27] Kapoor, A., Viraraghavan, T. (1997). Heavy metal biosorption sites in *Aspergillus niger*. *Bioresources and Technology* 61: 221–227.

[28] Schluter, A. (1988) Poly ([1,1,1]propellane). A novel rigid-rod polymer obtained by ring opening polymerization breaking a carbon-carbon A-bond. *Journal of the American Chemical Society* 21(5): 1208–1211.

[29] Mwangi, I.W., Ngila, J.C. (2012) Removal of heavy metals from contaminated water using ethylenediamine-modified green seaweed, *Caulerpa serrulata*. *Physics and Chemistry of the Earth* 50: 111–120.

[30] Singh, S.P., Ma, L.Q., Hendry, M.J.. 2006 Characterization of aqueous lead removal by phosphatic clay: equilibrium and kinetic studies. *Journal of Hazardous Materials* 136 (3): 654–662

[31] Benhima, H., Chiban, M., Sinan, F., Seta, P., Persin, M. (2008) Removal of lead and cadmium ions from aqueous solution by adsorption onto micro-particles of dry plants. *Colloids Surface Biochemistry: Biointerfaces* 61: 10–16.

[32] Godhane, I., Nouri, L., Hamdaoui, Q., Chiha, M. (2008) Kinetic and equilibrium study for the sorption of cadmium (II) ions from aqueous phase by eucalyptus bark. *Journal of Hazardous Materials* 152: 148

[33] Richardson, J.T. (1967) Crystal field effects in ion-exchanged faujasites. *Journal of Catalysis* 9(2): 178–181.

[34] Tripath, T., Ranjan-De, B. (2006) Flocculation: A New Way to Treat the Waste Water, *Journal of Physical Sciences*, 10:93 –127

[35] Li, X., Tang, Y., Cao, X., Lu, D., Luo, F., Shao, W. (2008) Preparation and evaluation of orange peel cellulose adsorbents for effective removal of cadmium, zinc, cobalt and nickel, colloids and surfaces. *Physicochemical and Engineering Aspects* 317: 512–521.

[36] Keskinkan, O., Goksu, M.Z.L., Basibu Yuk, M., Forster, C.F. (2004) Heavy metal adsorption properties of a submerged aquatic plant *Ceratophyllum demersum*. *Bioresource Technology* 92: 197

[37] Yang, J.B., Volesky, B. (1999) Cadmium biosorption rate in protonated Sargassum biomass. *Environmental Science and Technology* 33: 751.

[38] Khani, M. H. (2006) Biosorption of uranium from aqueous solutions by nonliving biomass of marine algae, *Cystoseira indica*. *Journal of Biotechnology* 9(2): 101–108.

[39] Ilhan, S.N.M., Kilicarslan, S., Ozdag, H. (2004) Removal of chromium, lead and copper from industrial waste by *Staphylococcus saprophyticus*. *Turkish Electronic Journal of Biotechnology* 2: 50–57.

[40] Puranik, P.R., Paknikar, K.M. (1997) Biosorption of lead and zinc from solutions using *Streptoverticillium cinnamoneum* waste biomass. *Journal of Biotechnology* 55: 113.

[41] Bradl, H.B. (2004) Adsorption of heavy metal ions on soils and soils constituents. *Journalof Colloid and Interface Science* 277: 1–18

[42] Drake, L.R., Lin, S., Rayson, G., Jackson, P.J. (1996) Chemical modification and metal binding studies of *Datura innoxia*. *Environmental Science and Technology* 30: 110–114.

[43] Agrawal, A., Sahu, K.K. (2006) Kinetics and isotherm studies of cadmium adsorption on manganese nodule residue. *Journal of Hazardous Materials* 137(2): 915–924.

[44] Topperwien, S.B.R., Xue, H., Asigg, A. (2007) Cadmium accumulation in *Scenedesmus vacuolatus* under fresh water conditions. *Journal of Environmental Science and Technology* 41:5383–5388.

[45] Comuzzi, C., Grespan, M., Melchior, A., Portanova, R., Tolazzi, M. (2001) Thermodynamics of complexation of cadmium (II) by open chain N-donor ligands in dimethyl sulfoxide solution. *European Journal of Inorganic Chemistry* 12: 3087–3094.

Water Quality

Aspects on the Accumulation of Trace Metals in Various Environmental Matrices (Water, Soil, Plant and Sediments): Case Study on Catchment Area of the Somes River, Romania

Cezara Voica, Andreea Maria Iordache,
Roxana Elena Ionete and Ioan Ștefănescu

Additional information is available at the end of the chapter

http://dx.doi.org/10.5772/65700

Abstract

A study was carried out to determine the accumulation of trace metals in water, sediments, and soil from several locations in the Transylvania region (Romania), using the inductively coupled plasma mass spectrometry (ICP-MS) technique. A significant number of metals (range of several $\mu g\ L^{-1}$) were identified, the toxic metal concentrations in mostly of the investigated waters being within the permissible limits. A seasonal variation in the metal content was also observed. Comparison of the metal concentrations to samples of sediment, soil, and vegetation coming from the surrounding areas of the same water reservoir revealed a higher accumulation of rare and toxic metals in sediments than in soil and vegetation.

Keywords: surface water, pollution, environmental quality

1. Introduction

The main strategic objective of Romania in the field of water is linked to European integration, which involves harmonization and implementation of the *acquis communautaire* in the field of water quality protection. National Water Law No. 107/1996 [1], updated in July 2015, legally enshrines a novel conception on the status of water (Article 1, paragraph 1):

renewable natural resource, vulnerable, and confined; indispensable for life and society (its physical dimension); raw material for productive activities, source of energy, and transport route (its economic dimension); critical element in maintaining the ecological balance (its environmental dimension). The European Union (EU) Water Framework Directive (WFD) [2] establishes a framework for the protection of waters and consists of a new vision for the management of water resources in Europe. Mainly sustained on ecological elements, the ultimate objective of the WFD is the achievement of at least *"a good ecological quality status"* for all surface waters.

According to the estimation of the World Health Organization, about two-thirds of diseases are caused by the polluted water. Through its accession to the EU, Romania has undertaken to comply with the European regulations on water quality [3]. Within the United Nations Environment Programme, which supports the surveillance of water quality in freshwater ecosystems worldwide, by its global system of environmental monitoring (GEMS)/Water Global Network [4], the determination of heavy metals concentration is mandatory when the water quality is assessed.

The proposed subject represents a highly current field; the concentration of heavy metals in water being an intensive subject researched worldwide [5–7]. Some researchers have determined metal concentrations in water, sediment, plant; others have studied the metals effect on live organisms [8–10]. Heavy metals are seen as potential hazard for human health and ecosystem as they cannot be degraded, being continuously deposited and incorporated in water, sediments, soil, and vegetation. Anthropogenic activities may lead to important accumulations of toxic metals into the environment; therefore, the assessment of contamination degree in the aquatic and terrestrial environments by means of elemental analysis became a common monitoring activity of our days.

2. Sample preparation and analytical methods selection for analyzing contaminant elements in environmental samples

2.1. Study area

Somes is a transboundary catchment located in the north-western Romania. Its main sources of water are the surface waters, reservoirs, and ground waters. The overall water resources in the basin are theoretical about 4.348 billion m^3 (of which 4.012 billion m^3 coming from surface waters and 336 million m^3 of groundwater), but only 21.7% are technically usable. In this region, there are 23 reservoirs with areas over 0.5 km^2 such as the Gilau Lake (0.67 km^2), Tarnita (2.2 km^2), Somes Cald (0.8 km^2), and Fantanele-Belis (8.15 km^2).

The main objective of this study was to characterize the surrounding areas of raw water accumulations from the catchment, in terms of its content in heavy metals and rare earth, correlated with the supply mechanism of these surface waters. Thus, an assessment of the quality of surface flowing waters, during 2009–2011, was performed. Waters from eight surveillance sections were sampled: the Gilau dam (Area 1); the Somes Cald River (Area 2); the end of Gilau Lake intersection with the Somes Cald River (Area 3); the Somes Cald Lake (Area 4);

the confluence of Somes Cald Lake with Somes Rece Lake (Area 5); the Somes Rece Lake (Area 6); Tarnita (Area 7); and the end of Tarnita Lake bottom (exit from hydro plant—Area 8). The measurements were made with a variable frequency. Once with establishing the sampling data, we envisaged the possibility to set any correlations between the values of the monitored parameters and the climatic factors. Samples of sediment, soil, and vegetation corresponding to the surface water sampling points were also taken.

2.2. Samples preparation

Water samples were collected in high-density polyethylene containers previously washed in a solution of 10% nitric acid in an ultrasonic bath for 15 min followed by repeated rinsing with bi-distilled water and finally rinsed with ultrapure water (resistivity 18.2 MΩ cm^{-1}). Until sampling, the containers were kept in sealed polyethylene bags. Water samples were stabilized with ultrapure nitric acid (0.5% HNO_3).

Samples of *sediment, soil, and vegetation* require a digestion process to bring them into solution. For inductively coupled plasma mass spectrometry (ICP-MS) analysis, the digestion process should satisfy some conditions: the entire quantity of sample has to be dissolved, only ultrapure reagents must be used, any loss of analyte has to be avoided, the vessels in which the sample is kept must not react with the sample or the reagents used for the digestion, and the entire process should be fast and reliable. The samples were lyophilized, and then, 0.1 g aliquot of each sample was weighed. These aliquots were then digested in an acid mixture (3 ml HNO_3 60% + 2 ml HF 40%) at high pressure and temperature. After cooling, the liquid was transferred through a semi-automatic pipette in a 50 cm^3 volumetric flask of high-density polyethylene and was brought to the required volume with ultrapure water.

2.3. Analytical procedure

The analytical methods used in environmental monitoring have greatly evolved. Inductively coupled plasma mass spectrometer offers fast multielement capabilities, a high dynamic range, and excellent detection limits in a large number of matrices. It can be used for a variety of aqueous samples like natural waters or environmental samples that can be dissolved [11, 12]. The mass spectrometry with inductively coupled plasma (ICP-MS) is among the most success-ful existing methods applied when concentrations of trace and ultra-trace elements (under ppb) are envisaged [13].

For this work, measurements were performed with a mass spectrometer ICP-MS model Elan DRC-e, Perkin Elmer, with the following characteristics: detection limit 0.001–0.01 µg L^{-1}; resolution <0.5 at 10% peak height; abundance sensitivity 1–5 × 10^{-7}; and precision: <2% at 20 min. The performance of an ICP-MS instrument strongly depends on the operating conditions. Working parameters for plasma were chosen so as to obtain a good compromise between high sensitivity and low oxide levels. The following instrumental parameters of Elan DRC-e spec-trometer were set: 0.92 L min^{-1} nebulizer gas flow (NEB); 1.2 L min^{-1} auxiliary gas flow (AGF); 15 L min^{-1} plasma gas flow (PGF); 1100 W. ICP RF Power; 0.0 quadrupole rod offset; and 70.00 discriminator threshold.

2.4. Reagents and standards

Ultrapure deionized water (18.2 MΩ cm^{-1}) from a Milli-Q analytical reagent-grade water purification system (Millipore) and ultrapure HNO$_3$ 60% were used. All the plastic lab ware employed for sampling was either new or cleaned by soaking 24 h first in 10% HNO$_3$ then in ultrapure water. A 10 mg L^{-1} solution of Mg, Cd, Cu, In, Ba, Ce, Pb, and U (in 1% HNO$_3$, PerkinElmer Atomic Spectroscopy Standard–Setup/Stab/Masscal Solution) was used as external standard. The calibration solutions for quantitative measurements were prepared from a multi-element standard purchased from PerkinElmer (standard ICP-MS containing 29 elements, matrix: 5% HNO$_3$, PerkinElmer Life and Analytical Sciences), of 10 mg L^{-1}. To determine the rare metals in the water samples, a multi-element standard (multi-element calibration standard 2: 10 mg L^{-1}: Ce, Dy, Er, Eu, Gd, Ho, La, Lu, Nd, Pr, Sm, Sc, Tb, Th, Tm, Y, Yb), PerkinElmer (atomic spectroscopy standard), was used.

2.5. Parameters of performance

A method for determining the concentrations of heavy metals in water by ICP-MS was developed and validated. The performance parameters were within specifications of SR EN ISO 17294-2: water quality—application of inductively coupled plasma mass spectrometry (ICP-MS). Limit of detection recorded by the validated method for the elements under study provides the minimum limit of quantification required for quantitative determinations of the concentrations of these elements in the investigated waters, very good linearity (with correlation factors R > 0.999) for most elements.

The minimum detection limit is the lowest concentration or quantity of analyte which can be measured with reasonable statistical certainty. To determine the limit of detection 3SD, a method developed by PerkinElmer (Estimating Instrument Detection Limits, Elan version 3.4., and Software Guide) was used. Ultrapure water of 18.2 MΩ cm^{-1} was aspired, and signal intensities for blank were recorded. The limit of detection was calculated by Eq. (1) where SD$_{blank}$ is the standard deviation for the signal recorded on the blank for the element studied, conc.$_{sample}$ is the concentration (μg L^{-1}) of the analyte in the sample, and I$_{sample}$ and I$_{blank}$ are the signal intensities recorded for the sample and blank, respectively:

$$LOD = 3 \cdot SD_{blank} \cdot conc_{sample} / (I_{sample} - I_{blank}). \tag{1}$$

The limit of quantitation (LOQ) is the lowest concentration that can be quantitatively determined with an acceptable level of repeatability and accuracy. It is generally considered to be approximately ten times the minimum detection limit (LOD). The analytical quality control included daily analysis of standards and triplicate analysis of samples and blanks. The accuracy and precision of the analytical technique were evaluated by analyzing a certified standard reference material. Precision of the instrument was determined by introducing the same quantity of one sample ten times, and then, the relative standard deviation (RSD) was calculated. RSD values ranged from 0.4 to 6.4% confirming the high precision of the method. Accuracy expresses the correlation between the arithmetic mean of the measured values and the accepted reference value. RE ranged from −0.3 to 13% confirming the accuracy of the implemented method (**Table 1**).

Elements				Parameters								
				Intra-day studies				Inter-day studies				
	R^1	LOD (ng L^{-1})	LOQ (ng L^{-1})	Added (µg L^{-1})	Found (average) (µg L^{-1})	SD^2 (%)	RSD^3 (%)	Added (µg L^{-1})	Found (average) (µg L^{-1})	SD^2 (%)	RSD^3 (%)	RE^4 (%)
Al	0.9994	1.7	16.7	25	25.59	0.23	0.9	25	25.69	0.18	0.72	2.76
Cr	0.9999	1.3	13.2	2.5	2.53	0.01	0.6	2.5	2.56	0.05	2.22	2.6
Mn	0.9995	2.6	26.5	2.5	2.51	0.04	1.5	2.5	2.60	0.07	2.93	4.0
Ni	0.9999	1.2	12.4	2.5	2.40	0.03	1.2	2.5	2.58	0.14	5.36	3.4
Cu	0.9997	1.3	12.7	2.5	2.57	0.03	1.2	2.5	2.64	0.07	2.61	5.5
As	0.9999	3.5	34.5	2.5	2.73	0.03	1.0	2.5	2.71	0.02	0.66	8.5
Sr	0.9989	0.5	5.3	25	25.32	0.11	0.4	25	25.24	0.26	1.02	0.95
Cd	0.9999	1.1	10.9	2.5	2.58	0.05	1.9	2.5	2.67	0.06	2.22	7.0
Ba	0.9998	0.9	8.7	2.5	2.67	0.04	1.7	2.5	2.83	0.18	6.48	13.3
Pb	0.9993	0.5	4.9	2.5	2.31	0.07	1.7	2.5	2.49	0.14	5.59	−0.3
Fe	0.9989	6.5	64.7	25	25.43	0.14	0.6	25	25.53	0.48	1.88	2.76
Ti	0.9998	15.8	157.8	25	25.82	0.08	0.3	25	26.19	0.54	2.06	2.76
Zn	0.9968	4.6	45.7	25	24.87	0.13	0.5	25	24.74	0.23	0.94	−1.06
La	0.9993	5.1	50.9	2	2.03	0.02	0.9	2	1.87	0.16	8.79	−6.4
Ce	0.9995	5.3	52.7	2	2.16	0.02	1.1	2	2.06	0.20	9.88	3.2

^{1}R—correlation coefficient; ^{2}SD—standard deviation; ^{3}RSD—relative standard deviation, RSD (%) = SD/[metal]$_{mean}$ × 100; ^{4}RE—relative error, RE (%) = ([metal]$_{found}$ − [metal]$_{added}$/[metal]$_{added}$) × 100.

Table 1. Performance parameters obtained for heavy metals studied.

3. Experimental

3.1. Element concentrations in surface waters

Quantitative determination of the elements content in studied surface waters showed an increase, from 2009 to 2011, of Al and Mn concentrations in three sampling areas (Area 2, Area 4, and Area 5); an increase of Zn and Pb to Area 7; of Fe, Ti, Zn, and Pb concentrations to Area 2; of Cu and Pb to the Area 5; and of Fe and Cu to Area 6 (**Table 2**). High values were also registered for Zn and As in all sampling areas, during 2011 campaign, but below the permissible minimum level [14]. Increased concentrations of Cu, Zn, As, and Pb were observed in samples of surface water collected in 2010 compared with those collected in 2009 and 2011. These concentrations can be correlated with the registered rainfalls that were much higher in 2010. Seasonal changes in the concentrations of the analytes of interest observed in surface waters result from the dilution effect that occurred during rainfall. The content of heavy metals from the sampled river waters reduces due to the mixing with large volumes of uncontaminated water draining from the slopes. When the river flow decreases, the reverse phenomenon occurs. The contaminant concentrations increase both due to the evaporation and of the bacterial activity of sulfides oxidation once with the increase of temperatures.

	Sampling area/concentration ($\mu g\ L^{-1}$)											
Elements	*Area 7*			*Area 2*			*Area 5*			*Area 6*		
	2009	2010	2011	2009	2010	2011	2009	2010	2011	2009	2010	2011
Al	2.74	4.47	18.16	4.84	5.86	15.58	9.81	9.88	17.86	3.98	8.16	12.89
Mn	0.24	8.39	5.68	0.72	1.58	12.41	1.31	2.036	4.41	1.23	4.29	6.81
Fe	6.36	28.15	24.20	7.48	24.43	62.86	51.22	45.99	31.12	29.70	35.75	39.28
Ni	0.38	0.33	0.49	0.38	0.76	0.70	0.50	0.29	0.58	0.56	0.29	0.61
Ti	13.24	12.54	15.63	14.56	17.29	28.94	15.61	11.89	11.09	30.02	16.97	10.15
V	0.12	0.12	0.07	0.05	0.01	0.24	0.072	0.10	0.096	0.17	0.06	0.09
Co	0.03	0.63	0.43	0.03	1.37	0.613	0.032	0.56	0.398	0.07	0.53	0.52
Cu	0.69	0.59	0.71	1.42	0.71	1.348	0.588	0.50	0.727	0.59	0.54	0.68
Zn	0.23	0.16	13.08	0.19	0.14	1.26	0.202	0.13	0.532	0.87	0.14	1.57
As	0.61	0.43	1.24	0.54	0.44	2.186	0.500	0.47	1.167	0.85	0.49	1.11
Pb	0.08	0.02	0.061	0.08	0.01	0.071	0.018	0.02	0.163	0.04	0.02	0.04

Table 2. Comparative situation of metal concentrations in water sampled during the monitoring campaign.

3.1.1. Metals in water: seasonal influence

Water quality is affected by the weather conditions. This was revealed by studying the seasonal influence on the content of metals in water samples taken from the same areas in different calendar periods. A variation of heavy metal concentrations was observed in samples of water, sediment, and soil, depending on the season and certain times of year.

Characterization in terms of Cr, Pb, and Cu concentrations in water samples taken from the accumulation of fresh water of Tarnita (Area 8), Somes Cald (Area 2), and Somes Rece (Area 6) in early 2010 showed that the highest concentrations of these elements (Area 7) were registered in June and the lowest in January. This could be explained by the fact that the pollution originates in the driving effect of heavy metals by precipitation water that "wash" the adjacent area before reaching the lake. In the summer months, it is normal that the effect is more pronounced than in the cold winter months when the ground is frozen and therefore less exposed to erosion. The occurrence in July of higher concentration levels for some metals (e.g., Co, Cu, As) could be explained by the elevated temperatures recorded in this month that led to partial oxidation and solubilization of the sulfides, including biotransformation processes. Moreover, the intense evaporation enabled the crystallization of minerals containing large amounts of constituents of the elements Co, Cu and As. These minerals may contain soluble salts, which dissolve when rainfall is recorded in higher amounts. **Table 3** shows the minimum and maximum values of the concentrations of elements in waters depending on the areas and sampling periods.

Elements/concentration range	Sampling period	Area of sampling					
		Area 1	*Area 2*	*Area 4*	*Area 5*	*Area 6*	*Area 7*
Al/2.6–31.78 μg L^{-1}	March						
	May						
	July	min.	min.	min.	min.	min.	min.
	September	max.	max.	max.	max.	max.	max.
Ti/0.18–50.37 μg L^{-1}	March	max.					max.
	May				max.	max.	
	July		min.	min.	min.		
	September	min.	max.	max.		min.	min.
Mn/0.16–25.5 μg L^{-1}	March	min.	min.			min.	
	May		max.		max.	max.	
	July			min.	min.		min.
	September	max.		max.			max.
Fe/5.6–86.8 μg L^{-1}	March	min.		max.			max.
	May						
	July		min.	min.	min.	min.	min.
	September	max.	max.		max.	max.	
Ba/7.8–40.1 μg L^{-1}	March						max.
	May	max.	max.	max.	max.	max.	
	July				min.	min.	min.
	September	min.	min.	min.			
Sr/15.6–101.2 μg L^{-1}	March	min.		min.	min.	min.	min.
	May		min.				
	July	max.		max.	max.	max.	
	September		max.				max.

Table 3. The registered minimum and maximum values of the metal concentrations in waters by sampling area.

Determination of rare earths from various types of environmental samples is particularly important because it can serve to establish a sample fingerprint, and thus, the results could be used in determining the origin of the concerned sample and to identify sources of pollution

[15, 16]. Mass spectrometry with inductively coupled plasma offers the possibility to measure the rare earths with an excellent accuracy, which cannot be achieved by any another method [17, 18]. In this context, systematic observations correlated with the climatic conditions were performed in this work for a prolonged period. Thus, by comparing the data on a sampling site, in different periods, higher concentrations of La, Ce, and Pr were observed in September.

Analyzing data obtained in the 3 years of monitoring (2009, 2010, and 2011) for the same sampling month, a constancy of the rare metal concentrations was revealed, except for Sc, which recorded a significant decrease in 2009 (**Table 4**). Determination of trace amounts of rare earth elements dissolved in water was correlated with their concentration in the soil; concentrations of rare metals can originate from the soils adjacent to waters that once washed by rainfalls reach these waters. Quality of surface water and sediments, in terms of chemical species concentrations of the metals, was discussed based on the results obtained for the surface water and sediment analyzed samples taken from locations.

	Sampling area/concentration (μg L^{-1})											
Elements	*Area 8*			*Area 7*			*Area 5*			*Area 6*		
	2009	2010	2011	2009	2010	2011	2009	2010	2011	2009	2010	2011
Sc	2.686	0.693	0.428	1.493	0.862	0.403	1.680	0.817	0.423	2.515	0.842	0.429
Y	0.033	0.056	0.035	0.028	0.058	0.031	0.017	0.065	0.030	0.065	0.064	0.034
La	0.037	0.036	0.057	0.025	0.047	0.02	0.019	0.037	0.052	0.082	0.038	0.036
Ce	0.073	0.103	0.047	0.029	0.053	0.012	0.038	0.069	0.07	0.142	0.041	0.093
Pr	0.017	0.009	0.018	0.005	0.008	0.005	0.004	0.011	0.014	0.023	0.010	0.01
Nd	0.039	0.056	0.03	0.016	0.032	0.011	0.013	0.039	0.047	0.090	0.045	0.046
Sm	0.013	<0.001	0.005	0.004	0.003	<0.001	0.003	0.007	<0.001	0.020	0.001	<0.001
Eu	0.002	0.001	0.002	0.003	0.002	<0.001	0.004	0.003	<0.001	0.009	0.004	<0.001
Gd	0.009	0.012	0.007	0.005	0.007	0.002	0.003	0.014	0.007	0.018	0.009	0.014
Tb	0.002	0.001	0.002	0.001	0.002	0.003	0.001	0.002	0.002	0.004	0.003	0.002
Dy	0.005	0.010	0.005	0.006	0.010	<0.001	0.002	0.004	0.005	0.018	0.012	0.003
Ho	0.001	0.002	0.001	0.001	0.003	<0.001	0.001	0.003	0.003	0.003	0.002	0.002
Er	0.003	0.004	<0.001	0.002	0.005	<0.001	0.001	0.008	<0.001	0.010	0.007	<0.001
Tm	0.001	0.001	0.001	<0.001	0.002	<0.001	0.001	0.001	0.001	0.002	0.001	0.003
Yb	0.001	0.005	<0.001	0.002	0.003	<0.001	<0.001	0.004	<0.001	0.008	0.004	<0.001
Lu	<0.001	<0.001	0.001	<0.001	<0.001	<0.001	<0.001	<0.001	<0.001	0.001	<0.001	<0.001
Th	0.022	0.013	0.057	0.028	0.010	0.02	0.020	0.010	0.052	0.027	0.033	0.036

Table 4. Comparative situation of the rare metal concentrations during the monitoring campaign.

3.2. Element concentrations in sediments samples

Valuable data regarding the distribution of pollutants in sediment [19–22] were published in recent years. However, there is a limited analysis of the sedimentological characteristics on the pollutant distribution, which can play a key role in distinguishing the different sources, transport processes, and conservation status of specific contaminants in the environment. All the physicochemical and biochemical processes in aquatic systems take place at interfaces water-atmosphere-lithosphere, and a particularly important role is assigned to chemical and biochemical reactions at the water-sediment interface, which adjust the composition of natural waters. Studying the influence of physicochemical factors on the distribution of metals in sediments and water is important when the retention of metals by various types of sediments is assessed, developing thus predictions that may assure data comparison, interpretation, and their extrapolation, as an important contribution to the sustainable management of the investigated area.

Trace elements are one of the main sources of pollution of the aquatic chain. Since they tend to be adsorbed in sediment, the study of the adsorption/desorption of heavy metals in sediment and the effect of sediment transport are of particular importance. Sediments from lakes are an excellent witness of the water quality since they preserve important information on the environment and are recognized as a source of contaminants in aquatic systems due to the local physicochemical conditions. It is worth mentioning that the sediments are the best environment for toxic metals due to their high absorption capacity; therefore, sediment plays an important role in storing and releasing metals.

Correlations of climatic factors with the sampling locations were performed in relation to pollution sources. The most important factor in determining the concentration values of chemical species of the pollutant metal proved to be the location of the sampling point relative to the anthropogenic sources of pollution. A correlation was also found between the concentrations values of chemical species of the metal and the climatic factors registered during the monitoring period. As shown in **Table 5**, the concentration values for the most elements (Cd, Cu, Pb, Hg, and Zn) in sediment samples collected from the study area are within the permissible limits.

Based on the analysis of experimental data, it can be concluded that there is a match between the total concentrations of chemical species of the metals studied in samples of surface water and sediments collected from the same locations. Possible transfers of metal compounds in both directions may occur, from the sediment into the surface water and vice versa.

The levels for most metals in sediment samples collected from the study area generally were within the limits of admissibility, except for As present in quantities that exceeded the mean admitted values (sediments sampled in 2010 from the Areas 2 and 5). The average concentrations of some metals such as As, Pb, Co, U, and Cd were higher in the dry season than in the wet season. This is possible due to the dilution by rainwater influencing the concentration and mobility of heavy metal. In contrast, the mean concentration of Fe in sediment samples taken from the driest month was lower than concentrations in samples collected within the wet period, probably due to the rainfall. Higher values of metal concentrations were found in sediments taken from lakes versus those sampled from rivers.

Elements/permissible limits [23]	Month	Sampling area/concentration (mg kg^{-1})							
		Area 1	Area 2	Area 3	Area 4	Area 5	Area 6	Area 7	Area 8
As/17 mg kg^{-1}	March	17.33	16.53	14.66	11.23	111.39	12.91	21.75	20.17
	May	35.43	34.69	55.95	19.41	133.79	51.06	76.34	48.22
	July	20.15	22.37	55.93	28.28	26.78	26.30	87.48	80.34
	September	19.07	25.43	35.78	36.12	172.32	159.34	169.63	34.49
Cd/3.5 mg kg^{-1}	March	0.16	0.11	0.07	0.10	0.12	0.15	1.12	0.96
	May	0.35	0.32	0.41	0.22	0.16	0.56	1.65	0.13
	July	0.10	0.14	0.25	0.15	0.12	0.15	2.97	1.54
	September	0.08	0.12	0.15	0.20	0.38	0.36	0.31	0.26
Cr/90 mg kg^{-1}	March	11.95	20.55	19.38	21.13	39.02	18.59	20.75	19.12
	May	31.23	36.74	42.56	82.38	44.24	88.55	72.90	32.28
	July	16.40	16.57	16.39	71.97	60.19	21.84	61.04	60.34
	September	11.19	32.92	49.64	53.23	102.47	91.17	89.18	35.83
Cu/200 mg kg^{-1}	March	11.03	10.62	9.91	10.59	20.08	11.21	11.65	10.15
	May	25.41	24.16	27.71	47.66	27.81	60.78	66.97	17.93
	July	10.29	11.62	14.30	41.52	30.23	17.76	85.25	80.56
	September	10.26	17.53	25.38	40.54	54.80	52.78	49.90	16.50
Pb/90 mg kg^{-1}	March	40.24	33.56	39.81	25.78	26.15	20.96	21.17	21.76
	May	39.27	39.77	39.87	13.14	13.78	20.12	22.24	22.60
	July	19.56	18.43	17.03	8.41	8.23	19.29	24.28	24.18
	September	17.54	17.13	27.44	25.43	23.53	22.93	22.92	26.80
Hg/0.5 mg kg^{-1}	March	<0.001	0.07	0.12	<0.001	0.09	<0.001	<0.001	<0.001
	May	<0.001	0.28	0.18	0.03	<0.001	<0.001	0.15	0.12
	July	0.01	0.01	<0.001	<0.001	<0.001	0.08	0.09	<0.001
	September	0.07	0.04	0.17	0.19	0.24	0.20	0.11	0.07
Zn/300 mg kg^{-1}	March	51.41	44.35	53.09	50.17	42.63	31.59	50.19	30.20
	May	81.56	79.43	86.77	80.40	51.30	85.03	167.01	66.27
	July	42.71	46.64	57.63	64.72	63.12	54.76	193.75	190.82
	September	62.76	64.44	81.60	90.53	103.85	98.82	96.48	134.67

Table 5. Metal concentrations in the analyzed sediment samples.

The predominance of metals in the free ionic form and as soluble compounds in water, in an acidic environment, favors both the exchange processes at the water-sediment interface and the bioabsorbtion processes with direct consequence on the toxicity of these metals. Hence, appears obvious the need to study the distribution of the metals contained in the sludge fractions where the metals are attached in combination with different abilities of participating to the heterogeneous equilibrium sediment-aqueous phase.

Thus, to study the heterogeneous equilibrium occurring at the water-sediment interface, the metal concentrations of sediment pore water were determined. The literature presents the calculation mode (Eqs. (1) and (2)) of trace metals distribution in the interstitial sediment-water interface following the formula:

$$P_{sed.} = (M_s/M_t) \; 100 \tag{2}$$

$$P_{int.water} = 100 - P_{sed}, \tag{3}$$

where $P_{sed.}$ (%) is the proportion of metal in the sediment, $P_{int.water}$ (%) is the proportion of the metal in interstitial water, M_s (mg) is the amount of metal in the sediment, M_t (mg) is the total amount of the metal (sediment + interstitial water) in the sample.

Experimental data showed that for the studied metals, the concentrations in pore water samples exceed those in surface water, which suggests that sediment through interstitial water may become a potential source to chemical species mobilization of the metals in water (**Table 6**). Characterization in terms of rare metal concentrations, comparing water samples with water resulting after settling sediment, revealed that the sediments accumulate rare metals (the concentrations both in water and in sediment are very low, but there is a slight increase in sediment). It was also highlighted that sediment accumulates a larger amount of rare earth metals than soils.

Sampling area	Sample	Metal concentrations ($\mu g \; L^{-1}$)					
		As	Zn	Pb	Co	Cu	Ni
Area 1	Water	1.396	0.583	0.137	0.529	1.422	0.586
	Water sediment	1.582	0.699	0.294	0.585	1.595	0.773
Area 4	Water	0.951	0.106	0.03	0.417	0.434	0.411
	Water sediment	1.100	0.131	0.044	0.532	0.512	0.587
Area 6	Water	1.131	0.084	0.031	0.535	0.462	0.521
	Water sediment	1.372	0.103	0.098	0.619	0.510	0.588
Area 7	Water	0.869	0.081	0.016	0.594	0.415	0.398
	Water sediment	1.187	0.102	0.044	0.827	0.546	0.448

Table 6. Trace element concentration in water/water from sediment, year 2011.

3.3. Metal concentrations in soil samples

The catchment of Someş River fits in the temperate continental area, the bioclimatic conditions in the area causing a moderately active or slow biological circuit, and a strong acidification of soils. Soil research and its quality assessment closely related to the dynamic use of land offer many possibilities in terms of approach. Chemical characteristics of the soil and organic carbon content, pH, the oxides forms, carbonates, and some physical properties such as clay content may influence the concentration of chemical elements.

This work has also proposed an assessment of heavy metals distribution in soils in the studied areas and an investigation of the extent to which heavy metal content in soil from the areas of waters accumulation may influence their concentration in water. **Table 7** highlights the concentrations of toxic elements in soils versus the limit values according to the Romanian laws [24].

Area	Metal concentrations (mg kg^{-1})						
	Cu	Pb	Zn	Cd	Ni	Cr	Mn
Area 1	37.46	33.59	99.23	0.45	43.91	37.05	249.73
Area 2	16.70	31.76	62.43	0.19	24.23	28.55	264.95
Area 3	11.14	22.81	44.37	0.12	14.95	23.81	151.71
Area 4	52.92	75.30	382.66	1.95	38.75	52.93	1027.20
Area 5	37.85	23.52	65.18	0.25	45.94	55.39	534.97
Area 6	58.72	26.50	85.57	0.97	53.84	63.55	807.63
Area 7	42.77	17.31	102.40	0.89	57.18	33.87	513.54
Area 8	71.07	22.59	77.05	0.22	19.57	30.26	207.05
NV	20	20	100	1	20	30	900
TA	100	50	300	3	75	100	1500
IT	200	100	600	5	150	300	2500

NV, normal values; TA, threshold alert; IT, intervention threshold [24].

Table 7. Heavy metal concentrations in soil samples collected in May 2011, in different areas.

Cd in soil ranged from relatively narrow limits (0.19–1.95 mg kg^{-1}), with an overwhelming dominance (>99%) of the contents lower than the alarm threshold. It can be considered as representative for the natural geochemical background of the investigated area. These abundances are consistent with the geological structure, relatively uniform, which is determined by sedimentary rocks, which are known for their low content in Cd. An exceeding of the normal values was observed to Area 4 (Somes Cald Lake), but below the alert threshold.

Mn content is between relatively wide limits (120–1100 mg kg^{-1}), with a similar distribution to that described by the normal distribution law. Therefore, 99.90% of the samples were below the maximum content of normal, and only 0.10% of the samples exceed the normal range, but not the alert threshold. The monitoring carried out on soil during year 2011 revealed an exceeding of normal values in almost all areas, but below the alert threshold.

Chromium is considered as one of the most harmful metals for human health. In aquatic environment, it is presented as Cr^{3+} ion as well as anionic species and CrO_4^{4-} si $Cr_2O_7^{2}$, these two forms being produced by human activities. The values determined for Cr and Ni in the soils of the studied areas fluctuate in the range between 12.1 and 86.0 mg kg^{-1} (for Cr) and 17.0–70.0 mg kg^{-1} (for Ni), with exceedings of normal values without reaching the alert threshold.

Pb ranges from relatively low limits of 20–80.4 mg kg^{-1}. Most samples showed an exceeding of lead concentrations than normal, except for samples collected from Somes Cald Lake where values above the alert threshold were reported. This clearly reflects the effects of human intervention in this area.

Zn content determined in 2011 on the analyzed soils varies widely (26–146 mg kg^{-1}) with a significant dominance (>95%) of the contents, not exceeding normal values for soils. In the Area 4 (Somes Cald Lake), Zn anomalies are spatially overlapping, largely over the Pb, suggesting the same originating polluting sources.

Comparing the content of heavy metals in soil and water samples to each sampling point, it can assert that there is a correlation only in certain areas (e.g., Al, Zn, Ti, and Pb for Area 1), where the concentrations of metals in water samples may result from soil washing during rainfall. Here are the results obtained for soil samples collected in different months (**Table 8**): in March, the highest concentrations recorded for Pb (Areas 1, 2, 3, 5), Cu, Co, Cr (Area 2), Zn (Areas 2, 7, 8), and Mn (Areas 1, 5); in May, for Zn (Area 4), Ni (Area 7); in July, for As (Areas 2, 3, 4, 5, 7), Cu (Areas 4, 7, 8), Pb, Cr, Mn, and Co (Area 7), Zn (Area 4); and in September, for As (Areas 1, 6, 8), Cu, Ni, Cr, Zn (Areas 1, 3, 5, 6), Co (Areas 6, 7), and Mn (Areas 2, 3, 4).

Following the values obtained for samples of soil taken from the same area, in different years, a growing trend 2009–2011 in the concentrations of Pb, As, Ni, Zn, Co, and Cu, was observed (**Table 9**).

3.4. Determination of metal concentrations in vegetation samples

The toxic effect of the metals in the tissues and plant cells varies according to the concentration, leading that at high concentrations the whole process of plant growth and development is inhibited. Plants heavily suffer due to the harmful action of impurities from polluted air, their behavior representing a good indicator of pollution.

The ability of removing metals is a general characteristic of the organisms tolerant to metals. Many organisms/plants instead to eliminate metals accumulate them in high concentrations, especially in roots and leaves. The manner in which plant metabolism responds to the exposure at different concentrations of heavy metals is an important step in determining the fate of plants, tissues, or cells and their ability to survive. Metal tolerance issues related to development are increasingly the focus of researchers [25].

The ability of plants to takeover chemical elements varies within a wide range. Elements such as Br, As, B, Cs, and Rb are easily taken, while others such as Ba, Ti, Zr, Sc, Bi, Ga, and Fe are less available; these aspects are being adjusted depending on the particularities of the soil plant. Specific to the fungi, these show an affinity for taking over metals like Hg, Cd, Se, Cu, and Zn. The problem of heavy metal pollution of soil and plants in the general context of ensuring the health of living beings (humans, animals, plants), avoiding disturbance on the balance of the ecosystem and the need to address and to elucidate issues related to quality greenhouse products in order to improve culture techniques without soil, requires intense monitoring differences in the content of pollutants in the plant.

Elements	Period sampling	Area sampling/metal concentrations (mg kg^{-1})							
		Area 1	*Area 2*	*Area 3*	*Area 4*	*Area 5*	*Area 6*	*Area 7*	*Area 8*
As	March	12.36	14.67	10.72	13.06	16.74	39.64	40.05	27.87
	July	16.95	26.27	28.62	93.30	169.74	53.41	90.18	33.73
	September	82.67	25.22	21.36	13.18	25.88	67.62	19.51	49.58
Cu	March	16.63	21.03	5.15	20.34	28.64	55.6	73.23	80.54
	July	37.21	15.37	13.4	60.53	39.01	52.86	74.87	121.06
	September	58.54	13.7	14.88	45.31	45.9	67.71	10.68	21.09
Pb	March	48.6	54.4	29.44	49.06	32.26	15.68	10.86	12.42
	July	32.56	23.56	21.27	73.06	18.82	20.13	19.84	21.06
	September	19.63	17.32	17.73	77.55	19.5	43.70	14.78	24.12
Zn	March	102.25	65.12	26.96	108.71	49.2	66.79	106.89	102.85
	July	53.85	61.03	55.7	432.00	59.87	77.18	146.37	55.27
	September	141.59	61.14	58.46	333.33	86.49	112.76	50.44	98.84
Cd	March	0.32	0.27	0.06	1.96	0.21	0.43	0.58	0.16
	July	0.27	0.23	0.20	1.95	0.27	1.70	1.64	0.14
	September	0.77	0.18	0.12	1.95	0.27	0.8	0.08	0.30
Ni	March	17.07	22.74	11.93	25.21	28.65	48.08	52.04	21.32
	July	21.09	15.13	15.03	32.81	33.18	40.31	100.92	17.39
	September	93.59	34.84	17.1	44.69	76	73.13	13.44	21.76
Cr	March	24.26	36.39	12.11	45.31	37.46	56.6	34.16	24.05
	July	20.95	23.03	27.28	50.73	59.72	47.51	59.76	24.04
	September	65.94	26.23	32.06	55.13	68.99	86.54	7.98	36.48
Mn	March	304.55	265.56	127.98	324.06	679.78	387.65	250.13	205.02
	July	220.33	207.68	235.08	376.52	289.36	1434.76	655.86	228.02
	September	224.33	321.63	220.06	1077.88	635.78	600.5	371.23	186.09
Co	March	8.75	11.00	3.57	4.25	12.4	24.55	20.71	17.12
	July	8.96	7.12	7.81	19.13	13.35	33.34	23.07	15.82
	September	11.93	7.25	8.96	18.55	16.45	29.68	6.69	9.01

Table 8. Comparative situation for metal concentration in soils.

We also considered in our work the characterization of vegetation samples from surrounding areas of raw water reservoirs. We noticed that higher concentrations were found in moss than in leaves, so they accumulate more quickly the toxic metals. Also, there was observed a concentration of heavy metals in aquatic plant tissues; in the same area of sampling, analyzing samples of terrestrial and aquatic vegetation, we observed higher concentrations values of Fe, Co, Ni, Zn, As, and Pb in the lacustrine vegetation. So there is the possibility of using certain plant species for "extracting" the potential toxic heavy metals from water. **Table 10** contains the metal concentrations of vegetation collected in May 2011, while **Table 11** presents the comparative data for soil-vegetation.

Elements	Sampling year	Area sampling			PA		PI	
		Area 4	*Area 6*	VN	Sensitive	Less sensitive	Sensitive	Less sensitive
Pb	2009	24.89	27.04	20	50	250	100	1000
	2010	39.89	36.72					
	2011	73.06	43.70					
As	2009	10.54	8.38	5	15	25	25	50
	2010	18.82	18.43					
	2011	13.18	67.62					
Ni	2009	14.28	8.61	20	75	200	150	500
	2010	28.77	26.55					
	2011	44.69	73.13					
Cd	2009	0.18	0.23	1	3	5	5	10
	2010	1.14	2.33					
	2011	1.95	0.80					
Zn	2009	46.47	48.24	100	300	700	600	1500
	2010	137.35	106.96					
	2011	333.32	112.76					
Cu	2009	17.00	28.65	20	100	250	200	500
	2010	58.12	34.12					
	2011	45.31	67.71					
Co	2009	11.63	11.26	15	30	100	50	250
	2010	12.85	25.50					
	2011	18.55	29.68					
Hg	2009	0.11	0.10	0.1	1	4	2	10
	2010	0.24	0.62					
	2011	0.05	0.16					

VN, normal values; PA, threshold alert; PI, intervention threshold [24].

Table 9. Element concentrations for 3 years of sampling (mg kg^{-1}).

	Sampling area/metal concentrations (mg kg^{-1})								
Elements	Area 1		Area 2	Area 3	Area 4	Area 5	Area 6	Area 7	
	Leaves	Moss	Leaves	Leaves	Leaves	Leaves	Leaves	Leaves	Moss
Ti	44.78	283.50	47.64	121.20	223.20	23.16	41.64	55.34	236.95
V	0.90	31.83	0.47	11.65	21.09	0.34	0.21	2.71	69.19
Cr	3.38	17.44	1.28	9.09	17.28	2.07	1.37	2.62	45.07
Mn	47.74	257.27	41.98	196.11	399.35	93.22	76.17	79.23	252.88
Fe	357.01	9104.51	161.19	3147.62	5580.78	109.56	98.10	631.14	11290.76
Co	0.24	5.62	0.33	2.10	4.47	0.24	0.67	0.86	7.87
Ni	2.05	11.04	2.89	4.72	10.09	3.13	2.15	5.88	24.75
Cu	8.52	20.02	19.34	10.50	19.11	14.51	8.32	20.73	18.75
Zn	44.78	79.84	33.91	32.75	44.88	27.43	27.38	51.31	54.50
As	0.66	11.47	0.36	7.40	13.30	0.66	0.14	2.01	45.97
Cd	0.04	0.72	0.11	0.36	0.35	0.22	0.45	0.95	1.05
Ba	34.23	132.58	64.13	117.8	121.47	26.64	21.40	130.86	703.31
Pb	1.12	35.90	1.88	4.02	8.20	1.84	1.11	1.80	12.25

Table 10. Metal concentrations in vegetation collected in May 2011, from the studied areas.

	Sampling area/metal concentrations (mg kg^{-1})			
Elements	Area 1		Area 3	
	Soil	Vegetation	Soil	Vegetation
Co	8.96	0.13	7.81	1.13
V	48.67	0.17	58.72	0.22
Ni	21.09	1.72	15.82	2.69
Cu	37.21	2.54	13.40	6.10
Zn	53.85	48.26	55.70	36.32
As	16.95	0.15	28.62	0.34
Pb	32.56	0.12	21.27	0.10

Table 11. Metal concentrations in soil and vegetation samples from July 2011.

3.5. Drinking water

The surface water pollution's main impact in the studied area is on the water processing plant in terms of providing potable water. The drinking water released from the water treatment plant at good quality parameters and distributed through the public network could reach the end users at a less quality level, in which the distribution network of water may represent a potential source of chemical contaminants in the processed water. All the investments undertaken for the rehabilitation and modernization of water supply chain have as final aim the improvement in water quality. Even if the public distribution of water from sources and treatment facilities to the networks and connections are upgraded, there is a gap in the rehabilitation of some buildings internal networks, and thus, water with appropriate quality to the branch may get degraded to the end user tap due to the poor state of the interior outdated pipes.

Due to the increasing influence of the anthropogenic factors on water sources, the ensuring of water quality is of primary importance. At global level, the environmental monitoring is assured by the IGBM (integrated global background monitoring of environmental pollution) and GEMS (global system of environmental monitoring) networks. IGBM deals with background monitoring (before the intervention of pollution), and GEMS follows the impact monitoring (after the intervention of pollution). Over 20 EEC Directives regarding the protection of aquatic environment and many guidelines laying down the quality standards regarding the use of water and the checking of wastewater discharge are in force to support these activities (e.g., Council Directive 76-464-EEC—pollution caused by the discharge of dangerous substances into the aquatic environment; Council Directive 88/20/EEC—limit values for discharges of certain dangerous substances). Metals with potentially harmful effect on groundwater are Zn, Cu, Ni, Cr, Pb, Se, As, Sb, Mo, Ti, Sn, Ba, Be, Bi, U, V, Co, Tl, Te, and Ag. The maximum permissible limits for metal concentrations in drinking water given by some international directives are shown in **Table 12**.

Elements	NS30	EU	USA	Romania
Ag	10	–	100	–
Al	200	200	200	200
As	50	10	10	–
Ba	1000	–	2000	–
Cd	5	5	5	5
Cr	50	50	100	50
Cu	3000	1000	1000	100
Fe	200	200	300	–
Hg	1	1	2	–
Ni	50	20	–	20
Pb	50	10	15	10
Sb	10	5	6	–
Se	10	10	50	10
U	–	–	30	–
V	–	50	–	–
Zn	5000	–	5000	5000
Mn	–	–	–	50

Table 12. Permissible limits, in $\mu g\ L^{-1}$, for drinking water in England (NS300), European Union (EU), United States of America (USA) and Romania.

In Germany, the maximum permissible limits for drinking water are the same as those for the member states in the EU, exception As (40 μg L^{-1}) and Pb (40 μg L^{-1}). In Spain, the quality criteria for wastewater that may affect quality of surface water has recently been amended, setting values of 50 μg L^{-1} for As$_{total}$, 5 μg L^{-1} for Cr (VI), and 1 μg L^{-1} for Se$_{dissolved}$. In Romania, certification of potable water is performed in accordance with the Laws 458/2002 and 311/2004, which are aligned with the European Framework water Directive 2000/60/EC (**Table 13**). They concern the organoleptic (sensory), physical, chemical (general and toxic), radioactive, bacteriological, and biological characteristics of water.

Area/sampling period			Metal concentrations (μg L^{-1})				
			As	Cd	Pb	Co	U
2011	Water treatment plant	Raw water	0.302	<0.001	0.002	0.338	0.102
		Decant water	0.250	<0.001	0.016	0.147	0.010
		Filtered water	0.165	0.007	0.033	0.284	0.041
		Chlorinated water	0.221	<0.001	0.005	0.255	0.026
	City water entry point		0.229	<0.001	0.003	0.230	0.076
2010	Gilău treatment plant	Raw water	0.220	<0.001	0.003	0.244	0.101
		Decant water	0.125	<0.001	0.002	0.159	0.058
		Filtered water	0.247	0.003	0.078	0.245	0.105
		Chlorinated water	0.234	<0.001	0.010	0.228	0.114
	City water entry point		0.184	<0.001	0.002	0.201	0.101
2009	Water treatment plant	Raw water	0.345	<0.001	0.007	0.453	0.078
		Decant water	0.236	<0.001	0.012	0.245	0.056
		Filtered water	0.188	0.101	0.051	0.387	0.074
		Chlorinated water	0.268	<0.001	0.003	0.368	0.066
	City water entry point		0.298	<0.001	0.005	0.392	0.069

Table 13. Characterization of waters from water treatment plant (entrance to city).

The waters investigated in this study ensure the drinking water sources of the cities from Cluj county, but also provide industrial water for energy purposes and water for a trout farm in the area downstream the dam. The trace elements present in the water supplied as drinking water to population but also in the wastewaters and water of sewage plants before discharging into the rivers were determined for this work. Quality of water at the inlet in the distribution networks versus the one at the household consumer was characterized to highlight potential sources of contamination due to the technical condition of pipe networks. The water quality monitoring consisted of three phases:

i.   Monitoring the water quality of Gilau Lake in order to establish a realistic image in terms of physicochemical parameters value of the water before entering the treatment plant — samples collected directly from the lake. In terms of toxic metals, the studied surface water fall in first quality class, with few minor exceptions. Overall, the overruns, in the absence of an organized source of pollution, would lead to the conclusion that it is a natural pollution, which could be confirmed geochemical.

ii. Water quality monitoring inside the treatment plant; water samples collected after each stage of the raw water treatment (coagulation, filtration, sedimentation, chlorination). Referring to their content in metal with toxic effect was characterized the water treatment plant from the point of entry in the city (located 10 km away from the water treatment plant) taken in 2009, 2010, 2011. The obtained concentration values for As were in the range of 0.1–0.6 µg L^{-1}, Cd <0.001 µg L^{-1}, Pb in the range 0.03–0.7 µg L^{-1}, and U between 0.007 and 0.2 µg L^{-1}. **Table 13** presents a quantitative characterization of metals from period 2009, 2010, 2011.

iii. Water quality monitoring in the drinking water distribution network.

The water samples were collected from many urban districts. Comparing the obtained data with the permissible values in Romania, we can state that in terms of toxic metals content the drinking water is adequate and is well below the admissible limits (**Table 14**).

City distribution area/sampling period			As	Cd	Pb	Ni	Co	Fe	Zn	U	Cu
2011	Area 1 city		1.271	0.007	0.909	2.944	0.288	42.890	143.930	0.133	43.94
	Area 2 city		1.422	0.015	0.070	0.590	0.239	39.671	4.000	0.137	1.600
	Area 3 city	Cold	1.330	0.017	0.074	0.661	0.238	43.880	16.201	0.124	1.991
		Warm	1.710	0.073	0.774	3.222	0.330	42.670	168.77	0.160	5.740
	Area 4 city		1.290	0.017	1.270	0.744	0.151	45.226	8.925	0.131	14.070
	Area 5 city		1.500	0.007	0.106	2.275	0.172	41.987	13.47	0.122	8.866
	Area 6 city	Cold	1.180	0.026	0.052	0.760	0.166	35.770	36.35	0.124	1.921
		Warm	1.103	0.027	0.126	0.369	0.204	46.905	20.544	0.093	4.659
	Area 7 city		1.435	0.005	0.034	0.594	0.288	45.223	5.226	0.132	1.696
	Area 8 city center		1.023	0.022	0.268	1.059	0.209	43.990	142.88	0.113	2.785
	Area 9		1.545	0.022	0.268	2.257	0.142	42.991	43.053	0.113	6.028
2010	Area city 1		0.247	<0.001	0.142	0.443	0.201	45.772	1.445	0.132	1.331
	Area city 2		0.195	<0.001	0.011	0.521	0.206	35.984	1.994	0.106	1.207
	Area city 3		0.167	<0.001	0.005	0.61	0.185	52.806	1.553	0.092	1.094
2009	Area city 1		1.060	0.006	0.143	0.592	0.032	36.536	4.162	0.062	1.408
	Area city 2		1.063	<0.001	0.043	0.528	0.047	46.900	0.831	0.069	1.209
	Area city 4		1.003	0.016	3.433	0.682	0.027	58.758	9.481	0.064	2.175
	Area city 3		1.013	0.002	0.047	0.528	0.030	54.270	21.290	0.074	0.931

Metals concentrations (µg L^{-1}) — spanning As through Cu.

Table 14. Characterization of metal concentrations in waters from urban distribution areas.

4. Conclusions

The metals are considered to be one of the main sources of environmental pollution, having a significant effect on the ecological quality. Potentially toxic metals derived from the anthropogenic activities may lead to severe disturbances in the ecosystems, and therefore, the identification of pollution sources along with the assessment of the long-term pollution potential is a must in order to take actions for reducing or stopping the pollution. The anthropogenic sources lead to an increase in the metals levels in the environment due to the industrial and atmospheric pollution and their accumulation in soil, affecting thus the ecosystems. Therefore, measurements of metals in soil, plants, and sediment are very important in monitoring environmental pollution.

The primary objective of this study investigation was to characterize the surrounding areas of raw water reservoirs in terms of the content in heavy metals and rare earth, linked to establishing the supply mechanism of these surface waters. High accuracy measurements for quality assessment were performed on water samples, in terms of toxic metals, over a period of 3 years: 2009, 2010, and 2011. An increase (2009–2011) in the levels of Al and Mn was registered in three sampling sites (Somes Cald River, the confluence Somes Cald/Somes Rece, and Somes Cald Lake); of Zn and Pb to Tarnita site; of Fe, Ti, Zn, and Pb to the Somes Cald River; of Cu and Pb to the confluence Somes Cald/Somes Rece; and of Fe and Cu to the Somes Rece Lake. A large increase in the content of Zn and As was reported in 2011 for all the areas, but below the admissibility limits.

Water quality is influenced by weather conditions, and this was revealed by studying their influence on the content of metals in water samples from the same area, but different calendar period. There is a variation in the concentrations of heavy metals observed in the studied matrices (water, sediment, soil) depending on season, for certain periods of the year.

Determination of rare earths in various types of environmental samples is particularly important because it can serve to establish a fingerprint sample, and so the results could be used in determining the origin of the sample in question and to identify sources of pollution. Comparing the data of the 3 years, 2009, 2010, 2011, from the same month of sampling, there is a constant concentration of rare metals, except for Sc, where there is a significant decrease from 2009. Heavy metals are one of the main sources of pollution of the aquatic environment, they tend to adsorb in sediment, and the study of the adsorption/desorption effect of heavy metals in sediment and sediment transport has a particular importance. The concentration levels for most metals in sediment samples collected from the study area were within the general admissibility limits, with exception of As which was present in sediment samples taken in 2010 in quantities that exceeded the mean admitted (in the sediment samples collected from Areas 4—Warm Somes, 5—confluence Somes Cald/Somes Rece); also in the Area 4—Somes Cald was registered a significant increase in the Cr content (150 mg kg^{-1}, the limit was estimated to 90 mg kg^{-1}).

We found higher values of metal concentrations in sediments taken from lake versus concentration values in sediment samples taken from the river water. In soil samples taken from the same areas, in the same calendar period (2009 and 2010), an increase in the concentrations of Pb and As was observed and a decrease in the content of Co (for soil sampled in 2010).

Correlations with the climatic factors and the location of the sampling points were performed in relation to the pollution sources. The most important factor in determining the concentration values of the pollutant metal chemical species has proved to be the location of sampling points relative to the anthropogenic sources of pollution. A correlation between the results obtained for the concentrations of metals chemical species and the climatic factors registered during the monitoring period was also noted. Comparisons were made between the water quality at the inlet into the distribution networks and the water quality to the end user, to highlight potential sources of contamination related to the technical condition of these networks.

There is our responsibility, now and certainly in the future, to balance and control the environmental quality for each component and as overall. Thereby improving environmental quality will become the action against disorder and the reaction against the inertia and of the compromises when considering human life environment. Maybe someday each item and parameter of the environment will be integrated "in a world of balance and harmony". This calls for an efficient management of the environmental resources and for mandatory preserving of a right balance between the nature and society. It means implementation of a scientifically based management, both for the exploitation of natural resources and for the recovery and recycling of waste so that the failures and discontinuities in the evolution of ecosystems can be eliminated. The results of our research can be used by the companies that manage water resources in improving the drinking water treatment technologies. The results may be also useful for the territorial environmental agencies and authorities that have as direct responsibility the water quality.

Author details

Cezara Voica[1], Andreea Maria Iordache[2]*, Roxana Elena Ionete[2] and Ioan Ştefănescu[2]

*Address all correspondence to: andreea.iordache@icsi.ro

1 National Institute for Research and Development of Isotopic and Molecular Technologies, Cluj-Napoca, Romania

2 National Research and Development Institute for Cryogenics and Isotopic Technologies, Ramnicu Valcea, Romania

References

[1] National Water Law No. 107/1996, published in The Official Gazette of Romania.

[2] EU Water Framework Directive-Directive 2000/60/EC of the European Parliament and of the Council establishing a framework for the Community action in the field of water policy.

[3] Council Directive 98/83/EC on the quality of water intended for human consumption, or Council Directive 91/271/EEC concerning urban waste-water treatment amended by the Commission Directive *98/15/EC*, www.apmbm.ro/MM/PPPRPC/Codificare_Directive.doc.

[4] UNEP GEMS/Water Programme Water Quality Outlook ISBN 95039-11-4 UN GEMS/Water Programme Office.

[5] Islama S, Ahmed K, Raknuzzamanb M, Mamunb H-Al, Islame MK: Heavy metal pollution in surface water and sediment: A preliminary assessment of an urban river in a developing country. Ecological Indicators. 2015;**48**:282–291. doi:10.1016/j.ecolind.2014.08.016

[6] Maher A, Sadeghi M, Moheb A: Heavy metal elimination from drinking water using nanofiltration membrane technology and process optimization using response surface methodology. Desalination. 2014;**352**:166–173. doi:10.1016/j.desal.2014.08.023

[7] Wu Q, Leung JYS, Geng X, Chen S, Huang X, Li H, Huang Z, Zhu L, Chen J, Lu Y: Heavy metal contamination of soil and water in the vicinity of an abandoned e-waste recycling site: Implications for dissemination of heavy metals. The Science of the Total Environment. 2015;**506–507**:217–225. doi:10.1016/j.scitotenv.2014.10.121

[8] Roychowdhury T, Tokunaga H, Ando M: Survey of arsenic and other heavy metals in food composites and drinking water and estimation of dietary intake by the villagers from an arsenic-affected area of West Bengal, India. The Science of the Total Environment. 2003;**308**(1):15–35. doi:10.1016/s0048-9697(02)00612-5

[9] Glasby GP, Szefer P, Geldon J, Warzocha J: Heavy-metal pollution of sediments from Szczecin Lagoon and the Gdansk Basin, Poland. The Science of the Total Environment. 2004;**330**:249–269. doi:10.1016/j.scitotenv.2004.04.004

[10] Özden Ö, Erkan N, Cengiz Deval M: Trace mineral profiles of the bivalve species *Chamelea gallina* and *Donax trunculus*. Food Chemistry. 2009;**113**(1):222–226. doi:10.1016/j.foodchem.2008.06.069

[11] Salomon S, Jenne V, Hoenig M: Practical aspects of routine trace element environmental analysis by inductively coupled plasma–mass spectrometry. Talanta. 2002;**57**:157–168. doi:10.1016/s0039-9140(01)00678-6

[12] Thomas R. Practical Guide to ICP-MS. 2004. NY, USA: Marcel Dekker.

[13] Ammann A: Inductively coupled plasma mass spectrometry (ICP MS): A versatile tool. Journal of Mass Spectrometry. 2007;**42**(4):419–427. doi:10.1002/jms.1206

[14] Law no. 311/2004, Drinking Water Quality, published in The Official Gazette of Romania.

[15] Weltje L, Heidenreich H, Zhu W, Wolterbeek HT, Korhammer S, de Goeij JJM, Markert B: Lanthanide concentrations in freshwater plants and molluscs, related to those in surface water, pore water and sediment. A case study in The Netherlands. The Science of the Total Environment. 2002;**286**:191–214. doi:10.1016/S0048-9697(01)00978-0

[16] Zhang Q, He M, Chen B, Hu B: Preparation, characterization and application of *Saussurea tridactyla* Sch-Bip as green adsorbents for preconcentration of rare earth elements in environmental water samples. Spectrochimica Acta Part B. 2016;**121**:1–10. doi:10.1016/j.sab.2016.04.005

[17] Dressler VL, Pozebon D, Matusch A, Becker JS: Micronebulization for trace analysis of lanthanides in small biological specimens by ICP-MS. International Journal of Mass Spectrometry. 2007;**266**:25–33. doi:10.1016/j.ijms.2007.06.018

[18] Kumar SA, Pandey SP, Shenoy NS, Kumar SD: Matrix separation and preconcentration of rare earth elements from seawater by poly hydroxamic acid cartridge followed by determination using ICP-MS. Desalination. 2011;**281**:49–54. doi:10.1016/j.desal.2011.07.039

[19] Di Toro DM, Mahony JD, Hansen DJ, Scott KJ, Hicks MB, Mayr SM, Redmond MS: Toxicity of cadmium in sediments: The role of acid volatile sulfide. Environmental Toxicology and Chemistry. 1990;**9**:1487–1502. doi:10.1002/etc.5620091208

[20] Simpson SL, Rosner J, Ellis J: Competitive displacement reactions of cadmium, copper, and zinc added to a polluted, sulfidic estuarine sediment. Environmental Toxicology and Chemistry. 2000;**19**(8):1992–1999. doi:10.1002/etc.5620190806

[21] Fan W, Wang WX: Sediment geochemical control on Cd, Cr, and Zn assimilation by the clam *Ruditapes philippinarum*. Environmental Toxicology and Chemistry. 2001;**20**:2309–2317. doi:10.1002/etc.5620201025

[22] Papaefthymiou H, Papatheodorou G, Christodoulou D, Geraga M, Moustakli A, Kapolos J: Elemental concentrations in sediments of the Patras Harbour, Greece, using INAA, ICP-MS and AAS. Microchemical Journal. 2010;**96**(2):269–276. doi:10.1016/j.microc.2010.04.001

[23] Romanian Ministerial Order No. 1888/2007 approving the list of organohalogenated substances and heavy metals, as well as the maximum allowable limits for the organohalogenated substances and heavy metals in water and sediment substrate.

[24] Order 756/1997-modificated in 592/2001. Regulations for environmental pollution evaluation.

[25] Baytak S, Kocyigit A, Turker AR: Determination of lead, iron and nickel in water and vegetable samples after preconcentration with *Aspergillus niger* loaded on silica gel. Clean. 2007;**35**(6):607–611. doi:10.1002/clen.200700124

Chapter 13

Formulating Specific Water Quality Criteria for Lakes: A Malaysian Perspective

Zati Sharip and Saim Suratman

Additional information is available at the end of the chapter

http://dx.doi.org/10.5772/65083

Abstract

Monitoring water quality of inland water bodies such as lakes, reservoirs and ponds throughout Malaysia is important to ensure that these water bodies can be managed sustainably for their ecosystem functioning and services. Determining the quality of these water bodies for different uses is limited due to the unavailability of specific criteria or standards for such water bodies in the country. The aim of this study is to develop national water quality criteria and guidance values for lakes to enhance the water quality of the water bodies in Malaysia. The work is based on a literature review and a consensus among experts from the various stakeholders' consultative sessions. The criteria were divided into four specific uses which aim at protecting the health of human and aquatic life. The criteria and standards are targeted for non-regulatory purposes to promote lake quality monitoring by various stakeholders. More than 20 parameters were identified in the lake criteria to determine the classification. The identification of parameters and limits for the standards, however, was limited by data availability and appropriate understanding of the water body characteristics. The role of the criteria and their limitation was also discussed.

Keywords: lakes and reservoirs, recreational, stakeholder consultations, standards, water quality criteria

1. Introduction

Eutrophication of inland water is a prevalent issue in Malaysia, one which threatens the functioning of lake and reservoir ecosystems throughout the country. A preliminary study of the status of lake eutrophication reported that more than 60% of the 90 lakes being studied

were nutrient rich [1] with a few of them especially the urban lakes experiencing algal blooms, which affect human uses of the water, such as recreation and aesthetic values, while other lakes faced macrophyte infestation problems [2]. To address these widespread challenges, developing specific standards for lakes is necessary before introducing a monitoring programme and identifying management measures for improving lake water quality. At present, there are only two national standards of water quality in Malaysia, namely the National Water Quality Standards (NWQS) and the National Drinking Water Quality Standards (NDWQS), which were developed for the purpose of river and drinking water protection. Currently, lake monitoring efforts have focused on using the NWQS due to the unavailability of specific water quality criteria for lakes. In an earlier work, it has been shown that the majority of the lakes studied were categorized as suitable for recreational purposes, although, they did experience eutrophication [2].

Numerous ambient water quality criteria and standards have been introduced and published in various countries throughout the world [3] and compiled based on different uses [4]. Many of these criteria were developed for different protection objectives, mostly for drinking and/or recreation, and derived using different methodologies. Some countries propose unified regulatory standards that classify all water bodies based on beneficial uses [5]. Other countries use general standards which are applied to the water bodies for multiple purposes, with different numerical limits set for different types of water bodies, such as rivers, canals, lakes and coastal waters [6]. In some countries, such as Japan, a specific law to preserve lake water quality, better known as the Clean Lake Law, has been established to protect and improve water quality in the water bodies. Individualized measures have had to be developed in Japan due to differing water quality conditions, sources and causes of pollution amongst lakes [7]. In many states in the United States, site-specific criteria incorporating lake basin features and maximum load have been proposed, such as in the cases of Lake Erie and Lake Tahoe [8, 9]. As national standards for river and marine waters are available, specific criteria and standards for lakes are also needed so that Malaysia can ensure the sustainable management and protection of this lentic water due to its inherent characteristics.

This work describes the efforts in developing a National Lake Water Quality Criteria and Standards (NLWQCS) for Malaysia. The paper will be presented in three ways: (i) a review of the lake water quality criteria and standards in existing literature; (ii) classification and criteria, based on multi-stakeholder consultations and published data and (iii) discussion of the role of this standard for lake management. The main objective in developing these criteria is to provide a standardized reference for the monitoring and management of lakes, based on a consensus of local experts and stakeholders, namely the lake managers and lake owners such the state authority, state water authority and government agencies.

2. Methodology

Various definitions of criteria and standards have been reviewed by Ward [10]. In this paper, the term 'water quality criteria' is defined as the level of constituents or state of the physical, chemical, radiological, biological and aesthetic properties of water that determine its suitability for specific uses. The term standard follows the definition by Streeter [11] which is 'an operational goal or objective', such as to enable concerted efforts towards sustainable management of lakes, rather than an established regulatory criterion with legal ramification. The methodology adopted in establishing water quality standards usually involves: (i) identifying the designated or beneficial water uses and (ii) selecting the water quality parameters and determining the necessary limiting criteria to protect the most vulnerable of the beneficial uses.

Reviews of available standards in existing literature were carried out, covering various water quality standards both in Malaysia and worldwide. The local literature evaluated includes the NWQS [12], Putrajaya Lake Water Quality Standards (PLWQS) [13] and the NDWQS [14], as well as the draft document of guidelines for recreational water [15]. Reference was also made to a compendium on water quality regulatory frameworks and selected international standards and guidelines in different continents such as the United States, European Union, Japan, Brazil and Australia as listed in **Table 1**. The reviews were conducted in order to determine the designated water use, and water quality parameters and values. In terms of threshold limits, emphasis for selecting the threshold limit was given to local criteria and lake data from literature. This is followed with criteria or threshold limits from other countries or states with similar tropical conditions such as Brazil, the Philippines, Florida State in the United States and northern Australia. Threshold values of some of the criteria involving human health are based on well-adopted criteria in the developed countries.

Additionally, three stakeholder consultations were organized in 2015 to obtain consensus from stakeholders and expert judgements on the criteria. In the first stakeholder consultation, identified stakeholders were presented with a preliminary idea of the classification and criteria followed by sharing of experience with other international lake standards namely the Japanese lake criteria. Stakeholders were identified randomly comprising selected water quality experts, government agencies and water supply companies. In addition to minutes of discussion, questionnaires were also distributed to stakeholders to identify the need and parameters most important for such standard and to record their input. Second stakeholder consultation was held during the national lake research committee forum to refine the classification of the proposed NLWQCS and the parameter threshold limits.

The national lake research committee consists of experts in various disciplines concerning lakes, and key representatives from federal government agencies and ministries in charge of lake management. In this forum, the role of the standards was also discussed and deliberated. The third stakeholder consultative session involved reviewing the proposed standard by various stakeholders in 45 federal and state jurisdictions, namely the lake owners, with regard to the development of the criteria. A separate engagement with selected experts was carried out to review the improvements made to the criteria resulting from the stakeholder consultations. The final NLWQCS was presented and disseminated in a seminar involving stakeholders. **Figure 1** shows the development process of the NLWQCS.

Country/region	Standards and/or guidelines	Purposes or classification
Australia & New Zealand	Australian and New Zealand Guidelines for Fresh and Marine Water Quality 2000	Recreation; aquatic ecosystems
Brazil	CONAMA Resolution 274, 2000	Beneficial uses
Canada	Guidelines for Canadian Recreational Water Quality 2012	Recreation
EU	EU Bathing Water Directive 2006	Bathing
	EU Water Framework Directive 2000	
Japan	Basic environmental law 1967; environmental quality standards regarding water pollution 1986	Beneficial uses (protection of natural environment, fishery, drinking, irrigation, industrial water, environmental conservation)
	Law concerning special measures for the preservation of lake water quality 1984	Lake
Malaysia	National Water Quality Standards	Beneficial uses
	National Drinking Water Quality Standards; Raw water quality guidelines	Drinking
	Putrajaya Lake Water Quality Standards	Recreational
South Africa	South African Water Quality Guidelines: recreational uses	Recreation
	South African Water Quality Guidelines: aquatic ecosystem	Aquatic ecosystems
	South African Water Quality Guidelines: domestic uses	Domestic uses
	South African Water Quality Guidelines: domestic uses	Irrigation
	South African Water Quality Guidelines: industrial uses	Industrial uses
	South African Water Quality Guidelines: agricultural uses	Aquaculture; livestock dewatering
UK & Ireland	UK The bathing water regulations 2013	Bathing
	Ireland Water Quality (Dangerous substances) Regulations, 2001	Dangerous substances
	Quality of Bathing Waters Regulations, 1992	Bathing
United States	2012 Recreational Water Quality criteria	Recreation
	State of Ohio Water Quality Standards	
	Lake Tahoe Basin Water Quality Plan	
	State of Michigan Water Quality Standards	
	Florida's surface water quality standards	
Various	WHO Recreational Water Quality Guidelines	Recreation

Table 1. List of selected water quality standards reviewed and compared.

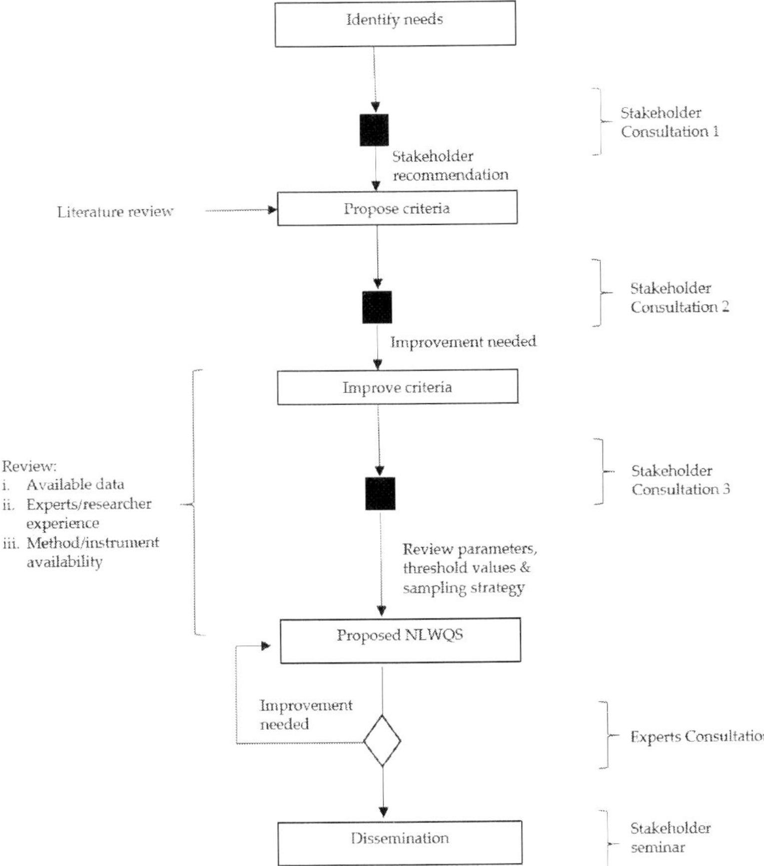

Figure 1. The development process of the NLWQCS.

3. Results and discussion

3.1. Comparison of different water quality criteria and standards

In general, most of the standards were developed to protect the use of water for human health purposes, such as drinking, bathing, and other resource efficiency such as industry, livestock dewatering and aquaculture [4]. The protection of aquatic health has been emphasized in the United States, Brazil, Australia and Japan. The criteria and number of parameters differ between the guidelines and standards. As drinking water standards are already established in Malaysia, emphasis in this work is placed on other human health applications, namely recreational, as well as aquatic life health. The commonly used criterion proposed by USEPA to measure human health is based on carcinogenicity and toxicity [10]. All recreational water quality criteria focus on microbiological parameters affecting human health. Microbial hazards are considered to be the primary concern by the WHO as they have the largest impact on health in terms of waterborne disease, especially when compared to chemical hazards which are usually associated with long-term exposure [16]. The Canadian recreational guidelines emphasize *Escherichia coli*, enterococci and cyanobacteria as indicators for microbial parameters [17]. In contrast, the South African guidelines identify nine microbiological parameters of concern related to recreational waters, namely algae measured by chlorophyll-*a* and blue-green algae counts, faecal coliforms, enterococci, *E. coli*, enteric viruses, coliphage and schistosoma/bilharzia [18]. Enterococci and *E. coli* are the only parameters monitored in the European Union bathing water directive, while microbiological parameters differ among states in the United States, with *E. coli* and faecal coliform mostly prescribed for monitoring. *E. coli* is the preferred indicator of faecal pollution for fresh recreational waters due to its strong correlation with the risk of gastrointestinal illness and the availability of a standardized method of detecting the organism within 24–48 h [17, 19]. *E. coli* is also a good indicator of human faeces, contributing 97% of the coliform organisms and recent faecal contamination due to its short survival rate [17]. Biochemical oxygen demand and chemical oxygen demand were not listed in many standards except for those in Japan and Malaysia. The impact on health from cyanobacteria is related to irritative dermal exposure to cyanobacteria genera such as Anabaena, Aphanizomenon, Nodularia and Oscillatoria, or the ingestion of cyanotoxin [16].

Within many standards and guidelines for recreational water, pH, clarity and colour are the most frequently prescribed physical and aesthetic parameters monitored (**Table 2**). Additionally, temperature and turbidity are important physicochemical and aesthetic parameters in Canadian Guidelines, while odour and floating objects are of concern within Australian and South African recreational guidelines, respectively. Turbidity is frequently used as a substitute for total suspended solids (TSS) and clarity, with numerical values specified in some standards or guidelines, such as in Australia and Malaysia.

Of common concern in terms of the chemical parameters for recreational water were total nitrogen (TN) and total phosphorus (TP), due to their ability to cause nuisance algal growth. Japan's regulations specified TN and TP of 0.2 mg/l and 0.01 mg/l, respectively for bathing purposes, while Lake Tahoe specifies TN and TP of 0.15 mg/l and 0.008 mg/l, respectively [8]. In Malaysia, the PLWQS and NWQS only list TP. In the United States, ammonia is also monitored in the various states' standards such as Ohio, Michigan and New York. In these US states, the narrative standard states that the water should be free of ammonia, or should contain amounts that cannot cause nuisance growths of aquatic weeds and algae. However, national recommendations on nutrient criteria were developed by USEPA and consider the region and waterbody approach by dividing the country into 14 nutrient ecoregions and four water body types, including lakes, reservoirs and wetlands [20]. Ranges of reference values were adopted in Australia to provide guidelines for managers to characterize ambient conditions [21]. In many regulations and guidelines, specific site studies are recommended to determine the appropriate concentrations for preserving the individual lakes [6, 9].

http://dx.doi.org/10.5772/65083

The number of chemicals criteria (specifically toxicants and pesticides) has increased greatly over the years [3]. More than 100 pesticide parameters have been specified in Australian guidelines [6]. The most typical pesticides mentioned due to health-related concerns are Aldrin, DDT, chlordane, lindane, endosulfan, malathion, paraquat and parathion.

	Units	Australia[1a]	Australia[1b]	Brazil[2]	Canada[3]	EU[4]	Japan[5]	Malaysia			South Africa			United States			UK	
								6	7	8	9	10	11	12	13	14	15	16
Temperature	°C	15–35			NR			±2		±2						NR		
Transparency	m	>1.6		NR	>1.2					0.6	>1					16		
Odour	–	NR		NR				NOO		NOO	1 TON				NR	NR	NR	
Taste	–			NR						NOT						NR		
floatables	–			NR						Nil	NR				NR	NR		NR
pH	–	5–9	6–8	6–9	5–9		6.5–8.5	6–9	6.5–9	6.5–9	6–9					6.5–8.5	6–8.5	
Colour	TCU	NR		<75	NR			150	15	150	NR				NR	NR		
Turbidity	NTU		2–200	<100	50			50	5	50	1					NR	≤29	
Hardness		500						250	500	250								
Oil & grease	mg/L	NR		NR												NR	≤5	
TSS	mg/L						≤5	50		50			<100		NR	NR		
DO	mg/L	>6.5		>6, >5			≥7.5	5–7		5–7								
TP	mg/L		0.01	0.025			≤0.01	0.2		0.05						0.01		
TN	mg/L		0.35	10			≤0.2						NR		NR	0.114		
BOD	mg/L			3–5				3		3			NR		NR			
COD	mg/L						≤3	25		25							NR	
Chlorophyll-a	µg/l			3						0.7	<15					0.9		
E. coli	Counts/100ml	35 (230)			≤70				0				≤130	126 (410)	126 (235)			900
Enterococci					≤400									35 (130)				330
Faecal coliform	Counts/100 ml	150 (1000)		1000				400	0	100	≤130				5000	200 (800)		
Total coliform	Counts/100 ml			5000		1000		5000	0	5000								
Enteric virus											0							
Cyanobacteria		>15,000			≤100,000													NR
Microcystin					≤20													
BGA											>6							
Clostridium perfringens											0	0						
Cadmium	mg/L	0.005		0.001		≤0.45	≤0.01	0.01	0.003	0.002		0.						

	Units	Australia[1a]	Australia[1b]	Brazil[2]	Canada[3]	EU[4]	Japan[5]	Malaysia			South Africa			United States				UK
								6	7	8	9	10	11	12	13	14	15	16
Lead	mg/L	0.05					≤0.01	0.05	0.01	0.05								
Arsenic	mg/L	0.05		0.05			≤0.01	0.05	0.01	0.05			0.01		0.01	≤0.05	0.05	
Mercury	mg/L	0.001		0.0002		≤0.07	≤0.0005	0.001	0.001	0.0001					#	#	0.012	
Chromium (IV)	mg/L	0.05		0.02			≤0.05	0.05	0.05	0.05					#	#		
Copper	mg/l	1						0.02	1	0.02					#	#		

Note: NR, narrative criteria or standard; #, a function of hardness; NOO, no obvious odour; NOT, no obvious taste.
[1] Australian and New Zealand Guidelines for Fresh and Marine Water Quality 2000, (a) recreation, (b) aquatic ecosystems.
[2] CONAMA Resolution 274, 2000.
[3] Canada, Recreation water quality 2012.
[4] EU Bathing Water Directive 2006.
[5] Basic environmental law; environmental quality standards regarding water pollution.
[6] National Water Quality Standards.
[7] National Drinking Water Quality Standards.
[8] Putrajaya Lake Water Quality Standards.
[9] South African Water Quality Guidelines: recreational uses.
[10] South African Water Quality Guidelines: domestic uses.
[11] South African Water Quality Guidelines: aquatic ecosystem.
[12] Quality criteria for water 1986/2012.
[13] State of Ohio Water Quality Standards.
[14] State of New York Nutrient standard plan.
[15] Florida's surface water quality standards.
[16] The bathing water regulations 2013.

Table 2. Comparison of selected parameters in different water quality standards and guidelines.

Water quality criteria for aquatic health are mostly chemical criteria that require knowledge of aquatic toxicity and environmental fate data. Most of aquatic ecosystems criteria are derived from toxicity data from multiple receptors in order to determine the ranges of tolerance for different organisms, including targeted protective species in relation to pollutants [3]. The commonly used criterion proposed by USEPA to measure the aquatic ecosystem health is usually the chronic effect value and the acute effect value. The minimum amount of data required for deriving water quality criteria for protecting freshwater aquatic environments varies between countries, as critically reviewed by Sha et al. [3]. In the United States and Europe, concentration approach is used with numeric criteria developed for cold water and warm water fish. In Australia, ecological health criterion is used compared to the contaminant limits approach adopted in the United States and Europe. Four biological criteria or indicators, namely species richness, species composition, primary production and ecosystem function, were recommended for assessing the ecosystem health [21]. When compared to developed countries where water quality criteria are well established for biota such as fish, specific information regarding the effects of pollutants that affect biota are very limited in tropical countries such as Malaysia. A recent work by Shuhaimi et al. has identified the acute test value for four heavy metals, namely iron, lead, nickel and zinc, on local biota [22].

Dissolved oxygen (DO), pH and turbidity levels are also important for aquatic life. Fish and other aquatic organisms could not survive if DO levels were less than 4 mg/l, and pH levels lower than 4 or above 11. For protecting specific wildlife such as salmonids, turbidity is specified in many standards in the United States and Alaska [23]. Various effects of turbidity and suspended solid concentrations include reduced growth, abundance, survival and feeding, fatality, altered behaviour and displacement [23]. The recommended criteria for turbidity vary as follows: <50 NTU for Canada and Malaysia, <100 NTU for Brazil and 2–200 NTU for tropical Australia. In the United States, numerical turbidity standards for protecting fish and aquatic habitats lie within 5–25 units above natural levels [23].

There are numerous criteria or limits for heavy metals due to their impact on both human and aquatic health. Almost all standards require criteria for lead, arsenic, mercury, cadmium and chromium. In the United States, criteria for most metals such as lead, cadmium, chromium and copper are based on natural logarithms of total hardness [24, 25] and are targeted for protecting human health. In Australia, numeric guideline values for maximum concentration are prescribed for about 17 parameters relating to heavy metals, while in South Africa a total of 12 parameters are listed [6, 26]. References on the lake conditions were made based on published literature [27–30].

3.2. Multi-stakeholder consultation on criteria and standards

A total of 32 local and four international experts were involved in the first stakeholder consultation. The findings from the first consultative process identified the need for the standards to be based on real data. They also stressed the need to consider the timing of data collection, the weather and the lake characteristics. Water quality parameter will differ with depth, and if sample is taken during daytime or night-time and during dry and wet weather conditions. The findings in the second consultative process further refined the classification of the proposed NLWQCS and the threshold limits of parameters that may affect water quality. The need to engage with various stakeholders in the state jurisdiction was also highlighted. The state authority and/or state water authority can decide whether water standard is to be used throughout the state for managing or controlling activities within lakes within the state. The application of water quality criteria in Malaysia can be regulatory, depending on regulation and law, or voluntarily such as to support in maintaining water quality for respective uses.

The findings of the third consultative process include further refinement of the threshold values based on various monitoring experiences by the respective stakeholders. In the consultation process, stakeholders were divided into four groups to discuss the individual classification. Discussions were focused on the parameters that need to be identified as criteria determining the class. Other toxicants criteria were also reviewed by the groups. The main parameters of concern in determining of threshold limits includes chlorophyll-*a*, turbidity, suspended solids, transparency, salinity, conductivity, total nitrogen, total phosphorus, coliform and *E. coli*. A separate group was formed to discuss the minimum sampling strategy to support the criteria and standard. This includes the minimum number of samples required and the compliance level of the criteria. The final engagement with experts on the criteria suggested to change the classification into category as the four classifications were not based on ranking of parameters. Refinement was also made in setting the parameters, such as BOD, which is defined as BOD_5 or BOD 5-day test consistent with other standards in Malaysia.

3.3. Proposed lake water quality criteria and standards

3.3.1. Classification and monitoring parameters

The development of the NLWQCS aims at providing a tool stakeholders and lake owners can use manage a lake or reservoir. NLWQCS applies to all lakes, reservoirs, ponds and wetlands. The proposed standard is divided into four categories, which were suggested as targets for the respective lakes to achieve [31] based on designated uses proposed by the respective authorities. Categories A and B were criteria to be applied for lakes used for recreational environment: Category A for primary body contact such as bathing, diving, skiing and wind-surfing activities, and Category B for secondary body contact such as boating, cruising and angling. Category A is the more stringent of the two and included criteria for parameters related to waterborne disease. The contaminant limits for certain parameters in Category A is more stringent than the raw water for drinking standard. Categories C and D were criteria to be applied for lakes that are productive and meant for fisheries and other economic activities: Category C for protection of aquatic health, and Category D for other limited uses. Category D is allowed to be more turbid, have higher concentrations of TSS and nutrients, and have lower transparency.

The main parameters identified for the different classes are categorized as (i) physical parameters; (ii) nutrients; (iii) biochemical and microbial constituents and (iv) other toxicants. Only approximately 23 physical-chemical parameters were identified to determine the categories. Factors considered in parameter selection include simplicity of measurement, availability of measuring devices and the reasonable cost of sampling and analysis [31]. Aesthetic parameters such as colour, odour, taste, floating objects, transparency, turbidity and suspended solids are emphasized in all categories for lake management. These parameters reflect aesthetic values of the lake and most can be measured easily. Normal temperature was set at $28 \pm 3°C$ [31]. Nutrient parameters relate to ammonia nitrogen, nitrate-nitrogen and TP. Hardness, which influences heavy metals, is deliberated to be around 50 mg/l.

The proposed criteria for lake water quality are given in **Table 3**. The criteria in this work are based on established criterion found in the literature assuming similar health and aquatic life effects will be experienced when levels of water quality parameters exceeding the criteria. In terms of aesthetic parameters, floating objects, colour and odour are rarely reported on in the literature. Suitability for swimming has been associated with the clarity or the depth of light penetration into water. The clarity of Malaysian lakes mentioned in the literature range between 0.3 m and 5.7 m [2, 27]. This parameter is associated with the colour of the water, levels of turbidity, algae and suspended solids. Guidelines for water clarity or Secchi depth were set at 1.2 m and 1 m in Canada and South Africa, respectively as being the minimum visibility level for water to be suitable for swimming [17, 18]. In Australia and New Zealand, the visual clarity level based on Secchi depth required to ensure swimmer safety in wadeable areas was recommended as 1.6 m [6]. A survey of perception for bathing in New Zealand found a Secchi depth of 1.5 m as the aesthetic consideration [32]. Currently, no survey has been carried out to measure perception for bathing in Malaysia. In the NLWQS, a lower threshold limit of 0.6 m was set following experience in Putrajaya. TSS in most lakes were generally found to be less than 20 mg/l, except in Sembrong and Aman Lakes which are known to experience algal bloom that can reach about 50 mg/l. DO concentrations in Malaysian lakes are highly variable and depend on the timing of the sampling. DO <5 mg/l is common for lakes located in peat swamp areas. In terms of nutrient parameters, TP was set to below <0.01 mg/l for primary contact and <0.05 mg/l for the other classes [31]. Nutrient values (specifically TP) were very high in many lakes and needed to be controlled to decrease eutrophication problems. Monitoring is necessary to ensure informed decision making regarding effective management measures to control nutrient inputs. These values are consistent with the limits set in Japanese regulations and at Lake Tahoe for bathing purposes, and are much more stringent than the PLWQS which specified higher TP value (0.05 mg/l) for its ambient recreational standard.

Parameter	Unit	Category A	Category B	Category C	Category D
Physico-chemical					
Temperature	°C	Normal ±3	Normal ±3	Normal ±3	Normal ±3
pH	-	6.5–8.5	6.5–8.5	6.0–9.0[a]	5.5–9.0[k]
Dissolved oxygen saturation	%	80–100	70–110	55–130	40–130
Dissolved oxygen	mg/L	6.3–7.8	5.5 – 8.7	4.5 – 10.3	3.3 – 10.3
Conductivity	µS/cm	1000[ai]	1000[ai]	2000	5000
Floatables	-	NV[i]	NV[i]	NV[i]	NV[i]
Odour	-*	NOO[i]	NOO[i]	NOO[i]	NOO[i]
Taste	-**	NOT[i]	NOT	NOT	NOT
Colour	TCU	100–200	150–300	300	300
Total suspended solid	mg/L	100	100–500	200	500
Turbidity	NTU	40	40–170	70	250
Transparency (Secchi depth)	m	0.6[i]	0.6[i]	0.3	0.3
Oil & Grease	mg/L	1.5[i]	1.5[i]	1.5[i]	1.5[i]
Salinity	psu	nvd	nvd	<1[a]	>1
Hydrogen carbonate	mg/L	<200	<200	**[a]	**[a]
Total dissolved solids	mg/L	1,000[>k]	1,000[>k]	1,000[>k]	1,000[>k]
Nutrient					
Ammoniacal Nitrogen	mg/L	0.1[a]	0.3[i]	1	2.7[a]
Nitrite-N	mg/L	0.04[i]	0.4[a]	0.4[a]	0.4[a]
Nitrate-N	mg/L	7[a]	7[a]	10	10
Total nitrogen (TN)	mg/L	0.35[f]	0.35[f]	0.35	0.35
Total phosphorus (TP)	mg/L	0.01[b]	0.035[b]	0.035[b]	0.05[b]
Biochemical and microbiological/water-borne disease					
Biochemical Oxygen Demand (BOD)	mg/L	3	6[a]	6	8

Parameter	Unit	Category A	Category B	Category C	Category D
Chemical Oxygen Demand (COD)	**mg/L**	**10[a]**	25[a]	25	50
Chlorophyll-a	**μg/L**	**10**	**3 – 15**	**15**	**25**
Cyanobacteria	**Cells/ml**	**15000[cf]**	15000[cf]	15000[cf]	15000[cf]
Faecal Coliform	MPN/ 100ml	150[fg]	<1000[g]	5000 (20000)*[a]	5000 (20000)*[a]
Total Coliform	**Counts/ 100ml**	**5000[a]**	5000[a]	50000[a]	50000[a]
E.coli	**cfu/100ml**	**100[d]**	1200[g]	3000[h]	3000[h]
Enterococci	**MPN/ 100ml**	**33[d]**	230[c]	nvd	nvd
Clostridium perfringens **(including spores)**	-	**nd[i]**	nd[i]	nvd	nvd
Cryptosporodium **sp.**	-	**nd[i]**	nd[i]	nvd	nvd
Giardia **sp.**	-	**nd[i]**	nd[i]	nvd	nvd
Leptospira **sp.**	-	**nd[e]**	Nd[e]	nvd	nvd
Enteroviruses	PFU/L	nvd	nvd	nvd	nvd
Microcystin–LR	μg/L	0	0	0	0
Heavy metals					
Arsenic	**mg/L**	**0.05[ai]**	**0.1[a]**	**0.15[d]**	**0.4[a]**
Aluminium	mg/L	0.1[ji]	0.1[ji]	0.05[i]	0.05[i]
Antimony	mg/L	0.03[i]	0.03[i]	0.03[i]	0.03[i]
Argentum	mg/L	0.05[k]	0.05>[k]	0.05>[k]	0.05>[k]
Barium	mg/L	0.1[g]	0.1	1[ai]	1[ai]
Beryllium	mg/L	0.004[i]	0.004[i]	0.004[i]	0.004[i]
Boron	mg/L	1[ai]	1[ai]	1[ai]	1[ai]
Calcium ion	mg/L	200	200	**[a]	**[a]
Cadmium	**mg/L**	**0.002[i]**	**0.01[a]**	**0.25[d]**	**0.25[d]**
Chloride	mg/L	250[j]	250	250	250
Chromium	mg/L	0.05[ac]	0.025[ac]	0.05[ac]	0.05[ac]
Cobalt	mg/L	0.05[c]	0.05[c]	0.05[c]	0.05[c]
Combined Chlorine	mg/L	>1.0	>1.0	>1.0	>1.0
Copper	mg/L	0.02[ai]	0.02>[ai]	0.02>[ai]	0.02>[ai]
Fluoride	mg/L	1	1	1.5[f]	1.5[f]
Iron	mg/L	1[ai]	1	1	1
Lead	**mg/L**	**0.05[a]**	**0.05[a]**	**0.0025[d]**	**0.05[a]**
Magnesium	mg/L	150[k]	150	150	150
Manganese	mg/L	0.1[ai]	0.1	0.1	0.1
Mercury	**mg/L**	**0.001[a]**	**0.002[a]**	**0.00077[d]**	**0.002[a]**
Nickel	**mg/L**	**0.05[a]**	**0.05[a]**	**0.052[d]**	**0.05[a]**
Potassium ion	mg/L	200	200	200	200
Silver	mg/L	0.05[ai]	0.05	0.05	0.05
Sodium	mg/L	200[k]	200[k]	200[k]	200[k]
Sulphur	mg/L	0.05[ai]	0.05	0.05	0.05
Zinc	mg/L	3[k]	3[k]	5[ai]	5[ai]
Organics or pesticides					
1,2-dichloroethane	μg/L	30[k]	30[k]	30	30
2,4-D	μg/L	30[k]	30[k]	70[ai]	70[ai]

Parameter	Unit	Category A	Category B	Category C	Category D
2,4-DB	µg/L	90[k]	90[k]	90	90
2,4-dichlorophenol	µg/L	90[k]	90[k]	90	90
2,4,5-T	µg/L	9[k]	9[k]	10[ai]	10[ai]
2,4,5-TP	µg/L	4[ai]	4[ai]	4[ai]	4[ai]
2,4,6-trichlorophenol	µg/L	200[k]	200	200	200
Acrylamide	µg/L	0.1[g]	0.1[>g]	0.1[>g]	0.1[>g]
Alachlor	µg/L	20[k]	20[>k]	20[>k]	20[>k]
Aldicarb	µg/L	10[k]	10[>k]	10[>k]	10[>k]
Aldrin / Dieldrin	µg/L	0.02[i]	0.02[>i]	0.02[i]	0.02[i]
Anionic Detergent MBAS	µg/L	1000[k]	1000[>k]	1000[>k]	1000[>k]
Atrazine	µg/L	nvd	nvd	nvd	nvd
BHC	µg/L	2[ai]	2[ai]	2[ai]	2[ai]
Benzene	µg/L	10[f]	10[f]	10[f]	10[f]
Benzo(a)pyrene	µg/L	0.01[c]	0.01[c]	0.01[c]	0.01[c]
Bromate	µg/L	10[>g]	10[>g]	10[>g]	10[>g]
Bromodichloro methane	µg/L	60[>k]	60[>k]	60[>k]	60[>k]
Bromoform	µg/L	100[>k]	100[>k]	100[>k]	100[>k]
Carbofuran	µg/L	7[>k]	7[>k]	7[>k]	7[>k]
Carbon Chloroform Extract	µg/l	500[ai]	500[ai]	500[ai]	500[ai]
Chlordane	µg/L	0.08[ai]	0.08[ai]	0.08[ai]	0.08[ai]
Chloroform	mg/L	0.2[>k]	0.2	0.2	0.2
Cyanide	mg/L	0.05[g]	0.05[g]	0.05[g]	0.05[g]
DDT	µg/L	2[>k]	2	2	2
Dibromoaceto nitrile	µg/L	100[>k]	100[>k]	100[>k]	100[>k]
Dibromochloro methane	µg/L	100[>k]	100[>k]	100[>k]	100[>k]
Dichloroacetic acid	µg/L	50[>k]	50[>k]	50[>k]	50[>k]
Dichloroaceto nitrile	µg/L	90[>k]	90[>k]	90[>k]	90[>k]
Endosulfan	µg/L	10[ai]	10[ai]	10[ai]	10[ai]
Epichlorohydrin	µg/L	4[f]	4[f]	4[f]	4[f]
Free Residual Chlorine	mg/L	1.5[i]	1.5[i]	1.5[i]	1.5[i]
Glyphosate	µg/L	200[k]	200[k]	200[k]	200[k]
Heptachlor	µg/L	0.05[ai]	0.05[ai]	0.05[ai]	0.05[ai]
Hexachloro benzene	µg/L	1[>k]	1[>k]	1[>k]	1[>k]
Lindane	µg/L	2[ij]	2[ij]	2[ij]	2[ij]
MBAS/BAS (Methylene Blue)	µg/L	200[f]	200[f]	200[f]	200[f]
MCPA	µg/L	2[>k]	2[>k]	2[>k]	2[>k]
Methoxychlor	µg/L	20[>k]	20[>k]	20[>k]	20[>k]
Mineral Oil	µg/L	300[>k]	300[>k]	300[>k]	300[>k]
Oil & Grease (Emulsified Edible)	mg/L	7:N[i]	7:N[i]	7:N[i]	7:N[i]
Oil & Grease (Mineral)	mg/L	0.04:N[i]	0.04:N[i]	0.04:N[i]	0.04:N[i]
Paraquat	µg/L	10[ai]	10[ai]	10[ai]	10[ai]
Parathion	µg/L	30[k]	30[k]	30[k]	30[k]
PCB	µg/L	0.1[ai]	0.1[ai]	0.0001	0.0001

Parameter	Unit	Category A	Category B	Category C	Category D
Pendimethalin	µg/L	20[>k]	20[>k]	20[>k]	20[>k]
Pentachlorophenol	µg/L	9[>k]	9[>k]	9[>k]	9[>k]
Permethrin	µg/L	20[>k]	20[>k]	20[>k]	20[>k]
Pesticides	µg/L	nvd	nvd	nvd	nvd
Phenol	µg/L	5[g]	5[g]	5[g]	5[g]
Polycyclic Aromatic Hydrocarbons	µg/L	nvd	nvd	nvd	nvd
Propanil	µg/L	20[>k]	20	20	20
Selenium	mg/L	0.01[fj]	0.01[f]	0.01[f]	0.01[f]
Simazine	µg/L	20[k]	20[k]	20[k]	20[k]
Sulphate	mg/L	250[j]	250[>j]	250[>j]	250[>j]
t-DDT	µg/L	0.1[ai]	0.1[ai]	0.1[ai]	0.1[ai]
Tetrachloroethene and Trichloroethene	µg/L	10[gj]	10[gj]	10[gj]	10[gj]
Total indicative dose	µg/L	nvd	nvd	nvd	nvd
Total organic carbon (TOC)	µg/L	nvd	nvd	nvd	nvd
Toxicants (heavy metal, organics)	µg/L	#	#	#	#
Trichloroacetic acid	µg/L	100[>k]	100[>k]	100[>k]	100[>k]
Trichloroaceto nitrile	µg/L	1[>k]	1[>k]	1[>k]	1[>k]
Trihalomethanes–Total	µg/L	1000[>k]	1000[>k]	1000[>k]	1000[>k]
Tritium	µg/L	nvd	nvd	nvd	nvd
Vinyl chloride	µg/L	5[>k]	5[>k]	5[>k]	5[>k]
Radiological parameters					
Gross-alpha	Bq/L	0.1[ai]	0.1[ai]	0.1[ai]	0.1[ai]
Gross-Beta	Bq/L	1[ai]	1[ai]	1[ai]	1[ai]
Radium-226	Bq/L	<0.1[ai]	<0.1[ai]	<0.1[ai]	<0.1[ai]
Strontium-90	Bq/L	<1[ai]	<1[ai]	<1[ai]	<1[ai]

Notes: Item in light grey—should be measured for categorization.
NV, not visible; NOO, no obvious odour; NOT, no obvious taste; nd, not detected; nvd, no value determined.
[a]DOE [12].
[b]Health Canada [17].
[c]ANZECC [6].
[d] USEPA [24, 25].
[e] WHO [16].
[f] Ministry of Health, unpublished report.
[g] EPA Ireland [40].
[h] Conversion using USEPA ratio (126 *E. coli* = 200 faecal).
[i] Perbadanan Putrajaya [13].
[j] CONAMA [5].
[k] NDWQS.
[*] maximum not to be exceeded.
[#] parameter not fully established.

Table 3. National Lake Water Quality Criteria and [C2]Standards 2015 [30].

Biochemical and microbiological parameters suggested in the standard include BOD, COD, total and faecal coliforms, *E. coli*, enterococci, cyanobacteria and three pathogens, namely *Cryptosporidium* sp., *Leptospira* and *Giardia* sp. [31]. Few researchers have debated the suitability of faecal indicator bacteria such as faecal coliform, enterococci and *E. coli* in tropical freshwater bodies [33, 34]. These studies show that the indicator bacteria can multiply to establish itself in the soil of tropical countries, and are thus inadequate to suggest an indicator of pollution from human and animal faeces [32, 33]. In another study, *Clostridium perfringens* has been suggested as a better indicator for recreational water quality standards in tropical countries due to its inability to multiply in the soil [35]. Despite this variability, enterococci and *E. coli* remain widely used as indicators of faecal coliforms [36]. In order to apply these standards to Malaysian lakes, *E. coli*, enterococci and *Clostridium perfringens* were identified as indicators. Additionally, pathogens such as *Cryptosporidium* sp., *Leptospira* and *Giardia* sp. were included as they can lead to potentially severe disease outbreaks in Malaysia, such as diarrhoea. Waterborne disease associated with *Cryptosporidium parvum* and *Giardia duodenalis* has emerged as an important public health concern in developed and developing countries [37]. The presence of *Giardia* sp., *Cryptosporidium* sp. and *leptospira* sp. have been reported in some of the urban lakes suggesting the use of this water body for recreational purpose is a major health concern [38, 39] and requiring monitoring and enforcement of source of pollutants.

Stakeholders recommended monitoring of five heavy metals in lakes, namely arsenic, cadmium, lead, mercury and nickel, due to their toxicity in human health. Some of these chemicals were frequently detected in many rivers throughout Malaysia [12]. Arsenic was classified as very toxic and widely associated with industrial pollution from the mining industry, dye manufacturers, the glass and ceramics industry, fertilizers and pesticides [40]. High arsenic levels were reported in post-mining lakes such as Blue Lake, due to its use in the gold mining process. This has led to the barring of the lake for any human activities [41, 42]. Both cadmium and lead are potentially hazardous to most forms of life and are considered to be toxic to aquatic organisms. The main environmental sources of cadmium are discharges from mining, metal smelters and agricultural uses of sludge, pesticides and fertilizers. For lead, the main sources are anthropogenic activities such as runoff associated with lead emissions from gasoline-powered motor vehicles, and industrial and municipal wastewater discharges [26, 40]. Mercury is of major concern in the natural aquatic environment due to its extreme toxicity to aquatic organisms, high concentrations of which in water bodies are associated with industrial pollution. Information on many toxicants, in particular pesticides in Malaysian lake water quality were not found in the literature. The threshold limit for many toxicants provided here presents as a starting point for monitoring of the pollutants. Future studies on threshold levels of toxicants in lakes are needed to confirm and validate the criteria.

3.3.2. Role of standards and limitation

The proposed standards were based on expert judgement and the best available information found in the literature. Part of the role of the NLWQCS is to provide the directional targets for research and management programmes. The standards can be used to guide rehabilitation measures and conservation efforts as well as to develop management measures to address eutrophication issues. The standards are proposed to be non-binding and to be gradually improved as more information and data are gathered by the stakeholders. As water quality monitoring in lakes in still in infancy, this NLWQCS does not provide a plan for enforcement to enable protection of the uses nor become an established regulatory criterion with legal ramification. However, the criteria can also be used by the respective agencies and stakeholders to assist in monitoring and to classify various lakes under their regulatory or management controls for their fitness for different uses.

In the proposed document, the sampling strategy was not described in detail. Water quality in lakes is known to differ temporally and spatially, both horizontally and vertically depending on lake depth. Seasonal and daily variations associated with irradiance, along with dissimilarity in surface waters' mixing related to weather patterns can induce variations in temperature, DO and the transportation of nutrients or pollutants. Thermal and chemical stratification are common features of deeper lakes which affect the water quality being monitored. Variations in water quality are also dependent on the composition of discharges over both short and extended periods. Discharges from housing, commercial buildings and industry can vary within a day, a week or a season. Domestic discharges depend on the homes' occupancy, which is usually higher during the early morning, at midday and in the early evening, while industrial discharges depend on operation hours. Higher home discharges and lower industrial discharges could happen during weekends and festive seasons. Exceedances of bacterial indicator to health risk were also temporally sporadic and geographically limited with the reduction of pollutant loading not necessarily reducing the health risk [43]. Narrative criteria were proposed in NLWQCS namely to consider the size and shape of the lake when choosing a sampling location, so that the site selection is representative of the whole lake. The choice of site for routine water monitoring sampling can make a significant difference to the classification of microbial water quality [44], owing to the hydrodynamics and proximity of pollutant sources.

With respect to the criteria for protecting aquatic life in the NLWQCS, many limitations exist due to unavailability of extensive data. The criteria in this are mostly based on chronic and acute effect values in temperate countries which have different species. Most of the water quality criteria were also based on a single pollutant model which mostly targeted single species instead of community response [3]. Some of the criteria for aquatic life protection adopted the same value for the human health protection criterion assuming that the effects to all types of aquatic life, all stages of their life cycle and the whole aquatic community are similar. Further research is much needed on deriving the chronic and acute effects of many pollutants on freshwater local species in order to establish more accurate criteria for protecting the Malaysian aquatic environment.

In these criteria, the sampling methodology aims at testing the ambient condition of the lakes via a minimum of three sets of samples, to be monitored at least twice a year, once in the dry and once in the wet season. The proposed depth is surface measurement, in accordance with standard methods [45]. However, adult chest depth at ~1.2–1.5 m is the most common sampling depth recommended in the United States and Canadian recreational guidelines, due to strong evidence in the form of the mathematical relationship between indicator organism density and swimmer illness. Intensifying the number of samples and frequency of monitoring may provide a better representation of the lake's overall water quality and trends. A minimum frequency of once per week during the swimming season was recommended by Canadian guidelines [17] and monthly by the UK and a few US states [9, 46, 47], in order to make more informed decisions regarding lake suitability for recreation. As bathing or swimming activities in lakes is not widely practiced in Malaysia, nor is there a specific swimming season, monitoring measures will depend on the authority's management budgets.

In order to determine the extent of violation or compliance levels, this standard proposes the 90th percentile of sampling results to be considered as acceptable for determining any class. Few guidelines or standards such as in the United States propose the use of dual limits, with the first being a maximum limit for the geometric mean concentration over an interval, and the second being a single-sample maximum or threshold limit set to better evaluate the water quality in both the short- and long-term. The short-term limit is usually set over a 30-day interval and aims at addressing immediate water quality issues, while the long-term limit is set over the duration of the swimming season and aims to address chronic contamination problems. Other standards such as that of the UK advocate the use of percentage compliance levels, mostly of 95% or 90% [46] and some others such as Japan use annual averages [48]. The 90th percentile was taken into account following stakeholder consensus, the values of which consider top end variability in the distribution of water quality and also in order to curb influence of possible small sample sizes. The water quality classification of risk or status was not described in the NLWQCS due to unavailability of reference data.

Financial issues relating to water testing were collectively identified by the stakeholders as major challenges for NLWQCS implementation. Water quality monitoring was usually performed based on the parameters to estimate the water quality and Carlson's trophic state index [49] due to limited funds. Pathogenic parameters such as *Leptospira* and *Giardia* sp. are rarely monitored. Measurements were usually undertaken by the relevant departments when there were reported cases. In this NLWQCS, focus is placed on the main physicochemical, microbial and selected toxicants in order to promote lake monitoring. The main aim is to monitor and control nutrient, microbial and organic pollutants, and improve aesthetic values of existing lakes. The various lists of pesticides and toxicants is encouraged to be monitored if funding is not limited and if testing methods are available at lower detection limits, such as for trace chemicals. The proposed criteria and standard serve as a reference and a starting point in the proper monitoring of lakes and in working towards sustainable management of the water bodies. Future review has been suggested upon the availability of more data from monitoring efforts, and possibly to expand the standard further from water quality-based management to ecologically based management to protect the whole aquatic system [21]. It is hoped that the criteria will evolve and be adopted as a regulatory monitoring program in the future to enable a better connection to standard compliance and violations.

4. Conclusion

The importance of this work is to develop criteria that can be used for sustainable management of lakes and reservoirs in Malaysia. The lake criteria and standards are proposed to be non-regulatory to promote monitoring efforts by various stakeholders. The development of such a criteria and standard, however, may be limited by time constraints, fund allocations and expert knowledge as well as the variability of environmental data. Future work will look into governing the standard with an appropriate methodology and regulatory framework to ensure an effective national standard for application.

Acknowledgements

This research was supported by the Ministry of Natural Resources and Environment of Malaysia (no. P23170009000421). We acknowledge the contribution of various stakeholders in the consultative sessions, meetings and surveys. We also would like to express our gratitude to the study team including Akashah Majizat, Zelina Zaiton Ibrahim, Normaliza Noordin, Yusof Ishak and Nurul Hasni Abdullah for their effort.

Author details

Zati Sharip[1*] and Saim Suratman[2]

*Address all correspondence to: zati@nahrim.gov.my

1 Lake Research Unit, Water Quality and Environment Research Centre, National Hydraulic Research Institute of Malaysia, Ministry of Natural Resources and Environment, Seri Kembangan, Selangor, Malaysia

2 National Hydraulic Research Institute of Malaysia (NAHRIM), Ministry of Natural Resources and Environment, Seri Kembangan, Selangor, Malaysia

References

[1]   NAHRIM. 2009. Study on the Status of Eutrophication of Lakes in Malaysia. NAHRIM, Seri Kembangan.

[2]   Sharip Z, Zaki ATA, Shapai MAH, Suratman S, Shaaban AJ. 2014. Lakes of Malaysia: water quality, eutrophication and management. Lakes Reservoirs: Res. Manage., 19: 130–141.

[3]   Yan S, Zhou Q, Gao J. 2012. Methodology for derivation of water quality criteria for protecting aquatic environment and future development. Crit. Rev. Env. Sci. Technol., 42: 2471–2503.

[4]   UN-Water. 2015. Compendium of Water Quality Regulatory Frameworks: Which Water for Which Use?

[5]   CNE (Council National Environment). 2000. CONAMA resolution 274, Brazil.

[6]   ANZECC (Australian and New Zealand Environment and Conservation Council). 2000. Australian and New Zealand Guidelines for Fresh and Marine Water Quality. Volume 1. The Guidelines. ANZECC and Agriculture and Resource Management Council of Australia and New Zealand, Canberra.

[7]   Kai-Qin X, Ebie Y, Jimbo Y, Inamori Y, Sudo R. 2009. Measures and Policies against the Eutrophication for Lake Water Quality in Japan. Proceeding of 13th World Lake Conference. Wuhan, China.

[8]   CRWQB (California Regional Water Quality Control Board). 1995. Water quality control plan for the Lahontan Region. CRWQB, California.

[9]   Ohio EPA. 2014. State of Ohio water quality standards. Ohio Environmental Protection Agency.

[10] Ward R. 2001. Development and use of water quality criteria and standards in the United States. Regional Environ. Change 2: 66–72.

[11] Streeter HW. 1949. Standards of stream sanitation. Sewage Work J. 21.

[12] DOE. 2013. Malaysia Environmental Quality Report 2013. Department of Environment Malaysia, Petaling Jaya.

[13] Perbadanan Putrajaya. 2000. Catchment Development and Management Plan for Putrajaya Lake. Perbadanan Putrajaya, Putrajaya.

[14] MOH (Ministry of Health). 2004. National Standard for Drinking Water Quality. Engineering Services Division, Putrajaya.

[15] MOH (Ministry of Health). 2013. National Water Quality Standards and Guidelines for Monitoring Recreational Water Quality Environment (marine and freshwater). Engineering Services Division, Putrajaya, Unpublished report.

[16] WHO. 2003. Guidelines for safe recreational water environment. WHO, Geneva.

[17] Health Canada. 2012. Guidelines for Canadian Recreational Water Quality, Third Edition. Water, Air and Climate Change Bureau, Healthy Environments and Consumer Safety Branch, Health Canada, Ottawa, Ontario.

[18] DWAF (Department of Water Affairs and Forestry). 1996a. South African Water Quality Guidelines (second edition), Volume 2: Recreational Use.

[19] Kinzelman J, Ng C, Jackson E, Gradus S, Bagley R. 2003. Enterococci as indicators of Lake Michigan recreational water quality: comparison of two methodologies and their impacts on public health regulatory events. Appl. Env. Microbiol., 69: 92–96.

[20] USEPA. 2000. Nutrient Criteria Technical Guidance Manual: Lakes and Reservoirs. Office of Water, US EPA, Washington.

[21] Hart B, Campbell I, Angehrn-Bettinazzi C, Jones M. 1993. Australian water quality guidelines: a new approach for protecting ecosystem health. J. Aquat. Ecosys. Health, 2: 151–163.

[22] Shuhaimi-Othman M, Nur-Amalina YNR, Umirah NS. 2012. Deriving freshwater quality criteria for iron, lead, nickel and zinc for protection of aquatic life in Malaysia. Sci. World J. 861576–861576

[23] Lloyd DS. 1987. Turbidity as a water quality standard for salmonid habitats in Alaska. N. Am. J. Fish. Manage., 7: 34–45.

[24] USEPA. 2012b. 2012 Recreational Water Quality Criteria. Office of Water, US EPA, Washington.

[25] USEPA. 2012a. Federal Water Pollution Control Act. Office of Water, US EPA, Washington.

[26] DWAF. 1996b. South African Water Quality Guidelines, Volume 7: Aquatic Ecosystems.

[27] ASM (Akademi Sains Malaysia). 2010. Managing lakes and their basins for sustainable use in Malaysia (Lake Briefs Report Series I), ASM, Kuala Lumpur.

[28] Furtado JI, Mori S. 1982. The Ecology of a Tropical Freshwater Swamp, the Tasek Bera, Malaysia. Dr W. Junk Publishers, The Hague.

[29] NAHRIM. 2012. Managing lakes and their basins for sustainable use in Malaysia. NAHRIM, Seri Kembangan.

[30] NAHRIM. 2013. Proceedings National Seminar on Research and Management of Lakes and their Basins for Sustainable Use: The Current Status of Selected Lakes in Malaysia. Seri Kembangan.

[31] NAHRIM. 2015. The Development of Lake Water Quality Criteria and Standards for Malaysia. NAHRIM, Seri Kembangan.

[32] Smith DG, Croker GF, McFarlane K. 1995. Human perception of water appearance. N. Z. J. Mar. Freshwater Res., 29: 29–43.

[33] Hazen TC. 1988. Fecal coliforms as indicators in tropical waters: a review. Toxicity Assessment 3: 461–477.

[34] Fujioka R, Sian-Denton C, Borja M, Castro J, Morphew K. 1998. Soil: the environmental source of Escherichia coli and Enterococci in Guam's streams. J. Appl. Microbiol. 85: 1365–2672.

[35] Fujioka R, Roll B, Byappanahalli MN. 1997. Appropriate recreational water quality standards for Hawaii and other tropical regions based on concentrations of *Clostridium perfringens*. Proc. Water Environ. Fed., 4: 405–411.

[36] Boehm AB, Sassoubre LM. 2014. Enterococci as indicators of environmental fecal contamination. In: Enterococci: From Commensals to Leading Causes of Drug Resistant

Infection (Gilmore MS, Clewell DB, Yasuyoshi I, Nathan S, eds.), Massachusetts Eye and Ear Infirmary, Boston.

[37] Carmena D. 2010. Waterborne transmission of Cryptosporidium and Giardia: detection, surveillance and implications for public health. In: Current Research, Technology and Education Topics in Applied Microbiology and Microbial Biotechnology (Mendez-Vilas A, ed.), pp. 1–14.

[38] Benacer D, Woh PY, Zain SNM, Amran F, Thong KL. 2013. Pathogenic and saprophytic leptospira species in water and soils from selected urban sites in peninsular Malaysia. Microbes Environ., 28: 135–140.

[39] Onichandran S, Kumar T, Lim YAL, Sawangjaroen N, Andiappan H, Salibay CC, Chye TT, Ithoi I, Dungca JZ, Sulaiman WYW, Ling LY, Nissapatorn V. 2013. Waterborne parasites and physico-chemical assessment of selected lakes in Malaysia. Parasitol Res., 112: 4185–4191.

[40] EPA Ireland. 2001. Parameters of Water Quality: Interpretation and Standards. Ireland Environmental Protection Agency.

[41] Sari SA, Ujang Z, Ahmad UK. 2006. Geospeciation of arsenic using MINTEQA2 for a post-mining lake. Water Sci. Technol., 54: 289–299.

[42] Harith MN, Hassan R. 2008. Preliminary investigation on vertical distribution of phytoplankton biomass and nutrients in Tasik Biru (Blue Lake), Bau, Sarawak. In: Proceedings of the 3rd Regional Symposium on Environment and Natural Resources "Conservation for a Green Future", pp. 446–454

[43] Soller JA, Eisenberg JNS, DeGeorge JF, Cooper RC, Tchobanoglous G, Olivieri AW. 2006. A public health evaluation of recreational water impairment. J. Water Health, 04.1: 1–19.

[44] Quilliam RS, Clements K, Duce C, Cottrill SB, Malham SK, Jones DL. 2011. Spatial variation of waterborne *Escherichia coli* – implications for routine water quality monitoring. J. Water Health, 09.4: 734–737.

[45] APHA (American Public Health Association). 2012. Standard methods for the examination of water and waste water. APHA, Washington.

[46] DEFRA & EA (Department for Environment, Food & Rural Affairs and Environment Agency). 2013. Water resources: The Bathing Water Regulations 2013 No. 1675, UK.

[47] FDEP (Florida Department of Environmental Protection). 2012. Surface Water Quality Standards. FDEC, Florida.

[48] Ministry of Environment. 1985. Law concerning special measures for the preservation of lake water quality. Law 61, Government of Japan, p. 15.

[49] Carlson RE. 1977. A trophic state index for lakes. Limnol. Oceanogr., 22: 361–369.

Chapter 14

Integrated Approaches in Water Quality Monitoring for River Health Assessment: Scenario of Malaysian River

Salmiati, Nor Zaiha Arman and Mohd Razman Salim

Additional information is available at the end of the chapter

http://dx.doi.org/10.5772/65703

Abstract

Current practice of determining river water quality in Malaysia is based mainly on physicochemical components. Perhaps, owing to the lack of information on habitat requirements and ecological diversity of aquatic macroinvertebrates and on unearthly taxonomic key of benthic macroinvertebrates in this region makes it less popular than conventional methods. The study took place in three rivers in the state of Johor, Southern Peninsula of Malaysia, which exhibited different degrees of disturbances and physical properties, namely Sungai Ayer Hitam Besar, Sg Berasau, and Sg Mengkibol. Benthic macroinvertebrates were sampled using rectangular dipnet with frame dimension 0.5 m × 0.3 m. Although physicochemical elements such as water temperature, pH, and dissolved oxygen (DO) were measured using a YSI Professional Plus handheld multiparameter instrument, other parameters such as biochemical oxygen demand (BOD_5), chemical oxygen demand (COD), total suspended solid (TSS), and ammoniacal nitrogen (NH_3N) were tested using the procedure of APHA Standard Method. The study found that the status of water quality varies among the three rivers. A multivariate analysis, the canonical correspondence analysis (CCA), was applied to elucidate the relationships between biological assemblages of species and their environment using PAST (version 2) software. The present findings reveal that human-induced activities are the ultimate causes of the alteration in macroinvertebrate biodiversity.

Keywords: benthic macroinvertebrate, ecosystem health, anthropogenic impact, water quality monitoring, biological assessment

1. Introduction

Water is a natural resource that is vital to all life-forms. Although nearly 70% of the world is covered by water, only 2.5% of the total is freshwater. The rest is ocean-based saline water. However, only 1% of the freshwater is easily accessible, with much of it trapped in glaciers and snowfields. In tandem with the growing global population and improvement of living standards, the increasing demand for freshwater has been said to overshadow the concerns of the warming effect of climate change [1].

Since time immemorial, rivers have played a major role in the development of human society, serving as transport routes and as a vital supply of water for domestic and agricultural use, while yielding an important source of protein for human consumption. Hence, it is not surprising that many major towns and cities are situated on the banks of rivers. For example, early urban settlements such as Uruk, Eridu, and Ur, established at the dawn of human civilization about 6000 years ago (4000 BC) in Mesopotamia and Babylon, were built in the fertile valley irrigated by the Tigris and Euphrates rivers [2].

2. Scenario of river in Malaysia

Rivers have similarly played an important role in the growth of towns and cities in Malaysia, with early settlements springing up along river banks and estuaries [3]. Many major cities and towns in such locations include Kuala Lumpur, Kuala Terengganu, Alor Setar, Kuantan, Kota Bharu, Kuching, and Melaka City [4, 5]. The discovery of tin deposits in the flood plains and river valleys also encouraged settlements to mushroom in these areas, leading to a booming tin-mining industry in the 1800s till 1980s, which made the country the largest producer of tin in the world.

Malaysia has grown rapidly over the last three decades, transforming from a rural economy based on agriculture and tin mining to an export-based, manufacturing economy. In the eighteenth century and the first half of the nineteenth century, large areas of land were cleared for coffee and sugarcane cultivation. This was followed by large-scale land clearing for rubber plantations, making Malaya the world's largest producer of natural rubber. In recent years, much of the rubber growing lands has been converted to oil palm cultivation, while further new areas have been cleared for this crop. Unfortunately, rapid changes of land use, especially of forested land and food crops to plantations as well as urban development, have triggered river erosion, surface runoff, and sedimentation of rivers, resulting eventually in overstressed river systems. River basins are frequently facing problems arising from flooding. Many rivers are gradually losing their ability to supply fresh water, and as a result, these rivers are now mainly used for transportation [6].

In Malaysia, the sources of raw fresh water are rivers, storage dams, and groundwater. Rivers supply 90% of the nation's water supply, providing water for various uses such as domestic, agricultural and industrial processes, power generation, besides serving as waterways for transport and communication. Aquatic harvests from rivers are also an important source of food. However, as the country develops, water pollution is becoming more serious, affecting the function of the river system as a source of raw water supply. Although raw water supply is not yet depleted, clean water that can be safely consumed by humans is becoming hard to come by.

The major causes of water pollution in Malaysia include effluent from wastewater treatment plants, discharge from agro-based industries and livestock farming, land clearing activities, and domestic sewage [7]. Rivers in both urban and rural areas are experiencing the same problems. Although environmental issues in Malaysia raise serious concerns, the measures taken to address the problem thus far have been fragmented and inadequate. An integrated and holistic approach that is required is now gaining recognition, and this is reflected in the government's latest policies.

3. River-related issues and river management in Malaysia

As a responsible authority for ensuring the sustainability of integrated river basin and water resources management, the Department of Irrigation and Drainage (DID) under the Ministry of Natural Resources and Environment (NRE) upheld the Integrated River Basin Management (IRBM) concept more than 10 years ago. IRBM, a subset of Integrated Water Resources Management (IWRM), is an effective method or approach to achieve the objectives of the IWRM-based river basin. In other words, IRBM is the management of river basin as an entity, not as a series of isolated individual rivers. It is geared towards integrating and coordinating policies, programs, and practices in addressing water and water-related issues. It also requires the improvement of professional and financial practices as well as legislative, managerial, and political capacity on water-related issues. The One State One River Program (1N1S), launched by DID in 2005, was an extension of the Love Our Rivers Campaign with the slogan "Sungaiku Hidupku" ("My river, My life"). This program is one of the pilot projects for the implementation of the IRBM concept. In this program, DID and the state governments selected 13 rivers, one river for each state. Among the main criteria of the river selection was that the polluted rivers should be running through major cities in the country. The main goals of the 1N1S were to achieve and maintain the status of clean and vibrant river within Class IIB of water quality by 2015 [8]. Under the RMK-9, a sum of RM57.5 million was allocated to each state, while in the RMK-10, an allocation of RM26 million was provided for a period of 2 years (2011–2012) for 13 selected rivers.

The results showed that the program had achieved some measure of success, especially in terms of improved water quality from Class V to Class III in some rivers, namely Sungai Petani, Kedah; Sungai Galing, Pahang and Sungai Pinang, Pulau Pinang. In addition, Sungai Kinta in Perak achieved an improvement in water quality index (WQI) from Class III to Class IIB. However, the water quality for Sungai Hiliran, Terengganu and Sungai Penchala, Kuala Lumpur remain unchanged [8]. Those river restoration programs have not only shown positive effects and significantly improved the quality of water, but also enhanced amenities and riverside landscape. Nevertheless, the positive effects of the measures on riverine biota are rarely observed or documented. Sungai Melaka, for example, has shown tremendous changes in water quality, from heavily polluted to slightly polluted after undergoing several rehabilitation efforts. However, in terms of faunal diversity and aquatic life, only tolerant and hardy species such as the tilapia fish have been found to inhabit the river. A similar situation also occurs in two rehabilitated rivers in Johor, namely Sungai Sengkuang and Sungai Sebulong, where only hardy, non-economic fish species have been observed.

However, taking into consideration the physicochemical aspects alone are not sufficient to indicate a healthy ecosystem as a whole. In fact, this does not guarantee health of aquatic life because it does not directly reflect the biological responses to pollution. Although physicochemical evaluation might be appropriate to particular circumstances at the time of sampling, it does not provide an insight into the effects of pollution on habitat and aquatic life. Aquatic communities respond to ecosystem changes in various ways. The distribution and abundance of certain species and changes in their behavioral, physiological, and morphological of individual organisms indicate whether that habitat has been adversely altered. High biodiversity of aquatic species and the presence of sensitive species are good signs of a healthy stream. Nature of the river as a collection point for water flowing from every corner reflects the health of the surrounding area. Therefore, any changes or modification on riparian vegetation and surrounding landscape may subsequently alter the composition and functional structure of aquatic life inhabiting it. Healthy water body shows ecological integrity, which represents the natural or undisturbed area. Ecological integrity is a combination of three components, namely chemical, physical, and biological integrities. When one or more of these components are degraded, the health of the water body is affected, and in most cases, aquatic life living in it will reflect the degradation. According to Gordon et al. [9], stream health measurement takes into consideration the water quality, habitat availability and suitability, energy sources, hydrology, and the biota themselves.

In order to achieve a comprehensive evaluation of healthy water bodies, biological assessment tool should be carried out simultaneously with the standard physicochemical method. Biological assessment, the primary tool to evaluate the biological condition of a water body, comprises surveys and other direct measurements through biological communities such as plankton, periphyton, microphytobenthos, macrozoobenthos, aquatic macrophytes, and fish. Among all, benthic macroinvertebrates are the most favored in freshwater monitoring and are widely used to evaluate the water body health and condition [10, 11]. The advantages of using biological indicators, particularly macroinvertebrates, are biological communities that reflect the overall ecological quality and provide a broad measurement of fluctuating environmental conditions. In addition, the result of biological monitoring is reliable and relatively inexpensive compared to toxicity testing [12]. Liebmann (1962) quoted that the history of biological monitoring methods for assessing water quality began more than a century ago by Kolenati (1848) and Cohn (1853) both quoted by [13]. However, such studies in Malaysia are still very limited and started relatively late with the earliest documented was in the early 90s [14, 15]. After year 2000, interest on this topic is gaining attention and grows, and example of studies can be seen in [16–20]. In the year 2009, DID in collaboration with Universiti Sains Malaysia produced a Guideline for Using Macroinvertebrates for Estimation of Streams Water Quality. The guideline provides simple, inexpensive, and easy approach to estimate water quality through the identification of freshwater macroinvertebrates. This government's effort is an initial step to the development of such studies in Malaysia and proving biological methods in the study of water quality began to be accepted.

4. Description of study area

This study focused on the description of the existing ecological environment of three rivers with different environmental gradient ecosystem, *viz.* Sg Ayer Hitam Besar (forest reserve), Sg Berasau (logged area), and Sungai Mengkibol (urban and rehabilitated rivers). Three main processes explored in this study were consisted of physical characteristics (general characteristics that are important in influencing the river's aquatic ecology such as channel forms, instream habitats, substrates, riverbank vegetation, and structure; additional habitat attributes such as anthropogenic alterations to the river were briefly described), biological characteristics (focusing on the composition and abundance of macroinvertebrates species) and chemical characteristics (documentation of existing conditions related to commonly observed water quality parameters). The study also investigated the correlation between the physicochemical attributes and variations in the macroinvertebrates assemblages.

4.1. Sg Ayer Hitam Besar

Sg Pontian Besar drains a total area of 362.047 km² to the Straits of Malacca. The main tributaries are Sg Ayer Hitam Besar and Sg Rambutan. The Sg Ayer Hitam Besar subcatchment is situated between Ulu Pontian and Kampung Seri Gunung Pulai settlement area. This river has a length of 11.2 km with a width of 35 m, before joining the main river and end up at the Straits of Malacca. The study was conducted at the Pulai Waterfall in Gunung Pulai Forest Reserve, in the southwestern part of Johor. Located within an 8-ha protected forest reserve, this former water catchment area for Singapore serves an important function as an area for habitat and biodiversity protection, recreation, tourism, research, and education. Gunung Pulai is a hill dipterocarp forest type on granite-based soil; its peak is about 700 m above sea level. The Gunung Pulai Forest Reserve is divided into 27 compartments to facilitate administration and management. However, only two compartments, namely Compartment 9 for recreational pursuits and Compartment 7 for educational purposes, are open to the public.

4.2. Sg Berasau

Sg Berasau is located in the Kota Tinggi District, about 42 km northeast of Johor Bahru. With an area of 54 km², this river basin is a sub-basin of Sungai Ulu Sedili Besar river basin. The study area was a first-order river situated about 8 km from the Kota Tinggi-Jemaluang main road and within the Ulu Sedili Forest Reserve. The sampling areas could be reached only by 4WD through forest plantations managed by Aramijaya Sdn. Bhd., which granted the study team access to the area for this study. Sg Berasau is surrounded by a permanent forest reserve.

However, there is active logging in the area planted with *Acacia mangium*, a tree species of the pea family, Fabaceae. *Acacia mangium* is suitable as raw material for sawn timber, and woodchips in the pulp and paper industry, and reconstituted wood for the furniture industry. In the Asia–Pacific region, Japan is among the larger importers of wood chips, while the pulp and paper industry finds markets in Taiwan and South Korea. Based on the macro-EIA for forest management units (FMU) in Johor, Sg Berasau supports a variety of fish species such as Terbul, Sebarau, Baung, Seluang, and Tapah.

4.3. Sg Mengkibol

Sg Mengkibol is a second-order river located within the Endau watershed. This river receives flows from Sg Melantai before joining Sg Semberong. This river basin is approximately 185 km² width and 20 km long. The study area is located in the middle section of Sg Mengkibol, starting from the Sg Mengkibol Riverine Park until the wet market. From December 2006 until January 2007, Malaysia experienced large-scale flood events that affected most of the state of Johor. This phenomenon was caused by extremely high rainfall attributed to Typhoon Utor that made landfall in the Philippines and Vietnam. A series of massive floods hit the states of Malacca, Pahang, and Negeri Sembilan, with Johor as the worst hit state. Among the major towns affected by the floods were Batu Pahat, Johor Bahru, Kluang, Kota Tinggi, Mersing, Muar, Pontian, and Segamat. Consequently, the Department of Irrigation and Drainage (JPS) allocated RM2 million to deepen Sg Mengkibol at its banks for flood mitigation in low-lying residential areas [21].

5. Sampling procedure and data collection

Most of that data used for analyses in the study were primary data obtained from sampling and laboratory analyses. Data of benthic macroinvertebrate assemblage and water quality assessment were obtained from *in situ* analyses and further analyses in the laboratory. River habitat and morphology data were measured on-site in field surveys. On the other hand, secondary data such as land use, rainfall data, and stream catchment maps were obtained from several agencies such as the Department of Environment, Department of Drainage and Irrigation, Department of Survey and Mapping and local authorities.

5.1. Benthic macroinvertebrate

The multihabitat approach of USEPA's Rapid Bioassessment Protocol (RBP) was adopted for this study as it was suitable for sampling a wide variety of stream types [22]. In this study, a rectangular dipnet with 500-μm mesh attached to a 0.5 m × 0.3 m frame and a long pole were used for this purpose. For multiple habitats, the habitat types were sampled in proportion to their relative surface area within the sampling reach. A total of 20 sample units were collected from all major habitat types by kicking the substrates or jabbing with a dipnet within a sampling station to obtain a composite of 60 sample units in total. The samples were washed and any detritus present was removed on-site as it would be impractical to wash large samples in the laboratory. Following this, benthic materials were sieved and rinsed before preservation in 70% ethanol. The sample containers were labeled to show all the essential information, including date and sampling location. Preprinted labels were preferably used using marker pens as ethanol would remove writing. Moreover, labeling on container lids was avoided in case they were interchanged. In laboratory, benthic macroinvertebrates were rinsed thoroughly in 500 μm-mesh sieves to remove preservative and sediment, while remaining debris was visually inspected and discarded. During the separation process, the samples were soaked for about 15 min in tap water to hydrate the preserved organisms and prevent them from floating on the water surface during sorting. The samples were then spread over an enamel tray and sorted out into major taxa. All organisms were identified to the lowest practical level using a dissecting microscope and taxonomic Key from Yule and Yong [23].

5.2. Water quality

Water quality assessment data were obtained by two methods, namely *in situ* and laboratory analyses. *In situ* measurements were made on temperature, pH, conductivity, and dissolved oxygen (DO) by using a YSI Professional Plus handheld multiparameter instrument. Meanwhile, other parameters such as biochemical oxygen demand (BOD_5), total suspended solid (TSS), ammoniacal nitrogen (NH_3N), and chemical oxygen demand (COD) were measured in the laboratory based on the Standard Methods for the Examination of Water and Wastewater [24]. Both *in situ* readings and water samples were collected from the same location. Water samples were collected in 1-L polyethylene bottles and chilled in a cold box filled with ice cubes (4°C) to minimize the metabolism of organisms contained in the water. The water samples were labeled in a manner similar to that used for the benthic samples.

The water quality index (WQI) was calculated to indicate the level of pollution and the corresponding suitability for use according to the National Water Quality Standards for Malaysia (NWQS). Water quality class was determined based on the water quality index (WQI), ascertained by the six parameters, *viz.* pH, DO, BOD_5, COD, TSS, and NH_3N, according to the DOE formula (1):

$$WQI = (0.22*SIDO) + (0.19*SIBOD) + (0.16*SICOD) + (0.15*SIAN) + (0.16*SISS) + (0.12*SIpH) \tag{1}$$

where SI is the subindex of the respective water quality parameters which is used to calculate the WQI (**Table 1**). The WQI classification based on water use is shown in **Table 2**.

Subindex for DO (in % saturation)	
SIDO = 0	for x ≤ 8
SIDO = 100	for x ≥ 92
SIDO = −0.395 + 0.030x² − 0.00020x³	for 8 < x < 92
Subindex for BOD	
SIBOD = 100.4 − 4.23x	for x ≤ 5

Subindex for DO (in % saturation)	
$SIBOD = 108 \times \exp(-0.055x) - 0.1x$	for $x > 5$
Subindex for COD	
$SICOD = -1.33x + 99.1$	for $x \leq 20$
$SICOD = 103 \times \exp(-0.0157x) - 0.04x$	for $x > 20$
Subindex for NH3-N	
$SIAN = 100.5 - 105x$	for $x \leq 0.3$
$SIAN = 94 \times \exp(-0.573x) - 5 \times |x - 2|$	for $0.3 < x < 4$
$SIAN = 0$	for $x \geq 4$
Subindex for SS	
$SISS = 97.5 \times \exp(-0.00676x) + 0.05x$	for $x \leq 100$
$SISS = 71 \times \exp(-0.0061x) - 0.015x$	for $100 < x < 1000$
$SISS = 0$	for $x \geq 1000$
Subindex for pH	
$SIpH = 17.2 - 17.2x + 5.02 \times 2$	for $x < 5.5$
$SIpH = -242 + 95.5x - 6.67 \times 2$	for $5.5 \leq x < 7$
$SIpH = -181 + 82.4x - 6.05 \times 2$	for $7 \leq x < 8.75$
$SIpH = 536 - 77.0x + 2.76 \times 2$	for $x \geq 8.75$

Table 1. Best fit equations for the estimation of various subindex values.

Class	Uses
Class I	Conservation of natural environment Water supply I—practically no treatment necessary Fishery I—very sensitive aquatic species
Class IIA	Water supply II—conventional treatment required Fishery II—sensitive aquatic species
Class IIB	Recreational use with body contact
Class III	Water supply III—extensive treatment required Fishery III—common, of economic value and tolerant species; livestock drinking
Class IV	Irrigation
Class V	None of the above

Table 2. Water classes and uses.

5.3. Characterization of river habitat

A visual-based habitat assessment based on the USEPA habitat assessment survey was carried out simultaneously with the biological sampling. Several features in habitat assessment include a general description of the site, a physical characterization and water quality assessment, and also a visual assessment of instream and riparian habitat quality. Physical characterization comprised documentation of general land use, description of the stream origin and type, and a summary of the riparian vegetation features that included measurements of instream parameters such as width, depth, flow, and substrates. The observed channel dimensions were carried out in the survey stretch, located either in the presence of riffle or at a suitable shallow section of the river. Channel dimension measurements were taken according to the river habitat survey (RHS) method.

5.4. Streamflow gauging

Streamflow gauging was conducted to measure the flow rate of the study area. The equipments used in this study were flow meter, measuring tape, staff level, hammer, and ropes/cables. The Cole Parmer Model BS 11000 flow meter used in this study was equipped with a propeller to allow it to rotate according to the velocity of the water. The mean section method was used to measure the river discharge using the flow meter and other tools, whereby cross-sectional area of the river was divided into several subsections. Stream velocity was measured at depths of 0.6d, 0.2d, and 0.8d depending on the need, where d represents the variable of depth from the water surface. This method was used to obtain the average velocity of the represented river. The general hydraulic formula for river discharge is as follows (2)

$$Q = AV \tag{2}$$

where Q is discharge (volume/unit time-e.g. m^3/s, also called cumecs), A is the cross-sectional area of the stream (e.g. m^2), and V is the average velocity (e.g. m/s).

6. Results and discussion

6.1. Benthic macroinvertebrate compositions

Overall, a total of 1081 fauna were recorded from Sg Ayer Hitam Besar. On the other hand, 610 individuals were caught from Sg Berasau, while sampling from Sg Mengkibol documented 1008 individuals. This brings the total number of macroinvertebrates assemblage based from all sampling events of 2699 individuals. Common netspinners larvae from caddisfly family dominate the research finding in Sg Ayer Hitam Besar, contributing approximately one-third of the total samples (30.52%). Some intolerant taxa represented by Plecopteran families were seen widely distributed in the study area. They are Perlidae (242 individual) and Chloroperlidae (194 individual). Interestingly, Decapods (Palaemonidae) has been found widely distributed with relatively high abundance and dominated in every sampling session. On the other hand, mollusks were found in very low composition restricted to Physidae and Pleuroceridae (**Table 3**). Trichoptera represents the percentage of the most abundant and adaptive species with the most dominant family of Hydropsychidae. This higher number is believed to be associated with the presence of algal biomass [25]. This insect tends to live in sheath made from organic debris and mineral fragments and makes the surface of the substrate as their habitat. This insect larvae are also often attached to rocks, facing the flow and feed on the particles trapped in their nets [26]. Several groups of aquatic insects favored rocky substrate as it offers habitat for protection and oviposition [27]. In this present study, sufficient numbers of oviposition sites were observed, including plenty of rocky substrates and riverbank vegetations, which could explain the high abundance of caddisfly larvae in this area.

Order	Family	Abundance	Percentage (%)
Decapoda	Palaemonidae	153	14.15
	Potamidae	5	0.46
Ephemeroptera	Heptageniidae	57	5.27
	Ephemeridae	3	0.28
Plecoptera	Perlidae	242	22.39
	Capniidae	3	0.28
	Chloroperlidae	194	17.95
Trichoptera	Hydropsychidae	305	28.21
	Limnephilidae	12	1.11
	Polycentropodidae	6	0.55
	Leptoceridae	7	0.65
Coleoptera	Elmidae	60	5.55
	Pyralidae	6	0.56
Odonata	Calopterygidae	1	0.09
	Lestidae	2	0.19
	Gomphidae	3	0.28
	Libellulidae	16	1.48
Gastropoda	Pleuroceridae	1	0.09
	Physidae	4	0.37
Hemiptera	Veliidae	1	0.09
	Total	**1,081**	**100**

Table 3. Benthic macroinvertebrate compositions in Sg Ayer Hitam Besar.

Decapoda exhibited the highest distribution with an abundance of 343 individual in Sg Berasau. Both families, Palaemonidae and Potamidae, contributed more than half (56.23%) from the total amount. Caridean prawn (Palaemonidae) were found in all sampling events with large numbers compared to others (291 individual). The second group which had the highest distribution was Odonata, with an abundance of 153 individual. Gomphidae, which belongs to Anisoptera suborders, were the second largest taxa found, consisting of 91 individuals. The least dominant families in Sg Berasau were Capniidae, Sialidae, and Pyralidae, contributing 0.16% each from total percentage (**Table 4**). Freshwater prawns of the genus *Macrobrachium* are free-living decapod crustaceans, present in almost all permanent water bodies. They inhabit a wide variety of habitat even in extreme condition, where waters can reach pH 3.3, and stagnant pool with daytime temperature may reach 35°C [28]. Their feeding habits are variable, with some are scavengers or being detritivorous [29]. As such, they are very important in recycling organic matter in the environment. The inclusion of organic matter into water bodies due to logging activities in Sg Berasau is beneficial to shrimp, as can be seen from the abundance of these organisms. However, they are prone to human disturbance and development and become extinct. This happened in Sg Gombak, whereby the populations of *Atyopsis* species are now very rare due to rapid development, resulting in water pollution.

Order	Family	Abundance	Percentage (%)
Decapoda	Palaemonidae	291	47.70
	Potamidae	52	8.52
Ephemeroptera	Heptageniidae	30	4.92
	Ephemerellidae	18	2.95
	Baetidae	13	2.13
	Leptophlebiidae	2	0.33
	Potamanthidae	3	0.49
Plecoptera	Perlidae	11	1.80
	Capniidae	1	0.16
	Nemouridae	8	1.31
	Leuctridae	2	0.33
	Perlodidae	2	0.33
	Chloroperlidae	4	0.66
Odonata	Calopterygidae	17	2.79
	Lestidae	3	0.49
	Gomphidae	91	14.92
	Libellulidae	42	6.89
Gastropoda	Pleuroceridae	8	1.31
Hemiptera	Belostomatidae	2	0.33
	Nepidae	2	0.33
Megaloptera	Sialidae	1	0.16
Coleoptera	Hydrophilidae	6	0.98
	Pyralidae	1	0.16
	Total	610	100

Table 4. Benthic macroinvertebrate compositions in Sg Berasau.

Sampling of macrobenthic assemblages from Sg Mengkibol consists of moderately intolerant to very tolerant families. Odonates are on top of the list with highest abundance (448 individuals). Chironomidae or blood worm dominated the overall findings with cumulated percentage 22.32%. Gastropods, physidae, are in the second place with slight difference of cumulated percentage (21.92%). Odonates represented by Lestidae and Libellulidae also donated a relatively high number of 260 individuals. Interestingly, sensitive taxa were found in this study area, although the percentage is very low. They are mayflies and stoneflies (**Table 5**). Fly larvae can be found in various aquatic habitat and survived in most conditions. According to Yule [30], Chironomidae is probably the most diverse and abundant group of all stream macroinvertebrates. Chironomus, for example, were widely distributed in polluted areas [20, 31]. Hemoglobin pigment helps *Chironomus* spp. to adapt to unfavorable condition, since hemoglobin helps to sustain aerobic metabolism under low oxygen conditions [32]. Most fly larvae eat dead or dying plant and animal materials.

Order	Family	Abundance	Percentage (%)
Decapoda	Palaemonidae	20	1.98
Ephemeroptera	Ephemerellidae	1	0.10
	Baetidae	1	0.10
Plecoptera	Leuctridae	15	1.49
	Perlodidae	1	0.10
Coleoptera	Psephenidae	2	0.20
Odonata	Calopterygidae	60	5.95
	Lestidae	149	14.78
	Gomphidae	77	7.64
	Libellulidae	111	11.01
	Aeshnidae	41	4.07
	Coenagrionidae	10	0.99
Gastropoda	Pleuroceridae	3	0.30
	Physidae	221	21.92
	Viviparidae	4	0.40
Hemiptera	Naucoridae	35	3.47
	Nepidae	1	0.10
Hirudinea	Hirundinidae	28	2.78
Diptera	Chironomidae	225	22.32
	Syrphidae	3	0.30
	Total	**1,008**	**100**

Table 5. Benthic macroinvertebrate compositions in Sg Mengkibol.

6.2. River habitat survey

The Sg Ayer Hitam riverbed comprises more than 70% of natural structures such as cobble (riffles), large rocks, fallen trees, logs, and branches. These optimal conditions allow colonization, refugia, feeding/ spawning sites for aquatic faunal. Both Sg Ayer Hitam Besar and Sg Berasau consist of all four velocity/depth regime present in their study reach. The occurrence of slow-deep, slow-shallow, fast-deep, and fast-shallow velocity patterns reflects of habitat diversity and ability of stream to provide and maintain balance aquatic habitat. No channel alteration or dredging works present at studied reach in both rivers. Sg Ayer Hitam Besar showed and optimal condition of vegetative protection, as it covers more than 90% of streambank surface with native vegetation including trees, understory shrubs, or non-woody macrophytes. An optimal condition of vegetative zone serves as a buffer to pollution and nutrient input to the stream runoff, other than erosion control. Meanwhile, around 50–70% of the streambank surface is covered by riparian vegetation in Sg Berasau. Logging activities leave an obvious disruption as cropped vegetation/ bare soil potentially prone to high potential of streambank erosion during heavy downpour (30–60%).

Meanwhile, Sg Mengkibol exhibits an unsatisfactory habitat quality. Historically, Sg Mengkibol was hit by massive flood event in late 2006. In relation to deal with the incident over and over again, upgrading the river system for flood mitigation project has been carried out. Among the works are dredging and sediment disposals, as well as strengthening the river channel. As a result, variety of natural structure less than desirable due to frequent disturbed of epifaunal substrate. Compared to Sg Ayer Hitam Besar and Sg Berasau, shallow pools are more prevalent than deep pools at these rivers. Percentage of deposition of sediments in the Sg Mengkibol is approximately 50–80%, composing of gravel, sand, or fine sediment on the old and new bar. Increasing level of sediment deposition is an indication of instability and changing environment, thus unsuitable for many organisms. Deepening and dredging works as a part of river rehabilitation and restoration process have changed the shape of the stream channel drastically. More than 80% of stream reach has been straightened with the construction of anti-erosion measure in both sides of the banks. Straightened channel decreases the stream length 1–2 times shorter than its natural state. Channel sinuosity provides diverse habitat and fauna, as well as being able to handle surges as a result of storm. The construction of slope stabilization is carried out to reduce the amount of erosion that is likely to occur. Nevertheless, those artificial structures prevent plants from growing on streambanks. Therefore, the natural habitat for aquatic organism is limited. Riparian zone serves as a buffer to prevent the entrance of nutrients and pollutants directly into rivers. However, for urban river, riparian vegetative zone width is usually <6 m, due to extensive use of impervious surfaces. Therefore, it increases the volume of runoff and decreases groundwater recharge.

6.3. Water quality index (WQI)

A water quality index representing a gradation number describes the overall water quality in particular location and time based on several water quality parameters. The use of this index is not intended specifically for human health or aquatic life regulation, but provides simple guidance on water quality based on some important parameters. In Malaysia, the assessment and classification of water quality status are based on the water quality index (WQI) and the National Water Quality Standards (NWQS), which eventually grouped into certain classes. Index developed for Malaysia, the WQI is ascertained by six parameters, *viz.* pH, DO, BOD_5, COD, TSS and NH_3N. As summarized in **Figure 1**, overall mean WQI for Sg Ayer Hitam Besar was at Class I, indicating as an excellent quality. The mean values for each sampling event were ranged 90.67–97.00. Based on this index, Class I is defined as naturally very clean and preserved river. Its water resources are suitable as drinking water with minimal treatment. In terms of ecology, habitats are able to accommodate very sensitive aquatic species. Different situation is observed for Sg Berasau, although surrounded by a natural environment, land use activities such as deforestation have greatly affected the ecosystem health. Its effect can be seen through water quality status, which categorized this river into Class II (Clean). Based on general rating scale of WQI, Class II of water resources still can be used as a source of drinking after conventional treatment method. It is also suitable for recreational use with body contact. On average, the mean values of WQI in Sg Berasau are between 75.00 and 89.33 . Result of this study coincides with the finding from [17] which proves that logging activities which comply with prescribed standards still have an adverse impact on the riverine ecosystem even in a small proportion. On the other hand, Sg Mengkibol exhibits a moderately clean river status. Eight out of ten sampling events showed water quality for this river was in Class III (slightly polluted). This type of river status requires an extensive treatment as a drinking water supply. This river also accommodates certain fish species that are more tolerant and low in economic value such as catfish (*Clarias batrachus*) and tilapia (*Tilapia mossambica*) [33].

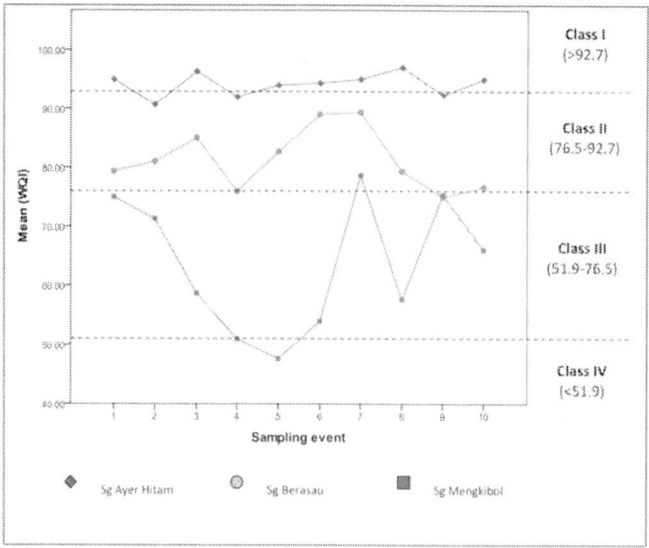

Figure 1. WQI values for all sampling sites.

6.4. Influence of hydrological, physicochemical, and habitat characteristics on the biological assemblage

In this section, a multivariate method, the canonical correspondence analysis (CCA), was applied to elucidate the relationships between biological assemblages of species and their environment using PAST (version 2) software. The CCA ordination biplot illustrated the relationship between several hydrological, physicochemical parameters, and distribution of the aquatic macroinvertebrates. The first two axes derived from CCA model accounted for 88.57% of the macroinvertebrates–environmental variations. CCA demonstrated that axis 1 was strongly correlated with DO, habitat quality (epifaunal substrate and vegetative protection), whereas COD, BOD_5, NH_3N, temperature, and velocity were negatively correlated with it (**Table 6**). Several taxa associated with three pollution-sensitive orders, EPT, showing good adaptation features and presenting highest score on the first axis. They consist of order Ephemeroptera (Heptageniidae, Ephemeridae), Plecoptera (Capniidae, Perlidae, Chloroperlidae), and Trichoptera (Hydropsychidae, Polycentropodidae, Leptoceridae, Limnephilidae). Low habitat quality, deterioration of DO and elevated concentration of nutrient and organic pollutant, suspended particulate, and temperature were positioned on the negative side of the first axis and were associated with moderate to tolerant taxa such as Odonata (Gomphidae, Libellulidae, Calopterygidae, Lestidae, Coenagrionidae, Aeshnidae), Hemiptera (Nepidae, Naucoridae), Coleoptera (Psephenidae), Gastropoda (Physidae, Planorbidae, Viviparidae), Diptera (Chironomidae, Syrphidae, Simuliidae, Tipulidae, Culicidae) and Hirundinidae (Hirundinidae). The second axis was positively related to TSS and pH. In particular, these species taxa (Pleuroceridae, Perlodidae, and Palaemonidae) showed moderate preference for velocity, water temperature, rainfall precipitation, and less vegetative cover. However, this group showed dependence on high suspended particulate and alkalinity.

Variable	Axis 1	Axis 2
Temperature	−0.891	−0.534
pH	0.281	0.953
DO	0.995	0.035
BOD	−0.977	−0.280
COD	−0.978	−0.260
TSS	−0.102	0.952
NH_3N	−0.849	−0.410
Rainfall	0.679	−0.424
Velocity	−0.682	0.639
Epifaunal	0.968	0.057
Vegetative	0.905	−0.307
Eigenvalue	0.842	0.447
Percentage of variance explained	57.85	30.72

Table 6. Summary statistic for the canonical correspondence analysis (CCA) relating aquatic macroinvertebrate–environmental variables (11 variables).

7. Conclusion

Rivers in Tropical Asia is closely related to the effect of seasonal flow imposed by unpredictable monsoon and seasonal rainfall [34, 35]. Several studies have analyzed the changes of benthic community dictated by seasonal rainfall [36–38], including those in Peninsular Malaysia [39–41]. In general, rainfall pattern in Malaysia is much influenced by wind flow pattern during the seasonal period. To some extent, it is also influenced by local topography. The present study indicated that seasonal rainfall has not significantly affected the distribution of benthic communities in the four studied rivers. This study is in line with the findings from Refs. [33, 42]. The anthropogenic impacts were more significant than the seasonal rainfall. However, the EPT populations were seen correlated to seasonal variation, as been reported earlier by Suhaila et al. [40] in Gunung Jerai Forest Reserve. According to Ref. [43], these types of species react quickly to the changes in environment.

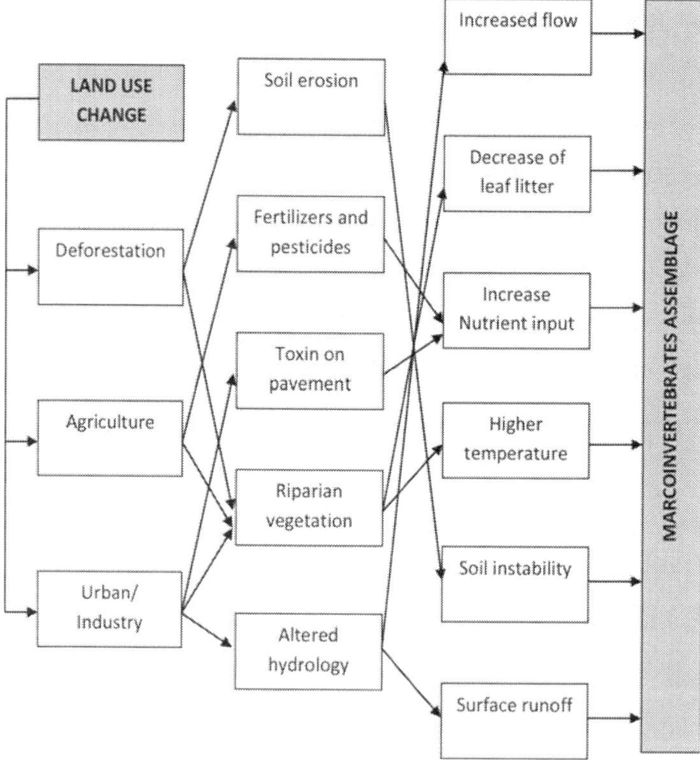

Figure 2. Illustrative schematic of the potential interactions between the threats, impact, and the response to the stream macroinvertebrate assemblage.

The present findings reveal that human-induced changes in natural habitat are the ultimate causes explaining the alteration in macroinvertebrates biodiversity. There is no doubt that anthropogenic disturbance impacted the structure of macroinvertebrate communities, either in the tropics or in the temperates [44–47]. Land use change is an integration of various human activities that negatively impact the river ecosystem. Among which, flow variability and sediments transport into the river by impervious surface and drainage in urban areas, stream channelization, agriculture, and deforestation. All of these threats will be manifested in changes in flows, benthic habitat conditions, and riffle-pool integrity. **Figure 2** presents an illustrative example of how macroinvertebrate communities can respond to land use change through a chain of indirect effects that lead to changes to the macroinvertebrate assemblage in both taxa richness and relative abundance.

Author details

Salmiati[1,2*], Nor Zaiha Arman[1] and Mohd Razman Salim[1,2]

*Address all correspondence to: salmiati@utm.my

1 Department of Environmental Engineering, Faculty of Civil Engineering, University of Technology, Johor Bahru, Johor, Malaysia

2 Center for Environmental Sustainability and Water Security (IPASA), RISE, University of Technology, Johor Bahru, Johor, Malaysia

References

[1] Nienhuis PH. Water and values: ecological research as the basis for water management and nature management. Hydrobiologia. 2006 Jul;565(1):261–75. doi:10.1007/s10750-005-1918-2.

[2] Macionis JJ, Parrillo VN. Cities and Urban Life. Upper Saddle River, NJ: Pearson Education; 2004.

[3] Chan Ngai Weng, Abdullah AL, Ibrahim AL and Ghazali S. River pollution and restoration towards sustainable water resources management in Malaysia. National Seminar on Society, Space and Environment in a Globalised World: Prospects & Challenges, 29–30 May 2003. The City Bayview Hotel, Penang, p. 208–219.

[4] Rahman HA. An Overview of River Pollution Issues in Malaysia. Retrieved on March. 2007;26:2014.

[5] Andaya BW, Andaya LY. A History of Malaysia. University of Hawaii Press; 2001.

[6]   Mohamad S, Toriman ME, dan Mokhtar Jaafar KA. River and Development: Waterfront City Malaysia. Bangi: UKM; 2015.

[7]   Department of Statistics Malaysia. Compendium of Environment Statistics Malaysia. Kuala Lumpur: Department of Statistics Malaysia; 2013.

[8]   Harian S. Favourable Initiative of River Conservation (2013 September). Retrieved 26 May 2016, from http://www.sinarharian.com.my/article/conservation inisiative-river-favourable-1.198987

[9]   Gordon ND, Finlayson BL, McMahon TA. Stream Hydrology: An Introduction for Ecologists. John Wiley and Sons; 2004 Jun.

[10]   Conti ME. Biological Monitoring: Theory & Applications: Bioindicators and Biomarkers for Environmental Quality and Human Exposure Assessment. WIT Press; 2008.

[11]   Wei M, Zhang N, Zhang Y, Zheng B. Integrated assessment of river health based on water quality, aquatic life and physical habitat. Journal of Environmental Sciences. 2009 Dec;21(8):1017–1027. doi:10.1016/S1001-0742(08)62377-3

[12]   Iliopoulou-Georgudaki J, Kantzaris V, Katharios P, Kaspiris P, Georgiadis T, Montesantou B. An application of different bioindicators for assessing water quality: a case study in the rivers Alfeios and Pineios (Peloponnisos, Greece). Ecological Indicators. 2003 Feb; 2(4):345–360. doi:10.1016/S1470-160X(03)00004-9

[13]   Liebmann H. Manual of Freshwater and Sewage Biology. In Manual of Freshwater and Sewage Biology. R. Oldenbourg; 1962.

[14]   Khan NISA. Assessment of water pollution using diatom community structure and species distribution—a case study in a tropical river basin. Internationale Revue der gesamten Hydrobiologie und Hydrographie. 1990 Jan;75(3):317–338. doi:10.1002/iroh.19900750305

[15]   Arsad A, Abustan I, Rawi CS, Syafalni S. Integrating biological aspects into river water quality research in Malaysia: an opinion. OIDA International Journal of Sustainable Development. 2012 May;4(02):107–122.

[16]   Ghani WM, Rawi CS, Hamid SA, Al-Shami SA. Efficiency of different sampling tools for aquatic macroinvertebrate collections in Malaysian streams. Tropical Life Sciences Research. 2016 Feb;27(1):115.

[17]   Zaiha AN, Ismid MM, Salmiati, Azri MS. Effects of logging activities on ecological water quality indicators in the Berasau River, Johor, Malaysia. Environmental Monitoring and Assessment. 2015 Aug;187(8):1–9. doi:10.1007/s10661-015-4715-z

[18]   Sharifah Aisyah SO, Aweng ER, Razak W, Ahmad Abas K. Preliminary study on benthic macroinvertebrates distribution and assemblages at Lata Meraung waterfall, Pahang, Malaysia. Jurnal Teknologi. 2015 Jan;72(5): 1–4

[19]   Al-Shami SA, Rawi CS, Ahmad AH, Hamid SA, Nor SA. Influence of agricultural, industrial, and anthropogenic stresses on the distribution and diversity of macroin-

vertebrates in Juru River Basin, Penang, Malaysia. Ecotoxicology and Environmental Safety. 2011 Jul;74(5):1195–1202. doi:10.1016/j.ecoenv.2011.02.022

[20] Al-Shami SA, Rawi CS, HassanAhmad A, Nor SA. Distribution of Chironomidae (Insecta: Diptera) in polluted rivers of the Juru River Basin, Penang, Malaysia. Journal of Environmental Sciences. 2010 Nov;22(11):1718–1727. doi:10.1016/S1001-0742(09)60311-19

[21] RM2 juta atasi masalah banjir di Mengkibol (2011 September). Utusan Online. Retrieved May 14, 2015, from http://www.utusan.com.my/utusan/info.asp?y=2011&dt=0923&pub=Utusan_Malaysia&sec=Johor&pg=wj_04.htm.

[22] Barbour MT, Verdonschot PF, Stribling JB. The multihabitat approach of USEPA's rapid bioassessment protocols: benthic macroinvertebrates. Limnetica. 2006;25(3):839–850.

[23] Yong HS, Yule CM. Freshwater Invertebrates of the Malaysian region. Akademi Sains Malaysia; 2004.

[24] Federation WE, American Public Health Association. Standard methods for the examination of water and wastewater. American Public Health Association (APHA): Washington, DC, USA; 2005.

[25] Quinn JM, Cooper AB, Davies-Colley RJ, Rutherford JC, Williamson RB. Land use effects on habitat, water quality, periphyton, and benthic invertebrates in Waikato, New Zealand, hill-country streams. New Zealand Journal of Marine and Freshwater Research. 1997 Dec;31(5):579–597. doi:10.1080/00288330.1997.9516791

[26] Triplehorn CA, Johnson NF. Study of Insect. Thompson Brooks: United States of America; 2005.

[27] Lancaster J, Downes BJ, Arnold A. Oviposition site selectivity of some stream-dwelling caddisflies. Hydrobiologia. 2010 Sep;652(1):165–178. doi:10.1007/s10750-010-0328-2

[28] Wowor D, Cai Y, Ng, PK. Crustacea: Decapoda, Caridea. In Yong HS, Yule CM (eds.). Freshwater Invertebrates of the Malaysian Region. Malaysian Academy of Sciences, 2004; p. 337–357.

[29] Wowor D, Muthu V, Meier R, Balke M, Cai Y, Ng PK. Evolution of life history traits in Asian freshwater prawns of the genus Macrobrachium (Crustacea: Decapoda: Palaemonidae) based on multilocus molecular phylogenetic analysis. Molecular Phylogenetics and Evolution. 2009 Aug;52(2):340–350. doi:10.1016/j.ympev.2009.01.002

[30] Yule CM. Insecta: Diptera. In Yule CM and Yong HS (eds.) Freshwater Invertebrates of the Malaysian Region. Academy of Sciences Malaysia, 2004; p. 610–612.

[31] Marziali L, Armanini DG, Cazzola M, Erba S, Toppi E, Buffagni A, Rossaro B. Responses of Chironomid larvae (Insecta, Diptera) to ecological quality in

Mediterranean river mesohabitats (South Italy). River Research and Applications. 2010 Oct;26(8):1036–1051. doi:10.1002/rra.1303

[32] Weber RE, Vinogradov SN. Nonvertebrate hemoglobins: functions and molecular adaptations. Physiological Reviews. 2001 Apr;81(2):569–628.

[33] Arman NZ, Salmiati, Said MI, Azman S. anthropogenic influences on aquatic life community and water quality status in Mengkibol River, Kluang, Johor, Malaysia. Journal of Applied Sciences in Environmental Sanitation. 2013 Sep;8(3):151–160.

[34] Dudgeon D. The ecology of tropical Asian rivers and streams in relation to biodiversity conservation. Annual Review of Ecology and Systematics. 2000 Jan;Vol. 31: 239–263.

[35] Gopal B. Conserving biodiversity in Asian wetlands: issues and approaches. The Asian Wetlands: Bringing Partnerships into Good Wetland Practices. Penerbit Universiti Sains Malaysia: Pulau Pinang, Malaysia; 2002; p. ID15–ID22.

[36] Silveira MP, Buss DF, Nessimian JL, Baptista DF. Spatial and temporal distribution of benthic macroinvertebrates in a southeastern Brazilian river. Brazilian Journal of Biology. 2006 May;66(2B):623–632. doi:10.1590/S1519-69842006000400006

[37] Joshi PC, Negi RK, Negi T. Seasonal variation in benthic macro-invertebrates and their correlation with the environmental variables in a freshwater stream in Garhwal region (India). Life Science Journal. 2007;4(4):85–89.

[38] Waite IR, Herlihy AT, Larsen DP, Urquhart NS, Klemm DJ. The effects of macroinvertebrate taxonomic resolution in large landscape bioassessments: an example from the Mid-Atlantic Highlands, USA. Freshwater Biology. 2004 Apr;49(4):474–489. doi: 10.1111/j.1365-2427.2004.01197.x

[39] Harun S, Al-Shami SA, Dambul R, Mohamed M, Abdullah MH. Water quality and aquatic insects study at the lower Kinabatangan River catchment, Sabah: in response to weak la niña event. Sains Malaysiana. 2015 Apr;44(4):545–558.

[40] Suhaila AH, Che Salmah MR, Nurul Huda A. Seasonal abundance and diversity of aquatic insects in rivers in Gunung Jerai Forest Reserve, Malaysia. Sains Malaysiana. 2014;43(5):667–674.

[41] Suhaila AH, Salmah MR, Al-Shami SA. Temporal distribution of Ephemeroptera, Plecoptera and Trichoptera (EPT) adults at a tropical forest stream: response to seasonal variations. The Environmentalist. 2012 Mar;32(1):28–34. doi:10.1007/s10669-011-9362-5

[42] Azrina MZ, Yap CK, Ismail AR, Ismail A, Tan SG. Anthropogenic impacts on the distribution and biodiversity of benthic macroinvertebrates and water quality of the Langat River, Peninsular Malaysia. Ecotoxicology and Environmental Safety. 2006 Jul; 64(3):337–347. doi:10.1016/j.ecoenv.2005.04.003

[43] Flecker A, Feifareck B. Disturbance and temporal variability of invertebrate assemblages in two Andean stream. Freshwater Biology. 1994;31:131–142. doi:10.1111/j.1365-2427.1994.tb00847.x

[44] Liow LH, Sodhi NS, Elmqvist T. Bee diversity along a disturbance gradient in tropical lowland forests of south-east Asia. Journal of Applied Ecology. 2001 Feb;38(1):180–192. doi:10.1046/j.1365-2664.2001.00582.x

[45] Hanski I, Koivulehto H, Cameron A, Rahagalala P. Deforestation and apparent extinctions of endemic forest beetles in Madagascar. Biology Letters. 2007 Jun;3(3):344–347. doi:10.1098/rsbl.2007.0043

[46] Sodhi NS, Lee TM, Koh LP, Brook BW. A meta-analysis of the impact of anthropogenic forest disturbance on Southeast Asia's Biotas. Biotropica. 2009 Jan;41(1):103–109. doi: 10.1016/j.tree.2004.09.006

[47] Sandin L, Solimini AG. Freshwater ecosystem structure–function relationships: from theory to application. Freshwater Biology. 2009 Oct;54(10):2017–2024. doi:10.1111/j.1365-2427.2009.02313.x

Assessment of Impacts of Acid Mine Drainage on Surface Water Quality of Tweelopiespruit Micro-Catchment, Limpopo Basin

Bloodless Dzwairo and Munyaradzi Mujuru

Additional information is available at the end of the chapter

http://dx.doi.org/10.5772/65810

Abstract

This research aimed to contribute to current literature for Tweelopiespruit micro-catchment, Limpopo Basin, by trending SO_4^{2-}, Cl^-, Ca^{2+}, Mg^{2+}, Na^+, K^+, Fe, pH and EC, for points F1S1, F2S2, W1S3, F6S7, F8S9, F10S11 and F11S12, as identified by the Department of Water and Sanitation, South Africa, for years 2003 to 2008. Results showed that pollutant concentrations generally increased downstream, which questioned their possible sources since pollution generally attenuates towards downstream. A possible explanation was that groundwater (polluted with the effluent) could be decanting from various places, thus contributing to the increase in concentrations, in places. This could potentially add value to existing efforts, which aim to halt and reverse impacts of acid mine drainage (AMD) in the micro-catchment and possibly in the Goldfields (a highly negatively impacted environment), which incorporates the Cradle of Humankind. Conclusions reached could provide invaluable options for alternative technological or methodological approaches that could be adopted for the treatment of AMD. This is critical to South Africa's water quality trending and sustainability of this ecosystem, especially because the Tweelopiespruit micro-catchment supports humans and a variety of wildlife like giraffe, within the preserve of the Krugersdorp Game Reserve (KGR) and also its outer boundaries.

Keywords: acid mine drainage, Limpopo Basin, Tweelopiespruit micro-catchment, water quality, pollution

1. Introduction

Acid mine drainage (AMD) is a pollutant that arises from exposure of metal sulphide minerals such as the abundantly available pyrite (FeS_2) to oxygen and water during the mining of metals and coals [1, 2]. Pyrite undergoes oxidation in a series of reactions, the first stage (trigger) of which results in production of sulfuric acid and ferrous sulfate as provided in

Eq. (1) [2]. The last stage results in formation of stable and soluble ferric iron (at pH lower than 3.5) or formation of the red precipitate ferric hydroxide (at pH greater than 3.5) [2]. Although AMD formation processes are accelerated by exposure to air [1], in oxygen-independent reactions, ferric iron becomes the main oxidant of the various other metal sulfides, which tend to associate with the pyrite in mineral formations. Naturally occurring bacteria can speed up the formation of AMD when they break down sulfide minerals [3]:

$$FeS_2 + \frac{7}{2}O_2 + H_2O \rightarrow 2SO_4^{2-} + Fe^{2+} + 2H^2{}^+ \tag{1}$$

Because pyrite is associated with gold and coal formations, mining of these minerals has subsequently resulted in very toxic and degraded environment, which are mainly highly acidic and usually contain excessive concentrations of metals, sulfides, sulfates, heavy metals, and salts [2, 4–9]. This is noted even in the South African content where it has been shown that coal formations of the Permian and Triassic-Permian ages, which lie in the E. Kalahari Precambrian Belt and the formations of the Permian, Permian–Carboniferous, and Triassic ages found in the Karoo Supergroup, are associated with gold deposits (**Figure 1** [10]). Indeed, this coformation means that large tracts of the South African environment are impacted by AMD.

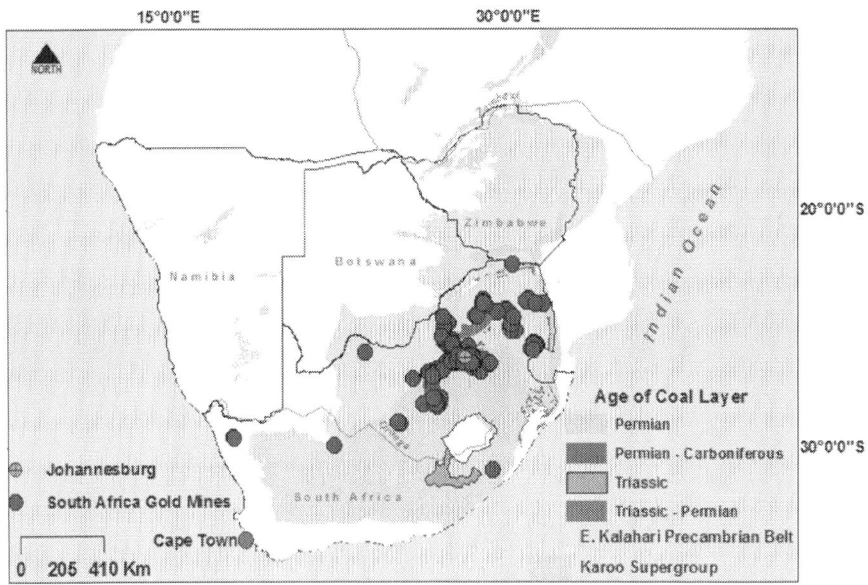

Figure 1. South Africa's gold mine locations and coal deposits (Software platform: ESRI [10]. Source of shapefiles: Internet).

At a global level, the latest Blacksmith's report by Harris and Andrew [11] has provided a tool in the form of a geospatially coded map (**Figure 2** [11]), to assist governments with prioritizing future resource allocation and pollution clean-up efforts. In this report it has been noted that mining activities occupy positions number one (artisanal gold mining), six and seven (mining and ore processing) in the top 10 of the world's 20 worst toxic pollution problems [11]. All three activities aforementioned are major sources of AMD, which Benedetto de Almeida [7] also described as one of the most serious environmental problems that the mining industry has ever created.

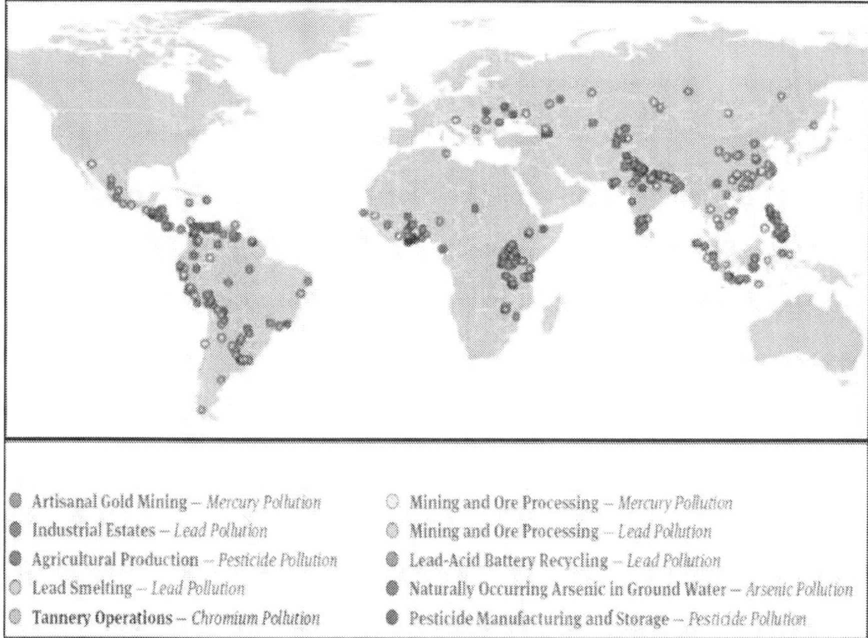

Figure 2. Geospatially coded map of top ten of the world's 20 worst toxic pollution processes [12].

Although Harris and Andrew [11] highlights that the South African environment is impacted by pesticide residues, Zilles Peccia [13] argued that AMD is the single most significant threat to the country's environment. For example, other researchers concurred that apart from the fact that mine dumps create harsh acidic and chemically toxic ecosystems in the country, a major environmental concern of pollution from AMD is the severe impact it has on productive land (e.g., agricultural land) as well as on groundwater, surface water, and aquatic life (e.g., the Vaal River Basin) as shown in (**Figure 3** [10, 14, 15]).

Therefore, treating AMD-impacted environments is a priority for South Africa as much as it is for the world, because if the environments are left as they are, the problem will just get worse, rendering more and more ecosystems uninhabitable. Evidently there are large tracts of land in South Africa, which are unusable because they are already impacted by AMD, examples having been documented in the East, Central, and Western basins of South Africa's Goldfields.

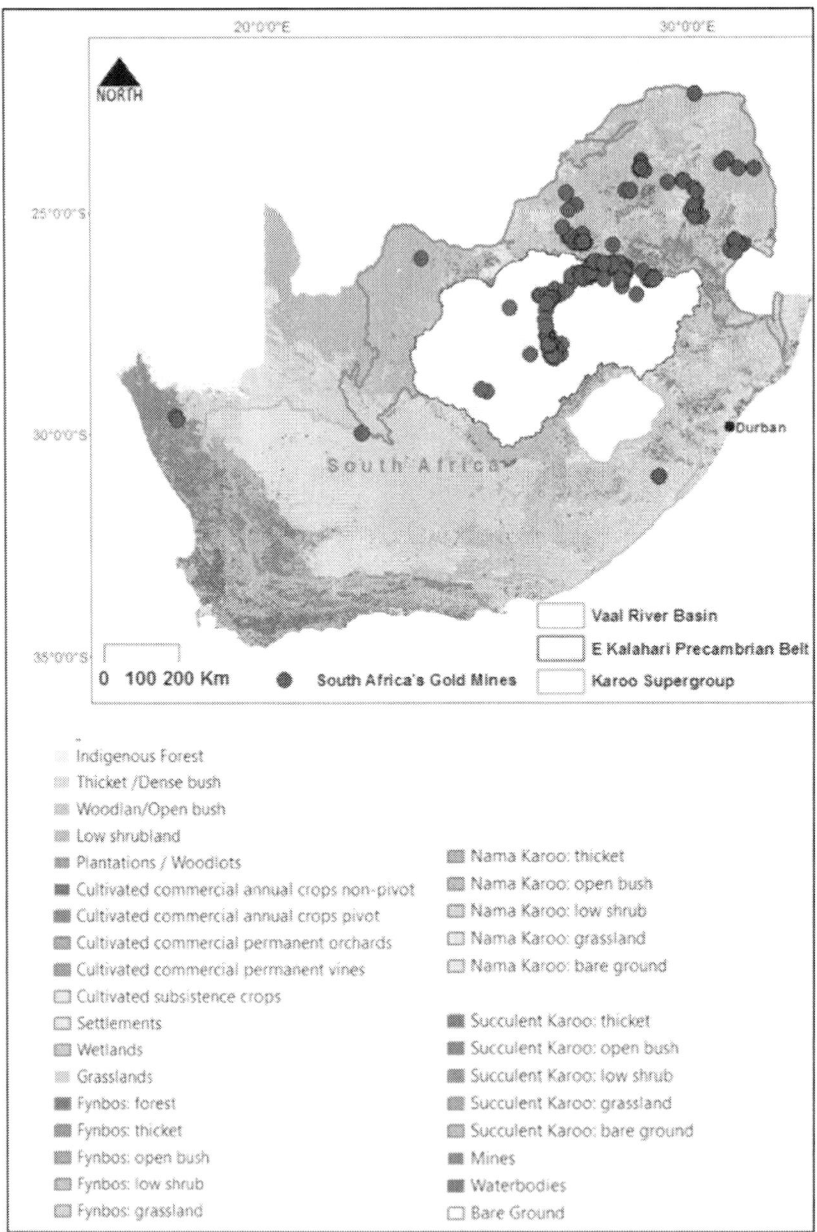

Figure 3. South Africa's land cover and the locations of gold mines (Software platform: ESRI [10]. Source of shapefiles: Internet).

Here, surface and groundwater are extremely polluted and unusable [11, 15–17] because gold and coal are mined largely from ores that also contain pyrite. The underlying hard-rock unit is made up of the Witwatersrand System in combination with others like the Transvaal System-Dolomite, The Ecca System, The Karoo, etc. It is noted that the Witwatersrand System, as represented by the Witwatersrand mines, is completely located in the Vaal Basin, a very strategic basin in South Africa, as indicated in **Figure 4** [10]. Therefore, due to the economic implications of polluting key livelihood environment, it has been suggested that where treatment processes are economically feasible and practical, it is necessary to reclaim the impacted environment and mitigate against pollution.

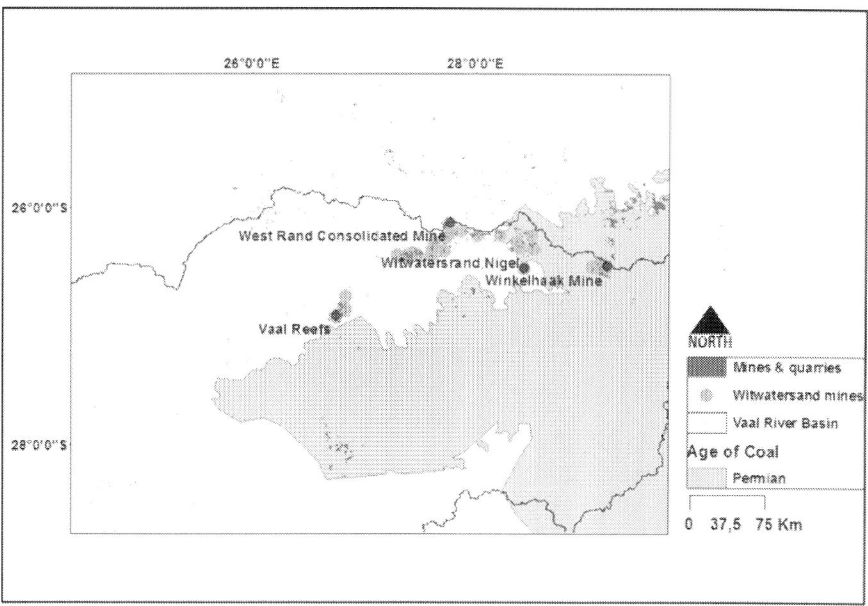

Figure 4. Witwatersrand System of gold mines located entirely in the Vaal Basin (Software platform: ESRI [10]. Source of shapefiles: Internet).

For example, Kruse, Bowman [15] reported on AMD treatment in a watershed near the village of Carbondale, Ohio, Hewett Fork subwatershed. The treatment process utilized involved neutralizing the AMD with lime. Results indicated that between the years 2000 and 2004, pH had improved from about four to around nine, with concomitant improvement in the biological communities in the study area. The major conclusion drawn from this intervention was that a 2-week interruption in treatment impacted on the fish community to a great extent while the macroinvertebrate community showed very minor perturbation. The reported community shift is a typical phenomenon for perturbed trophic structures [15].

Additionally, Wei, Wei [14] conducted a stream monitoring study in the United States for a period of 7 years. The objective was to evaluate the water quality trend and land cover in a

Mid-Appalachian watershed. The study area was a reclaimed former coal mining environment. GIS tools and multivariate analysis were applied to correlate the water quality trends and land cover. Results for pH, sulfate, and metals indicated that AMD was the major factor leading to overall poor water quality. It was concluded that water quality improvement was evident in subwatersheds which were originally heavily impacted but which were later reclaimed by reforestation. This indicated that good reclamation practices had positive impacts on water quality over time [14].

Benner, Gould [4] instead used bacterial populations and water chemistry to profile groundwater at Nickel Rim mine tailings impoundment in Ontario, Canada. The objective was to trace a plume of pollutants from the tailings impoundment and to find out if that plume was impacting groundwater in the vicinity. Results from groundwater analysis showed elevated populations of iron and sulfur oxidizing bacteria. These bacterial populations were restricted to hydrologically defined zones of recharge and discharge. It was concluded that active oxidation in the Nickel Rim tailings was occurring immediately above the water table, where water content was high in comparison to unsaturated zones further away from the water table. One plausible reason for this was that the water table interface provided continuous moisture gradient/potential difference enough to sustain ideal conditions for bacterial growth [4].

Despite these best efforts to try and reclaim impacted ecosystems, legal instruments have fallen short of implementing recommendations in order to deter further environmentally insensitive mining activities, challenges abound. For example, in South Africa, it has not been able to offer legal recourse for mine-related polluted environment because the situation is very complex, even though South Africa's Constitution [18] and related legal instruments support environmental sustainability. Many of the mines are closed off or dysfunctional, which should call for directors of these former mines to be answerable [19–22] for prosecution or jail terms, yet destruction continues. Implementation of the legal instruments seems to be the major stumbling block.

As part of on-going technology trials in South Africa's Witwatersrand Goldfields, Bologo, Maree [23] conducted experiments in order to understand the dynamics of reducing concentrations of Ca, Fe, SO_4^{2-}, and Mg from AMD-polluted effluent. The magnesium-barium-oxide process resulted in a reduction of pollutant concentrations. The technology also managed to recover the starter chemicals for reuse [23].

De Beer, Maree [24] used a CSIR ABC desalination process in a pilot plant to neutralise AMD samples from the Western Basin of the Goldfields. The process managed to remove total dissolved solids from 2600 to 360 mg/L. Metals were precipitated with CaS, $Ca(HS)_2$, or Ca$(OH)_2$ while SO_4^{2-} was reduced to 100 mg/L in a two-step process that employed gypsum crystallization followed by $BaCO_3$ treatment. The starting raw materials were recovered for reuse, making the process sustainable and cost-effective. It was demonstrated that the final treated water met the South African National Standard (SANS) 241 drinking water quality standard [24]. Motaung, Maree [25] used the same pilot plant to demonstrate that South Africa could produce sulfur from AMD treatment at a cost of ZAR2.21 m^{-3} of raw effluent. The potential value of the water and by-products amounted to R11.10 m^{-3} at a Rand value of US $1.00 = ZAR7.60 [25].

To support rehabilitation efforts, many studies have been done with the aim of characterizing and/or analyzing AMD as well as assessing its impact on ecosystems. Whichever rehabilitation process is chosen; the resultant treated effluent should be of good quality that is fit-for-purpose. Treated water may be channeled back into the mining operations or it could be released into the natural environment, while precautions should be taken not to transfer pollution from one stream to another. The receiving environment should be able to recover but if this is not possible, the effluent could be diverted elsewhere. Strict monitoring and evaluation of the effluent (treated or raw) could form part of the strategic long-term planning when mitigating against AMD impacts [17].

It has been documented that AMD has seriously impacted the surface water quality of the Eastern, Central and Western basins of the Witwatersrand and Goldfields [2, 9, 17, 24–29]. Because South Africa relies heavily on surface water for drinking and agricultural purposes, AMD thus threatens livelihoods of many as well as national economic returns from agriculture. Consequently, AMD impacts are expected to persist for the next centuries in a "do-nothing" scenario [17], which is unacceptable because while water quality is threatened directly, decanting AMD effluent also threatens to drown sensitive historical and wildlife sanctuaries around the City of Johannesburg Locus.

Researchers in South Africa and elsewhere, however, are continuously developing alternative interventions that require integrated implementation of a range of measures [5, 23–25, 30, 31] including neutralization, crystallization, and diversion (pumping the decanting effluent for reuse) in order to mitigate and rehabilitate affected environment. Various other example successes are reported for mitigating and treating AMD from polluted environment [30, 32–36]. However, these mitigatory activities impact on the environment and thus require monitoring in order to evaluate effectiveness of interventions.

Active and defunct gold and coal mines continue to pollute ecosystems through AMD and deposit of elements like radioactive material and heavy metals. First, the pollutant's acidity leads to a decrease in pH of the recipient water, should that water body have insufficient buffering capacity. Secondly, when pH in receiving water is lowered, some of the metals remain in soluble toxic form, thus making AMD a potent effluent for receiving watercourses [8].

In South Africa, although gold mining in the Witwatersrand System (see **Figure 5** [10]), is declining, massive closure in the 1990s caught the government unprepared for the environmental degradation, especially the rising of groundwater as it filled the voids, which were abandoned after mining activities had removed much of the precious element–bearing rocks.

Reactions of water exposed to pyrite and oxygen then subsequently created AMD whose postclosure decant is currently an enormous threat to the environment. Consequently, pollution could get worse if remedial activities are delayed or not implemented [17]. Additionally, polluted effluent from the mines and quarries that extend into the Limpopo Basin (**Figure 6** [10]) threatens to flood downstream environment including the Cradle of Humankind which continue to pollute ecosystems through AMD and speciation [9, 23, 37].

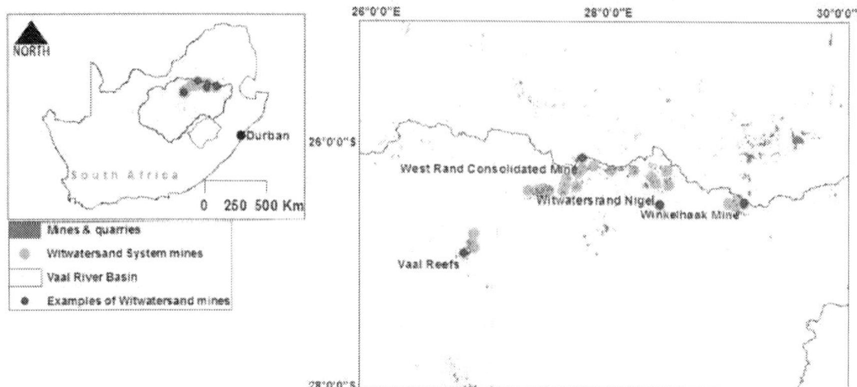

Figure 5. Witwatersrand System of gold mines in the Vaal Basin, South Africa (Software platform: ESRI [10]. Source of shapefiles: Internet).

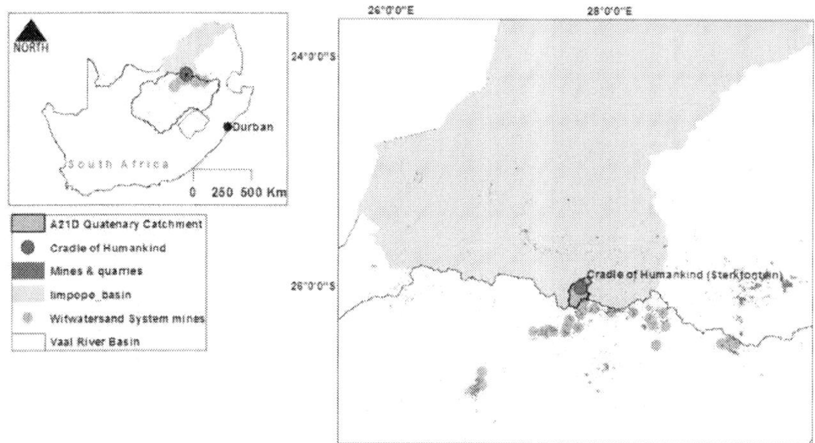

Figure 6. Threatened environments of the Cradle of Humankind, South Africa (Software platform: ESRI [10]. Source of shapefiles: Internet).

The Cradle of Humankind, which is located in the quaternary catchment of the Limpopo Basin called A21D, is one of South Africa's eight heritage sites and pollution threats of this magnitude are worrisome. Current initiatives are underway to either clean up the AMD before it reaches these vital and sensitive communities or stabilize it for reuse in fit-for-purpose situations [37–40]. These measures are crucial and strategic for the polluted environment where in 2010 about 60 ML/d AMD was decanting in the rainy season against a typical 20 ML/d in normal weather [37]. Decant polluted water flows via Tweelopiespruit and surrounding tributaries, through the Krugersdorp Game Reserve (KGR) and into Bloubankspruit that passes by the Cradle of Humankind, threatening this national treasure (**Figure 7** [10]). The KGR is home to a variety of wild animals which drink water from the polluted Tweelopiespruit.

To this end, solution development at Randfontein (**Figure 7** [10]) treatment plants include minimising the impact of waste (including AMD) from mining/AMD treatment, on the receiving aquatic environment by treating a portion of the effluent for re-use and release.

Using the Witwatersrand System effluent alone, researchers found out that the financial potential return of treating AMD was estimated at 350 ML/day (1ML = 1000 m^3) [29]. This calculation revealed that if the effluent was treated back to raw water quality guidelines, it could represent 10% of the daily potable water supplied by Rand Water Board to municipalities in Gauteng Province and surrounding areas, at a cost of R3000/ML, indeed a financial justification to treat the polluted effluent from these environments.

Concomitantly, the current paper reports on research that aimed to contribute to research literature for the Tweelopiespruit, Limpopo Basin, South Africa, by assessing impacts of treated effluent on the Tweelopiespruit micro-catchment as a receiving environment. This was envisaged to enhance understanding of the extent of the AMD problem in order to inform on possible mitigation measures in the quaternary catchment A21D and possibly in the wider Vaal and Limpopo hydrological primary basins, which are the basins that are majorly impacted by gold and coal mining activities.

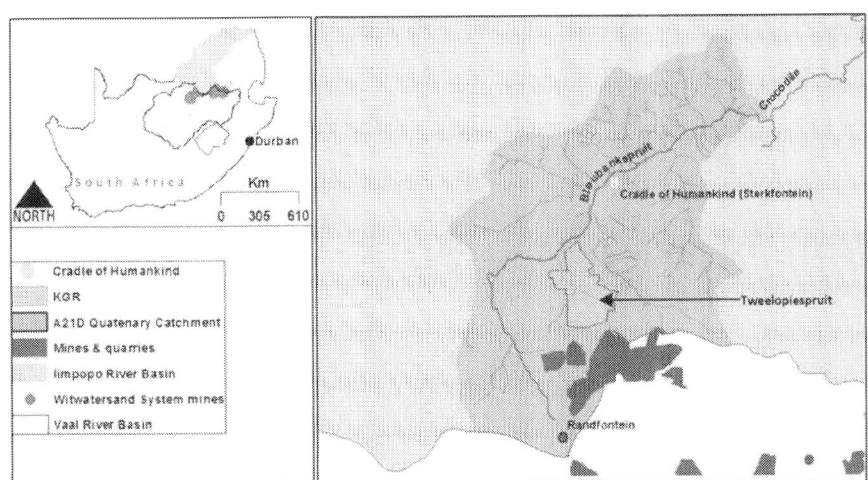

Figure 7. Tweelopiespruit, KGR and the Cradle of Humankind, South Africa (Software platform: ESRI [10]. Source of shapefiles: Internet).

Conclusions reached could provide information regarding whether treatment specifications could justifiably be continued as they were or could be enhanced to produce better quality treated effluent, especially as the Tweelopiespruit supports wildlife in the Krugersdorp Game Reserve (KGR).

2. Research problems and objectives

This research sought to answer the following questions:

- What are the trends in water quality of Tweelopiespruit and selected sampling sites around Randfontein plant, using the parameters SO_4^{2-}, Cl^-, Ca^{2+}, Mg^{2+}, Na^+ and K^+, Fe, pH, and electrical conductivity (EC)?

- What are the characteristics of the different water sample sources?

- What is the overall downstream water quality impact of the treatment plant intervention?

The overall aim of the research was to assess the impacts of treated AMD effluent on Tweelopiespruit's receiving and downstream ecosystem.

Specific objectives were:

- To evaluate the quality of water in the study area by analyzing and trending for SO_4^{2-}, Cl^-, Ca^{2+}, Mg^{2+}, Na^+, K^+, Fe, pH, and electrical conductivity (EC).

- To assess the impact of intervention (treatment plants) on Tweelopiespruit's health using the spatial and temporal trending patterns of the parameters.

The results from this study were to be submitted to the Team which was carrying out the overall neutralization process at the Randfontein AMD treatment plant. The report could assist them in assessing the impacts of the treated effluent on receiving waters, judging from the resultant water quality samples from the specified study monitoring sites.

3. Study area

In the Upper reaches of the Limpopo Basin's land use (see **Figure 8** [10]), polluted mine water decants from underground and flows from the Randfontein mining environment into Tweelopiespruit stream and the surrounding farming lands [17, 37, 39]. The massive discharge has altered the nature of this water course. Bologo et al. [21] estimated that about 50 ML is decanted into the Randfontein receiving environment each day. Some effluent enters the Tweelopiespruit wetland on the mine grounds via surface seepage [39].

Previous studies as reported by [25] identified the quality of water in the receiving karst groundwater environment as comprising a mixture of acid mine drainage and treated wastewater. This has compromised the sustainability of biodiversity (both plants and animals) in the KGR. This degeneration process currently (2013) poses a threat to the Cradle of Humankind, which has a geological formation of dolomitic rocks that are susceptible to attack and dissolution by AMD.

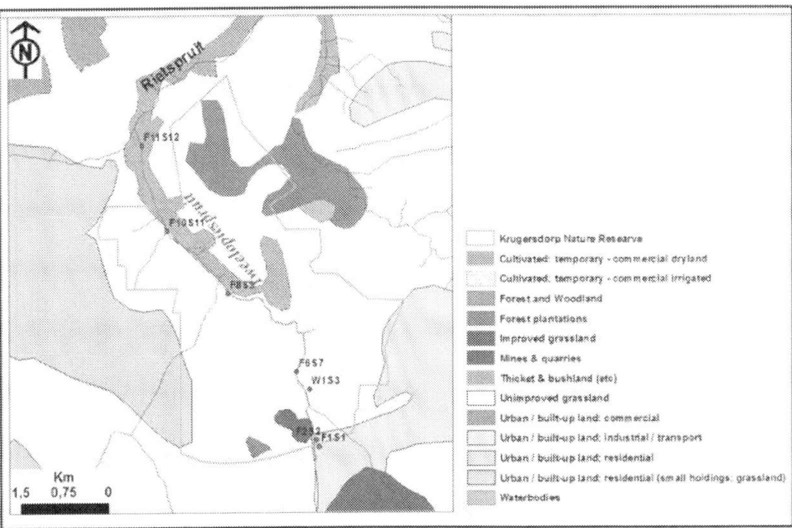

Figure 8. Tweelopiespruit land use (Software platform: ESRI [10]. Source of shapefiles: Internet).

Tweelopiespruit, which carries the sampling sites for this study, is a defined stream network thus it is identifiable as a hydrological unit and the seven sampling sites are clearly marked in **Figure 9**. Except for F1S1 and F11S12, five of the monitoring sites are located inside the KGR.

During the period of the study, the stream received both treated effluent AMD from three treatment pilot plants located at the Randfontein experimental site, as well as AMD that was decanting from underground within the surrounding environs. At one of the pilot treatment plants, AMD tertiary treatment was being employed, where a fraction of its volume (about 50% of decant volume by three treatment plants), was collected and treated in a dedicated neutralization plant that used lime ($Ca(OH)_2$) [35]. Two of the three pilot plants that treated AMD used this method while the third one employed a different treatment technology. The lime process was reported by Khorasanipour, Moore [41] as a preferred method to others like the alkaline method, claiming that it had a high removal efficiency for dissolved heavy metals, relatively low cost, and was insensitive to seasonal temperature fluctuations. In all the three treatment plants, AMD was pumped from underground shafts to the surface for treatment before the treated effluent was released to Tweelopiespruit. It was expected that by pumping and treating the AMD, the underground level of the mine waste water would fall below a critical level to allow stoppage of the decantation process.

For this research, the sampling points were chosen to represent the flow of effluent and treated water within the micro-catchment. This allowed for trending based on spatial locations of the sampling sites as well as temporal and spatial analysis of the data. **Table 1** describes the sampling sites using the same identification which is used by the Department of Water and Sanitation (DWS).

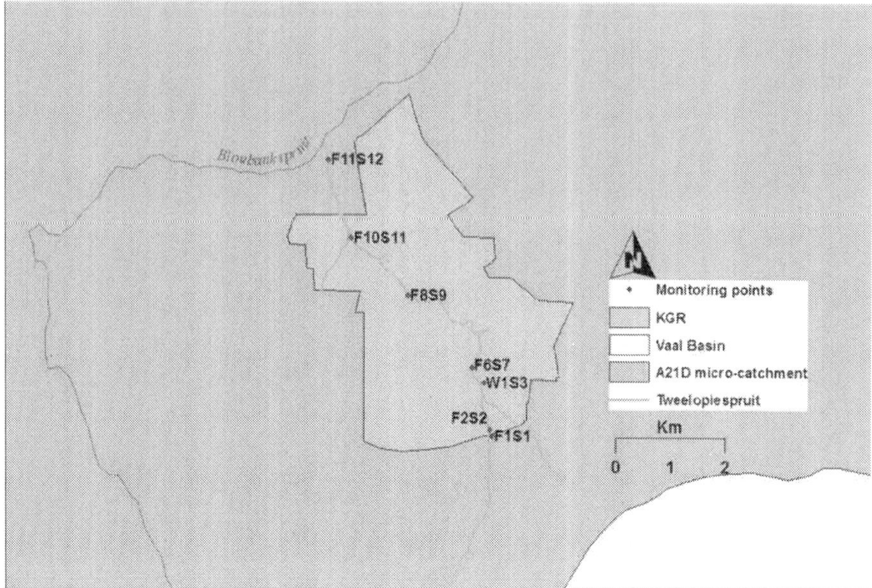

Figure 9. Sampling points along Tweelopiespruit (Software platform: ESRI [10]. Source of shapefiles: Internet).

Sampling point	Description	Latitude	Longitude
F1S1	Upstream of R24 at Randfontein Estates on Tweelopie	−26.10752	27.72268
F2S2	Willow tree in KGR on Tweelopie	−26.10653	27.72227
W1S3	Hippo Dam in KGR on Tweelopie	−26.09917	27.72128
F6S7	Cemetery Spring (1) (Spring 1) in KGR on Tweelopie	−26.09671	27.71932
F8S9	Lodge Spring (2) (Spring 2) at broad crest in KGR on Tweelopie	−26.08527	27.70886
F10S11	Northern fence in KGR	−26.07620	27.69963
F11S12	Tweelopie at the N14 intersection	−26.06374	27.69589

Table 1. Sampling sites on Tweelopiespruit.

4. Methods and materials

In this research, the quantitative research design was used. The research methods combined statistical analysis of retrospective (historical) data and batch analysis of water samples from the sites. Experimental analysis was performed on two batches of water samples, one in August and the other in September. Analysis was performed using the same analytical and standard methods which were used for the historical data in order to validate the historical data range that was employed in this study. The approach was flexible while mindful of the monitoring data which spanned hydrological years under a wide spectrum of hydrologic variability.

The experimental design had a random assignment where all samples from the monitoring sites had an equal chance of being assigned to a given experimental condition. Random assignment was used to ensure that experimental conditions did not differ significantly from each other.

An analytical approach was taken in order to determine the quality of water from the seven sampling points. The spatial and temporal changes were documented using tools that could aid in understanding the chemistry and the sources of the water, including geospatial mapping, and assessment of the impact of the treated effluent on the receiving water body.

- First, retrospective (historical) data were acquired from the DWS, which is a government entity that monitors chemical and biological quality of water as part of its mandate to provide environmental assessments and protection. It was treated for outliers before trending and geospatial mapping on ESRI ArcGIS 10.2.

- The second task involved conducting two sampling runs parallel with the regular monitoring exercise in the study area, in order to validate the retrospective data. The objective was to understand various aspects of the sampling sites in relation to the impacts of treated effluent on the quality of the receiving waters. Sampling and testing of water conformed to procedures for both the sampling and analysis of chemical parameters. All parameters were analyzed according to the standard methods of analysis.

5. Results and discussion

The results for chosen parameters, i.e., SO_4^{2-} (mg/L), Cl^- (mg/L), Ca^{2+} (mg/L), Mg^{2+} (mg/L), Na^+ (mg/L), K^+ (mg/L), Fe (mg/L), pH (pH units), and EC (m Sm^{-1}) are shown for all monitoring points in **Figures 10–20**; using trends that were plotted on MS Excel, for each sampling site according to their geocoordinates, starting with F11S12. While monitoring continues and is active at these sites since the 1970s, it is noted that there is a general trend toward increase in parameter concentration. For an environmentally sensitive micro-catchment, which also houses the Cradle of Humankind in its downstream ecosystem, efforts should be done to reduce the study area parameter concentrations in order improve the ecosystem. Because these are combination graphs for a mixture of parameters, individual units of measure could not be indicated on the y-axis of the graphs, hence the use of the label"parameter test value."

pH trends (**Figure 17**) for all monitoring points are lowest at the highest point (F1S1) but also is very unstable on all the other points because the environment itself is prone to nonpoint AMD pollution.

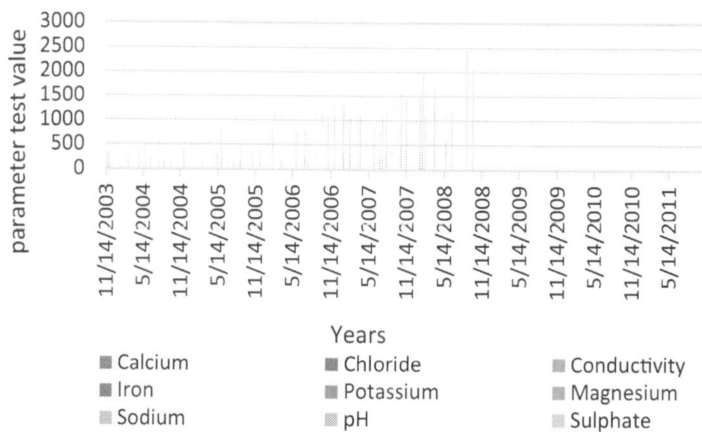

Figure 10. F11S12 monitored parameters for 2003–2011.

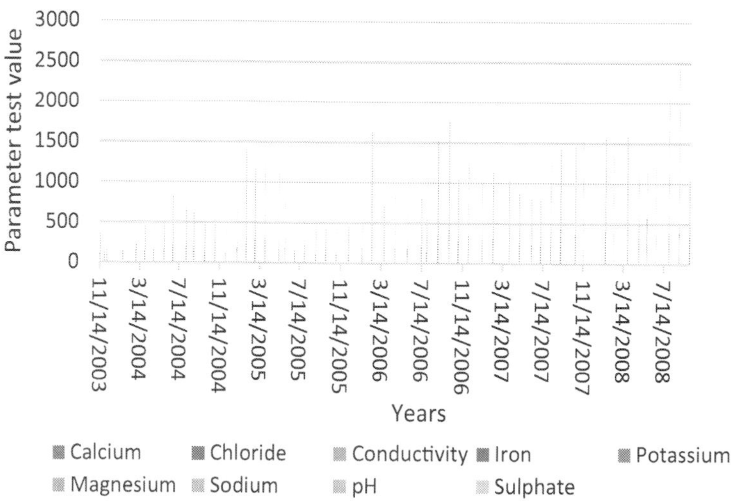

Figure 11. F10S11 monitored parameters for 2003–2008.

Just like the spiking values for pH, **Figure 18** indicates variable trends for sulfate along Tweelopiespruit.

However, the figure also indicates the conservative nature of sulfate as it shows lowest concentration at the sampling point farthest from the AMD sources along the river (F11S12 and

F10S11) and highest trends close to old mine shaft activities (F1S1 and F2S2). Sulfate does not mobilize easily and is therefore deposited along the stream soon after its entry. **Figure 19** shows that iron is typically high at the upper reaches of the river (F1S1), indicating the major source of AMD and a potential priority point for mitigation and management efforts to control AMD in the micro-catchment.

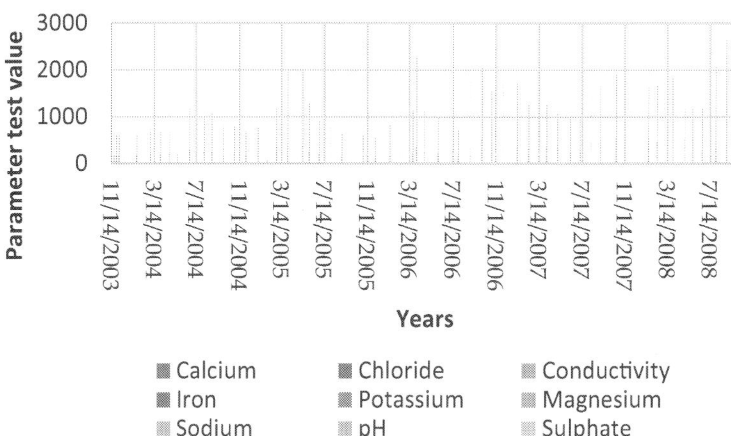

Figure 12. F8S9 monitored parameters for 2003–2008.

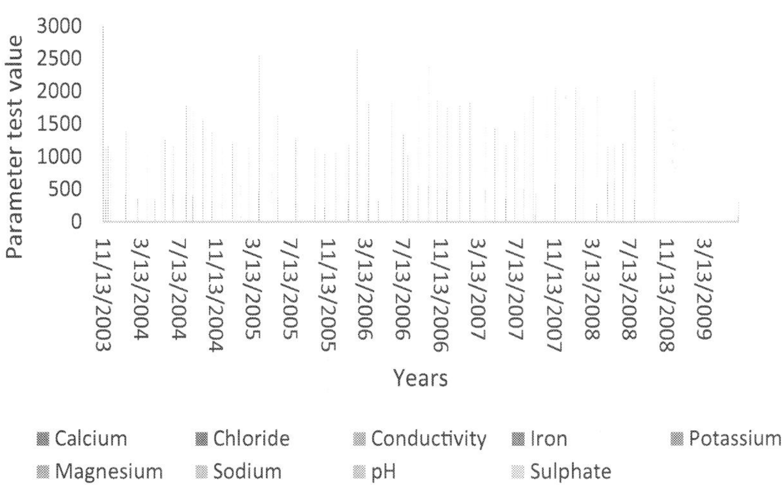

Figure 13. F6S7 monitored parameters for 2003–2009.

Calcium indicates a high concentration in the upper reaches of the river, too, as indicated in **Figure 20**.

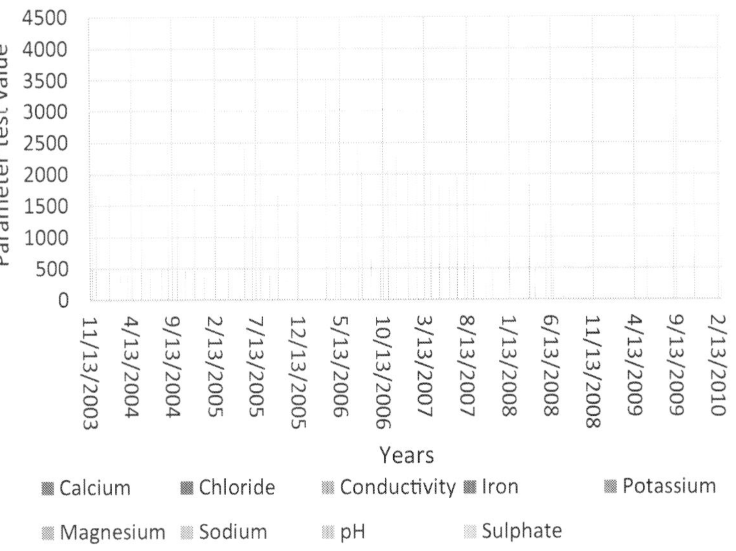

Figure 14. W1S3 monitored parameters for 2003–2010.

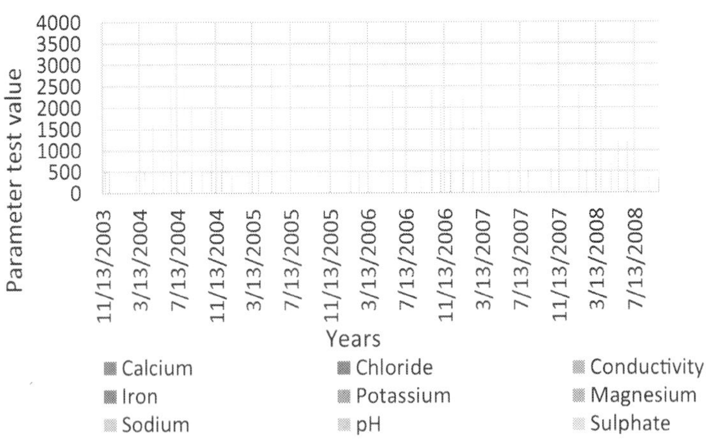

Figure 15. F2S2 monitored parameters for 2003–2008.

W1S3 indicates higher values for calcium and could be subject for further investigation regarding the type of water that passes by that monitoring point. For example, calcium contributes to water hardness which affects mobility of other related ions and anions in the water and sediments.

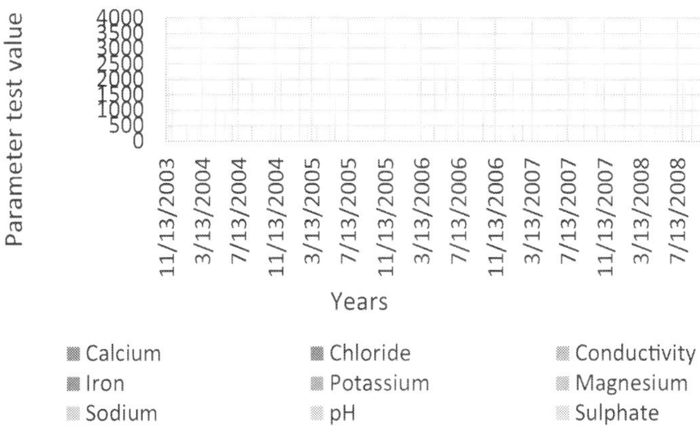

Figure 16. F1S1 monitored parameters for 2003–2008.

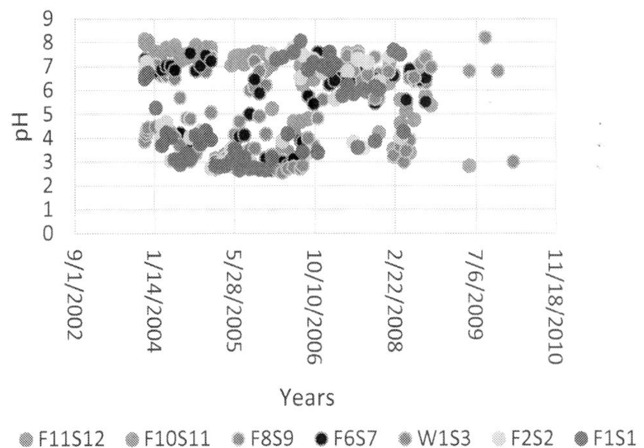

Figure 17. pH trending for all monitoring points.

The study area is critical in its strategic position because, apart from preserving its endowed environments, it is also because of its proximity to the Vaal Basin, which is the heartland of South Africa's economic activities as indicated in the digital elevation map (see **Figure 21**). Thus, measures and actions focusing on managing Tweelopiespruit pollution could also potentially benefit the Vaal System.

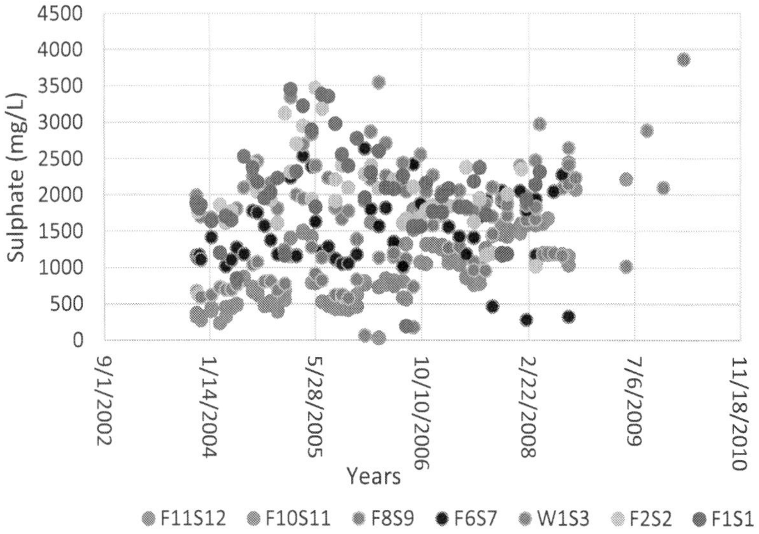

Figure 18. Sulfate trending for all monitoring points.

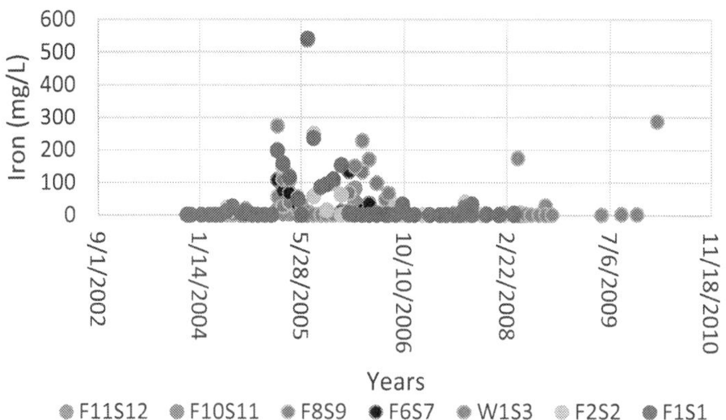

Figure 19. Iron trending for all monitoring points.

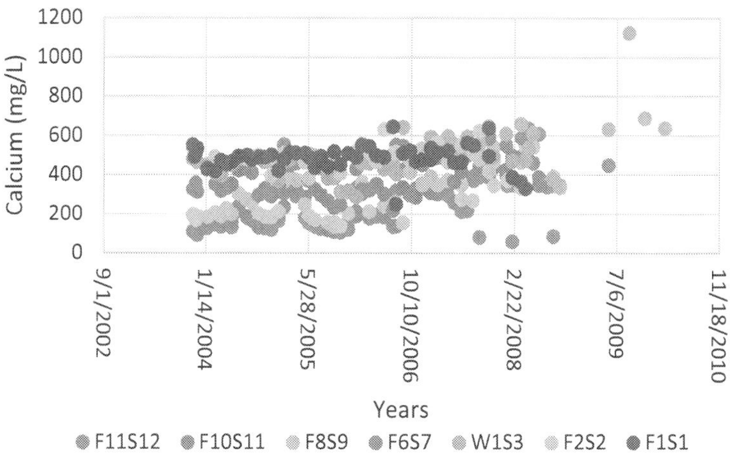

Figure 20. Calcium trending for all monitoring points.

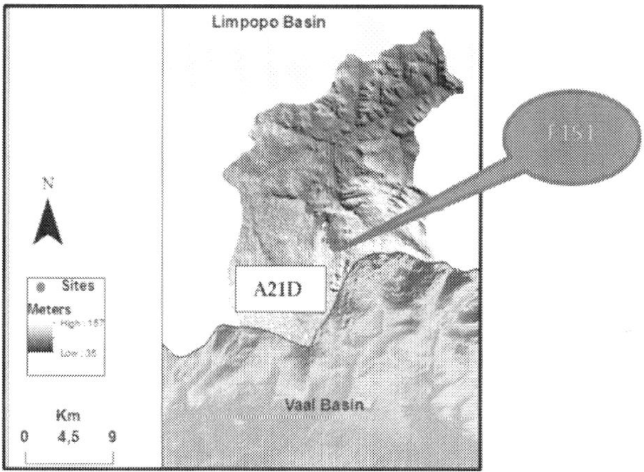

Figure 21. Overlaid monitoring points along Tweelopiespruit on micro-catchment A21D.

6. Conclusions

Pollution in the Krugersdorp Game Reserve is very significant as indicated by the chemical analysis' results for the monitoring points. Treatment of the polluted effluent does not seem to have an impact on the effluent for the study period up to 2008.

This research could benefit from the land use change detection, especially from satellite images which could show the devastating effects of the AMD on the environment within the Tweelopiespruit micro-catchment. These satellite images, freely available from the USGS website, could inform on the worsening situation in the micro-catchment.

It can be noted that peace-meal treatment works at the Randfontein site do not seem to have made a noticeable impact on pollution of the "dead" Tweelopiespruit, from AMD. The watershed, spanning the Witwatersrand System and the Upper reaches of the quaternary catchment A21D, is a hot spot for environmental disaster that is set to impact outer ecosystems for many years to come, and South Africa will have to pay for non-stringent environmental legislation, which was in place before the 1990s. The research results and conclusions aim to provide a baseline for critiquing ongoing research in the Tweelopiespruit micro-catchment in order to assist with answering the research questions that were initially raised against each objective. The use of satellite and remote sensing methods are recommended for further research.

Acknowledgements

The following people are most sincerely acknowledged for contributing to this paper in various but very valuable forms: Prof. Fred A.O. Otieno from Masinde Muliro University, Ms. Malebo D. Matlala, and Prof. Kevin Mearns – both from UNISA, Prof. Jannie Maree from Tshwane University of Technology, Mr. Phillip Hobbs, and the entire team at the Randfontein experimental treatment plant. The following organizations are acknowledged sincerely: Department of Water and Sanitation and Hydrology Section, USGS, CGIAR Consortium for Spatial Information, ESRI-South Africa, South African Weather Services and the South African Demarcation Board.

Author details

Bloodless Dzwairo[1]* and Munyaradzi Mujuru[2]

*Address all correspondence to: ig445578@gmail.com

1 Durban University of Technology, Department of Civil Engineering (Midlands), Durban, South Africa

2 University of Limpopo, Department of Water and Sanitation, Faculty of Science and Agriculture, Sovenga, Medunsa, Polokwane, South Africa

References

[1] Costello, C., Acid mine drainage: Innovative treatment technologies. National Network of Environmental Management Studies Fellow for US Environmental Protection Agency, Washington, DC, 2003.

[2] Tutu, H., T.S. McCarthy, and E. Cukrowska, The chemical characteristics of acid mine drainage with particular reference to sources, distribution and remediation: The Witwatersrand basin, South Africa as a case study. Applied Geochemistry, 2008. **23**(12): p. 3666–3684.

[3] Akcil, A. and S. Koldas, Acid Mine Drainage (AMD): causes, treatment and case studies. Journal of Cleaner Production, 2006. **14**(12–13): p. 1139–1145.

[4] Benner, S.G., W.D. Gould, and D.W. Blowes, Microbial populations associated with the generation and treatment of acid mine drainage. Chemical Geology, 2000. **169**(3–4): p. 435–448.

[5] Manders, P., L. Godfrey, and P. Hobbs, Acid Mine Drainage in South Africa, in Briefing Note 2009/02. 2009, CSIR Natural Resources and the Environment: Pretoria.

[6] Cole, M., et al., Development of a small-scale bioreactor method to monitor the molecular diversity and environmental impacts of bacterial biofilm communities from an acid mine drainage impacted creek. Journal of Microbiological Methods, 2011. **87**(1): p. 96–104.

[7] Benedetto, J.S., et al., Monitoring of sulfate-reducing bacteria in acid water from uranium mines. Minerals Engineering, 2005. **18**(13–14): p. 1341–1343.

[8] Hallberg, K.B., New perspectives in acid mine drainage microbiology. Hydrometallurgy, 2010. **104**(3–4): p. 448–453.

[9] McCarthy, T.S., The impact of acid mine drainage in South Africa. South African Journal of Science, 2011. **107**(5-6): p. 01–07.

[10] ESRI, Environmental Systems Research Institute (ESRI) ArcGIS for Desktop 10.2.

[11] Harris, J. and M. Andrew, The Top Ten of the Toxic Twenty, in The World's Worst Toxic Pollution Problems Report. 2011, Blacksmith Institute and Green Cross Switzerland: New York.

[12] Harris, J. and M. Andrew, The Top Ten of the Toxic Twenty, in The World's Worst Toxic Pollution Problems Report. 2011, Blacksmith Institute and Green Cross Switzerland: New York.

[13] Zilles, J.L., J. Peccia, and D.R. Noguera, Microbiology of Enhanced Biological Phosphorus Removal in Aerated-Anoxic Orbal Processes. Water Environment Research, 2002. **74**(5): p. 428-436.

[14] Wei, X., H. Wei, and R.C. Viadero, Post-reclamation water quality trend in a Mid-Appalachian watershed of abandoned mine lands. The Science of The Total Environment, 2011. **409**(5): p. 941–948.

[15] Kruse, N.A., et al., The Lasting Impacts of Offline Periods in Lime Dosed Streams: A Case Study in Raccoon Creek, Ohio. Mine Water and the Environment, 2012. **31**(4): p. 266–272.

[16] Donato, D.B., et al., A critical review of the effects of gold cyanide-bearing tailings solutions on wildlife. Environment International, 2007. **33**(7): p. 974–984.

[17] McCarthy, T., et al., Mine water management in the Witwatersrand gold fields with special emphasis on acid mine drainage. Report to the inter-ministerial committee on acid mine drainage. 2010: Pretoria. p. 1–128.

[18] Statutes of The Republic of South Africa - Constitutional Law: Act No. 108 of 1996. 18 December 1996: South Africa. p. 1241–1331.

[19] DEAT, No. 36 of 1998: National Water Act., in Government Gazette No. 19182, Vol. 398, August 1998. National Water Act, 1998. 1998, Department of Environmental Affairs and Tourism (DEAT): Cape Town.

[20] DEAT, No. 107 of 1998: National Environmental Management Act. Department of Environmental Affairs and Tourism, in Government Gazette No. 33306, Vol. 540, No. 9314 of 2010: National Environmental Management Act. 1998, Department of Environmental Affairs and Tourism (DEAT): Pretoria.

[21] DEAT, No. 53 of 2008: National Radioactive Waste Disposal Institute Act, in Government Gazette No. 31786, Vol. 523 number 19, January 2009. National Radioactive Waste Disposal Institute Act, 2008. 2008, Department of Environmental Affairs and Tourism (DEAT): Cape Town.

[22] DEAT, No. 28 of 2002: Mineral and Petroleum Resources Development Act, in Government Gazette No. 23922, Vol. 448, No. 1273 10 October 2002. Mineral and Petroleum Resources Development Act, 2002. 2002, Department of Environmental Affairs and Tourism (DEAT): Cape Town.

[23] Bologo, V., J. Maree, and F. Carlsson, Application of magnesium hydroxide and barium hydroxide for the removal of metals and sulphate from mine water. Water SA, 2012. **38**(1): p. 23–28.

[24] De Beer, M., et al., Acid mine water reclamation using the ABC process. 2010, Council for Scientific and Industrial Research (CSIR): Pretoria.

[25] Motaung, S., et al., Recovery of Drinking Water and By-products from Gold Mine Effluents. Water Resources Development, 2008. **24**(3): p. 433–450.

[26] Winde, F. and I. Jacobus van der Walt, The significance of groundwater-stream interactions and fluctuating stream chemistry on waterborne uranium contamination of streams–a case study from a gold mining site in South Africa. Journal of Hydrology, 2004. **287**(1-4): p. 178–196.

[27] Dzwairo, B., Modelling raw water quality variability in order to predict cost of water treatment, in Department of Civil Engineering. 2011, Tshwane University of Technology: Pretoria. p. 237.

[28] Naicker, K., E. Cukrowska, and T.S. McCarthy, Acid mine drainage arising from gold mining activity in Johannesburg, South Africa and environs. Environmental Pollution, 2003. **122**(1): p. 29–40.

[29] Liefferink, M., Greater intervention needed to tackle acid mine drainage, in Creamer Media's Mining Weekly. 2012.

[30] Kalin, M. and W.L. Caetano Chaves, Acid reduction using microbiology: treating AMD effluent emerging from an abandoned mine portal. Hydrometallurgy, 2003. 71(1–2): p. 217–225.

[31] Bologo, V., Treatment of acid mine drainage using magnesium hydroxide and barium hydroxide, in Department of Environmental, Water and Earth Sciences. 2012, Tshwane University of Technology: Pretoria.

[32] Baruah, B.P. and P. Khare, Mobility of trace and potentially harmful elements in the environment from high sulfur Indian coal mines. Applied Geochemistry, 2010. 25(11): p. 1621–1631.

[33] Martins, M., et al., Characterization and activity studies of highly heavy metal resistant sulphate-reducing bacteria to be used in acid mine drainage decontamination. Journal of Hazardous Materials, 2009. 166(2–3): p. 706–713.

[34] Benzaazoua, M., et al., The use of pastefill as a solidification and stabilization process for the control of acid mine drainage. Minerals Engineering, 2004. 17(2): p. 233–243.

[35] Luptakova, A. and M. Kusnierova, Bioremediation of acid mine drainage contaminated by SRB. Hydrometallurgy, 2005. 77(1–2): p. 97–102.

[36] Burford, M.A., et al., Correlations between watershed and reservoir characteristics, and algal blooms in subtropical reservoirs. Water Research, 2007. 41(18): p. 4105–4114.

[37] Hobbs, P. and J.E. Cobbing, The hydrogeology of the Krugersdorp Game Reserve area and implications for the management of mine water decant, in Groundwater Conference, 8-10 October 2007. 2007: Bloemfontein. p. 10.

[38] Davies, T.C. and H.R. Mundalamo, Environmental health impacts of dispersed mineralisation in South Africa. Journal of African Earth Sciences, 2010. 58(4): p. 652–666.

[39] Durand, J.F., The impact of gold mining on the Witwatersrand on the rivers and karst system of Gauteng and North West Province, South Africa. Journal of African Earth Sciences, 2012. 68(0): p. 24–43.

[40] Oberholster, P.J., et al., An ecotoxicological screening tool to prioritise acid mine drainage impacted streams for future restoration. Environmental Pollution, 2013. 176(0): p. 244–253.

[41] Khorasanipour, M., F. Moore, and R. Naseh, Lime treatment of mine drainage at the sarcheshmeh porphyry copper mine, Iran. Mine Water and the Environment, 2011. 30(3): p. 216–230.

Chapter 16

The Behaviour of Natural and Artificial Radionuclides in a River System: The Yenisei River, Russia as a Case Study

Lydia Bondareva, Valerii Rakitskii and Ivan Tananaev

Additional information is available at the end of the chapter

http://dx.doi.org/10.5772/65743

Abstract

The Yenisei River is one of the largest rivers in the world. There is Mining and Chemical Combine (MCC) of Rosatom located at Krasnoyarsk, on the bank of the River Yenisei, 50 km downstream of the city of Krasnoyarsk. Since 1958 MCC used river's water for cooling of industrial nuclear reactors for the production of weapon plutonium—^{238}Pu. Besides the pollution caused by industry-related radionuclides, pollution by natural radionuclide—uranium and its isotopes— are also investigated. Besides the natural uranium isotopes (^{234}U, ^{235}U, ^{238}U), exclusive artificial isotope—^{236}U was also found. Yenisei water was also polluted by high tritium content: from 4 Bq/L (back road value) to 200 Bq/L (some sample of water). The total amount of radionuclides investigated was about 20 radioisotopes. These radionuclides have different physical and chemical properties, different half-lives, and so on. Thus, the data on artificial radionuclides entering the Yenisei River water were obtained by long-term monitoring, which is likely to be connected with the activity of the industrial enterprises located on the river's banks of the studied area.

Keywords: Yenisei River, migration, radionuclide, Siberia, isotopes, Russia

1. Introduction

The major part of population of Krasnoyarskii region lives on the banks of Yenisei River. Yenisei is—one of the largest rivers in the World: its length from junction of Big Yenisei and Small Yenisei is 3487 km, from Small Yenisei's rise—4287 km and from Big Yenisei's rise—4123 km. The place of junction of Big and Small Yenisei near city of Kyzyl is considered as geographical centre of Asia. Rising in the south, in the mountain deserts of Mongolia, Yenisei flows in the

north direction for nearly 3000 km, crosses various latitudinal geographical zones, falls into the Arctic Ocean, forming estuary zone up to 30 km wide. Length of Yenisei exceeds the same of Danube River (2857 km), Mississippi (3770 km) and Indus (3180 km). Yenisei River is the most affluent river of Russia with a runoff rate of 624 km^3/year. Mean water consumption in the estuary is 19,800 m^3/s and the maximal value is 190,000 m^3/s. With respect to basin area (2580 thousand km^2) Yenisei holds second place (after the Ob) and the seventh place among all rivers of the world. The nominal border between Western and Eastern Siberia lies along Yenisei. There are three hydroelectric power plants (HPP) on the Yenisei River and on the rivers falling into it. River's waters are characterized by high transparency (up to 3 m) and low mineralization (mean value is 54 mg/l) and also by high oxygen concentration. Flow velocity and river width can change considerably: from 1.5 to 12–15 km/h and from 0.2–0.5 to 3–5 km, respectively. Solids of the channel in the uppers are faceted soils that are changed into gravelly sand in the middle course and into sandy-clay in the lower course near the fall into the Arctic Ocean.

There is a constant mixing of water layers because of hydroelectric power plant's activity, thus not affecting water temperature from the depth of water flow even on higher distances after HPP stanch. At the beginning of July, water temperature in Krasnoyarsk district and after 100–150 km further down the course is ~10°C, at the end of July–August it is 15–17°C. River's ecosystem is related to oligotrophy with fauna-rich river, there are more than 500 species of algae and diatoms [1].

There is Mining and Chemical Combine (MCC) in Rosatom, located at Krasnoyarsk, on the bank of the River Yenisei in 50 km downstream of the city of Krasnoyarsk. There are atomic reactors and radiochemical production in the MCC. Since 1958 MCC used water for cooling industrial nuclear reactors for the production of weapon plutonium $-^{238}$Pu. River water, while passing through the cooling system of reactors, returned to Yenisei. Effluent waters contained a great amount of radionuclides that were formed during neutron activation of traces (solid slurry and dissolved compounds), which are present in river water. Two direct flow reactors were withdrawn in 1992, because the activity level of the effluent waters of MCC was remarkably decreased.

2. Radionuclides (natural and artificial) in the streams of the Yenisei River

As a result of long-term activity of MCC, the Yenisei's ecosystem contains considerable amounts of industry-related radionuclides [2]. In particular, an increased level of radioisotope contents in bed deposits and alluvial soils was found [2–6] and distribution and migration of radionuclides both in near-field influence of MCC [7, 8] and in significant distance away from effluent zone, including estuary of Yenisei, were indicated. As early as in the beginning 1970, the pollution zone of Yenisei's bottom land by ^{137}Cs was found by airborne gamma survey. In district of Yeniseysk city (island Gorodskoi around 300 km downstream of MCC), the specific activity of ^{137}Cs reaches 16,300 Bk/kg in some places, power of exposure (PE)—270 μR/h. According to present standards, bottom sediments and alluvial soils at this region are related to solid radioactive wastes. ^{137}Cs is the main radionuclide polluting soils and bottom sediments are $^{152+154}$Eu and ^{60}Co [9].

In this chapter, the results of research conducted mainly in the middle course of Yenisei in the 15 km region (from fall place of Ploskiy river (0 km) to Bolshoy Balchug (15 km), **Figure 1**) are described. In this region, at a water flow rate $Q = 4085$ m³/s the depth and current velocity were defined as $H \approx 7$ m, $v = 1.25–1.8$ m/s, respectively. Jet with industrial wastes spends along the right bank not more than 0.1 of river's width, i.e. along bottom land, where current velocity and depth are several times lower.

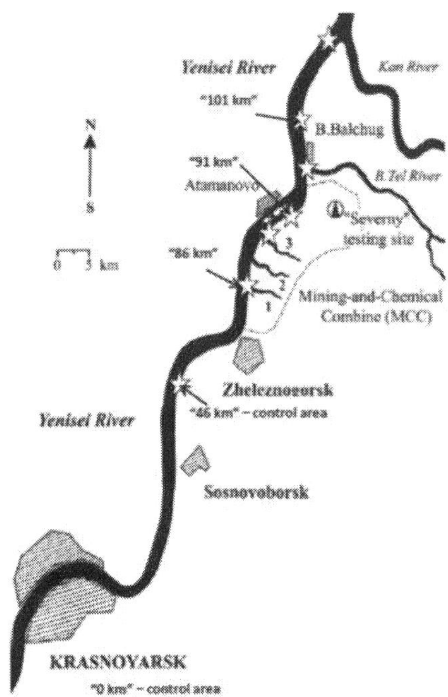

Figure 1. Sketch-map of the some region of the Krasnoyarsk Territory near the Mining—Chemical Combine of the Rosatom—surface water of the Yenisei River basin. 1: Shumikha River; 2: Stream No. 2; 3: the Ploskii Stream; – – – – : the boundary of the MCC sanitary-protective zone. ☆point of collection. Sampling points: '0 km'—56°27'05"N, 93°36'31"E; '2 km'—56°23'18"N, 93°37'13"E, '5 km'—56°23'40", '15 km'—56°27'05", 93°42'22"E.

2.1. Uranium: natural and artificial

Besides the pollution caused by industry-related radionuclides, pollution by natural radionuclide—uranium and its isotopes are also investigated.

The total uranium content is the main factor to determine the radiation level of water sources, its value is standardized and controlled by ecological services. Uranium in water is truly dissolved and found in the form of uranyl carbonate complex anions. In general, river waters contain 600 ng/l of dissolved uranium. Despite that main natural transport agents—water carries uranium in small amounts, one should not exclude that there can be local transfers of uranium in significant amounts [10].

The main feeders, contributing to the radioactive pollution of the Yenisei, are majorly the right bank feeders, situated near MCC outlet: river Kan, on the bank of which the electrochemical plant (ECP, Zelenogorsk city) is situated, and river Bolshaya Tel', flowing along the border of testing area 'Sverniy' MCC (Zheleznogorsk city).

According to data, presented in the monograph [9], the most of the region's waters, related to the bottomland of Kan, contain from 0.04 to 3 µg/l of uranium that is considered as highly pure with respect to natural radionuclide content. In addition, there was no trend in uranium content from the location of selection. Only in one place at the turn of Kan's course to the north vs. course of Bogunay river, it was revealed that all of the waters contain uranium from 1 to 3.3 µg/l. Industrial waters discharged by ECP into Kan near the plant administration were similar to natural uranium content and contained 0.05–0.08 µg/l of uranium.

Natural stream feeding Syrgyl river contained from 0.03–0.07 to 1.0–7.3 µg/l of uranium. The contents in the range 0.3–5.0 µg/l were shown to be natural geochemical background of uranium in the studied region, in particular, in the bottomland of Kan. All of the excesses are considered as abnormal.

The analysis data [1–12] shows that the geochemical background level of uranium in the Yenisei River is in agreement with the mean statistical level for the basins with major contribution of natural uranium resources, e.g. Baikal Lake and rivers of Altai region: from 0.15 to less than 2.0 µg/l.

Uranium content in waters which were collected from Bol'shaya Tel' in the September 2007 at the 1000 m place from the estuary is 3–60 times higher than values obtained for uranium (mean value 0.33 ± 0.08 µg/l) in background samples (Yenisei, tideway). Moreover, this period was indicated by significantly higher uranium concentrations as compared with other studied months. This increase becomes remarkable for the 1000 m place, where uranium concentration is 16 µg/l that is very close to the accepted in Canada and Australia standards for the minimal allowed uranium concentration—20 µg/l and by 8 times exceeds accepted by WHO standard—2 µg/l. Despite that obtained values are lower than the level of exposure (LE = 75 µg/l) accepted by in NRS of Russian Federation [9, 10], uranium concentration in some places of Bol'shaya Tel' in September is, in general, can be considered as abnormal. It is known that natural uranium is a mixture of three isotopes: ^{238}U—99.2739% ($T_{1/2}$ = 4.468 × 10^9 years), ^{235}U—0.7024% ($T_{1/2}$ = 7.038 × 10^8 years) and ^{234}U—0.0057% ($T_{1/2}$ = 2.455 × 10^5 years). In contrast to other isotope pairs, last two isotopes are in constant proportion, regardless of high migration activity of uranium and geography: $^{238}U/^{235}U$ = 137.88 [13, 14]. The presence of uranium was truly established in the waters of Bol'shaya Tel', it can only be originated artificially: in the sample from 1000 m (October 2006) ~0.05 ng/l and in the sample from Bol'shaya Tel' (March 2007) ~0.03 ng/l. In addition, the ratio of $^{236}U/^{234}U$ at these places is 1:0.8, respectively.

Besides, water samples obtained in September provided information about anion content of NO_3^- (~2 mg/l, while the maximum permissible concentration (MPC) is 45 mg/l), CH_3COO^- (~7 mg/l) in the waters of Bol'shaya Tel' (1000 m from estuary). It is considered that the presence of such anions can indicate the non-equilibrium conditions in basin solution. Such situation is considered rather usual for liquid radioactive wastes, where acetate and nitrate, due to kinetic limitations of the acetate oxidation by nitrate, can coexist even at high (about 100°C) temperatures [10].

Generalized information about the total uranium content in water samples of the Yenisei River is given in **Figure 2**.

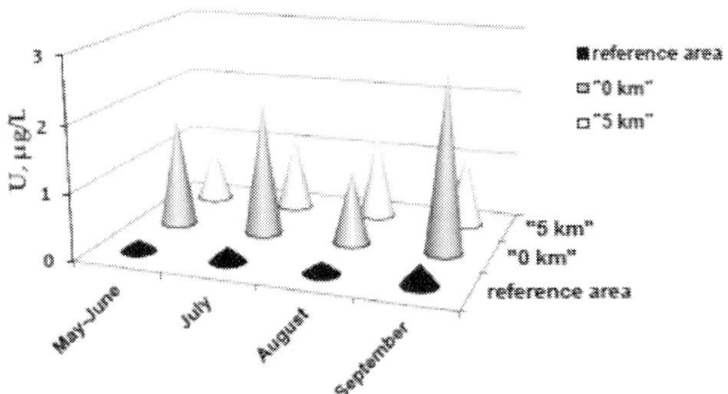

Figure 2. Results of determination of total uranium content in Yenisei water at distances from water discharge of MCC '0 km' and '5 km', taken 2006–2009, 'distance from water discharge MCC'.

Presented data indicate uranium content in the estuary of Ploskiy river '0 km' to exceed by 6–9 times background values of uranium typical for Yenisei. Further investigation of isotope composition of indicated water samples revealed that a ratio of uranium isotopes differ from natural isotopes and also the presence of ^{236}U can also evidence the industrial origin of high uranium concentrations as compared with background values. Isotope analysis of some samples has been carried out.

In water samples of Yenisei (pick point '0 km') the ratio of $^{238}U/^{235}U$ is 119:120. Besides, artificial uranium isotope ^{236}U ($T_{1/2} = 2.39 \times 10^7$ years) was found, the ratio of which to ^{234}U equals $^{236}U/^{234}U$ ~0.1–0.2. Thus, one can state that high uranium concentration in Yenisei waters is caused by MCC activity.

2.2. Tritium and other radionuclides

2.2.1. Tritium

Besides artificial radionuclides, Yenisei water was also polluted high tritium content. To prove this, the tritium content was determined in the picked water samples. Results are given in **Figure 3**.

Tritium content in the picking site '0 km' exceeds by 15–20 times the background tritium content obtained via long-term monitoring and typical for Yenisei (4 ± 2 Bk/l) [15–20].

To prove industrial origin of tritium in water samples it is recommended to control content of gamma-emitting radionuclides. There is significant amount of artificial radionuclides in the studied water.

2.2.2. Radionuclides without tritium

Depending on the state of radionuclides that can be present as simple ions to molecules and hydrolyzed forms, colloids and pseudocoolloids, organic and inorganic particles [21, 22] and, respectively, migrates over long distances and be sorbed by ecosystem immediately near the discharge area. Content of TUE in surface basins is extremely low and equals 10^{-10}–10^{-15} M, within limits of the most sensitive spectral techniques, e.g. mass-spectrometry [23, 24]. For the precise determination of TUE contents as well others radionuclides such as ^{90}Sr in water systems, the most frequently used methods are hybrid ones, combining preliminary concentrating and separating of radioisotopes with various detecting methods, e.g. alpha-, beta- and gamma-spectrometry [25–27].

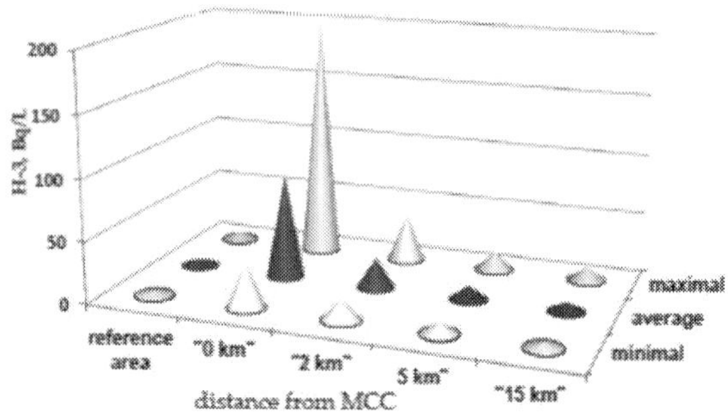

Figure 3. Average tritium content in water samples of Yenisei (distance down the stream from places of water discharge by MCC).

To increase the number of identified radionuclides, the method for concentrating the radionuclide from Yenisei water samples has been introduced [8]. Data obtained after concentration of water samples is given in **Tables 1** and **2**.

Water samples contain the bunch of artificial radionuclides. To increase the number of identified radionuclides, the method for concentrating the radionuclide from Yenisei water samples has been improved [8].

The method for concentrating the radionuclide was accepted on the basis of two widely known methods of co-precipitation with oxyhydroxide of Fe (III) and Mn (IV) oxide [28, 29].

Artificial radionuclides, which have different origin, have been found in water samples: induced (activated) radionuclides—^{24}Na, ^{46}Sc, ^{51}Cr, ^{54}Mn, ^{59}Fe, ^{60}Co, ^{65}Zn, ^{76}As and others; satellite radionuclides—^{99}Mo, ^{124}Sb, ^{131}I, ^{133}I, ^{141}Ce, ^{144}Ce and others. The most distinctive are trans-uranic radionuclides—^{239}Np, isotopes of Pu. In water samples, taken down the stream from MCC (5 km), besides the decreasing concentration of artificial radionuclides there were found some natural radionuclides: ^{210}Pb and ^{232}Th. There were included the presence of long-living satellite isotope ^{152}Eu ($T_{1/2}$ = 13.6 years) ~ 0.04–0.06 Bk/l and the presence of short-living activated radionuclide ^{58}Co ($T_{1/2}$ = 71.3 days) ~ 0.03–0.07 Bk/l in water samples.

2.3. Suspended matter of the Yenisei River: trucks for transport of radionuclides in the water flow

Because major part of radionuclides has been found in the suspended matter, transporting by water stream of Yenisei, more thorough studies of suspended matter of Yenisei have been conducted.

№	Isotopes	2007	2008	2009
1	^{24}Na	2.5 ± 1.4	1.9 ± 0.2	–
2	^{46}Sc	0.21 ± 0.01	0.136 ± 0.006	0.086 ± 0.06
3	^{51}Cr	6.0 ± 0.2	2.7 ± 0.1	3.4 ± 0.1
4	^{54}Mn	0.014 ± 0.003	0.014 ± 0.003	0.007 ± 0.002
5	^{59}Fe	0.16 ± 0.01	0.11 ± 0.008	0.07 ± 0.01
6	^{60}Co	0.13 ± 0.01	0.17 ± 0.008	0.09 ± 0.01
7	^{65}Zn	0.11 ± 0.01	0.055 ± 0.007	0.03 ± 0.004
8	^{76}As	8.5 ± 0.6	4.5 ± 0.2	4.7 ± 0.6
9	^{85}Sr	–	0.014 ± 0.003	0.003 ± 0.001
10	^{99}Mo	–	0.093 ± 0.008	0.04 ± 0.01
11	^{103}Ru	0.027 ± 0.004	0.026 ± 0.003	0.012 ± 0.006
12	^{106}Ru	0.078 ± 0.025	–	0.04 ± 0.01
13	^{124}Sb	0.016 ± 0.003	0.020 ± 0.003	0.012 ± 0.004
14	^{131}I	0.051 ± 0.013	0.031 ± 0.005	0.028 ± 0.008
15	^{133}I	–	–	0.14 ± 0.02
16	^{137}Cs	0.057 ± 0.005	0.142 ± 0.009	0.09 ± 0.02
17	^{141}Ce	0.048 ± 0.006	0.050 ± 0.006	0.021 ± 0.007
18	^{144}Ce	0.08 ± 0.02	0.13 ± 0.02	0.04 ± 0.01
19	^{239}Np	29.5 ± 1.4	17.1 ± 0.3	10.3 ± 0.8

Table 1. Radionuclide content in water samples after concentrating, taken in the place of MCC discharge ("0 km"), Bk/l.

N	Isotopes	2006	2007	2008	2009
1	^{24}Na	–	–	–	0.07 ± 0.025
2	^{46}Sc	0.11 ± 0.02	0.09 ± 0.02	–	0.002 ± 0.001
3	^{51}Cr	2.6 ± 0.2	1.4 ± 0.2	0.037 ± 0.013	0.057 ± 0.009
4	^{58}Co	–	–	–	–
5	^{60}Co	0.14 ± 0.02	0.11 ± 0.04	0.006 ± 0.001	0.002 ± 0.001
6	^{65}Zn	0.10 ± 0.03	0.07 ± 0.02	0.003 ± 0.001	0.005 ± 0.002
7	^{76}As	3.1 ± 0.3	0.08 ± 0.03	1.07 ± 0.08	0.103 ± 0.015
8	^{106}Ru	0.3 ± 0.1	0.4 ± 0.3	–	0.0064 ± 0.0061
9	^{131}I	0.04 ± 0.01	–	0.002 ± 0.001	0.0023 ± 0.0009
10	^{137}Cs	0.07 ± 0.02	0.04 ± 0.01	0.001 ± 0.001	0.0015 ± 0.0013
11	^{140}La	0.16 ± 0.03	0.08 ± 0.03	–	0.006 ± 0.002
12	^{144}Ce	0.25 ± 0.07	0.04 ± 0.02	–	–
13	^{152}Eu	0.06 ± 0.02	0.04 ± 0.02	–	–
14	^{239}Np	0.27 ± 0.02	0.32 ± 0.04	0.39 ± 0.02	0.261 ± 0.007

Table 2. Radionuclide content in water samples, taken from Atamanovo region, after concentrating (taken at 5 km down the stream from the place of discharge), Bk/l.

The investigations were carried out in the middle reach of the River Yenisei at the site 15 km (from the inflow of the Plosky stream (0 km) to the village Bolshoy Balchug (15 km) (**Figure 1**). The stream with technogenic admixtures propagates along the bank of the river not more than 0.1 one-tenth of the width of the river, i.e. along the flood plain where the river flow speed and the width are several times less.

As a result of ultra-filtration method, it was found that the main part of the suspended particles (up to 90%) was concentrated in the pelitic fraction of >5 µm. The filters with the suspensions were fixed on the specimen mount with the help of the conducting double-sided adhesive carbon type and placed into the electron microscope chamber. The precipitate was found to contain particles of quartz, mica and iron-containing minerals (limonitic and magnetic iron), mainly, with the size not exceeding 10–15 µm. Moreover, the precipitate revealed the presence of a considerable amount of various biological objects (diatoms, annelids, plant spores, etc.). All the mineral particles and biota were covered with a layer of fine limonitic-clayish particles. Spectral analysis of some parts of the sample (selected particles, characteristic details) was carried out. The suspended matter contains a large colony of diatoms, for example, *Meridion circulare, some cyclotellas* and *opyphoros, Cyclotella vor. Jacutca* (**Figure 4**).

The fraction with the size of '5-1 µm' uniformly covers the filter surface with a layer of fine particles. The precipitate mainly consists of mineral components (calcite, clays, clayish minerals, quartz and gypsum debris).

The fraction '1.0–0.2 µm' uniformly covers the filter surface with a layer of fine particles of the micron and submicron size, they are mainly aluminosilicate compounds having various structure and composition, limonite, calcite and gypsum.

Figure 4. Material composition of the water suspensions (separated by the ultra-filtration method). The fraction ≥5 µm. Magnification power of 2000×.

The material composition of the solid suspensions in the Yenisei River water generally corresponds to the mineral compositions of the rocks and the products of their hypergenesis which collected from the channel and the banks of the river. Occasionally, the admixture of the particles of technogenic origin (ash wastes from boiler stations) is observed.

Thus, it was shown that the suspended substance is similar to its geomorphology with the bottom sediments of the Yenisei River. However, the suspensions entering the river with the industrial discharge water significantly differ from the suspensions of the mainstream both in their composition and particle size.

At the sampling of the district runoff of radionuclides when the time of the discharge contact with the river water was insignificant, the radionuclides ^{3}H, ^{24}Na, ^{60}Co, ^{239}Np and ^{99}Mo (~90%) were mainly presented as a fraction <0.2 μm (filtrate). These can be both free ions in the molecular solution (e.g., ^{24}Na^{+}), and molecules or sorbed ions in colloid particles which managed to pass through a 0.2 μm filter. ^{46}Sc, ^{214}Bi, ^{103}Ru are mainly presented in solid phase, while the last two isotopes being in the coarsest fraction (more than 90% of them). ^{85}Sr and ^{131}I have less uniform phase distribution. ^{76}As is almost absent in the most coarse fraction (>5 μm). In the samples taken 5 km downstream, there is a decrease of the total activity, first of all, due to the coarse particle sedimentation. The radionuclide redistribution according to the size fractions was found: almost the whole amount of ^{60}Co is concentrated in the fraction with the size of >1 μm, a considerable amount of ^{214}Bi is transformed into a solution (the fraction <0.2 μm), almost 40% of ^{99}Mo and up to 70% of ^{24}Na are transformed into the fraction of 1–0.2 μm. With the total background level decrease there appear natural radionuclides ^{212}Pb and ^{234}Th in the solid phase as well as ^{65}Zn in the solution.

3. Mathematical calculations of the mass transport of technogenic radionuclides in the water flow of the River Yenisei in the impact zone of the Mining and Chemical Combine

In the chapter, the results radiation-chemical situation in the middle reach of the Yenisei River located in the nearest zone of the influence of the Mining and Chemical Combine of Rosatom have been described. It has been shown that a wide range of radionuclides, heavy metals and organic substances of different genesis flow into the waters of the Yenisei River. It has been demonstrated that radionuclides and other pollutants are transported by the water flow in the form of molecular solution or colloids or with suspended matter. In this case, the suspended matter consists of pelitic finely dispersed mineral particles, plant and organic detritus and amounts of living biological objects.

Calculations have been made according to the described method in the area of the River Yenisei from the estuary of the river Plosky up to the island Atamanovsky. Assuming the water discharge to be Q = 4085 m^{3}/s the river depth $H \approx 7$ m and the flow rate v = 1.25–1.8 m/s in the given section are estimated based on the hydraulic model. According to an earlier estimation, the stream with the technogenic admixtures propagates along the right bank, not far than one tenth of the river width, i.e. along the flood plain where the flow rate and depth are several times lower than those calculated based on the hydraulic model. According to the calculations: $H_n \approx 2.5$ m, $v_n \approx 0.38$–0.44 m/s.

Transport of radionuclide along the Yenisei River is based on a modified one-dimensional model proposed by Schnoor et al. [30]. For the whole length of the Yenisei, a homogeneous distribution of radionuclides over the cross-section is presupposed. It is assumed that both in the water column and in the active sediment layer the radionuclides are present in two forms: soluble and adsorbed forms. The most important processes influencing the behaviour of radionuclides include adsorption and desorption, sedimentation of suspended particles from the river water and resuspension from the active sediment layer, activity exchange between the pore water of the sediment and overlying water due to diffusion through the boundary and radioactive decay.

The calculations presented in this chapter are limited to the abiotic form of substance transport since the contribution of the biogenic component is considered to be insignificant [9].

Complex fresh water systems, such as large rivers, are assumed to be composed of a chain of interconnected 'elementary segments (ES)' that are comprised of: (a) the water column, (b) an upper sediment layer strongly interacting with water ('interface layer'), (c) an intermediate sediment layer below the 'interface layer' ('bottom sediment'), (d) a sink sediment layer below the 'bottom sediment', (e) the right and left sub-catchments of each ES.

Depending on the water discharge rate and geometry of the river bed the stream velocity varies which determines the transport of the sediment suspensions and sediment disturbance-sedimentation. To estimate the accumulation of radionuclides in the bottom sediments, a mathematical model described by Belolipetsky and Genova was used [31].

The concentrations of radionuclides on solid particles were assumed to be proportional to the area of the particle surface. We used the field data the fraction distribution of radionuclides in the initial solution. Then, the particle transport and sedimentation along the river bed was estimated. In the channels and floodplain (in the areas with small stream velocities) there occurs sedimentation of the sediment suspensions. During the periods of the increased water discharge rate (spring floods, increased volume of the hydroelectric station), the sediment disturbance is also possible as well as transport of impurities downstream (secondary pollution).

To describe the sediment suspension transport in a turbulent flow of non-compressible liquid a simplified equation is used:

$$\partial Si/\partial t + u_6\, \partial Si/\partial x \;=\; qSi/h + q/\omega \cdot Siq \tag{1}$$

where S_i is the concentration of the ith fraction [kq/m^3]; S_{iq} is the concentration of an impurity of the ith fraction, entering with the tributary on the way q; q_{Si} is sediment disturbance-sedimentation of the impurity of the i-th fraction; t is time; x is a coordinate directed along the current; Q is the discharge rate; ω is the cross-section area of the river bed; $u_6 = Q/\omega$ is the cross-section; average velocity h is the depth.

The bottom exchange is determined by the formula

$$qSi \;=\; (Si\,tr - Si0)\cdot wgi,\; Si\,tr = 0.01\cdot \alpha i \cdot Str,\; qS = \sum qSj \tag{2}$$

$$S_{tr} = \begin{cases} 0.2 \cdot u_{s}/gh\, w_{g}, \; if \; w_{g} \; < \; w_{*} \\ , \; w_{g} = (\rho_{S} - \rho_{s})/\rho_{s} \cdot g/18v \cdot d^{2}_{cp} \\ 0, \; if \; w_{g} \; \geq \; w_{*} \end{cases} \tag{3}$$

The transport capability of the flow S_{tr} depends on the depth-average flow velocity, depth and hydraulic coarseness; q_{s} is the mass exchange with the bottom.; S_{i0} is the concentration of the i-th fraction near the bottom; α_{i} is the percent content of the i-th fractions in the bottom sediments. When calculating $S_{i\,tr}$ using Eq. (3) it should be taken into account that $S_{i\,tr}$ cannot exceed the concentration of the i-th fraction in the bottom sediments ($S_{i\,day}$), therefore, when $S_{i\,tr} > S_{i\,day}$ it is assumed that $S_{i\,tr} = S_{i\,day}$. If the concentration of the i-th fraction in the bottom sediments is equal to zero, then $S_{i\,tr} = 0$.

The main change in the bottom sediment composition is assumed to be due to sediment disturbance and sedimentation. When $q_{s} > 0$, the bottom sediments enter the flow (washing out, sediment disturbance) and when $q_{s} < 0$ the silting of the river bed is observed (sedimentation of the suspended particles).

Let z_{*} be the thickness of the active layer of the bottom sediments. Assuming that the formation of the upper layer of the bottom sediments (with the thickness z_{*}) results in the sediment disturbance-sedimentation, the mass conservation equation for the i-th fraction in the bottom sediments $S_{i\,day}$ is written as follows:

$$\partial(z_{*} \cdot S_{i\,day}) / \partial t \; = \; -q_{Si} \tag{4}$$

Since $\sum q_{Si} = q_{S}$, $\sum S_{i\,day} = \rho$, from Eq. (4) one obtains the equation to find z_{*}:

$$\partial z_{*}/\partial t \; = \; -q_{S}/\rho \tag{5}$$

The calculation algorithm for the suspended and bottom sediment dynamics consists of the following stages:

Stage 1. The water flow rates u_{w} are determined as well as the depth h from the solution of the Saint-Venant equation.

Stage 2. Determination of the initial conditions. The granulometric composition of the bottom sediments in the section $X - X_{j}$ is taken to be $(d_{i}, a^{0}_{iday,j})$, where d_{i} is the diameter of the ith fraction particle (mm), $a^{0}_{iday,j}$ is the percentage of the i-th fraction in the bottom sediments, $i = 1, 2, \ldots, n$.

Stage 3. Establishment of the boundary condition in the initial section ($X = X_{0}$). In the initial section, $S^{n}_{iday,0}$ are determined using relations employed for the second stage, $S^{n}_{i,0}$ are estimated using the field data.

Stage 4. Estimation of the mass exchange between the bottom water and water flows. From the condition $w_{gi} \leq w_{*}$, $w_{*} = 0.4u_{*}$ one determines the fractions which are suspended. Let the suspended fractions be assigned the following index $i = 1, 2, \ldots, i_{*}$, $a_{i,j}$ is the percentage of the suspended fractions in the section. The percentage of all the suspended fractions is $r_{j} = a_{1,day,j} + a_{2,day,j} + \ldots + a_{i,day,j}$. Then, the percentage of the suspended ith fraction is

$$a_{i,j} = 100 \cdot r_{j}^{-1} \cdot a_{i,day,j}, \; i = 1, 2, \ldots.i \tag{6}$$

If $r_j = 0$ (the suspended fractions are absent), then, all $a_{i,j} = 0$.

Stage 5. Estimation of the concentrations of the suspended and bottom sediments as well as the location of the water-bottom interface.

Stage 6. Calculation of the granulometric composition of the bottom sediment:

$$a_{i,\ \partial H,j} = S^{n+1} \cdot \rho^{-1} \cdot 100 \tag{7}$$

Stage 7. Estimation of the bottom sediment radioactive contamination in the calculation sections.

Each fraction is assumed to be uniformly contaminated by radionuclides:

$$R^n_{i,j} = \lambda_i S^n_{i,j} \tag{8}$$

Knowing the contamination level in the initial section $R^n_{i,0} = \lambda_i S^n_{i,0}$, it is possible to estimate the level of the radionuclide contamination in the sections downstream the river

$$R^n_{i,j} = S^n_{i,j} \cdot (S^o_{i,j})^{-1} \cdot R^n_{i,0} \tag{9}$$

In the next time interval, the calculations are repeated (from stage 3 to stage 7).

The influence of the suspension-sedimentation processes on the admixture transport in the river flow close to the right bank of the River Yenisei in the studied area has been estimated.

The calculations made show that the concentrations of the lightest fraction in the calculation area almost do not change, while for the heavier fractions the decline of the suspended sediment concentrations is observed and the level of the radionuclide contamination also decreases (**Table 3**).

In the field data, the increase of the coarse fraction concentration is observed which is not connected with the suspension-sedimentation process. (S_nat, R_nat are the measured values, S_calc, R_calc are the calculated ones)

d, мм	0.00020	0.00045	0.005	0.01
District reset MCC				
S^0, г/л	0.0001	0.0005	0.0043	0.0031
R^0, Бк/кг	118.904	0.1728	0.1165	1.8224
Island Atamanovsky				
S_nat	0	0.0001	0.0009	0.0583
S_calc	0.0001	0.0005	0.0039	0.0021
Island Atamanovsky				
R_nat	2.4727	0.01101	0.01596	0.0853
R_calc	118.9009	0.1727	01044	1.2065

Table 3. Concentrations of particulate matter size fractions: real and calculated data.

Thus, the abiogenic mass-transport of the technogenic radionuclides, metals being among them, occurs mainly due to the coagulation of the suspended particles and contamination redistribution into bigger fragments.

Our calculations show that the concentration of the lightest fraction of the water on the current site remains virtually unchanged. However, we observed that concentrations of suspended sediment had decreased for heavier fractions and, consequently, decreased the level of contamination. In addition, our field data indicated an increase in the concentration of coarse fraction, which is associated not only with the resuspension-deposition, but also with the coagulation of suspended solids.

Thus, the data on artificial radionuclides entering the Yenisei River water obtained by long-term monitoring, which is likely to be connected with the activity of the industrial enterprises located on the river's banks of the studied area.

Acknowledgements

This investigation was made with financially supported by the Russian Bureau of fundamental researches N-16-05-00205

Author details

Lydia Bondareva[1]*, Valerii Rakitskii[1] and Ivan Tananaev[2]

*Address all correspondence to: lydiabondareva@gmail.com

1 Federal Scientific Center of Hygiene named after F.F. Erismana, Moscow, Russia

2 Far Eastern Federal University, Vlodivostok, Russia

References

[1] Sposito G. The Chemistry of Soils. Oxford: Oxford University Press; 1989. 347 p.

[2] Bolsunovsky A., Bondareva L. Actinides and other radionuclides in sediments and submerged plants of the Yenisei River. Journal of Alloys and Compounds, 2007; **444–445**: 495–499.

[3] Vakulovsky S.M. Radioactive contamination of surface water in the territory of Russia in 1961–2008. In.: Problems of hydrometeorology and environmental monitoring. Ed. Vakulovsky S.M., Obninsk, Roshydromet. Russia; 2010; 2: 115–127.

[4] Kuznetsov Y.V., Revenko Y.A., Legin V.K. et al. By the estimation of the Yenisey River contribution to the total radioactive pollution of the Kara Sea. Radiochemistry, 1994; **36**: 546–553.

[5]   Kuznetsov Y.V., Legin V.K., Shishlov A.A. et al. Investigation of behaviour of ^{239,240}Pu and ^{137}Cs in system Yenisei River - Kara Sea. Radiochemistry, 1999; **41**: 181–186.

[6]   Kuznetsov Y.V., Legin V.K., Strukov V.N. et al. Transuranic elements in the floodplain sediments of the Yenisei River. Radiochemistry, 2000; **42**: 470–477.

[7]   Bondareva L.G., Bolsunovsky A.Y. The study of modes of occurrence of man-made radio-nuclides ^{60}Co, ^{137}Cs, ^{152}Eu, ^{241}Am in the sediments of the Yenisey River. Radiochemistry, 2008; **50:** 475–479

[8]   Bondareva L.G., Bolsunovsky A.Y., Trapeznikov A.V., Degermedzhy A.G. Using the new technique concentrating of transuranic elements in the Yenisei river water samples. Doklady Chemistry, 2008; **423**: 479–482.

[9]   Sukhorukov F.V., Degermedzhy A.G., Belolypetsky V.M. et al. Patterns of distribution and migration of radionuclides in the valley of the Yenisei River. - Novosibirsk: SD RAS, GEO, 2004. 286 p. (in Russian)

[10] Broder J. Merkel, Britta Planer-Friedrich, Christian Wolkersdorfer (ed). Uranium mining and hydrogeology III, International Mine Water Association Symposium, 15–21 September 2002, Freiberg/Germany (Preventing uncontrolled spread of radionuclides into the environment, Novosibirsk: SD RAS NIC OIGGM, 1996).

[11] Fedorin M.A. Multiwave XRF-SR determination of U and Th in bottom sediments of Lake Baikal: Brunhes paleoclimatic chronology. GEOL GEOFIZ, 2001; **42**(1–2): 186–193 (in Russian)

[12] Egorova I.A., Puzanov A.V., Blokhin S.N. Natural radionuclides (^{238}U, ^{232}Th, ^{40}K) in the water of the northwestern Altai. World of Science, Culture, Education, 2007; **4**: 16–19 (in Russian)

[13] Tilton G.R. et al. Isotopic composition and distribution of lead, uranium, and thorium in a precambrian granite. Bulletin of the Geological Society of America,1956; **66** (9): 1131–1148.

[14] Rosholt J.N., et al. Evolution of the isotopic composition of uranium and thorium in Soil profiles. Bulletin of the Geological Society of America 1966; **77**(9): 987–1004.

[15] Bolsunovsky A.Y., Bondareva L.G. New data on the tritium content in one of the tributaries of the Yenisei River. Doklady Chemistry, 2002: **385**(5): 714–717.

[16] Bolsunovsky A.Y., Bondareva L.G. Tritium in surface waters of the Yenisei River basin. Journal of Environmental Radioactivity, 2003; **66**: 285–294.

[17] Bondareva L.G., Zharovtsova S.A. Determination of tritium content in the environment. Vestnik KrasGU, ser. Analytical Chemistry, 2003; **2**: 127–128.

[18] Bondareva L.G. New data on the ecological state of the River Yenisei. Russian Journal of General Chemistry, 2010; **3**: 153–161.

[19] Bondareva L.G. Mechanisms of tritium transfer in freshwater ecosystems. Vestnik of National Nuclear Center. Republic Kazachstan, Bulletin of the National Nuclear Center. 2011; **1**: 10–23 (in Russian)

[20] Bondareva L. Natural occurrence of tritium in the ecosystem of the Yenisei river. Fusion Science and Technology, 2015; **60**(4): 1304–1307.

[21] Pirkko H. Radionuclide migration in crystalline rock fractures. Academic Dissertation, Helsinki, 2002.

[22] Salbu B., Krekling T. Characterisation of radioactive particles in the environment. Analyst, 1998; **123**: 843–850.

[23] Solatie D. Development and comparison of analytical methods for the determination of uranium and plutonium in spent fuel and environmental samples: Academic Dissertation. Helsinki, 2002. 63 p.

[24] Solatie I. D., Carbol P., Betti M. et al. Ion chromatography inductively coupled plasma mass spectrometry (IC-ICP-MS) and radiometric techniques for the determination of actinides in aqueous leachate solutions from uranium oxide. Fresenius Journal of Analytical Chemistry, 2000; **368**: 88–94.

[25] Colley S., Thomson J. Particulate/solution analysis of ^{226}Ra, ^{230}Th and ^{210}Pb in sea water sampled by in-situ large volume filtration and sorption by manganese oxyhydroxide. Science of the Total Environment, 1994; **155**: 273–283.

[26] Meece D.E., Benninger L.K. The coprecipitation of Pu and other radionuclides with $CaCO_3$. Geochimica et Cosmochimica Acta, 1993; **57**: 1447–1458.

[27] Cochran J. K., Livingston H. D., Hirschberg D. J. et al. Natural and anthropogenic radionuclide distributions in the northwest Atlantic Ocean. Earth and Planetary Science Letters, 1987; **84**: 135–152.

[28] Romantchuk A.U. Laws of sorption behavior of actinide ions in the mineral colloidal particles. Russian Journal of General Chemistry, 2010; **3**: 120–128.

[29] Petrova A.B. Sorption of Np and Pu in colloids particles of Fe(III) and Mn(IV) oxides in the present of gumic acids. Dissertation, 2007. 117 p (in Russian)

[30] Schnoor, J.L., Mossmann, D.J., Borzilov, V.A., Novitsky, M.A. et al. Mathematical model for chemical spills and distributed source runoff to large rivers. In: Schnoor, J.L. (Ed.), Fate of Pesticides and Chemicals in the Environment. New York: Wiley Interscience, Environmental Science and Technology Series, 1992. pp. 347–370.

[31] Belolipetsky B.M., Genova C.H. Calculating algorithm for definition of dynamics of suspended and bed sediments in channel. Computational Technologies, 2004; **2**: 9–25 (in Russian)

Chapter 17

Processes and the Resulting Water Quality in the Medium-Size Turawa Storage Reservoir after 60-Year Usage

Marek Ruman, Żaneta Polkowska and
Bogdan Zygmunt

Additional information is available at the end of the chapter

http://dx.doi.org/10.5772/66226

Abstract

The characteristics of water in the Turawa reservoir, one of the important storage reservoirs in Poland, were thoroughly studied. The reservoir and also the rivers flowing into the reservoir were monitored in the period 2004–2006 with respect to the basic physico-chemical parameters determining the quality of water such as water temperature, specific conductance, pH, biochemical oxygen demand (BOD_5) and chemical oxygen demand (COD_{Mn}), water hardness, content of typical ions (sodium, potassium, sulphates, phosphates and chlorides), heavy metals, and so on. The observed seasonal and long-term changes of the parameters measured were discussed and the processes responsible for the changes suggested. The causes of the observed deterioration of the ecological status of the Turawa reservoir were given and the remedial operations proposed. The latter included improvement in the management of solid wastes and wastewater in the area, reducing the runoff of nutrients in the catchment, decrease in discharge of pollutants into the rivers flowing into the reservoir, removal of bottom sediments and also increasing the average water table and decreasing its fluctuations. The effect of the remedial operations will be further studied.

Keywords: Turawa reservoir, physico-chemical parameters, changes of water characteristics, remediation propositions, pollution sources

1. Introduction

The resources of surface waters in Poland are relatively low, not equally distributed and characterized by seasonal variability. The total annual runoff is about 57 km³ of water, which is equal to 1600 m³/person—this is a third of the European average runoff per person (statistic data). In order to improve the availability of water, quite a number of storage water reservoirs have been built in different parts of the country. Generally, water storage behind large dams is applied whereby the water is used for power generation, municipal water supply and flood control. Large storage reservoirs with controlled discharge provide larger amounts of water for periods of drought or low flow in the streams. In Poland, there are about 100 reservoirs characterized by a capacity of over 1 million m³ each. The total storage capacity of the Polish reservoirs is about 4000 million m³, which constitutes about 6% of the annual runoff (statistic data). Unfortunately, this storage capacity has not been sufficient up to now. Moreover, the strong anthropopressure still observed makes that the water quality is not satisfactory in many cases. So, the efforts are being undertaken to increase water retention and eliminate or, at least, strongly decrease water pollution.

One of the larger water reservoirs in Poland is the Turawa reservoir, operated since 1948. It is used to drive a hydroelectric power plant, as a source of water for municipal water supply systems, and also for recreation.

The Turawa reservoir is situated between 50°42′27″ and 50°44′32″ north latitude and 18°04′51″ and 18°10′59″ east longitude in Mała Panew Valley, which is the part of Równina Opolska (Opole Plain) (**Figure 1**).

Figure 1. Localization of the Turawa reservoir in Poland (Terra/Modis NASA).

It is situated on the Mała Panew river 16 km away from where it flows into the Odra river. The two other smaller rivers supplying the reservoir with water are the Libawa and the Rosa. The participation of these rivers in the surface water inflow to the Turawa reservoir is 87, 9 and 4%, respectively.

Morphometrical parameters of the Turawa reservoir are as follows: capacity, 95.5 hm³; surface area, 20.9 km²; length, 7.4 km; maximum width, 3.45 km; average width, 2.26 km; length of the shore line, 29.5 km; maximum depth, 13.68 m and average depth, 4.70 m. The water turnover time in the Turawa reservoir in the years 1996–2005 was 59 days on average: the shortest 37 days in 2001 and the longest 87 days in 2003. The turnover time was calculated as the quotient of the sum of water evaporating, water exchanged with catchment and water that flows out of the reservoir divided by the annual average capacity of the reservoir. The water turnover time is the theoretical value, which allows for the approximate assessment of the process. On the basis of this parameter, the Turawa reservoir can be classified as a high-flow water body.

The Turawa reservoir plays an important role in Poland, but due to the strong agricultural and industrial anthropopressure its water quality in the studied period was rather poor. The quality of water in the Turawa reservoir is determined by the composition of the water of the rivers flowing into the reservoir, that is, the Mała Panew river, the Libawa river and the Rosa river and also by the processes proceeding in the reservoir itself. The Mała Panew is the most polluted of the above-mentioned rivers; it carries industrial, municipal and agricultural pollutants. The largest amounts of wastewater are discharged from the several cities in the neighbourhood of the reservoir [1, 2].

The main source of nitrogen and phosphorous compounds is the application of fertilizers in arable land in the catchment of the Mała Panew and the Libawa rivers.

Any pollution prevention and remedial programmes need the determination of the large number of physico-chemical parameters characterizing water quality and the dynamics of the changes of these parameters.

2. Measured parameters and sampling points

All typical parameters generally used to classify surface waters were measured, mostly in the recognized laboratories of water analysis by means of recommended procedures and some on-site using specialized measuring instrumentation. Laboratory-measured parameters are given in **Table 1** (parameters typical for water purity classification).

The measurements were made for 20 sampling points situated in the selected sites of the reservoir, the rivers flowing into the reservoir, namely the Mała Panew, the Libawa and the Rosa, the river flowing out of the reservoir, that is, the Mała Panew, and the small lakes in the neighbourhood of the reservoir. The points for on-site measurements and for collecting samples for laboratory analysis are presented in **Figure 2**.

Specific conductance was measured on-site using conductance metre whose accuracy was 0.1 μS cm^{-1}. For oxygen content in water, the oxygen metre, enabling to carry out measurements up to the depth of 20 m, was used. Temperature was measured with the use of a mercury thermometer and also thermometers being integral parts of pH, oxygen and conductance metres. To determine water transparency, Secchi disk was applied. Bathymetric measurements were performed by means of a Lowrance sonar and an acoustic Doppler current profiler.

Parameter	Parameter (AAS)
Chemical oxygen demand (COD_{Mn})	Magnesium (Mg)
Biochemical oxygen (BOD_5)	Manganese (Mn)
Total hardness (TH)	Iron (Fe)
Calcium (Ca^{2+})	Cobalt (Co)
Magnesium (Mg^{2+})	Nickel (Ni)
Alkalinity (HCO_3^-)	Copper (Cu)
Chloride (Cl^-)	Zinc (Zn)
Sulphate (SO_4^{3-})	Cadmium (Cd)
Phosphate (PO_4^{3-})	Lead (Pb)
Nitrate (NO_3^-)	
Sodium (Na^+)	**Parameter**
Potassium (K^+)	Cyanides
Escherichia coli	PAH (total)
Total organic carbon	Total Phosphorous
Ammonia nitrogen	Total dissolved matter
Nitrogen by Kjeldahl	Phenol index
Total nitrogen	Anionic surfactants
Fluorides	Mineral oil index
Total alkalinity	

Table 1. Laboratory-measured parameters.

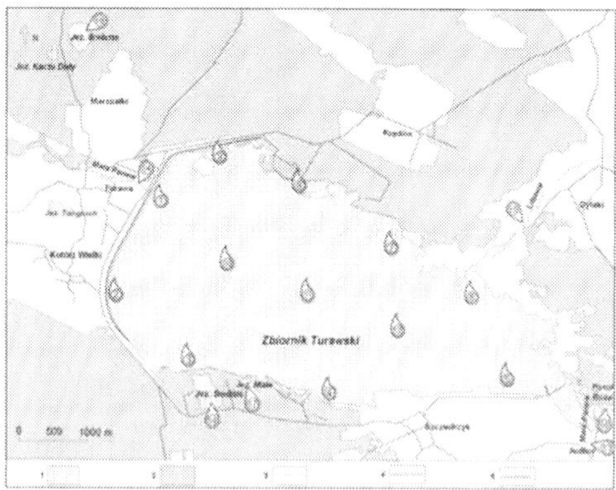

Figure 2. Location of the points for collecting samples for physico-chemical measurements: Geographical features: 1, water reservoirs: Jez. - lake and Zbiornik Turawski—the Turawa reservoir; 2, forests; 3, rivers; 4, roads; 5, dam.

Bathymetric measurements have shown that the reservoir is the deepest in the vicinity of the dam and the shallowest at the inflows of the rivers into the reservoir: the Mała Panew, the Libawa and the Rosa (**Figure 3**).

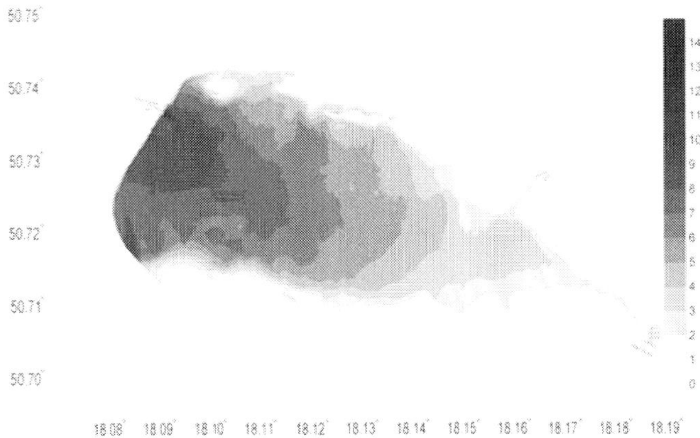

Figure 3. Bathymetric chart of the Turawa reservoir.

3. Water temperature in the reservoir

The water temperature and its fluctuations are important characteristics of storage reservoirs. In the areas characterized by well-defined cyclic temperature changes, the diurnal and seasonal temperature changes of surface water are typical. Thermal conditions of water reservoir are determined by climate (temperature of air, insolation, direction and speed of wind) and reservoir morphology, subsoil type and vegetation [3]. Water temperature can also be influenced by anthropogenic factors.

In the case of the Turawa reservoir, the water temperature changes in a wide range. It depends mainly on air temperature. In 2004–2006, the average water temperature was relatively well correlated with the changes of average monthly air temperature (correlation coefficient of 0.9).

In a year cycle, there are two periods of relatively stable temperature, winter season and summer season, and also two periods characterized by large water temperature changes, spring and winter seasons. The monthly minimum temperatures in the range of 0.6–0.8°C were observed in the period from December to February. The maximum temperatures ranging from 23.1 to 25.4°C were measured in July and August. In the period discussed, the lowest temperature was 0.3°C (January 2004) while the highest was 26.7°C in July 2006.

The changes of the limits of thermal layers and periods of water circulation and stagnation were also studied. Spring circulation starts after ice melts off and is characterized by temperature increase (from 0°C). Heavier water moves downwards while colder water from the bottom layer moves up resulting in water mixing and constant temperature in a vertical profile [4].

The period of summer water stagnation is characterized by temperature drop with depth. This is typical water stratification whereby the upper layer of the highest temperature close to air temperature is epilimnion and the coldest bottom layer is hypolimnion. The water between these two layers is metalimnion. Due to the small depth of the reservoir (in summer rarely more than 10 m in the deepest point) and intensive wind mixing in summer stagnation period thermal stratification was not observed—vertical temperature difference did not exceed 4°C (**Figure 4**).

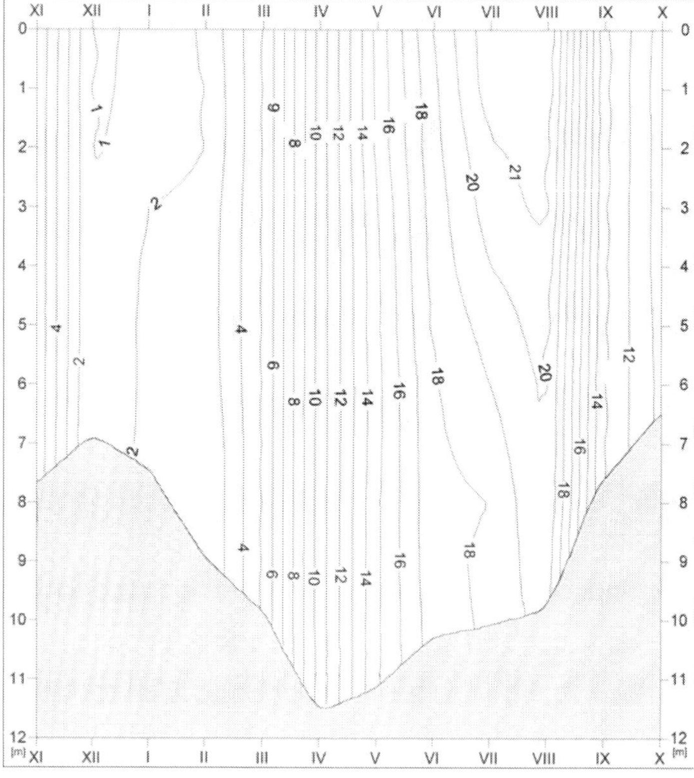

Figure 4. Changes in water temperature in the Turawa reservoir in the hydrological year 2004—vertical profile: 50°44'10.28"N, 18°05'30.22"E.

At the beginning of September, the period of autumn circulation starts in the Turawa reservoir. Due to cooling of surface layer, wind and convective mixing, the difference between the water temperature of epilimnion and hypolimnion in storage reservoirs decreases [5]. In the case of the Turawa reservoir, the process of decreasing difference in temperature was accelerated by intensive water outflow in the period from September to December. The water level dropped by 2.5 m and the water volume approximately by 30 million m^3. The process of water cooling ended when the temperature dropped to 4°C in the whole thermal profile, what for the Turawa reservoir was at the turn of November and December. From December to February (hydrological year 2004) or to March (hydrological years 2005 and 2006), winter stagnation occurred. In the deepest parts of the Turawa reservoir, water temperature was in the range of 2–2.7°C, while directly under ice layer it was 0.8°C.

In fact, summer stratification did not occur in the Turawa reservoir—the average water temperature was high and ranged from 20 to 21.8°C, whereas in winter it was in the range of 0.8–2.0°C. The annual average water temperature for the whole reservoir changed from 10.1 to 10.7°C. Thermal variability of the reservoir or the ratio of maximum temperature to minimum temperature reached the value of 26. In the hydrological years 2004–2006, the thermal conditions of the reservoir were strongly influenced by fluctuations of water level and the corresponding large changes of the volume of water from about 16 to 91 million m^3. The relatively small average depth of the Turawa reservoir favours wind mixing.

The interesting characteristics of the Turawa reservoir are water-freezing phenomena. The earliest time ice cover appeared was the end of November while the latest time its disappearance was the beginning of April; on average, it was the second half of December and the middle of March, respectively. The ice cover disappeared first in the coastal zone, most likely due to higher absorption of heat by land masses and intensive supply of surface and ground water.

The annual length of time the Turawa reservoir was covered with ice varied within a wide range. In the studied period, it changed from 18 days in 1994 to 106 days in 1984, on average 61 days. There were also large differences in the thickness of ice cover, an annual average cover ranged from 6 cm in 1988 to 26 cm in 1996. Due to the formation of ice cover, anthropogenic storage reservoirs, including the Turawa reservoir, can be used for recreation, too.

4. Oxygen saturation

The oxygen dissolved in surface water comes mainly from the atmosphere and photosynthesis processes; its content is one of the most important parameters characterizing the quality of water. Oxygen is necessary for the life of fishes and other water animals. The oxygen content depends on the physical (temperature) parameters and on biological processes. Water pollution generally results in drop of oxygen content. In the hydrological years 2004–2006, the Turawa reservoir water saturation with oxygen was typical (60–100%) from autumn to spring with the highest values in winter (**Figure 5**). In that period, the average oxygen saturation in vertical profile ranged from 58 to 102%.

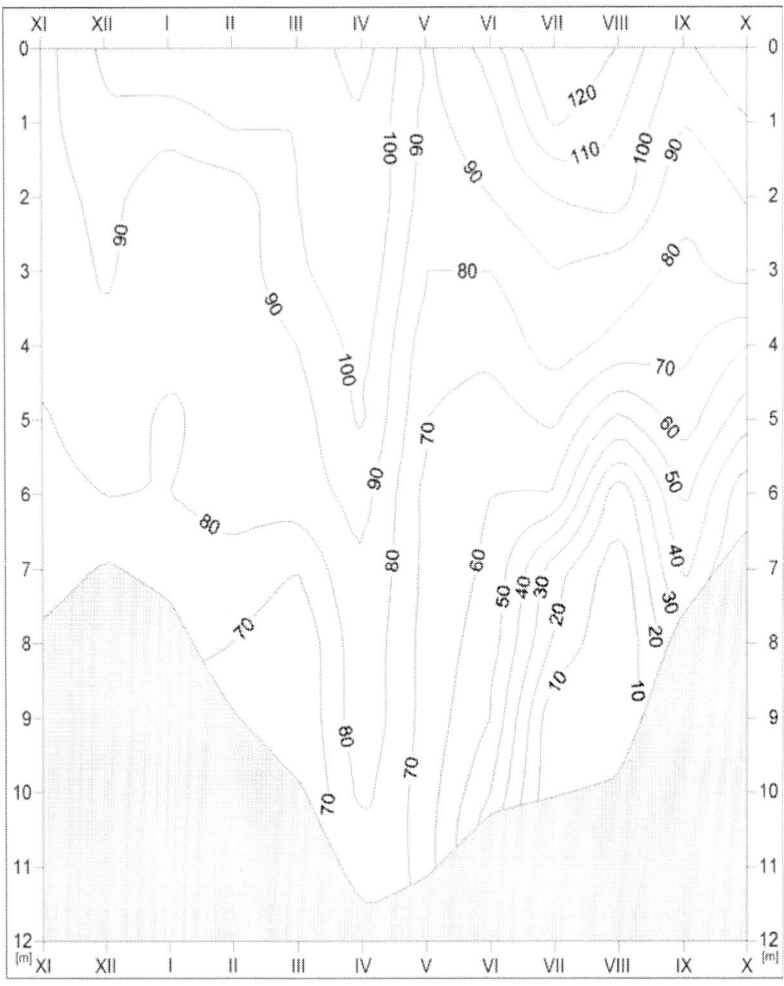

Figure 5. Oxygen saturation (in %) of the Turawa reservoir in the hydrological year 2004—vertical profile: 50°44′10.28″N, 18°05′30.22″E.

Due to increase of nutrients in limnic waters in summer, the oxygen saturation profiles are highly modified. In the Turawa reservoir, epilimnion water was oversaturated with oxygen (from 104 to 164%) because of intensive growth of phytoplankton which produces oxygen in the process. In hypolimnion, phytoplankton growth is limited and oxygen is consumed by decaying organic matter which results in drop of oxygen content. The oxygen deficit (saturation smaller than 60%) occurred from June to September. In July and August, the bottom layer of up to 6–7-m thick was completely depleted of oxygen. In fact, every year the oxygen saturation conditions of water in the Turawa reservoir were getting worse in summer period due to intensification of eutrophication processes. This is very unwanted process since oxygen deficiency is detrimental for the river fauna. With respect to oxygen content, water in the reservoir was classified to purity class III [6].

5. Conductance of water in the reservoir

Surface waters show large differences in water-specific conductance: from as low as 1µS cm^{-1} to 3000 µS cm^{-1}. Surface water conductance is strongly influenced by discharged wastewater, mainly industrial wastewater whose specific conductance can be as high as 10,000 µS cm^{-1}.

The concentration of mineral substances dissolved in water, which are responsible for water conductance, changes with time. The changes are caused by primary production, which diminishes the content of salts in water and also by the transfer of biogenic substances from bottom sediments to water which increases conductance. In the period 2004–2006, the monthly average specific conductance of water in the Turawa reservoir changed from 285 to 425 µS cm^{-1} (**Figure 6**). In the whole period studied, the lowest measured specific conductance was 272 while the highest was 495 µS cm^{-1} (average 339 µS cm^{-1}).

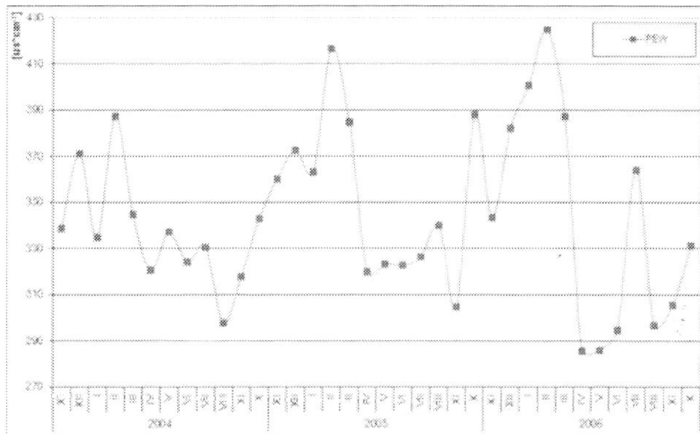

Figure 6. Changes in the average monthly specific conductance of the Turawa reservoir in the hydrological years 2004–2006.

In the season of snow melting (January and February), the conductance was the highest due to transfer to surface waters of inorganic compounds formed in the process of decay of organic matter present at the bottom and also due to runoff from spatial sources. The lowest specific conductance was measured in spring-summer season—most probably this resulted from maximum water level in the reservoir at that time. Statistically significant dependence (R^2 = 0.37) was found between the reservoir water level and conductance—the higher water level the lower the specific conductance. According to the regulations in force at that time, water in the Turawa reservoir was classified to purity class I with respect to that parameter in all measuring periods, in the hydrological years 2004–2006.

6. Water hardness

Water hardness is generally defined as the content of divalent cations, mainly calcium and magnesium in water. Hardness is divided into three categories: total hardness, temporary or bicarbonate hardness, and permanent or non-bicarbonate hardness. The total hardness or the total content of calcium and magnesium ions, generally expressed as mg $CaCO_3$ dm^{-3}, is most often used in the classification of surface water. Depending on geological characteristics of the catchment, it changes in a wide range, from several mg $CaCO_3$ dm^{-3} to several hundred mg $CaCO_3$ dm^{-3}. High hardness can be generally disadvantageous if water is to be used in household and industry. On the other hand, very low (smaller than 30 mg $CaCO_3$ dm^{-3}) hardness can be harmful for humans.

In the studied period, the monthly average total hardness in the surface water of the Turawa reservoir ranged from 106 to 191 mg $CaCO_3$ dm^{-3}. The lowest measured value of the total hardness was 95 mg $CaCO_3$ dm^{-3} while the highest was 210 mg $CaCO_3$ dm^{-3}; the monthly average hardness was 141 mg $CaCO_3$ dm^{-3} (**Figure 7**).

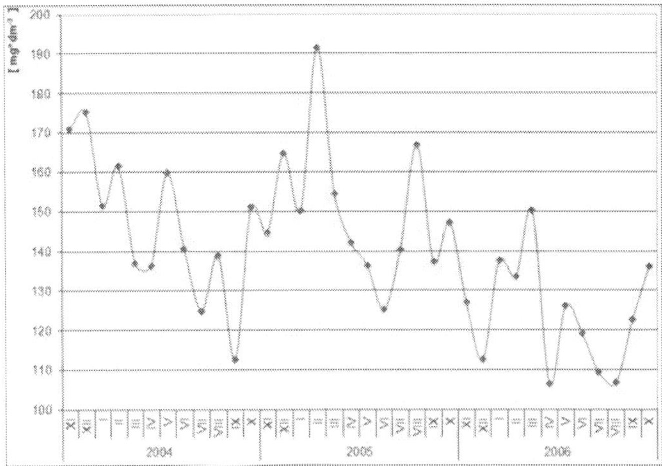

Figure 7. Changes in the average monthly total water hardness of the Turawa reservoir in the hydrological years 2004–2006.

In the studied period, the difference in the total hardness was relatively small. The water could be classified as soft or medium hard. Since the total hardness is determined by the concentration of calcium and magnesium, the changes of the monthly average concentrations of these ions were also measured. They are shown in **Figures 8** and **9**, respectively.

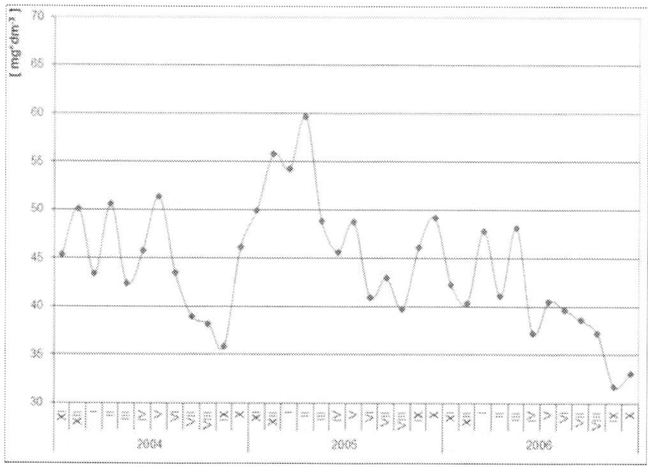

Figure 8. Changes in the average monthly calcium concentration in the waters of the Turawa reservoir in the hydrological years 2004–2006.

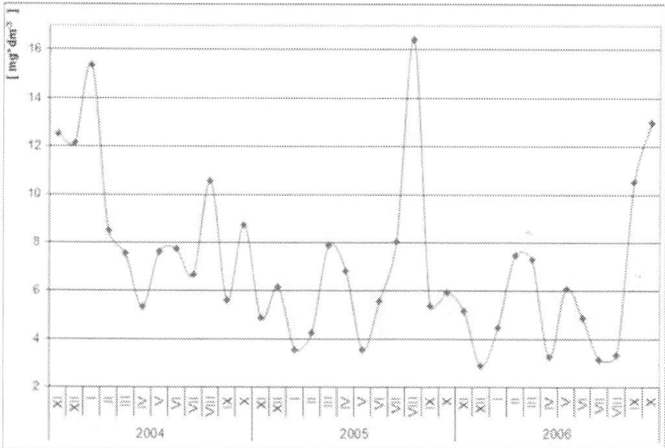

Figure 9. Changes in the average monthly magnesium concentration in the waters of the Turawa reservoir in the hydrological years 2004–2006.

Seasonal changes of calcium content were much higher than those of magnesium, which can be related to difference in solubility. The calcium content in non-polluted natural waters is generally four times higher than magnesium content. In the Turawa reservoir, it was about six times higher. The total hardness of water in the reservoir was the highest in winter, most probably due to mineralization of dead organic remains at that period.

7. Salinity changes

Water salinity is generally defined as the total content of the salts dissolved in water. In the case of storage reservoirs, usually the contents of chlorides, sulphates, potassium and sodium ions are determined in water. High salinity can be harmful to the environment and it should be monitored.

7.1. Chlorides and sulphates

The occurrence of chloride ions is common in the environment, including water. Their content in natural non-polluted water ranges from trace level to several hundred mg dm^{-3}. They are also present in vegetation and in animals. The presence of chlorides in water is responsible for the increased rate of corrosion. Chlorides are also harmful to fresh water vegetation [7].

In the hydrological years 2004–2006, the monthly average concentration of chlorides in the reservoir was in the range from 13.88 to 53.00 mg dm^{-3} (**Figure 10**). The concentration distribution was characterized by relatively low dispersion. The highest concentrations were found in early spring during snow cover melting. In winter, snow from roads was discharged into the neighbourhood of the reservoir and in the valley of the Mała Panew river. Generally, snow from the roads contains increased amounts of chlorides used in winter road maintenance. Chlorides with water from snow melting could reach the reservoir.

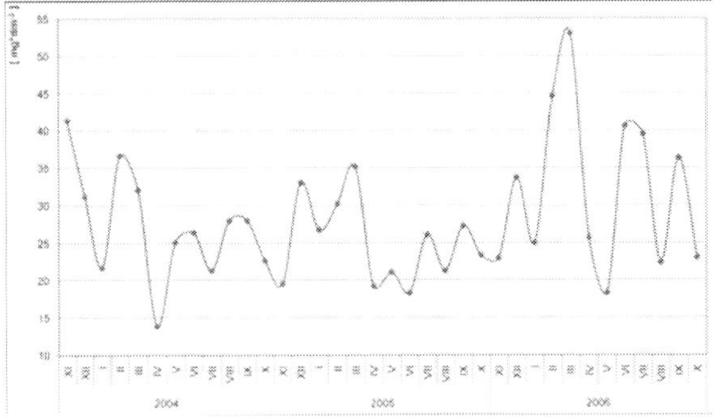

Figure 10. Changes in the average monthly chloride concentrations in the waters of the Turawa reservoir in the hydrological years 2004–2006.

The sulphates occur in natural waters in a wide range of concentrations. They are present at especially high concentrations, even up to 6000 mg dm^{-3}, in mine waters. The Turawa reservoir is surrounded by coniferous forests which assimilate larger amounts of SO_2 than deciduous forests. Sulphur in needle trees is in the form of sulphate ions, which can be leached out from fallen needles. However, anthropogenic factors had the most critical influence on the content of sulphates in the reservoir water. In the period studied, the average monthly concentration of sulphates in the reservoir water ranged from 28.9 to 76.7 mg dm^{-3} (**Figure 11**); the average for all the studied period was 52.2 mg dm^{-3}.

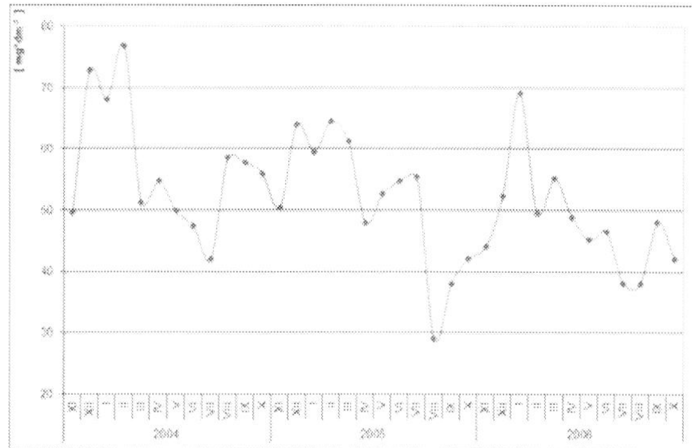

Figure 11. Changes in the average monthly sulphate concentrations in the waters of the Turawa reservoir in the hydrological years 2004–2006 (own elaboration).

The highest sulphate concentrations were found in winter due to decay of organic matter rich in sulphur compounds. In summer time, sulphate concentration dropped because living organisms were taking in sulphates and because, in anaerobic conditions in the benthic layer, they were reduced to hydrogen sulphide.

7.2. Sodium and potassium

The presence of sodium and potassium in natural waters is related to weathering of igneous rock and their leaching from sedimentary rock [8, 9]. Sodium salts are present in nearly all natural waters at the concentrations from several to 30 mg dm^{-3}. Similar to chlorides, sodium ions can come from rock salt and also from sea aerosols. There are also several anthropogenic sources of sodium, mainly wastewater from soda producing and processing industry, and from dye industry.

Potassium is usually present in natural waters in small quantity, generally at the level of several mg dm^{-3}. Strongly polluted industrial effluents and runoff water from fertilized arable land contain much higher concentration of potassium. Also in the leachate from municipal solid waste sites, potassium can be found at considerable concentrations.

In the Turawa reservoir, the average concentrations of Na$^+$ and K$^+$ ions were 12.4 and 4.5 mg dm^{-3}, respectively. In the period studied, the monthly average concentrations of sodium and potassium were in the range of 7.3–17.0 and 3.5–6.6 mg dm^{-3} (**Figures 12** and **13**), respectively. Fluctuations of Na$^+$ concentrations can be related to water level in the reservoir—the highest were in autumn-winter season when the water level was the lowest while the lowest were measured in spring during maximum filling of the reservoir. Seasonal changes of potassium concentration are not so distinct.

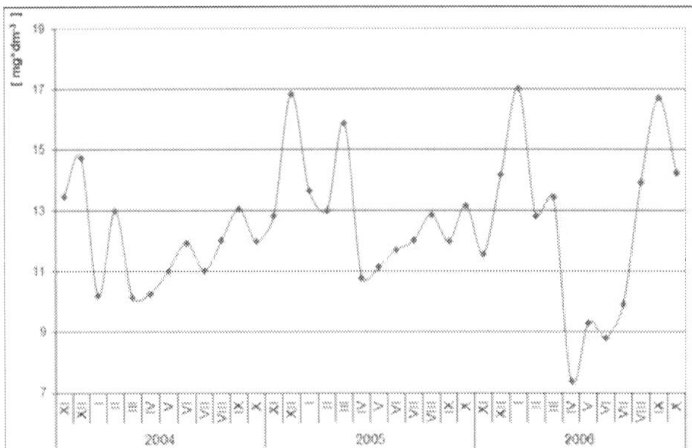

Figure 12. Changes in the average monthly sodium concentration in the waters of the Turawa reservoir in the hydrological years 2004–2006.

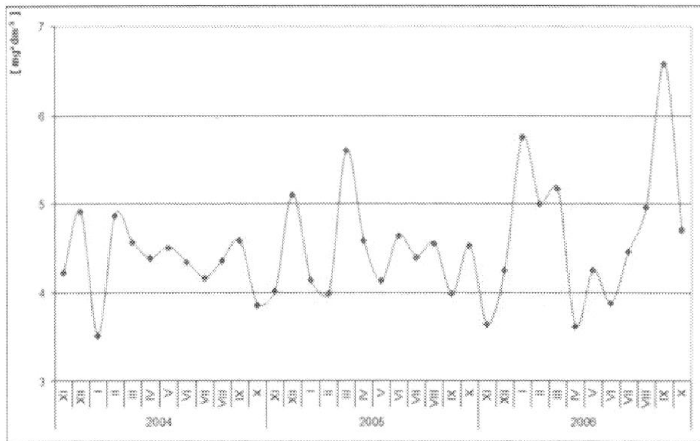

Figure 13. Changes in the average monthly potassium concentration in the waters of the Turawa reservoir in the hydrological years 2004–2006.

8. Concentration changes of organic substances

8.1. Classes of organic compounds

Surface waters can contain a wide spectrum of organic compounds belonging to different classes. The most important are carbohydrates, proteins, amino acids, esters, fats, organic acids, surfactants, soaps, ketones, alcohols, hydrocarbons, phenols, and also humic and fulvic substances. Most organic substances are of natural origin. Many are produced by animals and vegetation. They are also formed in the process of decay of dead remains of animal and vegetation. However, presently many organics found in surface waters come from anthropogenic sources. Except from humic substances, natural origin substances undergo biodegradation to simple inorganic compounds. Their presence is responsible for the deterioration of water smell and taste and can be harmful for living organisms. The presence of many anthropogenic compounds in surface water can disturb biological balance and make self-purification difficult [10].

Different human activities can really be a source of a large group of organic water pollutants. Chlorobenzenes are widely used in industry, in organic synthesis, and also as fungicides and insecticides. Some chlorophenols have fungicide and herbicide properties and some are used in producing other chemicals. Dioxins (polychlorodibenzodioxins and polychlorodibenzofurans), which are dangerous chemicals, are not produced on purpose, but they are contaminants of some chemical products. They are also formed in combustion of organic matter. Ethylenediaminetetraacetic acid is applied in steam generators to avoid precipitation of metals, in nuclear industry for surface treatment (decontamination), and so on. Glycols are used as hydraulic fluids and antifreeze agents. Polycyclic aromatic hydrocarbons are formed during forest fires, coal burning, and so on. Benzene, toluene, ethylbenzene and xylenes (BTEX) and aliphatic-chlorinated hydrocarbons are utilized as solvents. Volatile thiols can be present in the wastewater of petroleum industry. Anthropogenic phthalates are as common as natural chloride and sodium ions and presently can be found nearly everywhere; they are found in paints, inks, lacquers and solid organic polymers. Styrene can be found in plastics, synthetic rubber, copolymers; vinyl chloride is a residual monomer in polyvinyl chloride (PVC). Volatile organochlorinated compounds (VOX), mainly trihalomethanes (water chlorination), ethylene and ethane derivatives (solvents) can easily enter the environment. Some VOXs are produced by algae biomass in the process of chlorination. Pesticides include organochlorinated compounds (DDT, lindane, deldrin, etc.), organo-phosphorous compounds (parathion, malathion, etc.), sulphonates, carbamates and chlorophenoxy acetic acids. Phenols are formed in decomposition of vegetable products, can be present in industrial cellulose-containing wastewater, and were used as wood preservatives (pentachlorophenol). Sources of polychlorinated biphenyls (PCBs), pollutants widespread in the environment, are industrial wastes and leakage from transformers. PCBs undergo bioaccumulation easily. Being produced and applied or used in the production of other chemicals, organic compounds can enter the environment, including the surface water whereby they can be harmful in many ways.

8.2. Basic organic pollution parameters

Organic compounds differ greatly in physico-chemical properties and in threat to living organisms so they should be determined as groups or individual compounds. This could be and is costly and also labour- and time-consuming activity, and therefore as the first step of the determination of organic pollution, the total organic pollution parameters are used. The most typical and delivering important information on water quality are biochemical oxygen demand (BOD_5) and chemical oxygen demand (COD). The former is a measure of content in water of organic matter that can be decomposed biologically, while the latter shows the amount of organic matter which can be oxidized by strong oxidizer as potassium permanganate (COD_{Mn}) or potassium chromate (COD_{Cr}), respectively.

The basic water quality indicators mentioned above were monitored for the Turawa reservoir. In hydrological years 2004–2006, the monthly average BOD_5 for the whole studied period was 5.2 mg O_2 dm^{-3}; it was changing in a wide range from 0.6 to 16.7 mg O_2 dm^{-3} (**Figure 14**). The annual average BOD_5 increased from 3.6 mg O_2 dm^{-3} in 2004 to 8.6 mg O_2 dm^{-3} in 2006. Occasionally, very high concentrations were found. In the period from June to September 2006, the maximum BOD_5 values exceeded 40 mg O_2 dm^{-3}.

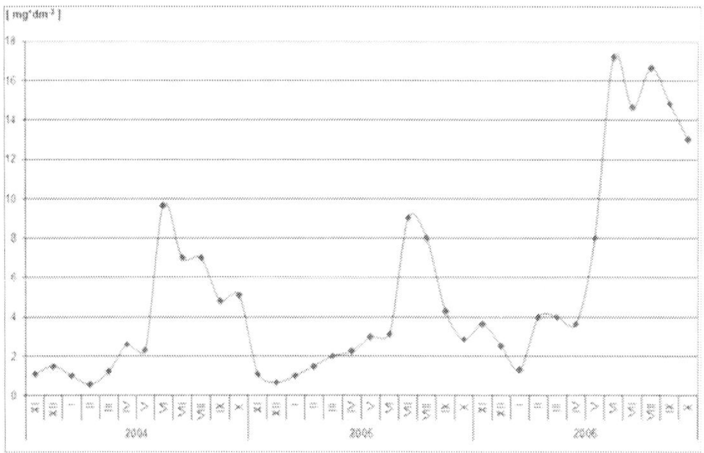

Figure 14. Changes in the average monthly BOD_5 concentration in the waters of the Turawa reservoir in the hydrological years 2004–2006.

In the above-measuring period, the monthly average COD_{Mn} parameter was 9.2 mg O_2 dm^{-3} and changed from 5.2 to 9.2 mg O_2 dm^{-3} (**Figure 15**). Occasionally, very high values of COD_{Mn} were measured, for example, from September to October 2006 the maximum COD_{Mn} values exceeded 25 mg O_2 dm^{-3}.

Both BOD_5 and COD_{Mn} values were season dependent; in winter were the lowest while in summer the highest (**Figures 14** and **15**) due to intensive eutrophication and resulting growth of algae and green algae during hot season. These organisms emit considerable amounts of organic compounds in metabolic processes and during decaying.

If water contains neither toxic substances nor organic matter resistant to biodegradation, the good correlation between BOD_5 and COD_{Mn} can be expected. For the water of the Turawa reservoir, correlation coefficient was 0.79.

In October 2006, a series of other pollution parameters were determined. They included total organic carbon (TOC), and indicators of industrial pollution such as free cyanides, phenolic index, anionic surfactants, mineral oils and polycyclic aromatic hydrocarbons. Fortunately, the indicators of industrial pollution were found quite low. This shows that the water in the Turawa reservoir was polluted mainly with organic matter of natural origin, which is caused by increasing eutrophication.

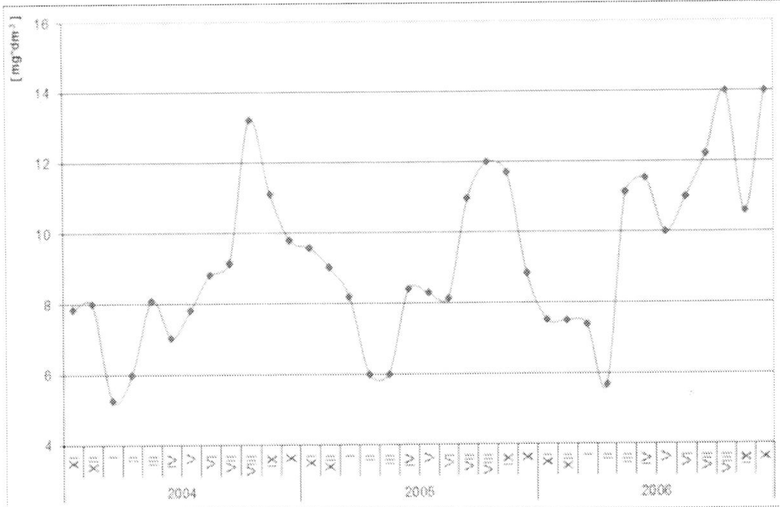

Figure 15. Changes in the average monthly COD_{Mn} concentration in the waters of the Turawa reservoir in the hydrological years 2004–2006.

9. Eutrophication processes

Eutrophication is a system response to the addition of nutrients, mainly phosphates, resulting in the 'bloom' or great increase of phytoplankton in a water body. Oversupply of nutrients, inducing explosive growth of phytoplankton and algae, results in the consumption of oxygen when these species die. The oxygen depletion level may lead to fish kills and a number of other effects reducing biodiversity. Generally, low oxygen content has many negative consequences [11, 12].

The eutrophication of the Turawa reservoir was mainly of anthropogenic origin. Nutrients were delivered by inflowing rivers, mostly the Mała Panew river; they come mainly from municipal wastewater, from arable land and forests surrounding the reservoir, and from tourist resorts. Considerable participation in eutrophication had nutrients released from bottom sediments gathering in the reservoir for tens of years. In 2004–2006, intensive release of phosphates from bottom sediments was observed during spring periods caused by water mixing. Phosphates in the form of compounds with iron or with aluminium were also released during summertime due to anaerobic conditions in the benthic layer. The intensive eutrophication in summer is favoured by hydrological and meteorological conditions such as small stream flow of inflowing rivers, low water level and high water temperature.

10. Content of heavy metals

There is no widely agreed criterion-based definition of heavy metals. Quantitative criteria used to define heavy metals included density, atomic mass and atomic number. Often applied density criteria range from above 3.5 g cm^{-3} to above 7 g cm^{-3}. In aquatic environment, heavy metals occur as micro-pollutants. Their natural sources include rock weathering, volcano eruptions, sea aerosols, forest fires and geological processes. Most important anthropological sources are transport, power industry, chemical industry, solid waste sites, fertilizers, flue gases, and so on. Heavy metals are stable pollutants and once introduced into a given environment can remain there for a long time [9].

In aquatic environment, heavy metals can be in the form of ions and dissolved complexes, and bonded to inorganic suspended matter and also to dead remains of vegetation and animals. Heavy metals are toxic species, even at rather low concentrations, and toxicity is strongly dependent on the metal. Their presence can influence the bloom of green algae.

In 2004–2006, the quarterly measurements of such heavy metals as manganese (Mn), iron (Fe), cobalt (Co), nickel (Ni), zinc (Zn), lead (Pb) and cadmium (Cd) in the water of the Turawa reservoir were conducted. They included the following: 12 measuring series were performed, 130 determinations of each metal in each series. The measurements were made for different reservoir filling to achieve the most reliable results. The program of sampling must have been precisely planned since heavy metals undergo multiple processes of desorption and adsorption during turbulent movement of water. The main source of heavy metals in the reservoir was the Mała Panew river. In the catchment of this river, zinc and lead ores were mined for about 100 years in the sixteenth/seventeenth century. At the end of the eighteenth century, mining and metallurgy of heavy metals were developed in the area again. Mines and numerous processing plants were discharging polluted industrial effluents into the Mała Panew river and its tributaries resulting in heavy pollution of river water and bottom sediments in which heavy metals accumulate resulting in steady increase in concentration. The heavy metals accumulated in sediments in the past can re-enter the liquid phase of the river water. Both forms of the pollutants, that is, dissolved in water and bonded to sediments, can be transported to the Turawa reservoir. In the years studied, there were 46 point sources discharging the pollutants to the Mała Panew river, 36 (26 industrial) of which were situated in the catchment of the reservoir. Due to the tendency to accumulate, the heavy metals were present at high concentrations in the sediments deposited in the reservoir, for example, the permissible level for cadmium was exceeded 10 times and for lead and zinc seven times. The heavy metals present in sediments can be reintroduced into the liquid phase during water rippling and flow which enhance the fluctuation of water level and hence uncovering the reservoir bottom.

10.1. Manganese (Mn)

The occurrence of manganese in surface water is quite common. It is leached out of rocks and soil. According to some researchers, the concentration of 6 $\mu g\ dm^{-3}$ is taken as a background for river water. The main anthropogenic sources include Mn metallurgy and burning coal and petrol. In small amounts, manganese is needed and present in vegetation and animals. It plays the role of catalyst in biochemical processes.

In the Turawa reservoir, the average content of manganese was 0.0590 mg dm^{-3} and changed from below detection limit to 0.2043 mg dm^{-3}. The highest concentration was determined in winter due to advantageous conditions of leaching from subsoil. In summer, the content dropped due to adsorption on suspended matter.

According to surface water classification, the reservoir water, with respect to Mn content, was classified most often as the second class of purity, and occasionally as the first class in summer and as the third in winter.

10.2. Iron (Fe)

Iron is quite common in the lithosphere; its content is about 5%. It is present in mineral waters due to the process of leaching out of rocks and soil. Considerable amounts are released into surface water with wastewater from the plants processing iron and with mine water. Its typical concentration in surface water is on the level of several mg dm^{-3}. Depending on the content of organic matter, oxygen, carbon dioxide, microorganism activity and pH, iron can be present in water in a dissolved form, as colloids and suspended mater. Some iron compounds can be assimilated by aquatic organisms.

In the water of the Turawa reservoir, the average iron concentration was 0.1338 mg dm^{-3} and changed in the range of 0.009–0.4145 mg dm^{-3}. According to surface water classification in the years the studies were conducted, the reservoir water with respect to Fe content was classified as purity class I in summertime, as class II in autumn-winter period when, at the low level of water and the increased bottom exposure, iron compounds are leached out of sediments.

10.3. Cobalt (Co)

Cobalt concentration in surface water is quite small, most often single to several $\mu g\ dm^{-3}$. The main anthropogenic sources are industrial waste and coal burning. In non-polluted surface waters, its typical concentration is 0.05 $\mu g\ dm^{-3}$ while the natural average content is about 0.02 $\mu g\ dm^{-3}$. Cobalt undergoes the process of bioaccumulation in phytoplankton and accumulates in bottom sediments.

In the water of the Turawa reservoir, the average cobalt concentration was about 12.0 $\mu g\ dm^{-3}$ and ranged from below detection limits to 62.6 $\mu g\ dm^{-3}$.

10.4. Nickel (Ni)

In water, nickel can be present in dissolved form as a two-valent cation, or in the complexes, most often as cyanide complex or in undissolved form as cyanide, sulphide, carbonate or hydroxide. In pure waters, its concentration is about 5 µg dm^{-3}, while in strongly industrialized areas, surface waters can contain up to about 0.020 mg dm^{-3}. Considerable amounts of nickel are emitted during combustion of diesel fuel [13].

In the water of the Turawa reservoir, the average cobalt concentration was about 6 µg dm^{-3} and ranged from below detection limits to 16.5 µg dm^{-3}. According to surface water classification in force in the years studied, the reservoir water with respect to Ni content was classified as purity class I. Only occasionally, the limiting concentration for class I was slightly exceeded.

10.5. Copper (Cu)

Copper is quite common in the earth crust; its content is about 0.02%. In water, it is in the form of quite mobile complexes with humic and fulvic acids. Copper compounds are easily bonded to sediments. Copper is present in mineral waters due to the process of leaching out of rocks and soil. The natural concentration of copper in surface waters is about 0.002 mg dm^{-3}. The main anthropogenic sources of copper are metallurgy, copper-processing factories, corrosion of products made out of this metal and its numerous alloys. Copper compounds are used as biocides to kill blooming algae [14].

In the water of the Turawa reservoir, copper concentration ranged from non-detectable to 0.0202 mg dm^{-3} with an average of 0.0111 mg dm^{-3}. According to surface water classification in force in the years studied, the reservoir water with respect to Cu content was classified as purity class I. The highest concentration was found in summertime during intensive water blooming since algae are capable to accumulate heavy metals, including copper. Algae were able to assimilate up to 90% of copper present in the reservoir.

10.6. Zinc (Zn)

Zinc in the earth is present in the form of minerals though its occurrence is not common. Small amounts of zinc can be found in water due to leaching from soil. However, the main sources of zinc in surface water are effluents of zinc smelters, zinc processing factories, chemical industry and coal burning. Pipe corrosion can also be a source of zinc in water, especially in drinking water. Polluted waters contain zinc at a concentration of 0.005–0.015 mg dm^{-3}.

In the water of the Turawa reservoir, the average zinc concentration was about 35.4 µg dm^{-3} and ranged from below detection limits to 169 µg dm^{-3}. According to surface water classification obligatory in the years studied, the reservoir water with respect to Zn content was classified as purity class I.

10.7. Lead (Pb)

Despite common occurrence in the earth crust, lead content in natural waters is low. However, due to common pollution of surface waters with lead, it is rather difficult to assess its natural content. Lead is present at 0.003 mg dm^{-3} in non-polluted waters. The main anthropogenic sources of lead are chemical industry, mining and ore-processing industry. In industrial areas whereby non-ferrous metal smelters are present, atmospheric precipitation can be a source of lead in surface waters [14].

In the water of the Turawa reservoir, the average lead concentration was 0.0116 mg dm^{-3}; it changed from non-detectable to 0.412 mg dm^{-3}. According to surface water classification obligatory in the years studied, the reservoir water with respect to Pb content was most often classified as purity class III and occasionally as class IV. The high concentration resulted from the fact the Mała Panew river delivering 87% water to the reservoir was strongly polluted with heavy metals, including lead. Most lead in water bodies collects in bottom sediments where from it can later move back to water.

10.8. Cadmium (Cd)

Cadmium, in the free form, is not present in the earth crust, but occurs in ores of other heavy metals, mainly zinc. It is very toxic and accumulates in living organisms. It is one of the most dangerous substances on the earth. The main sources of surface water pollution with cadmium are industrial effluents from cadmium mining and metallurgy and corrosion of galvanic covers. It is also present in some phosphate fertilizers. In non-polluted waters, it is present at 0.02 µg dm^{-3}. Cadmium remains in water for relatively long time, then precipitates as carbonate and undergoes adsorption on suspended matter and sediments. In the process of bonding to sediments, bacteria participate converting cadmium compounds to CdS [14].

In the water of the Turawa reservoir, the average Cd concentration was 0.0025 mg dm^{-3}; it changed from non-detectable level to 0.0164 mg dm^{-3}. According to surface water classification, the reservoir water with respect to Cd content was most often classified as purity class IV and occasionally as class V. The high concentration in the water of the reservoir resulted from the process of chemical denudation of this element in the catchment area of the Mała Panew river. Cadmium present at high concentration in bottom sediment did not re-enter water layer since it is sparingly soluble in neutral and alkaline solution. Leaching of cadmium took place only at the lower level of water (in winter) when some part of the reservoir bottom is uncovered. The additional reason of cadmium pollution is runoff from arable land whereby phosphate fertilizers and dolomite fertilizers produced out of waste of lead and zinc mining and metallurgy are applied. Lower concentration of cadmium was observed during intensive blooming of algae which can absorb up to 80% of cadmium in the reservoir. It was found out that vegetation and animals take in cadmium at the rate proportional to its content.

11. Conclusions

Conducting the studies on the conditioning and effects of physico-chemical properties of the water in the Turawa reservoir in the hydrological years 2004–2006 confirmed the hypothesis that its ecological status was deteriorating due to strong agricultural and industrial anthropopressure.

The changes of physico-chemical properties of the reservoir were determined by the quality of inflowing river water, mainly the Mała Panew river, and inflow of pollutants from localities and resorts without the appropriate sewage systems. Moreover, the biological and chemical processes proceeding in the reservoir resulted in the changes of physico-chemical properties of water.

According to the guidelines of monitoring programme of lakes, the Turawa reservoir is characterized by moderate susceptibility for anthropopressure according to morphometric, hydrological and basin characteristics. However, the water of the reservoir could have been classified as the third class of purity, that is, as water strongly polluted.

According to the qualitative classification of surface water, the reservoir was classified as unsatisfying (class IV) or bad (class V) with respect to oxygen-related parameters (BOD_5 and COD_{Mn}), total organic carbon, microbiological parameters and also to the concentration of heavy metals such as lead and cadmium.

In 2004–2006, due to strong eutrophication the following physico-chemical parameters of water worsened: colour, odour, oxygen saturation, transparency, pH, BOD_5 and COD_{Mn}. Positive tendency of decreasing water hardness and concentrations of calcium, magnesium and sulphates took place.

One of the largest threats of ecological status of the reservoir was water eutrophication which threatens ecosystem operation. The most important threats were oxygen deficit in deep areas of the reservoir which is dangerous for aerobic organisms, oxygen oversaturation of epilimnion, decrease in water transparency, pH increase, accumulation of organic matter, reservoir overgrowing, deterioration of taste and smell, and release of phosphorous compounds in anaerobic conditions. Progressing degradation of aquatic environment decreases the usable value of water, lowers the aesthetics of the environment and management of all the water bodies.

Deteriorating ecological status of the Turawa reservoir was required to take immediate comprehensive measures to prevent it from total degradation. Remedial propositions were as follows:

- Proper management of solid wastes and wastewater and building sewage system around water body.

- Reduction of runoff of nutrients out of the catchment to the reservoir (building of wastewater treatment plant).

- Increase in year average water table elevation.

- Reducing fluctuations of water table level.

- Decreasing the content of phosphorous and heavy metals in bottom sediments.

- Removal of bottom sediments.

After 10 years, the physico-chemical parameters are being measured again to assess the results of remedial actions.

Author details

Marek Ruman[1*], Żaneta Polkowska[2] and Bogdan Zygmunt[2]

*Address all correspondence to: marek.ruman@us.edu.pl

1 Faculty of Earth Sciences, KNOW (Leading National Research Centre) University of Silesia, Sosnowiec, Poland

2 Department of Analytical Chemistry, Faculty of Chemistry, Gdansk University of Technology, Gdańsk, Poland

References

[1] Tsakovski S, Kudlak B, Simeonov V, Wolska L, Namiesnik J. Ecotoxicity and chemical sediment data classification by the use of self-organising maps. Analytica Chimica Acta. 2009; 631: 142–152.

[2] Simeonov V, Wolska L, Kuczyńska A, Gurwin J, Tsakovski S, Protasowicki M, Namieśnik J. Sediment-quality assessment by intelligent data analysis. TrAC Trends in Analytical Chemistry. 2007; 26: 323–331.

[3] Wang S, Huang Z, WuYu-S, Winterfeld PH, Zerpa LE. A semi-analytical correlation of thermal-hydraulic-mechanical behavior of fractures and its application to modeling reservoir scale cold water injection problems in enhanced geothermal reservoirs. Geothermics. 2016; 64: 81–95.

[4] Bocaniov SA, Ullmann Ch, Rinke K, Lamb G, Boehrer B. Internal waves and mixing in a stratified reservoir: Insights from three-dimensional modeling. Limnologica – Ecology and Management of Inland Waters. 2014; 49: 52–67.

[5] Jorgensen SE; Loffler H; Rast W; Straskraba M; Evaluating Lake and Reservoir Water Quality. Chapter 3, Developments in Water Science. 2005; 54. Elsevier Science, ISBN: 9780080535340.

[6] Hudson JJ, Vandergucht DM. Spatial and temporal patterns in physical properties and dissolved oxygen in Lake Diefenbaker, a large reservoir on the Canadian Prairies. Journal of Great Lakes Research. 2015; 41: 22–33.

[7] Bernhardt-Römermann M, Kirchner M, Kudernatsch T, Jakobi G, Fischer A. Changed vegetation composition in coniferous forests near to motorways in Southern Edition Germany: The effects of traffic-born pollution. Environmental Pollution. 2006; 143: 572–581.

[8] Bartarya SK. Hydrochemistry and rock weathering in a sub-tropical Lesser Himalayan river basin in Kumaun, India. Journal of Hydrology. 1993; 146: 149–174.

[9] Gaillardet J, Viers J, Dupré B. Trace Elements in River Waters, Treatise on Geochemistry. 2003; 5: 225–272

[10] Gerbersdorf SU, Cimatoribus C, Class H, Engesser KH, Helbich S, Hollert H, Lange C, Kranert M, Metzger J, Nowak W, Seiler TB, Steger K, Steinmetz H, Wieprecht S. Anthropogenictrace compounds (ATCs) in aquatic habitats—Research needs on sources, fate, detection and toxicity to ensure timely elimination strategies and risk management. Environment International. 2015; 79: 85–105.

[11] Liu Y, Wang Y, Sheng H, Dong F, Zou R, Zhao L, Guo H, Zhu X, He B. Quantitative evaluation of lake eutrophication responses under alternative water diversion scenarios: A water quality modeling based statistical analysis approach. Science of The Total Environment. 2014; 468–469: 219–227.

[12] Munyati C. A spatial analysis of eutrophication in dam reservoir water on the Molopo River at Mafikeng, South Africa. Sustainability of Water Quality and Ecology. 2015; 6: 31–39.

[13] Mirabi A, Shokuhi RA, Nourani S. Application of modified magnetic nanoparticles as a sorbent for preconcentration and determination of nickel ions in food and environmental water samples. TrAC Trends in Analytical Chemistry.2015; 74: 146–151.

[14] Bilal M, Kazi TG, Afridi HI, Arain MB, Baig JA, Khan M, Khan N. Application of conventional and modified cloud point extraction for simultaneous enrichment of cadmium, lead and copper in lake water and fish muscles. Journal of Industrial and Engineering Chemistry. 2016; 40: 137–144.

Chapter 18

Impact of Wastewater on Surface Water Quality in Developing Countries: A Case Study of South Africa

Joshua N. Edokpayi, John O. Odiyo and
Olatunde S. Durowoju

Additional information is available at the end of the chapter

http://dx.doi.org/10.5772/66561

Abstract

Wastewater effluents are major contributors to a variety of water pollution problems. Most cities of developing countries generate on the average 30–70 mm^3 of wastewater per person per year. Owing to lack of or improper wastewater treatment facilities, wastewater and its effluents are often discharged into surface water sources, which are receptacles for domestic and industrial wastes, resulting to pollution. The poor quality of wastewater effluents is responsible for the degradation of the receiving surface water body. Wastewater effluent should be treated efficiently to avert adverse health risk of the user of surface water resources and the aquatic ecosystem. The release of raw and improperly treated wastewater onto water courses has both short- and long-term effects on the environment and human health. Hence, there should be proper enforcement of water and environmental laws to protect the health of inhabitants of both rural and urban communities. This study reports major factors responsible for the failing state of wastewater treatment facilities in developing countries, which includes poor operational state of wastewater infrastructure, design weaknesses, lack of expertise, corruption, insufficient funds allocated for wastewater treatment, overloaded capacities of existing facilities, and inefficient monitoring for compliance, among others.

Keywords: developing countries, health risk, pollution, surface water, wastewater effluents, wastewater treatment facilities

1. Introduction

Freshwater availability is one of the major problems facing the world, and approximately, one-third of drinking water requirement of the world is obtained from surface sources like rivers, dams, lakes, and canals [1]. These sources of water also serve as best sinks for the discharge of

domestic and industrial wastes [2, 3]. The biggest threat to sustainable water supply in South Africa is the contamination of available water resources through pollution [4]. Many communities in South Africa still rely on untreated or insufficiently treated water from surface resources such as rivers and lakes for their daily supply. They have no or limited access to adequate sanitation facilities and are a high risk to waterborne diseases [5]. Since 2000, there has been a dramatic increase in the episodes of waterborne diseases in South Africa [6, 7].

Surface water has been exploited for several purposes by humans. It serves as a source of potable water after treatment and as a source of domestic water without treatment particularly in rural areas in developing countries. It has been used for irrigation purposes by farmers, and fishermen get their occupation from harvesting fish in so many freshwater sources. It is used for swimming and also serves as centers for tourist attraction. Surface water, therefore, should be protected from pollution. Major point sources of freshwater pollution are raw and partially treated wastewater. The release of domestic and industrial wastewater has led to the increase in freshwater pollution and depletion of clean water resources [8]. Most quantities of wastewater generated in developing countries do not undergo any form of treatment. In few urban centers, various forms of wastewater treatment facilities (WWTFs) exist but most of them are producing ill-treated effluents, which are disposed of onto freshwater courses.

In some developed countries of the world, adequate supply of potable water and improved sanitation facilities have been achieved. Strict environmental laws and monitoring for compliance prevent undue pollution to freshwater sources. Good waste management technologies and increased environmental protection awareness have contributed immensely to the success story. This has resulted in fewer cases of waterborne diseases reported compared to developing countries.

Many people in developing countries of the world still rely on untreated surface water as their basic source of domestic water supply. This is so because either there is an incessant supply of potable water or inadequate water supply systems. This problem is exacerbated in rural areas. Surface water is increasingly under undue stress due to population growth and increased industrialization. The ease of the accessibility of surface water makes them the best choice for wastewater discharge. Wastewater which comprises of several microorganisms, heavy metals, nutrients, radionuclides, pharmaceutical, and personal care products all find their way to surface water resources causing irreversible damage to the aquatic ecosystem and to humans as the aesthetic value of such water is compromised. These pollutants decrease the supply of useable water, increase the cost of purifying it, contaminate aquatic resources, and affect food supplies [9]. Pollution combined with the human demand for water affects biodiversity, ecosystem functioning, and the natural services of aquatic systems upon which society depends on.

Urban areas in most developing countries do have several wastewater management systems some of which are very effective and meet international standards, but many others are plagued with poor designs, maintenance problems, and expansion including poor investment in wastewater management systems. Most rural and poor communities often do not have any form of wastewater management systems. Effluents from large- and small-scale industries are usually channeled to surface water courses, which often result in pollution, loss of biodiversity in the aquatic ecosystem, and possibly health risk to humans.

Environmental quality and antipollution legislations are the most widely used interventions to control and reduce environmental pollution [10, 11]. In most countries, environmental laws have been enacted by the government and enforced through its administrative structures [12]. The use of criminal sanctions has also limited pollution but the enforcement of these environmental laws remains inadequate [12, 13]. Enforcement of environmental laws in South Africa like other developing countries suffers major setbacks due to inadequate technical experts, insufficient funds, corruption, and low deterrent effects of sanctions [12, 13].

2. Surface water quality

Surface water is one of the most influenced ecosystems on earth, and its alterations have led to extensive ecological degradation such as a decline in water quality and availability, intense flooding, loss of species, and changes in the distribution and structure of the aquatic biota [14], thus, making surface water courses not sustainable in providing goods and services [14, 15]. For instance, the health of a river system is influenced by various factors, which include the geomorphology and geological formations, physicochemical and microbial quality of the water, the hydrological regimes, and the nature of instream and riparian habitats [15].

Water quality is described by chemical, physical, and biological characteristics of water that determine its fitness for a variety of uses and for the protection of the health and integrity of aquatic ecosystems [16]. Each aquatic ecosystem has the natural tendency to adapt and compensate for changes in water quality parameters through dilution and biodegradation of some organic compounds [17]. But when this natural buffering capacity of the aquatic ecosystem is exceeded due to the introduction of various classes of contaminants from point and nonpoint sources on a continuous basis, water pollution sets in.

In South Africa like most other developing countries in the world, surface water is usually used for domestic, recreational, and agricultural purposes mostly in the rural areas [9, 18]. Water quality is affected by both natural processes and anthropogenic activities. Generally, natural water quality varies from place to place, depending on seasonal changes, climatic changes, and with the types of soils, rocks, and surfaces through which it moves [5, 19]. A variety of human activities such as agricultural activities, urban and industrial development, mining, and recreation significantly alter the quality of natural waters and change the water use potential [16, 19].

Decrease in water quality can lead to increased treatment costs of potable and industrial process water [19]. The use of water with poor quality for agricultural activities can affect crop yield and cause food insecurity [4]. The presence, transport, and fate of heavy metals and organic compounds (which are toxic and persistent) in water bodies are a cause for serious concern globally [4]. Groundwater can be polluted through the release of chemicals contained in wastewater. Riverbeds and wetlands are threatened with increased sediment impoundments and the presence of toxic and persistent chemicals. Such pollution can persist long after their original sources have ceased.

The health of the aquatic ecosystem can be negatively affected by the presence of toxic substances. This is further exacerbated with high population of pathogens in the water. The use of microbiologically contaminated water for domestic and other purposes is detrimental to human health and the society at large [20]. These conditions may also affect wildlife, which uses surface water for drinking or as a habitat. Generally, for measuring water quality, the physical (turbidity, electrical conductivity, temperature, total dissolved solids, color, and taste), chemical (pH, COD, BOD, nonmetals, metals, and persistent organic pollutants, POPs), and biological (fecal coliform, total coliform, and *enterococci* count) analyses are usually performed [19].

3. Wastewater treatment

Wastewater comprises of all used water in homes and industries including storm water and runoffs from lands, which must be treated before it is released into the environment in order to prevent any harm or risk it may have on the environment and human health. The major types of wastewater are shown in **Figure 1**.

The major aim of wastewater treatment is to protect human health and prevent environmental degradation by the safe disposal of domestic and industrial wastewater generated during the use of water. One of the objectives of wastewater treatment is to recycle wastewater for reuse in irrigation, thereby preserving water resources, which is scarce in arid and semiarid regions of the world [21, 22]. In ancient times, there was no specific treatment given to wastewater. Instead, wastewater was channeled from buildings into waterways through gutters and canals, which eventually ended up in rivers, streams, lakes, and oceans, which were used by people [23]. This natural treatment process based on dilution was adequate presumably due to a smaller population and low population density as well as human activities, resulting in lower pollution load as compared to the present times [23].

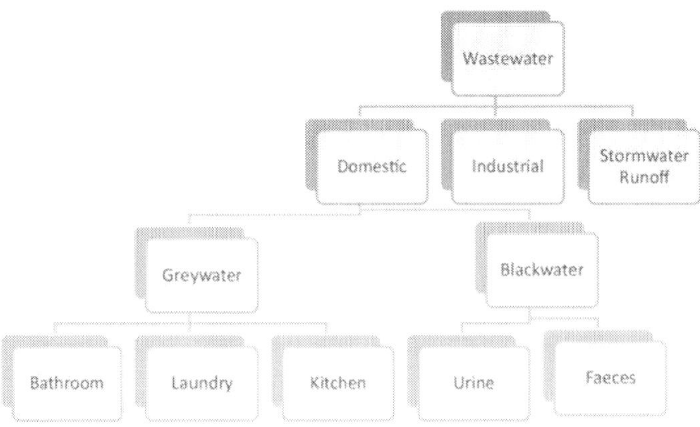

Figure 1. Types of wastewater.

Increase in population and industrial growth led to the generation of a high quantity of untreated wastewater channeled to water bodies as raw water [24]. Eutrophication, fish kill, and cholera outbreaks have commonly been reported in communities that use contaminated water for domestic and other purposes [24]. This necessitates the consideration of a more advanced technology in treating wastewater. Wastewater treatment facilities were initially designed to remove/decrease conventional pollution parameters (BOD_5, COD, total suspended solids, and nutrients) from the wastewater stream so that the final effluents do not constitute new sources of pollution [25]. However, it has been discovered that the wastewater organic load contains high levels of a variety of hazardous organic pollutants, and thus, additional treatment steps and control measures become very necessary [21, 25].

The quality of wastewater varies according to the types of influents the WWTFs receive such as domestic wastewater, dry and wet atmospheric deposition, urban runoff containing traffic-related pollution, or agricultural runoff [25]. The range of contaminants becomes broader when industrial wastewater is included into the raw water stream that enters a WWTF [25–27]. Recently, it has been shown that WW effluents contain emerging organic contaminants such as persistent organic pollutants (POPs), brominated flame retardants, per-fluorinated compounds, and pharmaceuticals, which are not removed during the treatment process [25, 28]. Wastewater treatment technology is fast changing so as to meet the current day challenge.

In many countries, urbanization is growing at an unprecedented rate, and such development is often unbalanced with much of the disposable municipal expenditure devoted to high-profile infrastructure with waste disposal and management coming well down in the list of priorities in terms of allocation of funds [29]. A study conducted by Saving Water South Africa [30] showed that less than half of the South African wastewater treatment plants (WWTPs) treat wastewater they receive to a safe and acceptable level [31]. The health risk from wastewater usually comes from microbial pathogens, nutrient loads, heavy metals, and some organic chemicals [31, 32]. Bacteria are the most common pathogens usually found in treated wastewater and cause several infections and diseases particularly to young, pregnant, immune-compromised and aged people [31, 33].

Most wastewater treatment facilities in South Africa dispose their effluents directly to nearby rivers or streams, which are used by the surrounding villages for their various water needs. Ogola et al. [34] demonstrated from their study on wastewater treatment facilities in Limpopo Province of South Africa that wastewater is rarely treated to acceptable standards, and this was further confirmed by Edokpayi [35] and Pindihama et al. [36]. Their findings suggest that inadequate investment in wastewater treatment infrastructure, shortage of skilled manpower, poor planning or corruption [31] could have resulted in the poor performances of wastewater treatment facilities in that region.

Wastewater needs to be adequately treated prior to its disposal or reuse in order to protect receiving water bodies from contamination [31]. The discharge of poorly treated wastewater usually affects water users downstream and contaminates groundwater [31]. Waste stabilization ponds (WSPs) are usually used to provide an effective and low-cost means of handling domestic wastewater for smaller towns and communities [34]. The use of WSPs is advantageous over the conventional WWTPs because they are very simple to design, operate, and maintain and do not necessarily need skilled manpower [37]. However, Jagals et al. [38] conducted a study on WSPs in the Free State Province of South Africa; their results revealed that most of the municipal waste stabilization ponds were performing less than the required standard. They recommended the implementation of a frequent monitoring program. Wastewater has also been implicated as a possible source of heavy metals, polycyclic aromatic hydrocarbons (PAHs), and microbial contamination to soils, surface water, sediment and groundwater [39–41]. The inadequate storage facility in most WWTPs often gives room to untreated wastewater loss into the surrounding lands and rivers, especially during heavy rains and flooding.

Weber et al. [21] showed from their studies the presence of a wide range of contaminants of emerging concerns in wastewater even after the conventional treatment process. Such contaminants include pesticides, polycyclic aromatic hydrocarbons, and pharmaceutical and personal care products. Bundschuh et al. [22] reported on the impacts of wastewater, that concerns have been largely associated the presence of microorganisms, while other toxic and persistent components like heavy metals and POPs have not been given appropriate consideration. Treated wastewater often contains some pollutants such as POPs and heavy metals which are not removed in the treatment processes. The use of such water for irrigation may lead to the accumulation of those contaminants in soils which can be bioavailable for uptake to plants and animals. Thus, providing a pathway through the food chain to man. They also have biological effects on soil fauna and flora after long-term application [39, 42, 43].

The cultivation of vegetables in soils irrigated with wastewater containing high concentrations of toxic metals usually take up such metals and accumulate them in edible and nonedible parts of the vegetables in quantities large enough to cause potential health risks both to animals and humans consuming these metal-rich plants [44–48]. Heavy metals have special features that make them toxic even at very low concentrations. They are nonbiodegradable and persistent in various environmental media and can accumulate in plants and animals [45]. The oral route has been identified as the major pathway through which heavy metals enter into the human body. Consumption of food crops from farmlands irrigated with wastewater and ill-treated wastewater effluents could make people who feed on them at risk to several diseases, some of which only become evident after many years of exposure [49, 50].

Wang et al. [43] reported the accumulation of polycyclic aromatic hydrocarbons (PAHs) in soils irrigated with wastewater such that the concentration of PAHs was found to be higher in soils very close to the main entrance of the wastewater and decreased gradually with distance away from the plant. Once these pollutants are released into the environment, they are capable of persisting for a long period of time. PAHs can affect humans and animals both externally and internally. On the skin, they cause several inflammations which are often associated with itching and irritation. They can also cause cancer and are endocrine disruptors [51]. Most studies on the impact of wastewater on the receiving watershed in South Africa are often limited to the microbiological quality of the discharged effluent [38, 39].

3.1. Impact of wastewater discharge onto surface water in South Africa

The release of raw and ill-treated wastewater onto water courses has both short- and long-term effect on the environment and human health. Freshwater sources have been negatively impacted by wastewater. Such impacts are dependent on the composition and concentration of the wastewater contaminants as well as the volume and frequency of wastewater effluents entering surface water source [52]. Eutrophication of water sources may also create environmental conditions that favor the growth of toxin producing cyanobacteria, and exposure to such toxins is hazardous to human beings.

3.1.1. Environmental impact

Poorly treated wastewater can have a profound influence on the receiving watershed. The toxic impacts may be acute or cumulative. Acute impacts from wastewater effluents are generally due to high levels of ammonia and chlorine, high loads of oxygen-demanding materials, or toxic concentrations of heavy metals and organic contaminants. Cumulative impacts are due to the gradual buildup of pollutants in receiving surface water, which only become apparent when a certain threshold is exceeded [18, 21, 34, 39]. All aquatic organisms have a temperature range for their optimum function and survival [51]. When there are sudden changes within those ranges, their reproductive cycle, growth, and life can be reduced or threatened. Owing to the organic load of wastewater, discharged effluents from wastewater treatment facilities usually contribute to oxygen demand level of the receiving water. There is increased depletion of dissolved oxygen (DO) in surface water that receives ill-treated wastewater. From previous studies, the levels of DO in the effluent of various wastewater treatment facilities in South Africa are usually lower than the required standard of 8–10 mg/L [53, 54]. DO level below 5 mg/L would adversely affect aquatic ecosystem. DFID [55], Momba et al. [39], and Morrison et al. [56] stated that the effect of ill-treated wastewater on surface water is largely determined by the oxygen balance of the aquatic ecosystem, and its presence is essential in maintaining biological life within the system.

Osuolale and Okoh [57] reported that DO concentration in two WWTPs in Eastern Cape province of South Africa was in the range of 3.9–9.6 mg/L and 6.9–9.4 mg/L, respectively, from September 2012 to August 2013. For most of the study period, the levels of DO measured in one of the WWTPs were lower than the concentrations of 8–10 mg/L, which is characteristic of unpolluted water except in December 2012 (9.6 mg/L). Momba et al. [39] recorded DO levels in the range of 3.26–4.57 mg/L in their investigation of the impact of inadequately treated effluents of four wastewater treatment facilities in Buffalo City and Nkonkobe Municipality of Eastern Cape Province of South Africa. Concentrations below 5 mg/L can have a negative effect on aquatic organisms in the water resource [39]. Igbinosa and Okoh [58] reported a DO concentration in the range of 4.15–6.26 mg/L in autumn, 4.99–5.38 mg/L in summer, 4.85–11.22 mg/L in winter, and 4.96–6.69 mg/L in spring. This shows that seasonal variations have significant influence on the levels of DO in surface water. The presence of degradable organics in wastewater is responsible for the low levels of DO determined when compared to surface water sources. Low DO values can lead to the malfunctioning of some fish species and can eventually lead to the death of fish [58].

BOD and COD usually give an estimate of organic pollution in water and wastewater. They are important wastewater quality parameters as they are used to measure the efficiency of most wastewater treatment facilities. Surface water is expected to have low BOD/COD values to sustain aquatic life. High levels of BOD and COD can cause harm to aquatic life, especially fish. Low levels of BOD and COD in river systems indicate good water quality, while high levels indicate polluted water. There is an inverse relationship between the BOD/COD levels and DO concentrations. When large biodegradable organics are present in water as it is the case with most wastewater, DO is consumed by bacteria. When this happens, the DO level drops below a threshold point, with negative impact on life as they are unable to continue their normal life sustaining processes such as growth and reproduction. Such decrease affects fish and other aquatic life. The levels of COD reported for the effluent of several WWTFs in South Africa is presented in **Table 1**.

WWTF's location	COD (mg/L)	References
Eastern Cape Province I	4.6–211	[56]
Eastern Cape Province II	10.33–88.33	[56]
Alice WWTP, Eastern Cape Province III	7.5–248.5	[57]
Thohoyandou WWTP, Limpopo Province	50–105	[31]
Siloam WSPs, Limpopo Province	82–200	[58]

Table 1. COD levels of the effluent from wastewater treatment facilities in South Africa.

The South African guideline value for COD in wastewater is 75 mg/L but this level was exceeded for most of the sampling months in the WWTFs. From **Table 1**, wastewater effluent is a major contributor to organic pollution in surface water of South Africa.

The influx of nutrients such as nitrites, nitrates, and phosphorus into water bodies can induce eutrophication. Generally, nitrogen-containing compounds are abundant in many wastewater streams, and the inadequate treatment of them can lead to their introduction on the receiving watershed with their attendant consequences. Eutrophication can result when nutrient-rich wastewater effluents are discharged onto water courses. This can lead to algae blooms and growth of plants in the aquatic ecosystem. When this happens, turbidity of the water increases, plant and animals' biomass increases, sedimentation rate increases, species diversity decreases, and anoxic conditions may develop, and this could give rise to change in dominant species of the aquatic biota [35]. Nitrate nitrogen and phosphorus levels capable of inducing eutrophication have been reported by several authors for wastewater effluents in South Africa [31, 53, 58].

3.1.2. Health impacts

Contamination of surface water with pathogenic organisms in wastewater could result in the transmission of waterborne diseases for people who use the water resource for domestic and other purposes downstream [59, 60]. About 25% of all deaths worldwide are the result of infectious diseases caused by pathogenic microorganisms [61]. Scientists have identified about 1400 species of microorganisms that can cause ill health, including bacteria, protozoa, protozoan parasites, parasitic worms, fungi, and viruses [4]. The major concern of wastewater discharge onto freshwater courses is the impact they have on public health. Wastewater consists of various classes of pathogens which are capable of causing diseases of various magnitude to man. Unlike some of the environmental impacts that can take a long time before they manifest, pathogens cause immediate negative health impact on people that use contaminated surface water resource for domestic, agricultural, and recreational purposes. Some common pathogens found in untreated and ill-treated wastewater are presented in **Table 2**.

Agent	Species	Disease
Bacteria	*Campylobacter jejune*	Gastroenteritis (possibly long-term sequelae e.g., arthritis)
	Escherichia coli	Gastroenteritis
	E. coli O157:H7	Bloody diarrhea, hemolytic uremic syndrome
	Helicobacter pylori	Abdominal pain, peptic ulcers, gastric cancer
	Salmonella spp.	Salmonellosis, gastroenteritis, diarrhea (possibly long-term sequelae e.g., arthritis)
	Salmonella typhi	Typhoid fever
	Shigella spp	Dysentery (possibly long-term sequelae e.g., arthritis)
	Vibrio cholera	Cholera
Helminths	*Ascaris lumbricoides* (roundworm)	Ascariasis
	Ancylostoma duodenal and *Necator americanos* (hookworm)	
	Clonorchis sinensis (liver fluke)	Clonorchiasis
	Fasciola (liver fluke)	Fascioliasis
	Fasciolopsis buski (intestinal fluke)	Fascioloidiasis
	Opisthorchis viverrini	Opisthorchiasis
	Schistosoma (blood fluke)	Schistosomiasis (Bilharzia)
	Trichuris (whipworm)	Trichuriasis
	Taenia (tapeworm)	Taeniasis
Protozoa	*Balantidium coli*	Balantidiasis (dysentery)
	Cryptosporidium parvum	Cryptosporidiosis
	Cyclospora cayetanensis	Persistent diarrhea
	Entamoeba histolytica	Amoebiasis (amoebic dysentery)
	Giardia lamblia	Giardiasis
Viruses	Adenovirus	Respiratory disease, eye infections
	Astrovirus	Gastroenteritis
	Calicivirus	Gastroenteritis
	Coronavirus	Gastroenteritis
	Enteroviruses	Gastroenteritis
	Echovirus	Fever, rash, respiratory and heart disease, aseptic meningitis
	Poliovirus	Paralysis, aseptic meningitis
	Hepatitis A and E	Infectious hepatitis
	Parvovirus	Gastroenteritis
	Norovirus	Gastroenteritis
	Rotavirus	Gastroenteritis
	Coxsackie viruses	Herpangina, aseptic meningitis, respiratory illness, fever, paralysis, respiratory, heart and kidney disease

Table 2. Pathogens found in untreated wastewater (adapted with permission from WHO [60]).

Several episodes of disease outbreaks such as diarrhea and cholera have been reported in various provinces of South Africa with wastewater effluents as the major contributor. In 2004, Mail and Guardian [62] reported a cholera outbreak in Delmas region of Mpumalanga Province of South Africa where 380 cases of diarrhea and 30 cases of typhoid fever were recorded. Similarly, sickness and death were recorded in KwaZulu-Natal and Eastern Cape Provinces of South Africa where sewage spills occurred on surface water sources [53, 63]. South Africa suffered a cholera outbreak in 2003 when 3901 cases were reported in Mpumalanga Province, the Eastern Cape Province, and Kwazulu-Natal Province, and 45 deaths were confirmed. In 2004, 1773 cases of cholera were reported in Mpumalanga's Nkomazi region, which borders Mozambique, and 29 people died. Also in the same year, 738 people were diagnosed with cholera in the Eastern Cape Province, of which 4 died [63]. And 260 more cases were reported in the North West Province of which two people died. In early 2014, a diarrhea outbreak was reported in Limpopo province [64]. Forty-five people were admitted to hospital for treatment after contracting diarrhea. In almost all the cases stated above, the use of contaminated water as a source of domestic water was implicated to be the major cause of the epidemics. Several studies have shown that wastewater effluents still contain high amount of fecal coliforms which do not conform, to the 1000 cfu/100 mL in the DWA guideline for wastewater discharge [6, 31, 38, 39, 58, 59, 65].

4. Conclusion

Surface water will remain as an alternative source of water to meet domestic water demand mostly in rural areas of the world if potable water is not supplied on a regular basis. Wastewater effluents should be treated efficiently so as not to pose a health risk to the users of surface water resources. The major cause for the failing state of wastewater treatment facilities in South Africa as well as other developing countries includes inadequate coverage of wastewater treatment facilities in both urban and rural areas, poor operational state of wastewater infrastructure, design weaknesses, expertise, corruption, insufficient funds allocated for wastewater treatment, overloaded capacities of existing facilities, and inefficient monitoring for compliance with recommended guidelines. Enforcement of water and environmental laws must be in place to protect the environment and the health of numerous people that still depend on surface water as their major source of water supply.

Acronyms

WWTP	Wastewater treatment plant
WWTF	Wastewater treatment facilities
DO	Dissolved oxygen
COD	Chemical oxygen demand
BOD	Biochemical oxygen demand
WHO	World Health Organization
DWAF	Department of Water Affairs

Author details

Joshua N. Edokpayi*, John O. Odiyo and Olatunde S. Durowoju

*Address all correspondence to: Joshua.Edokpayi@Univen.ac.za

Department of Hydrology and Water Resources, School of Environmental Sciences, University of Venda, Thohoyandou, South Africa

References

[1] Jonnalagada SB, Mhere G. Water quality of Odzi River in the eastern highlands of Zimbabwe. Water Resources. 2001; **35**:2371–2376. doi: 10.1016/S0043-1354(00)00533-9

[2] Das J, Acharya BC. Hydrology and assessment of lotic water quality in Cuttack City, India. Water, Air and Soil Pollution. 2003; **150**:163–175. doi: 10.1023/A:1026193514875

[3] Tukura BW, Kagbu JA, Gimba CE. Effects of pH and seasonal variations on dissolved and suspended heavy metals in dam surface water. Chemistry Class Journal. 2009; **6**:27–30.

[4] CSIR. A CSIR perspective on water in South Africa. CSIR Report No. CSIR/NRE/PW/ IR/2011/0012/A. [Internet]. 2010. Available online: http://www.csir.co.za/nre/docs/ CSIR%20Perspective%20on%20Water_2010.PDF [Accessed 2015/03/4].

[5] DWAF. Water Quality Management Series Guideline Document U1.5. Guideline for the management of waterborne epidemics, with the emphasis on cholera, 1st ed. 2002.

[6] Bessong PO, Odiyo JO, Musekene JN, Tessema A. Spatial distribution of diarrhea and microbial quality of domestic water during an outbreak of diarrhea in the Tshikuwi Community in Venda, South Africa. Journal of Health Population and Nutrition. 2009; **27(5)**:652–659.

[7] DWAF. Guideline for the Management of Waterborne Epidemics, with the Emphasis on Cholera–Co-ordination, Communication, Action, and Monitoring. 2002; pp. 1–3. Published by Department of Water Affairs and Forestry, Pretoria, South Africa.

[8] Avalon Global Research. Water and Waste Water Treatment Opportunity in India. An Overview. [Internet]. 2012. Available online: www.export.gov.il/uploadfiles/02_2012/ indiawater.pdf?loaded=true [Assessed 2016/06/27].

[9] Edokpayi JN, Odiyo, JO, Olasoji SO. Assessment of heavy metal contamination of Dzindi River, in Limpopo Province, South Africa. International Journal of Natural Science Research. 2014; **2**:185–194.

[10] Agarwal VK. Environmental laws in India: challenges for enforcement. Bulletin of the National Institute of Ecology. 2005; **15**:227–238.

[11] Gregory LR. First Preparatory Meeting of the World Congress on Justice, Governance and Law for Environmental Sustainability. Kuala Lumpur, Malaysia. [Internet]. 2011; pp. 1–30. Available online: http://www.unep.org/delc/Portals/24151/FormatedGapsEL. pdf [Accessed 2015/8/19].

[12] Faure MG. Enforcement issues for environmental legislation in developing countries. [Internet]. 1995. Available online: http://archive.unu.edu/hq/library/Collection/PDF_files/INTECH/INTECHwp19.pdf [Accessed 2015/8/19].

[13] Feris LA. Compliance notices—a new tool in environmental enforcement. Potchefstroom Electronic Law Journal. 2006; **9(3)**:53–70.

[14] Oberdorff T, Pont D, Hugueny B, Porcher J. Development and validation of a fish-based index for the assessment of 'river health' in France. Freshwater Biology. 2002; **47(9)**:1720–1734.

[15] Poff LN, Allan D, Bain MB, Karr JR, Prestegaard KL, Richter BD, Sparks RE, Stromberg JC. The natural flow regime: a paradigm for river conservation and restoration. Bioscience. 1997. **47**:769–784.

[16] DWAF. South African Water Quality Guidelines: Volume 7: Aquatic Ecosystems, 2nd ed. Pretoria, South Africa. 1996; p. 161. Published by Department of Water Affairs and Forestry, Pretoria, South Africa.

[17] Dallas HF, Day JA, Musibono DE, Day EG. Water quality for aquatic ecosystems: tools for evaluating regional guidelines. WRC Report No. 626/1/98. 1998; p. 240.

[18] Osode AN, Okoh AI. The impact of discharged wastewater final effluent on the physicochemical qualities of a receiving watershed in a sub-urban community of the Eastern Cape Province. Clean Soil, Air, Water. 2009; **37**:938–944. doi: 10.1002/clen.200900098

[19] Rainwater Harvest.co.za. Water Quality & Water Quality Management in SA. [Internet]. 2010. Available online: http://www.rainharvest.co.za/2010/10/water-quality-water-quality-management-in-sa/ [Accessed 2014/8/20].

[20] DWA. Wastewater limit values applicable to the discharge of wastewater into a river resource. National Water Act, Government Gazette No. 20528. [Internet]. 1999. Available online: http://hwt.co.za/downloads/NWA%20General%20and%20Special%20Authorisations.pdf [Accessed 2013/4/7].

[21] Weber S, Khan S, Hollender J. Human risk assessment of organic contaminants in reclaimed wastewater used for irrigation. Desalination. 2006; **187**:53–64. doi:10.1016/j.desal.2005.04.067

[22] Bundschuh M, Gessner MO, Fink G, Ternes TA, Sogding C, Schulz R. Ecotoxicological evaluation of wastewater ozonation based on detritus-detritivore interactions. Chemosphere. 2011; **82**:355–361. doi:10.1016/j.chemosphere.2010.10.006

[23] Tarr JA. McCurley J, McMichael, FC, Yosie T. Water and wastes: a retrospective assessment of wastewater technology in the United States, 1800–1923. Technology and Culture. 1986; **25(2)**:226–263. doi:10.2307/3104713

[24] Burian SJ, Nix SJ, Pitt RE, Durrans SR. Urban Wastewater Management in the United States: past, present, and future. Journal of Urban Technology. 2000; **7(3)**:33–62.

[25] Ratola N, Cincinelli A, Alves A, Katsoyiannis A. Occurrence of organic microcontaminants in the wastewater treatment process. A mini review. Journal of Hazardous Materals. 2012; **239–240**:1–18. doi:10.1016/j.jhazmat.2012.05.040

[26] Katsoyiannis A, Samara C. Persistent organic pollutants (POPS) in the sewage treatment plant of Thessaloniki, northern Greece: occurrence and removal. Water Research. 2004; **38**:2685–2698. doi:10.1016/j.watres.2004.03.027

[27] Katsoyiannis A, Samara C. Persistent organic pollutants (POPS) in the conventional activated sludge treatment process: fate and mass balance. Environmental Research. 2005; **97**:245–257. doi:10.1016/j.envres.2004.09.00

[28] Muller K, Magesan GN, Bolan NS. A critical review of the influence of effluent irrigation on the fate of pesticides in soil. Aquaculture, Ecosystems and Environment. 2007; **120**:93–116. doi:10.1016/j.agee.2006.08.016

[29] Klinch BA, Stuart ME. Human health risk in relation to landfill leachate quality; British Geological Survey, Technical Report; Overseas Series; British Geological Survey. Key Worth, Nottingham; United Kingdom. 1999.

[30] Saving Water SA. Over half of wastewater treatment plants are well below standard. [Internet]. 2011. Available online: http://www.savingwater.co.za/tag/wastewater-treatment-plants/ [Accessed 2012/8/9].

[31] Edokpayi JN, Odiyo JO, Msagati TAM, Popoola EO. Removal efficiency of faecal indicator organisms, nutrients and heavy metals from a Peri-Urban wastewater treatment plant in thohoyandou, Limpopo Province, South Africa. International Journal of Environmental Research and Public Health. 2015; **12**:7300–7320. doi:10.3390/ijerph120707300

[32] Toze S. Microbial pathogens in wastewater. Literature review for urban waste systems in multi-divisional research program. CSIRO Report No. 1/97. 1997.

[33] Ashbolt NJ, Dorsch MR, Cox PT, Banens B. Blooming *E. coli*, what do they mean? In Coliforms and *E. coli*, Problem or Solution. The Royal Society of Chemistry, Cambridge. 1997; pp. 78–85.

[34] Ogola JS, Chimuka L, Maina D. Occurrence and fate of trace metals in and around treatment and disposal facilities in Limpopo Province, South Africa (a case of two areas). In: Proceedings of the IASTED International Conference on Environmental Management and Engineering. 2009.

[35] Edokpayi JN. Assessment of the efficiency of wastewater treatment facilities and the impact of their effluent on surface water and sediments in Vhembe District, South Africa. A Ph.D. thesis submitted to the University of Venda, South Africa. 2016.

[36] Pindihama GK, Gumbo JR, Oberholster PJ. Evaluation of a low-cost technology to manage algal toxins in rural water supplies. African Journal of Biotechnology. 2011; **10(86)**:19883–19889. doi:10.5897/AJBX11.024

[37] Phuntsho S, Shon HK, Vigneswaran S, Ngo HH, Kandasamy J. The performance of waste stabilization ponds: experience from cold climatic conditions of Bhutan. In: Proceedings of 8th IWA Specialized Conference on Small Water and Wastewater Systems. 2008.

[38] Jagals C, Mackintosh G, Jack U, van Der Merwe J. Waste stabilization ponds and the health-related risks: the need for monitoring programs. [Internet]. 2006. Available online: http://www.ewisa.co.za/literature/files/285%20Jagals.pdf [Accessed 2013/3/21].

[39] Momba MNB, Osode AN, Sibewu M. The impact of inadequate wastewater treatment on the receiving water bodies—case study: Buffalo City and Nkokonbe Municipalities of the Eastern Cape Province. Water SA. 2006; **32**:5. doi:10.4314/wsa.v32i5.47854

[40] Song M, Chu S, Letcher RJ, Seth R. Fate, partitioning, and mass loading of polybrominated diphenyl ethers (PBDEs) during the treatment processing of municipal sewage. Environmental Science and Technology. 2006; **40**:6241–6246. doi:10.1021/es060570k

[41] Wojtenko I, Ray A. Treatment of stormwater by natural organic materials. In: Proceedings of the Engineering Foundation Conference, vol. 263. 2001; pp. 549–553.

[42] Chen Y, Wang C, Wang Z. Residues and source identification of persistent organic pollutants in farmland soils irrigated by effluents from biological treatment plants. Environment International. 2005; **31(6)**:778–783. doi:10.1016/j.envint.2005.05.024

[43] Wang M, Leung A, Chan J, Choi M. A review on the usage of POP pesticides in China, with emphasis on DDT loadings in human milk. Chemosphere, 2005; **60**:740. doi:10.1016/j.chemosphere.2005.04.028

[44] Naser HM, Sultana S, Haque MM, Akhter S, Begum RA. Lead, Cadmium and Nickel accumulation in some common spices grown in industrial areas of Bangladesh. The Agriculturists. 2014; **12(1)**:122–130. doi:10.3329/agric.v12i1.19867

[45] Ahmad JU, Goni MA. Heavy metal contamination in water, soil, and vegetables of the industrial areas in Dhaka, Bangladesh. Environmental Monitoring and Assessment. 2009; **166(1–4)**:347–57. doi:10.1007/s10661-009-1006-6

[46] Srinivasan JT, Reddy VR. The impact of irrigation water quality on human health: a case study in India. Ecological Economics. 2009; **68**:2800–2807. doi:10.1016/j.ecolecon.2009.04.019

[47] Arora M, Kiran B, Rani S, Rani A, Kaur B, Mittal N. Heavy metal accumulation in vegetables irrigated with water from different sources. Food Chemistry. 2008; **111**:811–815. doi:10.1016/j.foodchem.2008.04.049

[48] Alam MGM, Snow ET, Tanaka A. Arsenic and heavy metal contamination of vegetables grown in Santa village, Bangladesh. Science of the Total Environment. 2003; **308**:83–96. doi:10.1016/S0048 9697(02)00651-4

[49] Bahemuka TE, Mubofu EB. Heavy metals in edible green vegetables grown along the sites of the Sinza and Msimbazi rivers in Dar es Salaam, Tanzania. Food Chemistry. 1999; **66**:63–66. doi:10.1016/S0308-8146(98)00213-1

[50] Liu H, Probst A, Liao B. Metal contamination of soils and crops affected by the Chenzhou lead/zinc mine spill (Hunan, China). Science of the Total Environment. 2005; **339(1–3)**:153–166. doi:10.1016/j.scitotenv.2004.07.030

[51] Noth EM, Hammonda SK, Biging GS, Tager IB. A spatial-temporal regression model to predict daily outdoor residential PAH concentrations in an epidemiologic study in Fresno, CA. Atmospheric Environment. 2011; **45**:2394–2403. doi:10.1016/j.atmosenv.2011.02.014

[52] Akpor OB, Muchie M. Environmental and public health implications of wastewater quality. African Journal of Biotechnology. 2011; **10(13)**:2379–2387. doi:10.5897/AJB10.1797

[53] Mema V. Impact of poorly maintained wastewater and sewage treatment plants: lessons from South Africa. Pretoria: Council for Scientific and Industrial Research. [Internet]. 2010. Available online: http://www.ewisa.co.za/literature/files/335_269%20Mema.pdf [Assessed 2012/03/21].

[54] Watson IM, Robinson JO, Burke V, Grace M. Invasiveness of *Aeromonas spp.* In relation to biotype virulence factors and clinical features. Journal of Clinical Microbiology. 1985; **22(1)**:48–51.

[55] DFID. A simple methodology for water quality monitoring. G. R. Pearce, M. R. Chaudhry and S. Ghulum (Eds.), Department for International Development, Wallingford. 1999. p. 100.

[56] Morrison G, Fatoki OS, Persson L, Ekberg A. Assessment of the impact of point source pollution from the Keiskammahoek Sewage Treatment Plant on the Keiskamma River– pH, electrical conductivity, oxygen demanding substance (COD) and nutrients. Water SA. 2001; **27(4)**:475–480. doi:10.4314/WAS.V27I4.4960

[57] Osuolale O, Okoh A. Assessment of the physicochemical qualities and prevalence of Escherichia Coli and Vibrios in the final effluents of two wastewater treatment plants in South Africa: ecological and public health implications. International Journal of Environmental Research and Public Health. 2015; **12**:13399–13412. doi:10.3390/ ijerph121013399

[58] Igbinosa EO, Okoh AI. The impact of discharge wastewater effluents on the physio-chemical qualities of a receiving watershed in a typical rural community. International Journal of Environmental Science and Technology. 2009; **6(2)**:175–182.

[59] Chigor VN, Sibanda T, Okoh AI. Studies on the bacteriological qualities of the Buffalo River and three source water dams along its course in the Eastern Cape Province of South Africa. Environmental Science and Pollution Research 2013; **20**:4125–4136. doi:10.1007/ s11356-012-1348-4

[60] WHO. WHO Guidelines for the Safe Use of Wastewater, Excreta, and Greywater. World Health Organization, Geneva. 2006.

[61] UNEP (United Nations Environment Programme). Water quality for ecosystem and human health. United Nations Environment Programme Global Environment Monitoring System (GEMS)/Water Programme. 2006.

[62] Mail and Guardian. Officials Work to Curb Delmas Typhoid Outbreak. Johannesburg, South Africa, 8 September, 2005.

[63] Health news.com. Recent cholera outbreaks in SA. [Internet]. 2015. Available online: http://www.health24.com/Medical/Cholera/Recent-cholera-outbreaks/Recent-cholera-outbreaks-in-SA-20120721 [Accessed 2016/08/20].

[64] News24. Diarrhea outbreaks in Limpopo Province. [Internet]. 2014. Available online: http://www.news24.com/SouthAfrica/News/Diarrhoea-outbreak-in-Limpopo-20140127 [Accessed 2016/07/20].

[65] DWA. Wastewater risk abatement plan: a guideline to plan and manage towards safe and compliant wastewater collection and treatment in South Africa. [Internet]. 2010. Available online: https://www.dwa.gov.za/dir_ws/GDS/Docs/.../DownloadSiteFiles.aspx? id [Assessed 2016/03/ 21].